Foundations of Environmental Sustainability

Foundations of Environmental Sustainability

The Coevolution of Science and Policy

EDITED BY
Larry L. Rockwood
Ronald E. Stewart
Thomas Dietz

OXFORD
UNIVERSITY PRESS
2008

OXFORD
UNIVERSITY PRESS

Oxford University Press, Inc., publishes works that further
Oxford University's objective of excellence
in research, scholarship, and education.

Oxford New York
Auckland Cape Town Dar es Salaam Hong Kong Karachi
Kuala Lumpur Madrid Melbourne Mexico City Nairobi
New Delhi Shanghai Taipei Toronto

With offices in
Argentina Austria Brazil Chile Czech Republic France Greece
Guatemala Hungary Italy Japan Poland Portugal Singapore
South Korea Switzerland Thailand Turkey Ukraine Vietnam

Copyright © 2008 by Oxford University Press, Inc.

Published by Oxford University Press, Inc.
198 Madison Avenue, New York, New York 10016

www.oup.com

Oxford is a registered trademark of Oxford University Press

Library of Congress Cataloging-in-Publication Data
Foundations of environmental sustainability : the coevolution of science
and policy / edited by Larry L. Rockwood, Ronald E. Stewart, and Thomas Dietz.
     p. cm.
Includes bibliographical references and index.
ISBN 978-0-19-530945-4
1. Environmental policy—United States. 2. Environmental policy—Economic
aspects—United States. 3. Environmental sciences—United States.
I. Rockwood, Larry L., 1943– II. Stewart, Ronald E. (Ronald Eugene), 1943–
III. Dietz, Thomas.
GE180.F68 2007
333.72—dc22   2007015381

9 8 7 6 5 4 3 2 1

Printed in the United States of America
on acid-free paper

Preface

The purpose of this book is to describe and analyze a critical period in the history of human relationships with the environment (roughly from the 1950s to the present), assess where we stand today, and consider the challenges that lie ahead. The authors have all played key roles in the application of ecological science to conservation and to the development of public policy in the United States and throughout much of the world. The chapters that follow present a unique assembly of such leaders, each speaking from his or her own area of expertise.

This book originated when its authors, among others, gathered for a symposium arranged by George Mason University and organized around the life and career of Lee M. Talbot. His work over the last half of the twentieth century is of global distinction. Few others' careers have so fully spanned environmental science and policy, pioneering both research and application. Lee Talbot's career marks, and substantially helped bring about, the transition from the concept of conservation to the concept of sustainability.

Those who attended the symposium in Talbot's honor were an unusually distinguished group; such a gathering is extremely rare. It provided a unique opportunity for us, the editors, to obtain a firsthand overview of the evolution, status, and trends in environmental science and policy. So we invited the participants from the symposium to prepare chapters for this book. Additionally, in order to give greater depth and breadth to the topics covered, we invited another group of equally distinguished scholars (with whom Talbot had also developed working relationships) to prepare chapters on appropriate subjects.

Each author was asked to review the evolution of environmental science and policy in the past half century, to assess the present and consider the challenges of the future. As requested, they have provided critical assessments of both the current situation and the challenges for the future. Although the authors came from a variety of starting points, they followed paths that have generally converged on the concept of sustainability. Consequently this book is titled *Foundations of Environmental Sustainability: The Coevolution of Science and Policy*.

We wish to thank H. Paige Tucker for her dedication and for the long hours she devoted to this project. Paige well understands the old saying that trying to manage a group of scientists is akin to "herding cats." Without her efforts, the book never would have been possible. We also thank Megan M. Draheim for her heroic efforts in bringing the last stages of the book to fruition. Robert Jonas helped us transfer hard copy materials into appropriate electronic format. R. Christian Jones, chair of the Department of Environmental Science and Policy at George Mason University, provided financial support and a steady hand. And, as always, the contributions of Suzanne Stewart, Jane Rockwood, and Linda Kalof—our collaborators in science, public policy, and life—should never be underestimated.

Larry L. Rockwood, Ronald E. Stewart, and Thomas Dietz

Contents

Contributors

Charles Victor Barber is environmental advisor with the Office of Environment and Science Policy at the U.S. Agency for International Development (USAID), where he focuses on coordinating and enhancing the agency's participation in multilateral environmental conventions and processes related to biodiversity and natural resources. Prior to joining USAID, he worked as an independent consultant on environment and natural resources policy projects for the World Bank, the Asian Development Bank, IUCN, the UN Environment Programme, the Nature Conservancy, and the UN University Institute of Advanced Studies. He has also served as a representative of the International Marinelife Alliance, as a senior research associate at the World Resources Institute, and as a consultant on forest and land use policy for the Ford Foundation and other organizations. A prolific author of works on conservation policy and resource management, Barber holds an M.A. in Asian studies, a Ph.D. in jurisprudence and social policy, and a J.D. from the University of California, Berkeley.

Daniel B. Botkin is research professor in the Department of Ecology, Evolution, and Marine Biology at the University of California, Santa Barbara, and president of the nonprofit Center for the Study of the Environment. He was formerly a faculty member at George Mason University and Yale University and a scientist at the Woods Hole Marine Biological Laboratory. He is the author of numerous books, including *No Man's Garden: Thoreau and a New Vision for Civilization and Nature* (2000) and *Discordant Harmonies: A New Ecology for the Twenty-First Century* (1990). Botkin received his B.A. in physics from the University of Rochester, his M.A. in English literature from the University of Wisconsin, and his Ph.D. in plant ecology from Rutgers University.

James A. Burchfield is associate dean of the College of Forestry and Conservation at the University of Montana. Prior to becoming associate dean, he served as director of the Bolle Center for People and Forests at the University of Montana. He has also worked for the U.S. Department of Agriculture's Forest Service in several U.S. locations, conducted assessments of social conditions in the Columbia River basin, served in the international division of the Forest Service in Washington, D.C., and helped implement forest management operations for national forests in Michigan, Ohio, Oregon, and Washington. Burchfield's academic training was conducted at the University of Washington and the University of Michigan, where he received a

Ph.D. His recent work examines the principles of social acceptability in forest management, the effects of wildfires on rural communities, and the implications of stewardship contracting on public lands.

Leif E. Christoffersen is senior fellow at the Agricultural University of Norway, chairman of the GRID-Arendal (UN Environment Programme) Foundation in Norway, board chairman of Scandinavian Seminar College, committee chairman of the Norwegian Research Council, and president of Christoffersen Associates. He has also spent twenty-eight years working with the World Bank, serving in positions including chief of the African Environment Division and personal assistant to the president. He received the St. Olav Medal in 2006 from Norway for his contributions to the environment. Christoffersen did his undergraduate work at the University of Edinburgh and received his M.A. and M.A.L.D. in international economics from Tufts University.

Jason Clay is senior vice president at the World Wildlife Fund–U.S. For over twenty years, he has worked with human rights and environmental organizations; he helped invent green marketing in the 1980s, and he also established a trading company that developed markets for rainforest products with nearly 200 U.S. and European companies. In 1999 Clay created the Shrimp Aquaculture and the Environment Consortium (World Wildlife Fund, World Bank, Food and Agriculture Organization of the United Nations, and Network of Aquaculture Centres in Asia-Pacific) to identify and analyze better management practices that address the environmental and social impacts of shrimp aquaculture. The founder and editor of *Cultural Survival Quarterly*, he studied anthropology and Latin American studies at Harvard University and economics and geography at the London School of Economics. He holds a Ph.D. in anthropology and international agriculture from Cornell University.

Megan M. Draheim received a B.A. in fine arts from George Washington University and is currently pursuing graduate work in environmental science and policy at George Mason University, where she received her M.S. in 2007. Her research involves the ecology of urban foxes and coyotes as well as human attitudes toward these animals.

Mohamed T. El-Ashry is former chairman and CEO (1991–2002) of the Global Environment Facility, also known as GEF. Under his leadership, the GEF grew from a pilot program with less than thirty members to the largest single source of funding for the global environment, with 173 member countries. Additionally, during his tenure, the GEF helped finance more than 1,000 environmental projects in over 140 countries. El-Ashry came to the GEF from the World Bank, where he was chief environmental advisor to the president as well as director of the environment department. Prior to joining the World Bank, he served as senior vice president of the World Resources Institute and as director of environmental quality for the Tennessee Valley Authority. He has received numerous international awards and honors and is the author of three books and more than 200 papers. El-Ashry received his B.S. from the University of Cairo and a Ph.D. in geology from the University of Illinois.

Robert Goodland is former chief environmental advisor (1978–2001) to the World Bank Group. He has drafted, and persuaded the World Bank to adopt, most of its social and environmental safeguard policies. He has also served as chief environmental advisor (2001–2004) to the independent Extractive Industries Review of the World Bank Group. Goodland has authored or coauthored more than ten books, including *The Social and Environmental Impacts of Oil and Gas Pipelines: Best Practice and State of the Art* (2005). He holds a B.S. in biology and an M.S. and Ph.D. in ecology, all from McGill University.

Robert J. Hofman is former scientific program director (1975–2000) of the Marine Mammal Commission, a federal agency established by the 1972 Marine Mammal Protection Act to oversee all federal activities affecting the conservation and protection of marine mammals. In this position, he managed a small research program, organized workshops and reviews of other agencies' mammal research programs, facilitated commission reviews of domestic and international policies and programs affecting marine mammals, and represented the commission at both national and international meetings. Hofman received his B.S. and M.S. from Indiana University of Pennsylvania. He holds a Ph.D. from the University of Minnesota, where he conducted dissertation research on the biology and ecology of Antarctic seals.

Sidney J. Holt served, and led, UN organizations for over twenty-five years. He was the first director of the International Ocean Institute at Malta, and he is a former advisor to the UN Environment Programme, former UN advisor on Mediterranean Marine Affairs, and former director of the Food and Agriculture Organization's Department of Fisheries. In addition, he has been a member of numerous university faculties and of the International Whaling Commission Scientific Committee and the International Fund for Animal Welfare. The founder and current executive director of the International League for the Protection of Cetaceans, Holt has written more than 400 works on marine mammals, international whaling, and sustainability. He received his university training at Reading University in England, where he earned a B.Sc. in botany, chemistry, and zoology and a D.Sc. in zoology.

Stephen R. Kellert is Tweedy/Ordway Professor of Social Ecology at the Yale School of Forestry and Environmental Studies, where he also serves as codirector of the Hixon Center for Urban Ecology. His current projects include studies of basic values and perceptions relating to the conservation of biological diversity, sustainable environmental design, and biophilia. He has written and edited numerous books, including *The Value of Life: Biological Diversity and Human Society* (1995), *The Biophilia Hypothesis* (1995), and *Kinship to Mastery: Biophilia in Human Evolution and Development* (2003). He received his B.S. from Cornell University and his Ph.D. from Yale University.

Agnes Kiss joined the World Bank in 1985, working first in its Office of Environmental and Scientific Affairs and then in the African region. In Africa, she became interested in environmental policy and biodiversity conservation. She lived in Kenya for five years; during this time, she oversaw a major project that supported

Richard Leakey's efforts to establish the Kenya Wildlife Service and rehabilitate the country's national parks. Kiss is a prolific author not only of technical publications but also of environmental fiction books for children. She now works in the Sustainable Development Department for Europe and Central Asia at the World Bank. She holds an M.S. and Ph.D. from the University of Michigan.

Gene E. Likens retired in 2007 as director of the Institute of Ecosystem Studies in Millbrook, New York. He has been a faculty member at Yale University; Cornell University; the University at Albany, State University of New York; and Rutgers University. He has also served as president of the American Institute of Biological Sciences, the Ecological Society of America, the American Society of Limnology and Oceanography, and the International Association of Theoretical and Applied Limnology. Currently vice president of the New York Botanical Garden and a member of the National Academy of Sciences, Likens received the National Medal of Science in 2001. He received his B.S. in zoology from Manchester College and his M.S. and Ph.D. from the University of Wisconsin.

Thomas E. Lovejoy is president of the nonprofit H. John Heinz III Center for Science, Economics and the Environment. Founder of the public television series Nature, he formerly served as chief biodiversity advisor and lead specialist for the environment for the Latin American region for the World Bank, as senior advisor to the president of the UN Foundation, as assistant secretary for environmental and external affairs for the Smithsonian Institution, and as executive vice president of World Wildlife Fund's U.S. division. (He retains his link with the Smithsonian as a research associate of the Smithsonian Tropical Research Institute.) In 2001, Lovejoy received the John and Alice Tyler Prize for Environmental Achievement. He is past president of the American Institute of Biological Sciences, past chairman of the U.S. Man and Biosphere Program, and past president of the Society for Conservation Biology, and he continues to serve on numerous scientific and conservation boards and advisory groups, including those of the American Museum of Natural History, the New York Botanical Garden, the Institute for Ecosystem Studies, the Wildlife Trust, the Woods Hole Research Center, and the Yale Institute for Biospheric Studies. He holds a B.S. and Ph.D. in biology from Yale University.

Walter J. Lusigi is senior advisor for the Global Environment Facility. He is also a member of the Norwegian Academy of Science and Letters, professor of conservation biology at the University of Oslo, and an affiliate member of the range science faculty at Colorado State University. Lusigi previously worked with the World Bank and UNESCO, and he has been an ecologist for the Office of the President of Kenya. He received his B.Sc. and M.Sc. in rangeland ecology and wildlife management from Colorado State University and his Ph.D. in landscape ecology from the Technical University of Munich, Germany.

Kenton R. Miller recently retired as vice president for international development and conservation at the World Resources Institute (WRI). He has an extensive background in wildlands management; prior to joining WRI he served as director general of IUCN, and after joining WRI he directed its biological resources program for ten years. Miller has also served as international coordinator for the joint WRI/IUCN/UN Environment Programme's Biodiversity Program, which produced the global biodiversity strategy. He has published extensively on wildlands management, national parks and protected areas, biodiversity conservation, bioregional planning, and decentralization. In 2005 Miller received the Bruno H. Schubert Award for his dedication to conservation issues. His undergraduate degree is from the University of Washington's school of forestry, and he holds a Ph.D. in forestry economics from New York University.

Russell W. Peterson is director of research and development for the DuPont Corporation, as well as president emeritus of the International Council for Bird Preservation. He has previously served as governor of Delaware, chairman of the Council on Environmental Quality, president of the National Audubon Society, vice president of IUCN, president of New Directions, and president of the Better World Society. Additionally, he has held visiting professorships at various universities. Peterson received his Ph.D. in chemistry from the University of Wisconsin.

John E. Reynolds III is senior scientist and manager of the Manatee Research Program at Mote Marine Laboratory in Sarasota, Florida, as well as cochair of the IUCN Sirenian Specialist Group. He was previously professor of marine science and biology and chair of the Natural Sciences Collegium at Eckerd College (1980–2001), where he was integral in establishing the marine science major. In 1989, Reynolds became a member of the Marine Mammal Commission's Committee of Scientific Advisors on Marine Mammals; in 1990 he became chair of the committee. In 1991 he was appointed chair of the Marine Mammal Commission, and he served on the commission during the George H. W. Bush, William Clinton, and George W. Bush administrations, retiring in 2001. Reynolds holds a B.A. from Western Maryland College and an M.S. and Ph.D. in biological oceanography from the University of Miami's Rosenstiel School of Marine and Atmospheric Sciences, where he conducted thesis and dissertation research on behavioral ecology and functional morphology of manatees. He has published nearly 200 papers, abstracts, and books.

Nicholas A. Robinson is the Gilbert and Sarah Kerlin Distinguished Professor of Environmental Law at Pace University School of Law, where he has founded a program in environmental law, as well as the legal advisor and chairman of IUCN—The World Conservation Union's Commission on Environmental Law. He has practiced environmental law in legal firms for various municipalities. In 1969, Robinson was named to the Legal Advisory Committee of the Council on Environmental Quality to develop environmental law. He later served as the general counsel for the New York State Department of Environmental Conservation, a position in which he drafted New York's wetlands and wild bird laws; he is also former chairman of the statutory Freshwater Wetlands Appeals Board and Greenway Heritage Conservancy for the Hudson River Valley. The author of several books and numerous articles relating to legal aspects of environmental issues, Robinson holds an A.B. from Brown University and a J.D. from Columbia University.

V. Alaric Sample has been president of the Pinchot Institute for Conservation since 1995. He is also an affiliated researcher for the Yale School of Forestry and Environmental Studies, as well as a fellow of the Society of American Foresters. Sample's professional experience spans public, private, and nonprofit organizations; he specialized in resource economics and national forest policy as senior fellow of the Conservation Foundation in Washington, DC, and later as vice president for research at the American Forestry Association. He has served on numerous national task forces and commissions, including the Commission on Environmental Quality's task force on biodiversity on private lands, and as cochair of the National Commission on Science for Sustainable Forestry. He holds a Ph.D. in resource policy and economics from Yale University.

Thayer Scudder is professor emeritus of anthropology at the California Institute of Technology and a founding director of the Institute for Development Anthropology in Binghamton, New York. For the past forty-five years his primary research and policy work has dealt with the impacts of river basin development projects on local communities in Africa and Asia, including the Kariba Dam Project (Zambia-Zimbabwe), the High Dam at Aswan (Egypt-Sudan), and the Three Gorges Project (China). A commissioner on the World Commission on Dams during the commission's lifetime, he is currently a member of the environment and resettlement advisory panel for the Nam Theun 2 project in Laos. He holds an A.B. and Ph.D. from Harvard University.

John Seidensticker is senior scientist at the Smithsonian's Conservation and Research Center and at the National Zoological Park. He is also chairman of the National Fish and Wildlife Foundation's Save the Tiger Fund Council. His Smithsonian research efforts have focused on understanding and encouraging landscape patterns and conditions where large mammals can persist, training future conservation leaders, and promoting environmental understanding through writing, public appearances, and museum and zoo exhibits. Previously, Seidensticker pioneered the use of radio telemetry to study mountain lions in North America. As founding principal investigator of the Smithsonian-Nepal Tiger Ecology Project, he was the coleader of the team that captured and radio-tracked the first wild tigers in Nepal. He has written and edited numerous works on animals and animal conservation, and he holds a B.A. and M.S. from the University of Montana and a Ph.D. from the University of Idaho.

James Gustave Speth is dean of the Yale School of Forestry and Environmental Studies. He previously served as administrator for the UN Development Programme, president of the World Resources Institute (which he founded), chair of the Council on Environmental Quality, senior staff attorney for the Natural Resources Defense Council, and professor at the Georgetown University Law Center. He has a B.A. from Yale University, was a Rhodes Scholar at Oxford (where he studied economics), and earned a J.D. from Yale Law School.

Lee M. Talbot is professor of environmental science, international affairs, and public policy at George Mason University, as well as senior environmental advisor to the World Bank Group and to the United Nations. He was formerly director general of IUCN, chief scientist and director of international affairs for the Council on Environmental Quality, and head of environmental sciences at the Smithsonian Institution. Talbot has conducted environmental research and advising in 131 countries, and he has produced over 270 scientific publications, many of which have received national and international awards. Talbot holds an A.B. and M.A. in zoology and an interdisciplinary Ph.D. in geography and ecology, all from the University of California, Berkeley.

James G. Teer is professor emeritus at Texas A&M University, where he has engaged in evaluation and development of policy and best management practices. He previously served as professor and head of the Department of Wildlife and Fisheries Sciences at Texas A&M University and as director of the Welder Wildlife Foundation; he has also served as a research biologist with the U.S. Fish and Wildlife Service. He has conducted research on large mammals and worked as a teacher of, and advisor on, wildlife management and conservation biology throughout the world, including the Americas, Africa, Europe, and Asia. Teer has earned a B.S. from Texas A&M University, an M.S. from Iowa State University, and a Ph.D. from the University of Wisconsin.

Jack Ward Thomas, retired, was the Boone and Crockett Professor of Wildlife Conservation at the University of Montana. He is also former chief (1993–1996) of the U.S. Forest Service. A wildlife biologist from Oregon and a thirty-year veteran of the Forest Service, Thomas worked in a variety of capacities (including as chief research wildlife biologist) before taking over as the head of the agency. Thomas chaired the Forest Ecosystem Management Team that developed the plan to protect the northern spotted owl; moreover, he instituted the concept of ecosystem management in the Forest Service. He currently serves on the Board of Agriculture and Natural Resources within the National Research Council. Thomas received a B.S. in wildlife management from Texas A&M University, an M.S. in wildlife ecology from West Virginia University, and a Ph.D. in forestry from the University of Massachusetts.

Russell E. Train is chairman emeritus of the World Wildlife Fund–U.S. He has also served as judge on the U.S. Tax Court, president of the Conservation Foundation, president of the World Wildlife Fund's U.S. division, undersecretary of the interior, chair of the Council on Environmental Quality, administrator of the Environmental Protection Agency, head of the U.S. delegation to the Stockholm UN Conference on the Human Environment, and representative to the International Whaling Commission. Train holds a B.A. from Princeton University and a J.D. from Columbia Law School.

H. Paige Tucker is an environmental specialist with the U.S. Department of Homeland Security, Federal Emergency Management Agency. She received her B.S., B.A., and M.A. from the University of Maryland. Currently she is a candidate for a Ph.D. in environmental science and public policy at George Mason University, where she has served as a teaching assistant (Environmental Science and Policy Department) and as an adjunct instructor (George Mason Freshman Center).

John R. Twiss Jr. is former executive director (1974–2000) of the Marine Mammal Commission. He worked in Antarctica and the Southern Ocean area before joining the National Science Foundation's International Decade of Ocean Exploration program in 1970. He has twice chaired the board of the Student Conservation Association, has served on the boards of the Ocean Conservancy and the Cape Eleuthera Island School, and currently serves on the board of the Marine Conservation Biology Institute. Twiss received his B.A. from Yale University.

Frederic H. Wagner is professor emeritus in the Department of Forest, Range, and Wildlife Science at Utah State University. He specializes in ecology and management of animal populations and arid ecosystems, natural resource policy, and the role of science and policy. He previously served as director (1979–1998) of the Utah State Ecology Center, and he has cochaired a congressionally mandated nine-state assessment of climate change in the Rocky Mountains and Great Basin (1998). His most recent book is *Yellowstone's Destabilized Ecosystem: Elk Effects, Science, and Policy Conflict* (2006). Wagner received his B.S. in biology from Southern Methodist University and his M.S. and Ph.D. in wildlife management from the University of Wisconsin.

Michael L. Weber is marine advisor to the California Fish and Game Commission. He previously served as vice president of the Center for Marine Conservation and special assistant to the director of National Marine Fisheries Service. His published works include *From Abundance to Scarcity: A History of U.S. Marine Fisheries Policy* (2001), *The Wealth of Oceans* (1995), and *Fish, Markets, and Fishermen: The Economics of Overfishing* (1999). Weber earned an A.B. in classical languages and an M.A. in Greek literature from the University of California, Berkeley.

George M. Woodwell is founder, director emeritus, and senior scientist of the Woods Hole Research Center at Woods Hole in Falmouth, Massachusetts, as well as a founder and current board member of the National Resources Defense Council. A botanist by training, he helped create and direct the Ecosystems Center of the Marine Biological Laboratory in Woods Hole. He also served as a senior scientist at Brookhaven National Laboratory. Woodwell is currently supervising the construction of a new headquarters for research into global climate change and terrestrial ecology. He received the 2001 Volvo Environmental Prize in recognition of forty years of research into the role humans play in the environment. Woodwell holds an A.B. from Dartmouth College and an A.M. and Ph.D. from Duke University.

EDITORS

Larry L. Rockwood is associate professor of environmental science and policy at George Mason University, where he has taught general ecology, population ecology, and tropical ecology for over thirty years. He also chaired the biology department for over twelve years. Rockwood's research interests include population ecology and plant-animal interactions. He is the author of a recent textbook, *An Introduction to*

Population Ecology (2005), and of a laboratory manual on general ecology. He holds a B.S. and Ph.D. from the University of Chicago.

Ronald E. Stewart recently retired from his positions as visiting professor and graduate coordinator in the Department of Environmental Science and Policy at George Mason University. A fellow of the Society of American Foresters, Stewart came to George Mason from the U.S. Department of Agriculture after almost thirty years of service, during which he rose in the ranks from research forester and project leader to deputy chief of programs and legislation. His research and professional interests include the integration of science and policy, natural resource policy, and forest ecology; he has over forty-two publications in these areas. Stewart received his B.S. and Ph.D. in forestry management from Oregon State University.

Thomas Dietz is assistant vice president for research and graduate study, director of the environmental science and policy program, and professor of sociology and crop and soil sciences at Michigan State University. He is a fellow of the American Association for the Advancement of Science; he is also coauthor or coeditor of seven books and over eighty peer-reviewed papers and book chapters, including *The Drama of the Commons* (2002) and *New Tools for Environmental Protection: Education, Information, and Voluntary Measures* (2002). He has received numerous awards, including the Sustainability Science Award of the Ecological Society of America and the Distinguished Contribution Award from the Section on Environment, Technology and Society, American Sociological Association. Dietz holds an undergraduate degree from Kent State University and a Ph.D. in ecology from the University of California, Davis.

Foundations of Environmental Sustainability

Introduction: The Quest for Environmental Sustainability

The Coevolution of Science, Public Awareness, and Policy

Lee M. Talbot

Throughout history there has been "a dominant geographic theme which deals with the growing mastery of man over his environment. Antiphonal to this is the revenge of an outraged nature on man. It is possible to sketch the dynamics of human history in terms of this antithesis" (Sauer 1938, 765). This is how geographer Carl Sauer summed up the relationship between humans and their environment. The words are as true today as when they were written seventy years ago, but now that revenge is truly global.

The past half century has seen the most rapid and profound changes in human relationships with the environment in all of history. The population of the world has increased by two and a half times, standing now at about 6.5 billion. Global demands for food, energy, water, land, and other resources have expanded at an even faster rate. Significant developments in technology have given the growing population an ever greater leverage over the environment, and at the same time have provided us with the capabilities to see and measure these changes. There has been unprecedented recognition of the human impacts on the environment, ranging from pollution and loss of biodiversity to acid rain and global warming; likewise, recognition of human relationships with and reliance upon the environment has expanded from a few perceptive individuals to a worldwide phenomenon. Scientific understanding of the environment, ecosystem ecology, and global ecological processes has expanded exponentially.

With recognition has come demand for action. The nongovernmental environmental community has evolved from a few scattered, largely single-issue conservation organizations to a worldwide movement composed of tens of thousands of organizations, many with broad environmental concerns and sophisticated approaches. Armed with new and increasingly effective technologies such as remote sensing, satellite communications, and information handling abilities, the scientific community

3

has gained and presented comprehensive understanding of what we are doing to Earth's environment and of its significance for our present and future welfare. On a global basis there has been a dramatic evolution of environmental science and, at least partly as a result, an equally dramatic evolution of environmental policy.

Under pressure from the general public, nongovernmental organizations (NGOs) and the scientific community, as well as individual governments and intergovernmental bodies, have responded with an array of laws, institutions, and other policy instruments intended to protect various aspects of the environment. In this last half century the focus on the environment and environmental resources has radically expanded. Concerns with individual issues—for example, conservation of a particular species or area, maximization of a particular resource harvest, or coping with specific sources of pollution—have evolved into a more integrated recognition of the interrelationships involved, and toward the overarching concept of sustainability. The purpose of this book is to describe and analyze this critical period in history (roughly from the late 1950s to the present), assess where we stand today, and consider the challenges that lie ahead. Although the book's initial focus is the past half century, this period does not exist in isolation. Some historical perspective is useful for understanding this period and how it came about. Consequently, a brief consideration of its historical context is in order.

THE HISTORICAL PERSPECTIVE

First, it is important to realize that the human role in modifying the earth's environment is ancient and pervasive. Human prehistoric activities modified the vegetation, soils, waters, and wildlife of much—if not most—of the earth. Domestication of fire enabled them to make profound changes in vegetation, which in turn altered soil structure and location, water regimes and wildlife, and, at least in some cases, climate.

Sauer postulated in various publications that most of the earth's grasslands were the result of fire, for which human beings were largely responsible (see, e.g., 1947, 1950, 1962). He also emphasized the human role in affecting plant evolution during the Pleistocene. My own ecological studies on savanna ecology (Talbot 1964; Talbot and Kesel 1975; Talbot and Talbot 1963), and my other ecological studies throughout the world, have convinced me of the soundness of Sauer's basic conclusions. With few exceptions, the present location and composition of tropical savannas and most other grasslands are largely anthropogenic, through the use of fire and, in some cases, subsequent grazing of domestic livestock.

Humans have played a major role in many of the areas that are now desert. And the vegetation composition of many of the forests and woodlands of the temperate zones, the drier tropics, and even the moist tropics also appears to have been influenced, and in some cases, determined, by early humans.

Humans exerted a major impact on wild fauna as well as flora, both by altering vegetation and through direct hunting. It is now generally accepted that they played an important role in the extinction of some of the large fauna of the late Pleistocene. The more recent but still prehistoric spread of agriculture further altered the faunal scene. For example, based on my work in Southeast Asia, I am convinced that shifting cultivation allowed the spread of that area's rich variety of large wild animals (including many species of wild cattle, deer and deerlike animals, elephants, rhinos, and pigs) into areas that otherwise would have been closed tropical forests in which such animals cannot thrive or, often, even survive.

It has long been my conviction that human activities involving the clearing of forests and other vegetation changed the albedo and moisture balance and resulted in local and possible regional or global climatic changes. George Perkins Marsh postulated this view in 1864 but he lacked solid data, so for many years the view was not accepted. However, in 1979 Carl Sagan, Owen Toon, and James Pollack asserted that humans have substantially contributed to global climate changes during the past several millennia, and the evidence of early human modification of the climate continues to accumulate.

The domestication of livestock added another dimension to the leverage humans exerted over the vegetation, soils, waters, and wildlife. The results of overgrazing by goats and sheep are particularly noticeable in the Middle East and on the edges of the Sahara, but it should be remembered that domestic livestock have had a significant impact on virtually all environments, from reindeer in the Arctic to water buffalo and pigs in the tropics, and from camels, cattle, and sheep in the lowlands to yaks in the alpine zones.

And, of course, the exponential growth of human numbers has brought a corresponding exponential rise in environmental impacts which, in turn, have been amplified by increasing technology. Continued technological innovation, ranging from Clovis spear points and digging sticks to high-powered rifles and internal combustion engines, has had the effect of constantly increasing the leverage over the environment—the potential environmental impact—of each additional human.

My key point here is that human activities have played a major role in the evolution of the earth's ecosystems. In a real sense, human beings have been changing the face of the earth since their earliest times.

AWARENESS AND RESPONSE TO ENVIRONMENTAL CHANGE

Most people believe that concern with the environmental impact of human activities is a relatively new phenomenon. Many associate the beginning of this phenomenon with the first Earth Day in 1970, or the events that led to it. In addition, there is often the perception that concern with environmental conservation is a Western phenomenon that arose in North America or Europe.

It is true that much of the current public awareness of environmental change dates from the 1960s and 1970s; however, concern with the environment has an ancient history that long predates modern North American and European nations and cultures. Concern with environmental degradation, for example, was expressed by Plato in about 300 B.C., when he compared the deterioration of the deforested hills of Attica to the "skeleton of a body wasted by disease" (Taylor 1929).

The first recorded environmental policies calling for the creation of protected areas and protection of wildlife were created in India. Around 300 B.C. the *Kautiliya*

Arthasastra referred to establishing an animal park where all animals would be welcomed as guests and given full protection (Kangle 2003). At roughly the same time (about 250 B.C.), India's Emperor Asoka issued the Pillar Edicts, the earliest policies to provide total protection to certain species of wildlife (Maharajah of Mysore 1952). It should be noted that these policies must have derived from observations of the impacts of human activities on the environment, specifically the decreasing number of certain wildlife species due to hunting. They provide early examples of the coevolution of environmental science (in this case, presumably observation) and policy.

The first recorded environmental policy in Europe comes from Charlemagne, who, around 789 A.D., promulgated instructions for guarding forests, primarily as a refuge for wild animals. The same instructions also detailed how to kill wolves and clear land, and they may indicate both a groping for balance between forest and agriculture, and an acknowledgment that growing human populations were interrelated with multiple demands for forest use and demand for cropland (Glacken 1967).

There was systematic clearance of European forests at least as early as the sixth century a.d.; the process was sufficiently effective that by the twelfth century there was a need felt for forest protection (Glacken 1967). Throughout the world, primarily for their own use, rulers set aside preserves to protect wildlife and proclaimed laws to regulate hunting. For example, in 1066 William of Normandy brought well-established Continental hunting laws to Britain. In fact, in Europe, the threat to wildlife from hunting eventually led to policies of pure protection for some species (as it had in India centuries before). In 1423 Poland's King Ladislas Jagellon issued a law restricting hunting of wild horses and aurochs (both of which later became extinct) as well as elk; in 1569 the Swiss canton of Glarus prohibited hunting on Karpfstock Mountain, initiating a reserve that still exists; and in 1576 the Prince of Orange, in Holland, established the Wood of Hague as a pure reserve (Glacken 1967).

The idea of man's stewardship over nature also became increasingly codified in Europe, especially in the seventeenth century—for example, by Francis Bacon, Matthew Hale, John Ray, and others (Glacken 1967). But at the same time, except perhaps where hunted species were concerned, there was a predominant idea that changes in the environment were the work of God, not man. This belief was fostered by the slow rate of environmental change, and it persisted into the nineteenth century, when the seminal writings of George Marsh (1864) and others clearly identified man as the cause of these environmental impacts.

1700s–1800s

Throughout most of history the rate of anthropogenic environmental change was very slow. Then "in the late 18th century[,] the progressively and rapidly cumulative destructive effects of European exploitation became marked . . . In the space of a century and a half—only two full lifetimes—more damage has been done to the productive capacity of the world than in all of human history preceding" (Sauer 1938, 767).

This process was particularly marked in North America. From the late eighteenth century throughout most of the nineteenth century, energetic European settlers made profound changes to the continent. Vast areas of forest were cut and burned; agricultural lands from coast to coast were opened, abused, and frequently abandoned; wildlife was decimated, with endless herds of buffalo destroyed and the seemingly inexhaustible passenger pigeons exterminated. All of these changes took place over a few short decades. Such a visible disaster, in which cause and effect occur within individuals' memory spans, is often required in order to attract public and governmental attention and thereby spur action. I refer to this as "instant environmental disaster syndrome."

In the late nineteenth century people experienced the impacts of environmental degradation, and they recognized what was happening—demonstrating instant environmental disaster syndrome in operation. Unprecedented actions were taken to conserve forests, lands, wildlife, and other natural resources. Worried citizens formed NGOs focused on various environmental resources: the American Forestry Association (1875), the American Ornithologists' Union (1883), the Boone and Crockett Club (1887), the Sierra Club (1892), and the Audubon Society (1905). The federal government established environmental agencies, including the forerunners of the U.S. Forest Service, the Fish and Wildlife Service, and the National Park Service. For the first time concern for the environment, in the form of policies to conserve natural resources, became a priority issue for a national government.

Recognition of environmental degradation was also increasing in other parts of the world, focused primarily on larger forms of wildlife that were disappearing rapidly, and by the late 1800s a number of conservation initiatives were being undertaken. In Europe these included establishment of more reserves, promulgation of additional game laws, and proposals for national and international bird protection. In Africa, establishment of game reserves began with the Kruger Reserve (now Kruger National Park) in 1892. In Asia, the Bombay Natural History Society was created in 1883, becoming that continent's first environmental NGO.

1890–1909 IN THE UNITED STATES

In the late decades of the nineteenth century through the first decade of the twentieth century, American efforts to protect environmental resources continued at a high level. In some cases this was due to charismatic and influential people outside government, often scientists, who identified environmental problems and, through writing and speaking, brought public attention to them. These individuals sometimes also organized groups of supporters to pressure the government to take action on the problems. John Muir, a founder and first president of the Sierra Club, was such an individual, as was George Bird Grinnell, editor of *Forest and Stream* magazine and founder of the first state Audubon Society. In other cases, environmental efforts were sparked by leaders within government itself, such as President Theodore Roosevelt and Gifford Pinchot.

1909–1920s IN THE UNITED STATES

After this period of a high level of concern and activity, the environment dropped off the public's radar and ceased to be a national policy priority. Roosevelt's departure from the presidency doubtless played a role in the shift, but it was due at least in part to the nation's focus on World War I and on the period of prosperity, then depression, that succeeded it. However, although significant environmental actions were less nationally visible than before, they were still taking place. The National Park Act, for example, was passed by Congress in 1916 and the National Parks Association (an NGO, now the National Parks Conservation Association) was founded in 1919. Additionally, in 1924, Aldo Leopold and his Forest Service colleague M. W. Talbot helped create the nation's first official wilderness area, the Gila, in New Mexico.

1930s IN THE UNITED STATES

Then, in the 1930s, there was another spurt of heightened environmental concern, with resultant environmental policies and other actions. It was due, as had happened in the past, to another instant environmental disaster. Misuse of farmland and unwise attempts to expand cultivation into unsuitable grasslands in the Midwest had combined with drought conditions, leading to massive erosion and creating the Dust Bowl. The resultant clouds of dust blew as far as the nation's capital, and delegations of desperate farmers descended on Congress seeking assistance. Direct experience of the problem led to national action; new federal agencies were created to deal with the problems, including the agency that became the U.S. Soil Conservation Service (now the National Resources Conservation Service), as were various state agencies.

During the same period other environment and resource issues received substantial attention. Wildlife management was established as a distinct science-based profession, thanks in large part to Leopold, whose landmark *Game Management* was published in 1933—the same year President Franklin Roosevelt's establishment of the Civilian Conservation Corps further focused attention on environmental matters. A federal grazing service was established, as were several NGOs (among them the Wilderness Society in 1935 and the National Wildlife Federation the following year).

1940s–1950s

In the following two decades environmentalism's profile once again declined, for worldwide public and governmental concern was focused on World War II and its aftermath. The United States emerged from the war as a superpower with unprecedented industrial and technological might that it exercised enthusiastically. Internationally, major efforts were made to rebuild areas damaged by the war and to assist in the development of newly independent countries. New technologies such as DDT and nuclear power were developed and applied with little or no consideration for their possible environmental impacts. Although evidence of environmental problems was mounting, it went largely unnoticed by the general public.

During this period a number of individuals did recognize the problems, writing eloquently about them. In 1948 two extraordinarily perceptive books, *Our Plundered Planet* by Fairfield Osborn and *Road to Survival* by William Vogt, described the accelerating environmental impacts of human actions and their implications, as did the writings and exhortations of a number of other distinguished scientists and conservationists such as David Brower, Sigurd Olson, and Harrison Brown. Leopold's *A Sand County Almanac, and Sketches Here and There,* published posthumously in 1949, is recognized today as the "bible" of environment and especially of environmental ethics. Yet all this had little immediate effect on the public as a whole or, consequently, on public policy.

In the post–World War II period, environmental concerns had an even lower profile internationally than in the United States. There were some global environmental accomplishments, though. In 1930 Harold J. Coolidge, often called the "father of international conservation," started the American Committee for International Wild Life Protection, which he chaired until the 1970s. His efforts and those of others, including national and international colleagues, led to the establishment of the International Union for Protection of Nature in 1948. (Later known as the International Union for the Conservation of Nature and Natural Resources [IUCN], and now known as the World Conservation Union, this unique organization brings together more than 1,000 members including national governments, government agencies, and NGOs literally from all over the world.) And in 1949 the still-new United Nations held the international Scientific Conference on the Conservation and Utilization of Resources at its temporary headquarters at Lake Success, New York.

THE PAST HALF CENTURY: UNIQUE IMPACTS AND RESPONSES

The environmental impacts of the postwar period continued to accelerate well into the 1950s, while the public and governmental response remained muted. Then, starting in the early 1960s, the situation changed markedly. The impacts and subsequent responses reached unprecedented levels; indeed, the magnitude and global significance of the environmental changes of the past fifty years have far surpassed those of all humankind's previous history. This situation can be illustrated by a brief review of some of the changes since the 1955 Symposium on Man's Role in Changing the Face of the Earth (Thomas 1956). This landmark conference, held by the Wenner-Gren Foundation in Princeton, New Jersey, involved over seventy international scholars and authors. It provides us with a unique baseline from which to view the subsequent half century.

Changes Affecting the World Environment since the 1955 Conference

1. The human population has increased to two and a half times its 1955 size.
2. Well over half the world's tropical forests are believed to have been lost, and deforestation is considered to be a critical problem throughout the developing world.
3. The processes of overgrazing and desertification have greatly increased, resulting in dramatic spread of deserts or desertlike conditions, accelerated loss of agricultural lands, degradation of most of the world's rangelands, and untold human suffering.

In the 1967–1973 Sahelian drought, for example, anthropogenic land degradation led to the deaths of over 100,000 people and twelve million head of livestock, as well as the collapse of the agricultural bases of five countries (Mainali 2006).

4. Overfishing, pollution, and conversion of estuarine habitats have notably altered the world's marine habitats. Around 90% of the world's larger predatory fish have been lost to overfishing, with drastic—and, in some cases, probably irreversible—changes to their ecosystems. This is in stark contrast to the conclusion of the only paper on marine harvests in the 1955 symposium, which held that "the more one learns of marine ecology the less likely it seems that man would have any effect upon it" (Graham 1956, 488).

5. Biodiversity conservation has become recognized as a global environmental problem. In 1955 relatively few species were known to be endangered; the issue of endangered species received little attention, and when it did it was considered to be of local concern at best. Now we realize that loss of biological diversity is of major worldwide concern. Roughly one-quarter of the world's mammal species and one-eighth of its birds are recognized as endangered (Adams 2005), and many biologists believe that the human-caused extinctions today represent an extinction event as great as the one sixty-five million years ago that wiped out the dinosaurs.

6. Chemical pollution, hardly even recognized in 1955, has become all-pervasive and few species or lands escape its effects. The residues of pesticides, for example, are found in terrestrial and aquatic organisms around the world, often many thousands of miles from where they were originally applied. Thousands of new chemicals are released into the environment each year, mostly with unknown impacts. Commonly used chemicals or combinations of them are being found to have endocrine-disrupting impacts or other serious impacts on living organisms, even in minuscule amounts.

7. With increasing energy needs throughout the world, the combustion of fossil fuels has accelerated, leading to a host of local, regional, and global problems. Acid rain, not yet discovered in 1955, is now recognized as a major environmental problem. Production of CO_2 and other greenhouse gases is causing global warming, which presents humans with the most daunting environmental challenge in history. It has already led to climate and weather changes, loss of glaciers, changes in biodiversity, and alterations in terrestrial and aquatic ecosystems worldwide, especially at high altitudes and high latitudes. (In the 1955 symposium, the principal air pollution problem noted was from hydrocarbons, limited to the vicinity of their production from industrial and automotive sources.)

8. Assisting third world countries with economic development has become one of the world's largest industries. The annual flow of resources to developing countries is in the hundreds of billions of U.S. dollars, but much of this development effort has not resulted in sustainable benefits for the people and countries involved. Environmental impacts of development projects are many. Such projects—particularly those involving lumbering, resettlement, or displacement of people—have had numerous environmental impacts, leading to deforestation, loss of critical habitats and biodiversity, and the multiple problems involving dams. In other words, development projects that have not taken environmental factors into account have had the effect of reducing rather than enhancing the capacity of the lands involved to support people.

9. Alien or invasive species were mentioned in the 1955 symposium, but not as a major environmental problem. Since then they have become a principal environmental problem around the world, causing hundreds of millions of dollars of damage annually, as in the case of the zebra mussel in the United States. Alien species represent the second or third major cause of threat to endangered species in the United States and elsewhere.

Public and Governmental Responses to the Environmental Changes since 1955

During this past half century, as the nature and seriousness of environmental problems have been recognized, there has been a proliferation of governmental and nongovernmental actions to deal with them.

1. In 1955 there were only a few dozen environmental (or conservation) NGOs, mostly in the United States. Today there are tens of thousands of such NGOs throughout the world.

2. In 1955 few governments had environmental institutions (i.e., agencies, departments, or other governmental structures) as such. Today virtually all nations have such institutions, most of them supported by comprehensive bodies of environmental legislation developed since the 1970s.

3. In 1955 no government had legislation requiring assessments of the environmental impacts of governmental or other actions. Today almost all national governments and many states or other subunits of government have such legislation, as do all international development assistance institutions (such as the World Bank).

4. Environment has become an important concern of much of the UN system, led by the UN Environment Programme (which was established as a result of the UN Conference on the Human Environment, held in 1972).

5. In 1955 there were few international agreements or conventions that addressed environment, and most of these dealt with specific environmental resources such as fisheries or migratory species such as waterfowl. Today there is a comprehensive body of international environmental law, and a number of nations have negotiated bilateral agreements for cooperation on environmental matters.

1960s: DRAMATICALLY EXPANDING ENVIRONMENTAL AWARENESS WORLDWIDE

In this past half century the public and governmental responses to environmental changes have dwarfed all previous efforts. There are several reasons for this dramatic change. Certainly one important factor was instant environmental disaster syndrome, sparked in the 1960s by a series of devastating localized environmental disasters. The Cuyahoga River burst into flame at Cleveland, Ohio, when its heavy load of pollution was ignited. The Great Lakes were grossly degraded to the extent that Lake Erie was declared dead or dying (Moody 1994). There were massive spills of oil, including the *Torrey Canyon* tanker spill in the English Channel and the blow-out at the Union Oil platform off Santa Barbara, California. There were increasing numbers of serious smog incidents. Mercury pollution led to impacts on fisheries and human health in Scandinavia and elsewhere; the most widely recognized incident was an outbreak of Minimata disease, a toxic syndrome caused by mercury poison-ing in Japan's Minimata Bay, that caused about 3,000 deaths, deformities, and other pathologies.

These incidents were caused by pollution, but there were also growing numbers of environmental problems involving land and natural resources. The devastation caused by clear-cut logging was more and more visible and contentious. Endangerment of species such as elephants, rhinos, alligators, and leopards because of trade in their products (ivory, horn, skins, and hides) was increasingly becoming recognized. Valuable agricultural or scenic lands were being lost at climbing rates to unplanned urban sprawl, mining, and other uncontrolled development. The impact on wildlife (and humans) of the release of massive amounts of pesticides and other chemicals into the environment was becoming obvious, and was brought into sharp relief by Rachel Carson's book *Silent Spring* (1962).

The Key Role of the Media

Individually these incidents represented severe environmental impacts. But since the incidents and their effects were in general fairly localized, they would not have led to dramatic public awareness in previous years. However, by the 1960s, news media were greatly changing and expanding, television news broadcasts were ubiq-uitous, and television was available throughout much of the world. As a consequence, incidents that previously would have been of local concern were instead brought to national or global attention. In earlier decades the *Torrey Canyon* oil spill would have been of direct concern to people and local governments along the English Channel; with the television coverage of the 1960s it was brought into the living rooms of people throughout the world. Therefore the 1960s became a period of constantly increasing instant environmental disasters.

One result of the coverage was that Americans and others in ever larger numbers became aware of the existence and danger of environmental impacts of their actions. The public began to realize that government action was needed to protect the envi-ronment and human health, and in growing numbers these concerns were conveyed to elected officials.

The Role Played by the NGOs

This process was frequently led and encouraged by the NGO community. Prior to the 1960s the environmental (or conservation) NGOs were few in number. They were largely scattered, single-issue organizations, often working independently because they considered each other to be competitors for a limited supply of sup-porters and funding. In the 1960s the NGO community responded to the explosion of public environmental concern by broadening its focus, increasing its membership and resource bases, and spawning new NGOs to fill the new niches.

In the United States the principal pre-1960s NGOs were the Sierra Club, the National Audubon Society, the National Parks Conservation Association, the Wilderness Society, the National Wildlife Federation, the Izaak Walton League (founded in 1922), the Defenders of Wildlife (founded in 1947), and the Nature Conservancy (founded in 1951). Another important early NGO was the World Wildlife Fund (WWF), founded at the start of the decade. These NGOs had been "conservation organizations," focusing largely on protection of wildlife or areas. But in the 1960s most of the old-line conservation NGOs expanded their foci to include broader environmental issues, and from 1960 to 1969 the total membership of these organizations increased from 123,000 to 819,000. Late in the decade a series of new NGOs was founded. These included the Environmental Defense Fund (founded in 1967) and the Natural Resources Defense Council (founded in 1970), both of which provided legal leadership and assistance to the environmental community and aimed to use the nation's legal system to support environmental causes. Other new NGOs included Friends of the Earth (founded in 1969), Environmental Action (founded in 1970), and the League of Conservation Voters (founded in 1970).

The NGOs' Confrontational Approach

These new NGOs epitomized the new militancy of the environmental movement. For decades, the relatively few people concerned with environmental issues had felt, correctly, that they were fighting an uphill battle with potent opponents and an unresponsive government. The established industrial, agricultural, and other economically powerful interests were accustomed to proceeding with no concern for the environment, and with disdain for the environmental community—if they recognized that such a thing existed. As the environmental movement grew in the 1960s, it began to be forcefully attacked by some industries that felt threatened by it. The most publicized example was the chemical industries' personal and professional attacks on Rachel Carson in response to *Silent Spring,* which effectively brought to public attention the dangers to wildlife and humans of overuse of pesticides and other chemicals.

The executive branch of government was also widely regarded as an enemy by the environmental community. The agencies that managed or impacted environmental resources focused on their narrow missions and often excluded conservationists and their concerns. These agencies usually had come into being because the individuals or industries that would become their principal clients convinced Congress to create them. Congress then established committees to provide oversight of the new agencies, and their resultant missions became set in concrete in the form of the "iron triangle"—clients, congressional committees, and the agencies themselves. The initial mission of agencies such as the Forest Service, the Bureau of Commercial Fisheries, and the Fish and Wildlife Service was to support the client (lumbering industry, commercial fisheries, and hunters, respectively) and facilitate that client's exploitation of the resource involved. The Forest Service, for example, was established to bring about scientific management of timber harvesting and grazing. Accordingly, it was often the object of scorn from the timber and grazing industries and their representatives in Congress. It took federal court decisions to establish that the Forest Service even had the authority to regulate grazing on the national forests. Conservationists were frequently stonewalled or rebuffed by these agencies; until the 1970 enactment of the National Environmental Policy Act (NEPA) of 1969, conservationists usually did not have a legal basis to challenge the agencies or even find out what they were doing.

The environmental movement gained adherents, momentum, and strength during the 1960s, and by the latter part of the decade the movement had become very strong. The long-standing underdog status of the conservationists and their deep frustrations with industry and government led to an "us versus them" mentality and a confrontational approach that characterized much of environmentalism well into the 1970s. Only in recent decades has environmentalism moved from confrontation to a philosophy of cooperation.

The U.S. Congress in the 1960s

Congress showed interest in occasional environmental issues early in the decade, passing, for example, the Wilderness Act in 1964 and the first Endangered Species Act in 1966. But by the latter part of the decade Congress had developed a real focus on environmental problems and was actively seeking policy solutions. This concern resulted from a combination of communication from constituents and NGO lobbyists and the personal interests of some congressional leaders. Significantly, congressional concern with the environment was totally bipartisan at that time.

Congress established a Joint Senate-House Committee on the Environment with an outside scientific advisory board to provide guidance, and specific policy issues were taken up by the relevant Senate and House committees that prepared legislative proposals. In the fall of 1969 Congress almost unanimously passed NEPA. This was probably the most critical American environmental legislation ever enacted. Among other things, it established the nation's environmental policy, instructed the executive branch of government to adhere to the policy, required environmental impact statements (EIS) as an action-forcing mechanism to enforce the policy, established the Council on Environmental Quality (CEQ), and gave the public access to agency processes and standing to take legal action against agencies if their actions did not adhere to the policy.

Developments in Environmental Science

The 1960s was also a period of important growth for environmental science. Ecology became both more comprehensive and more recognized. By the end of the decade, most universities had established curricula in ecology and other environmental sciences. Additionally, research carried out in the 1950s under the International Geophysical Year effort had showed the benefits of international cooperation in scientific endeavor, and the International Biological Program (IBP) in the 1960s brought these benefits to ecology and related sciences. The IBP's biome programs, in particular, involved interdisciplinary research on selected ecosystems, leading to much improved understanding of the environment and of human impacts on it. The IBP's conservation program resulted in scientific acceptance of conservation issues and needs as well as a higher recognition of the importance of conservation, especially in developing nations.

International Efforts in This Period

Beyond the IBP, other international efforts reflected the increased environmental concerns of the 1960s and further raised awareness of environmental issues and the need for action. For example, in 1968 the UN Educational, Scientific and Cultural Organization convened an Intergovernmental Conference on the Scientific Basis for Rational Use and Conservation of the Resources of the Biosphere. This led, later that year, to a UN Resolution on the Problems of the Human Environment, which established environmental protection as a major policy thrust of the United Nations. Also in 1968, the Conservation Foundation and the Center for the Biology of Natural Systems held an international Conference on the Ecological Aspects of International Development (Farvel and Milton 1972), a preview of what would become the major international theme of environment and development.

THE 1970S: UNPRECEDENTED ENVIRONMENTAL ACTIONS

U.S. Environmental Policies

The 1970s were a period of unprecedented activity and accomplishment in terms of environmental science, policy, and activism. On January 1, 1970, President Richard Nixon signed NEPA into law as his first symbolic action of what he termed "the Environmental Decade." As mandated by NEPA, the CEQ shortly came into being under the chairmanship of Russell Train, who was followed as chair in 1973 by Russell Peterson and in 1978 by James Gustave Speth. During the 1970s the CEQ played a key role in developing a wide range of national and international environmental policies and other actions, in addition to its roles as federal environmental coordinator, overseer of the EIS process, and preparer of annual reports on the state of the American environment. In the early 1970s, when public and bipartisan congressional support for environmentalism were at their height, more environmental legislation and other expressions of environmental policy were accomplished than in any comparable period in U.S. history. A list of a few of the many environmental laws passed in this period illustrates the range of their coverage: the Clean Air Act (1970), the Marine Mammal Protection Act (1972), the Clean Water Act (1972), the Coastal Zone Management Act (1972), an Endangered Species Act (1973), the Forest and Rangeland Renewable Resources Planning Act (1974), the National Forest Management Act (1976), the Fishery Management and Conservation Act (1976), and the Federal Land Policy and Management Act (1976).

The principle established by NEPA that actions should not be undertaken without considering their environmental impacts, and the requirement for EIS or environmental analyses, were rapidly adopted by individual American states, many of which legislated "little NEPAs," as well as by foreign governments and international development agencies. Today virtually all nations and international agencies require environmental impact assessments.

Another governmental environmental milestone in 1970 was the creation of the Environmental Protection Agency (EPA). This was basically a reorganization intended to combine a number of existing environmental regulatory components of other agencies and departments into an independent environmental agency where they could be more effective.

The New Look for Resource Management Agencies and Personnel

During this period, many U.S. government agencies had to make very significant changes in their cultures and in the way they dealt with environmental issues and the public. NEPA and its EIS requirements gave the NGOs and public unprecedented access to agency decision making and other processes. And passage of the Marine Mammal Protection Act and the Endangered Species Act established the precautionary principle and changed the agencies' focus. The "iron triangles" became severely bent or recast as the agencies often were forced to recognize that their constituencies had changed—they were different and much broader than before, and consequently their missions were substantially widened. For example, the Bureau of Commercial Fisheries, by then renamed the National Marine Fisheries Service (NMFS), had to shift its focus from generally unquestioned promotion of and assistance to the fishing industry toward management of the impact of fishing and conservation of the fish and the marine ecosystem. W. W. Fox Jr., a director of the NMFS, noted that "this was a complete reversal of natural resource policy practice up to that point. Traditional practice was to allow fishing to occur unless scientists could prove that it was harming fish populations" (Weber 2002, 127).

These changes logically and profoundly affected the resource management personnel of the agencies, their professional societies, and ultimately the universities and other training institutions that produced the professionals. Consider, for example, the field of wildlife management, which developed primarily in response to the needs of hunters. Most of the funding for wildlife management, research, and training came from hunters, either directly from licenses and other fees or indirectly from dedicated excise taxes on hunting equipment. As an understandable consequence, game birds and animals were the initial focus of wildlife management, research, and training. But the growth of broad public interest in and concern with the environment was accompanied by an increase of public interest in wildlife that did not involve hunting. People actively involved in activities like bird-watching, wildlife photography, and simple study and enjoyment of nature soon greatly outnumbered those involved in hunting. Gradually, the wildlife profession adapted to this changed clientele. Into the 1960s, the articles in the Wildlife Society's *Journal of Wildlife Management* and the presentations at the North American Wildlife Conferences were largely hunting-related. Now nongame subjects predominate in those forums; in fact, the index of a recent issue of the journal (October 2005) lists forty-two articles, of which only three focus on hunting or hunted species as such. And university wildlife curricula in the 1960s focused almost exclusively on game species, whereas nongame and ecosystem-oriented studies are now the norm (with hunting-related subjects still present but no longer predominant).

The adaptation of the resource management agencies and their personnel to the new environmental realities was not a smooth or easy process, and that process has not yet been wholly completed. It originally represented a fundamental change, and so it was at times strongly resisted. As an assistant secretary of agriculture noted about the agencies, "It is as though they circled the wagons and sought to valiantly fight off the Indians of change" (R. Cutler, pers. comm., 1977).

Earth Day

Earth Day was first held on April 22, 1970. The event involved mass rallies in major U.S. cities, along with environmental speeches and other events at schools and universities and throughout communities nationwide. Sponsored by Senator Gaylord Nelson and Representative Paul McCloskey, the event (now an annual observance) was intended to increase public awareness and mobilize citizen participation in environmental issues. Earth Day 1970 is often regarded as the culmination of the growth of environmentalism of the 1960s, or as the start of the increased environmental activism of the 1970s. Doubtless it was both.

NGOs in the United States

The early 1970s arguably represented the zenith of the effectiveness of the NGO community in the United States in terms of its ability to influence and promote the development of new environmental policy. In subsequent decades many more NGOs have been established at national, international, and local levels; they also have become much larger, much better funded, and more professional. But they have not exceeded, in terms of the visible impacts of their efforts on policy, the work of the early 1970s NGOs. The success of those organizations stemmed from reasons including the high levels of public environmental awareness at the time, the consequent bipartisan congressional concern with environment, and good access to government, especially through the CEQ. The governmental policy processes were generally receptive to the NGOs and their message. However, all of this changed. By the later 1970s, environment had become a partisan issue; there was a backlash against environmental regulations by industry and some of the public, especially westerners with the "sagebrush rebellion"; and energy had become a dominant political and economic concern. One result was that NGOs (as well as those in government concerned with environment) had to spend much of their energy simply to keep the environmental gains of the early 1970s from being eroded.

Since the 1960s most of the larger American NGOs have developed international programs, and some have established branches or offices in other countries. By 1975 so many U.S. NGOs had international activities that the American Committee for International Conservation was created (from Harold Coolidge's original American Committee for International Wild Life Protection) as an umbrella, voluntarily coordinating organization for them. It also served, at first, as a national committee for IUCN.

International Developments

Internationally the key environmental event of the 1970s was the UN Conference on the Human Environment, held in Stockholm, Sweden, in 1972. Some 1,200 participants representing 113 governments were present, along with a similar number of representatives of NGOs. The conference produced a set of environmental principles and a long action plan of recommendations. One of these recommendations resulted in the establishment of a new UN agency, the UN Environment Programme (UNEP), intended to be the environmental focal point and conscience of the UN system. Under the leadership of Maurice Strong, the conference succeeded in placing the environment in a prominent position on the agendas of most nations throughout the world, and this may be its most significant result. The conference also established the principle that the NGO community should be involved along with governments, and this pattern was continued in subsequent environmental theme conferences convened by the United Nations in later years on topics including desertification, population, human settlements, and water.

This decade also saw the adoption of a number of international environmental agreements. The subjects of multilateral conventions included control of ocean pollution in 1972 and 1975, establishment of the World Heritage Trust in 1972, control of trade in endangered species in 1973, and protection of wetlands in 1975. There were also a number of bilateral environmental agreements, such as those between the United States and the (then) Soviet Union, and the United States and Japan.

1980–2007

Struggle in the United States

After the heady days of the 1970s, those concerned with environmental protection in the United States faced a radically different governmental environment. President Ronald Reagan's Republican administration was ideologically and actively antienvironment. His agency heads, for example at the Department of the Interior and the EPA, were so openly and aggressively antienvironment and so successful in starting to dismantle the environmental protection institutions created in the previous decade that they created a public and congressional backlash. NGOs such as the Sierra Club saw their memberships rise to unprecedented numbers, and the House of Representatives went so far as to cite Ann Gorsuch, the head of the EPA, for contempt of Congress. Reagan's administration was forced to backpedal on some of its antienvironment actions and remove some of the most egregious agency heads, even replacing a few with strong environmental leaders, such as William P. Mott as director of the National Park Service and William Ruckelshaus as administrator of the EPA.

The subsequent years have been characterized by partisan politics dominated by largely antienvironment conservatives. There have been some successes in terms of environmental policy and action, but for the most part the period up to 2001 can be described as a holding action. Much, if not most, effort by environmentalists has been directed at maintaining existing environmental policies and institutions in the face of attempts to reduce or dismantle them, rather than looking ahead and developing new environmental initiatives. The period from 2001 to date has seen unprecedented and often successful efforts by the George W. Bush administration to dismantle the nation's environmental policies, regulatory structures, and other environmental institutions. It appears that much of the American public remains unaware of the magnitude and significance of these efforts, for public attention is more focused on issues related to wars and security.

In the 1960s and 1970s the United States was the world leader in environmental policy and actions. The nation's role was epitomized by the almost universal adoption by other nations of the NEPA mechanisms and of many U.S.-led measures for wildlife management, national parks, endangered species protection, and pollution control. Largely through the CEQ, the United States played a leading role in developing and negotiating a series of international conservation and pollution control conventions; the United States also played a very critical role in the preparations for the Conference on the Human Environment and its outcomes. Additionally, the United States provided substantial environmental assistance to other nations through active international programs in government environmental agencies such as the Forest Service, the EPA, the Fish and Wildlife Service, and the National Park Service, and through the bilateral assistance provided by the U.S. Agency for International Development. However, the administrations in power from 1980 on either have not supported or have actively emasculated most of these programs. Not only did the United States lose its position of leadership in international environmental affairs, but its government became widely perceived as the opponent of international cooperation on environmental issues epitomized by global warming and biodiversity conservation.

Regardless of the government's stance, polls indicate that the support of the American public for environmental protection has remained fairly strong since the 1970s, although it has never reached the predominant levels of Earth Day and the early 1970s. Since then, in terms of public attention, environment has frequently taken a backseat to other issues including wars, the economy, security, and energy.

Since the 1970s the NGO community has continued to expand. Many new local NGOs have been created throughout the country, and some new national ones have joined those that existed in 1970. The 2005–2006 *Conservation Directory* lists hundreds of environmental organizations (National Wildlife Federation 2005). Listed alphabetically, under the letter *A* alone there are 220 nongovernmental, nonprofit environmental organizations. Most of the major NGOs have increased substantially in size and funding. They often have headquarters in Washington, DC, to give them better access to government, and numerous organizations have substantially increased their lobbying staffs. Many also have branches or offices elsewhere in the United States.

Continued International Progress

Internationally there has been relatively steady progress in environmental policy and actions since the 1970s. With the impetus of the Conference on the Human Environment, nations throughout the world took on environment as a governmental concern. By now nearly all nations have adopted some version of the NEPA approach—that is, a formalized national environmental policy, some form of government environmental institutions (agencies, departments, ministries, etc.), and the requirement that environmental assessments be made of major actions that may impact the environment.

Since the Conference on the Human Environment, environmental NGOs have pro-
liferated in virtually all countries to the point where there are now literally tens of
thousands of them. While the majority of the NGOs are local, many are effectively national and some international. The truly international NGOs, such as Flora
and Fauna International (formerly the Fauna Preservation Society), Conservation
International, the WWF, and BirdLife International (formerly the International
Council for Bird Preservation), have significantly increased their funding and activities and often have established regional offices or branches in addition to their central ones. The same is true for IUCN (although it is not a true NGO since it is an
organization combining both governmental and nongovernmental members). Many
formerly national NGOs from both North America and Europe have become international, and NGOs have become so important that most UN organizations and
international development institutions have established sections or departments
specifically to work with NGOs.

Environment and Development: The Goal of Sustainability

International considerations of environment have been intimately linked with considerations of development. Development is a main priority of developing nations.
Initially, many such nations perceived that environmental considerations might limit
or impede development. Some spoke of wanting pollution because they saw it as a
sign of industrialization and therefore development, and some spoke of environmental concerns as being a neocolonial attempt to keep them from developing. The
organization of the Conference on the Human Environment sought, with some success, to show that development and environment should be mutually supportive, not
antagonistic. But development activities (e.g., those funded by the World Bank or
conducted by bilateral aid agencies) often had very negative environmental impacts,
as well as frequently negative impacts on the indigenous peoples involved. In many
cases, failure to prevent environmental impacts led to the failure of the original
development's objectives.

So the formidable challenge was to convince developing nations, international
development institutions, and environmentalists that environment and development
were two sides of the same coin, essential to each other. Beginning in 1975 IUCN,
in collaboration with the WWF and UNEP, undertook a five-year effort to develop
the World Conservation Strategy. This strategy was undertaken to help achieve sustainable development through conservation, the management of human use of the
biosphere so that it may yield the greatest sustainable benefit to present generations
while maintaining its potential to meet the needs and aspirations of future generations (IUCN, WWF, and UNEP, 1980). It emphasized the inseparable mutual roles
of environment and development and introduced to a worldwide audience the concept of sustainability as a principal objective. It was launched at high-visibility events
worldwide in 1980.

In the same year, the Independent Commission on International Development
Issues, informally known as the Brandt Commission, completed two years of work
and released a report called *North-South, a Programme for Survival*. The report analyzed the world's economic and social predicament as it affects the third world and

concluded with a set of proposals for the reform and restructuring of the world system. It emphasized that few threats to peace and the survival of the human community are greater than those posed by the prospects of cumulative and irreversible degradation of the biosphere on which human life depends and stated that "it can no longer be argued that protection of the environment is an obstacle to development. On the contrary, the care of the natural environment is an essential aspect of development" (Independent Commission on International Development Issues 1980, 114).

Subsequently the United Nations set up the World Commission on Environment and Development to further examine how to bring environment and development together for the benefit of all. Sustainability was a central focus of the commission's final report, *Our Common Future* (World Commission on Environment and Development 1987), which also emphasized that sustainable development was not possible without environmental protection. This was the central focus, too, of the 1992 UN Conference on Environment and Development, informally called the Earth Summit, held in Rio de Janeiro, Brazil. It was attended by representatives of 172 nations and, among other things, occasioned the signing of major treaties concerning climate change and conservation of biological diversity.

Beginning to recognize the role of environment in development, the World Bank established a very small environment office in the 1970s. For many years Robert Goodland was the sole environmental scientist on a staff of about 10,000. However, he succeeded in drafting a number of policies and procedures to incorporate environmental and social concerns into the World Bank's development work. In the mid-1980s a major reorganization included greatly expanding the group's environmental staff, establishing a central environment department along with environment divisions in each of the group's geographic operational units. The World Bank's environmental policies and procedures have been outstanding in the international development field, and subsequently the other international development institutions (including the regional development banks) have all adopted somewhat similar approaches.

Environmental Science since the 1970s

There have been considerable developments in environmental science since the 1970s. First, new capabilities have been provided by significant developments in technology and methodology. Advances in computer technology have revolutionized information handling, analysis, and calculation capabilities, and the development of personal computers has made computer capabilities accessible to most researchers and many managers, regardless of where they are, throughout the world. Satellite technology and other remote sensing developments have provided another dimension to the types of information that we can obtain, and they have helped to create a much more global perspective on environmental issues. The development of geographic information systems has provided another computer-based technology for handling and analyzing a variety of data; related new technologies include improvements in telemetry and global positioning systems. The result has been wide accessibility of the capabilities for obtaining, handling, and storing vastly more information than ever before. Computer advances also allow analysis of complex databases and creation of models and simulations that permit us to avoid many of the simplifying assumptions that were required before fast computing was generally available.

Moreover, understanding of ecosystems has been dramatically improved. The IBP biome studies of the 1960s, preceded by pioneering East African ecosystem research in the late 1950s, showed how cooperative, interdisciplinary research could provide the information that leads to understanding complex ecosystem processes. Subsequent research—particularly that of the Institute of Ecosystem Studies in Millbrook, New York, under president and founder Gene Likens, and the multi-disciplinary programs of the National Science Foundation and the International Council of Scientific Unions—has added greatly to ecosystem knowledge and understanding.

One result has been a profound change in the way ecosystems are perceived. Formerly the dominant paradigm was that of an ecosystem that was stable, closed, and internally regulated, and that behaved in a deterministic manner. This paradigm was the foundation for management of living resources such as forests, rangelands, wildlife, and fisheries, as well as for conservation of parks and endangered species. Now we realize that an ecosystem is a much more open system, one that is in a constant state of flux, usually without long-term stability, and one that is affected by a series of human and other, often stochastic, factors, many originating outside the ecosystem itself. As a result the ecosystem is recognized as probabilistic and multi-causal rather than deterministic and homeostatic; it is characterized by uncertainty rather than the opposite.

This new understanding of ecosystem processes is leading to profound changes in conservation and management of living resources. It is now realized that management of a protected area must take the ecosystem dynamics and uncertainties into account; simply putting a fence around an area and assuming that the area will remain unchanged will not work. Similar considerations apply to species conservation and harvests of living resources. Scientists and some managers realize that it is necessary to take the ecosystem into account, not just a single species; to recognize that ecosystems are dynamic, not static; and to understand that impacts on the ecosystem, such as those caused by logging, fishing, hunting, or grazing, may lead to irreversible changes in the status of the ecosystem. Furthermore, it is now widely recognized that social and economic factors often lead to impacts on ecosystems, in addition to playing a fundamental role in determining management goals and actions, and that they must be taken into account in nearly any conservation situation. In the belief that the traditional scientific disciplines were not in a position to use this new knowledge to deal effectively with the scientific complexity, and the urgency, of conserving biological diversity, a group of scientists established the Society for Conservation Biology in 1986 and thereby developed conservation biology as a vigorous new discipline.

THE QUEST FOR SUSTAINABILITY

Environmental changes caused by human activities have multiplied gradually over time, reflecting the growing human population and the developing technologies with which the changes are closely linked. Since the nineteenth century, the pace of environmental change has greatly accelerated; the most rapid increase has occurred post–World War II. In the United States, public concerns with these environmental problems and governmental responses to them have risen and fallen over time, with notable periods of higher activity from roughly the 1880s through 1909, during the 1930s, and in the 1960s and 1970s. These periods coincided with heightened public awareness of the negative impacts of anthropogenic environmental change.

The 1960s, culminating in Earth Day 1970, were a period of dramatically rising environmental awareness. The 1970s saw remarkable environmental actions at both governmental and public levels worldwide. Internationally there has been steady progress in environmental policy and action since the 1970s. However, in the United States, the 1980s and 1990s were characterized by partisan struggles between anti-environment and proenvironment efforts, and since 2001 there has been increasingly effective dismantling of environmental policies and institutions.

In this past half century, both the environmental changes and the public and governmental responses to them have dwarfed all previous levels. The environmental changes—notably including global warming—are simply a continuation of Sauer's "revenge of an outraged nature." The environmental responses are the result of the coevolution of environmental science, public awareness, and policy. Increasingly the stated goal of this process is environmental sustainability, but while much progress has been made, that goal is not yet in sight.

References

Adams, W. M. 2005. *Against extinction: The story of conservation.* London: Earthscan.

Carson, R. 1962. *Silent spring.* Boston: Houghton Mifflin.

Farvel, M. T., and J. P. Milton, eds. 1972. *The careless technology: Ecology and international development.* Garden City, NY: Natural History Press.

Glacken, C. J. 1967. *Traces on the Rhodian shore: Nature and culture in western thought from ancient times to the end of the eighteenth century.* Berkeley: University of California Press.

Graham, M. 1956. Harvests of the seas. In *Man's role in changing the face of the earth,* ed. W. L. Thomas, 487–503. Chicago: University of Chicago Press.

Independent Commission on International Development Issues. 1980. *North-south, a programme for survival: Report of the Independent Commission on International Development Issues.* Cambridge, MA: MIT Press.

International Union for the Conservation of Nature and Natural Resources, United Nations Environment Programme, and World Wildlife Fund. 1980. *World conservation strategy: Living resource conservation for sustainable development.* Morges, Switzerland: IUCN/WWF; Nairobi, Kenya: UNEP.

Kangle, R. P. 2003. *The kautiliya arthasastra.* Pt. 2, 2nd ed. Delhi, India: Motilal Banarsidass.

Leopold, A. 1933. *Game management.* Madison: University of Wisconsin Press.

Leopold, A. 1949. *A Sand County almanac, and sketches here and there.* New York: Oxford University Press.

Maharajah of Mysore. 1952. *Indian Wild Life Bulletin* (December).

Mainali, K. P. 2006. Grazing causes desertification in Sahel. *Frontiers in Ecology and the Environment* 4(5): 232.

Marsh, G. P. 1864. *Man and nature*. New York: Scribner. Repr., Belknap Press, 1965.

Moody, D. W. 1994. Pollution, water: Case studies. In *Encyclopedia of the environment*, ed. R. A. Eblen, and W. R. Eblen, 555–557. New York: Houghton Mifflin.

National Wildlife Federation. 2005. *Conservation directory 2005–2006*. Washington, DC: Island Press.

Osborn, F. 1948. *Our plundered planet*. Boston: Little, Brown.

Sagan, C., O. B. Toon, and J. B. Pollack. 1979. Anthropogenic albedo changes and the earth's climate. *Science* 206(4425): 1363–1368.

Sauer, C. O. 1938. Theme of plant and animal destruction in economic history. *Journal of Farm Economics* 20(4): 765–775.

Sauer, C. O. 1947. Early relations of man to plants. *Geographical Review* 37(1): 1–25.

Sauer, C. O. 1950. Grassland climax, fire, and man. *Journal of Range Management* 3(1): 16–21.

Sauer, C. O. 1962. Fire and early man. *Paideuma* 7:399–407.

Talbot, L. M. 1964. The biological productivity of the tropical savanna ecosystem. In *The ecology of man in the tropical environment*, 88–97. IUCN New Series, no. 4. Morges, Switzerland: IUCN.

Talbot, L. M., and R. N. Kesel. 1975. The tropical savanna ecosystem. *Geoscience and Man* 10:15–26.

Talbot, L. M., and M. H. Talbot. 1963. *The wildebeest in western Masailand, East Africa*. Wildlife Monographs, no. 12. Washington, DC: Wildlife Society.

Taylor, A. E. 1929. *Plato, Timaeus and Critias: Translation from the Greek*. London: Methuen.

Thomas, W. L., ed. 1956. *Man's role in changing the face of the earth*. Chicago: University of Chicago Press.

Vogt, W. 1948. *Road to survival*. New York: Sloane.

Weber, M. L. 2002. *From abundance to scarcity: A history of U.S. marine fisheries policy*. Washington, DC: Island Press.

World Commission on Environment and Development. 1987. *Our common future*. Oxford: Oxford University Press.

PART I

SETTING THE STAGE: PERSPECTIVES ON ECOLOGY AND THE ENVIRONMENTAL MOVEMENT

Ronald E. Stewart

The authors of part I examine the development of the environmental movement. They detail events and advances in conservation thought and policy worldwide, focusing particularly on the United States. Such advances, they argue, often result from a combination of (1) the emergence in the public arena of a real or perceived environmental crisis and (2) a catalyst or charismatic leader who frames the issue in a way that captures public attention and galvanizes public opinion. Though scientists can often best define an issue and simultaneously carry the credibility to speak, few have become advocates for public policy change, for fear of damaging their scientific objectivity. Those who have been successful in bridging the gap between science and policy, such as Lee M. Talbot, have had such stature that they were able to make a difference.

The marriage of environmental science and public policy has been rocky; policy frequently trails science by years or even decades. The trend in availability of, and support for, knowledge of how ecosystems respond to human intervention has generally been upward. However, like the historic progression of policy, it is subject to reversals. Both science and policy are affected by changes in national and international economic conditions and politics. In general, though, reversals in progress are never complete so that forward progress continues, albeit often at a less than satisfactory pace.

A number of historic and contemporary examples illustrate the power of the convergence of an issue with a human catalyst for change. For example, timber famine, an issue that arose in the late nineteenth century, stemmed in part from the overharvesting of forests in the eastern United States and in the Great Lakes states. This problem was described and framed by George Perkins Marsh in *Man and Nature* (1864). Marsh's work helped Americans recognize that continued overharvesting of natural resources could lead to natural disasters. Franklin Hough then published a series of studies in the 1870s and 1880s on the condition and extent of U.S. forests; these studies called for the establishment of forest reserves to end the decades-long policy of disposal of lands from the public domain. This resulted in the 1891 establishment of the first forest reserves. Other leaders in the fight to protect and conserve natural resources included private citizens (such as John Muir, founder of the Sierra Club, in the late

1800s and early 1900s) and public officials (such as Gifford Pinchot, first chief of the Forest Service, and President Theodore Roosevelt in the early 1900s; and Senator Hubert Humphrey in the 1960s and 1970s).

In another example, the issue of overuse of pesticides and their effects on wildlife and humans arose in the 1960s. It was brought into the public arena by Rachel Carson's *Silent Spring* (1962). Carson's work, and a general environmental awakening following the excesses of the 1950s, catalyzed action and resulted in a phenomenal growth of environmentalism and environmental organizations throughout the 1960s. This paved the way for significant changes in U.S. environmental policy with passage of the Clean Air Act, the Clean Water Act, the National Environmental Policy Act, the Freedom of Information Act, and the Endangered Species Act in the late 1960s and 1970s.

The authors of part I identify many other issues, individuals, and organizations within and outside the United States that together brought about significant changes in perception of environmental problems and in public policy. (In several cases, key leaders were the authors themselves.) However, such advances cannot always be sustained. Each generation must step forward and continue to make individual and collective decisions to improve the earth's environment. Environmental concerns can be subsumed by economic or other national and international crises.

Improvements in environmental conditions have resulted from the combination of visionary individuals and organizations with growing public awareness of critical environmental issues. While significant improvements have been made, diligence is needed to maintain progress. Changes in economic conditions, political leadership, and continuing increases in human populations and demands on resources may result in backsliding.

In chapter 1, Russell W. Peterson depicts the excesses of the Industrial Revolution and the explosion of the human population as the issues that "opened our eyes to the devastating impact that human activities could have on the earth." But citizens didn't begin organizing to do something about this threat until the late nineteenth century; as with other important movements in history, events and individuals converged to catalyze change. Specifically, predicted timber famine, catastrophic floods, reductions in important species, and other issues and events raised the public consciousness, but it took persons of vision and influence to bring about change. Peterson identifies pioneer environmentalists John Muir and President Theodore Roosevelt, along with the Audubon Society, as key to environmental change around the turn of the twentieth century in the United States.

Peterson traces the environmental movement through the twentieth century, noting that it gained an ethic and a mission with two important midcentury publications: Aldo Leopold's *Sand County Almanac* in 1949 and Rachel Carson's *Silent Spring*. Many governmental and nongovernmental environmental organizations around the world were created shortly thereafter. These included the International Council for Bird Preservation (now BirdLife International), the International Union for the Conservation of Nature and Natural Resources (IUCN, now the World Conservation Union), and the World Wildlife Fund. Increasing involvement of the United Nations in environmental issues resulted in the first UN Conference on the Human Environment in Stockholm, Sweden, in 1972. The United States was a major supporter of this conference, having established from 1970 to 1972 the Environmental Protection Agency, the Council on Environmental Quality (CEQ), environmental impact statements, and, as previously mentioned, a number of innovative environmental laws. (Leaders during this period included Peterson himself, as well as Russell Train and Lee Talbot, who impacted environmental issues on both the local and international levels.) Peterson calls the National Environmental Policy Act "America's Magna Carta for the environment."

While much has been accomplished by individuals and environmental organizations through direct efforts and the rallying of public pressure, affirms Peterson, all must remain diligent. For changes in economic conditions, and, often, accompanying political changes, can result in backsliding. Peterson also calls for the embracing of sustainable development principles and for adoption of what he calls a "soft energy" path, which involves using energy conservation and developing renewable forms of energy. Although it will take several decades to make the transition to a soft-energy future, Peterson believes that doing so will provide us with one of our best paths to resolving the global predicament.

Finally, Peterson reviews the life and accomplishments of Lee Talbot, a key player in the development of U.S. environmental policy and practice. Peterson's thirty-year friendship with Talbot provides a unique perspective of an individual who continues to influence environmental thinking worldwide today.

In chapter 2, Russell E. Train identifies the 1960s and 1970s as a period of groundbreaking efforts in environmental protection and awareness, and traces the history of environmental conservation in Africa and the United States during this period. In Africa, the development of important private organizations, such as the African Wildlife Leadership Foundation and others, resulted in improvements in conservation of African wildlife. In the United States, environmental organizations and helpful politicians brought about changes in environmental policy that continue to guide the nation's environmental conservation efforts.

According to Train this period was one of unprecedented accomplishment on the environment front. As discussed earlier, such progress involved "a combination of presidential leadership; bipartisan support in Congress; strong international cooperation; active involvement by private, nonprofit environmental organizations; and a deeply concerned public." Today's problems, such as global warming, are even more urgent, and have tended to become more complex and more contentious over time. It will take strong U.S. leadership, especially presidential leadership, at home and abroad to resolve these problems.

In chapter 3, James Gustave Speth discusses the *Global 2000 Report*. This report was prepared at the request of President Jimmy Carter in the late 1970s and originally suggested by Lee Talbot, then chief scientist for the CEQ. The report was a conceptualization of the future, circa 2000. It presented trends that might unfold between 1980 and 2000 in population and the environment if societies were to continue on their current paths. *Global 2000* effectively moved global-scale environmental challenges into American politics. Many of its projections proved correct, at least approximately.

The report had numerous shortcomings and got some projections, such as those for prices of food and minerals, wrong. But on most of the big issues of population and environment, the report accurately indicated the trends and predicted consequences and were echoed in reports from the UN Environment Programme, the Worldwatch Institute, the National Academy of Sciences, and others. Clearly, the nature of emerging global-scale environmental concerns was known to political leaders a quarter century ago. These key reports heralded "a new environmental agenda—more global, more threatening, and more difficult than the predominantly domestic agenda that spurred the environmental awakening of the late 1960s and the first Earth Day in 1970."

Global 2000 also identified the link between environmental decline and human security and social stability, noting that environmental threats "are inextricably linked to some of the most perplexing and persistent problems in the world—poverty, injustice and social conflict" and concluding that "vigorous, determined new initiatives are needed if worsening poverty and human suffering, environmental degradation, and international tensions and conflicts are to be prevented" (CEQ and U.S. Department of State 1980, I:4). The 1980 report of the International Commission on International Development Issues further called attention to these linkages: "War is often thought of in terms of military conflict, or even annihilation. But there is a growing awareness that an equal danger might be chaos—as a result of mass hunger, economic disaster, environmental catastrophes, and terrorism, so we should not think only of reducing the traditional threats to peace, but also of the need for change from chaos to order" (1980, I:3).

Not since President Carter have global environmental challenges been given such priority. But the failure to act has been bipartisan. "These issues, more than most, require true political leadership, which we have not yet had. But they also require changes that are far more sweeping and difficult than voting in a new slate of political leaders, as useful as that may sometimes be."

In chapter 4, Gene E. Likens identifies the conservation and protection of ecosystems and landscapes as a critical need in the largely destabilized world of the early twenty-first century. Likens views this threat as coming from several factors, including "(1) war and terrorism, which seriously threaten and degrade natural and human-dominated ecosystems and the ecosystem services they provide; (2) increased globalization of the marketplace, which markedly accelerates the invasion of alien species and spread of infectious disease; (3) the accelerating rate of suburban sprawl and fragmentation of landscapes and shorelines; and (4) the deficiency of leadership on these issues from what is currently the most powerful country in the world—the United States."

Many within the conservation movement see acquiring land or marine areas as an end in itself, believing that such areas will be "preserved" once acquired. Acquisition is an obvious first step for conservation and protection. But ecosystems are dynamic and will change with time through normal ecological and evolutionary processes. Such processes include succession, disease, alien invasions, mutations, immigration, emigration, and local extinctions. Beyond the initial acquisition of key areas, clear goals of conservation must be established early on for a site. Innovative adaptive management must then be used toward attaining and/ or protecting these goals.

From 1970 to 1973, Likens explains, the United States was "the world leader in environmental concern and protection. During this time, the National Environmental Policy Act, the Clean Water Act, and the Clean Air Act were passed, and the Environmental Protection Agency was established; the United States also led international agreements on ocean dumping, endangered species, and the World Heritage Trust, providing models that were adopted throughout the world. And how does the United States fare today as a role model for conservation and environmental protection? It is no longer a leader." The inequalities created by its disproportionate wealth, disproportionate utilization of resources, and uneven consumption of energy "can no longer be kept from the rest of the world, particularly the poor and disadvantaged." Likens develops his argument through two brief examples: the ecology of Lyme disease and the effect of the global marketplace on the invasion of alien species.

S. T. A. Pickett, R. S. Ostfeld, M. Shachak, and Likens (1997) have summarized a framework for the scientific practice of conservation focused on biodiversity, ecosystem function, and heterogeneity. This approach replaces the old paradigm for ecology and conservation of the "balance of nature" with a "flux of nature" paradigm. Given the enormous importance of protecting and sustaining large, unfragmented natural areas, inaction or inadequate action cannot be tolerated. "We need to find ways to show convincingly the ecological importance of wild living resources to the public, to demonstrate their connections to our life support, and then to change the public's short-term and sometimes greedy view into a long-term, more holistic (ecosystem) view about these vital resources' value."

References

Carson, R. 1962. *Silent spring.* Boston: Houghton Mifflin.

Council on Environmental Quality and U.S. Department of State. 1980. *The global 2000 report to the president: Entering the twenty-first century.* 3 vols. Washington, DC: Government Printing Office.

Independent Commission on International Development Issues. 1980. *North-south, a programme for survival: Report of the Independent Commission on International Development Issues.* Cambridge, MA: MIT Press.

Leopold, A. 1949. *A Sand County almanac, and sketches here and there.* New York: Oxford University Press.

Marsh, G. P. 1864. *Man and nature.* New York: Scribner.

Pickett, S. T. A., R. S. Ostfeld, M. Shachak, and G. E. Likens, eds. 1997. *The ecological basis of conservation: Heterogeneity, ecosystems, and biodiversity.* New York: Chapman and Hall.

1

The Environmental Movement in the United States

Russell W. Peterson

THE MODERN ENVIRONMENTAL MOVEMENT

The modern environmental movement has been a long time coming. This is surprising in light of the importance of the quality of our environment to the quality of human life. It took many years for us to recognize that human life is interdependent with all plant and animal life and with the sun, air, water, and land. We now appreciate that we are but one part of the interconnected whole of planet Earth.

The excesses of the Industrial Revolution, coupled with the explosion of the human population, opened our eyes to the devastating impact that human activities could have on the earth. In the late nineteenth century, citizens finally began to organize to do something about the threat. Here in the United States, the Audubon Society stopped the slaughter of Everglades birds by hunters who were selling the birds' plumes to the New York millinery trade. John Muir established the Sierra Club to save choice natural areas in the United States from threatened development. And Theodore Roosevelt joined the fight as our first "environmental president," setting aside natural areas for future generations.

In 1949 Aldo Leopold's *Sand County Almanac* gave rise to the powerful concept of the "land ethic." Conservationists, scientists, and managers of natural resources were inspired, studying, writing, and teaching about the natural world and its interconnections.

In 1962 Rachel Carson's *Silent Spring* awakened the world to the grave environmental threat of the massive production and use of hazardous chemicals. Environmental organizations inside and outside government sprang up around the world, including the International Council for Bird Preservation (now BirdLife International), the International Union for the Conservation of Nature and Natural Resources (IUCN, now the World Conservation Union), and the World Wildlife Fund. This action increased UN involvement in environmental issues, resulting in the first Conference on the Human Environment in Stockholm, Sweden, in 1972. The United States was a major force in the creation of this conference, having established between 1970 and 1972 the National Environmental Policy Act, the Environmental Protection

Agency, the Council on Environmental Quality, the Clean Air Act Extension, and environmental impact statements.

At roughly the same time (1971), we in the small state of Delaware passed legislation to save our unspoiled coastal zone from the massive industrialization planned for it by the U.S. Department of Commerce, thirteen international oil and transportation companies, and Delaware's own Chamber of Commerce and construction unions. Today this waterfront, free from the deep sea port, the string of oil refineries, and the mountains of coal and iron ore planned for it, attracts many visitors to its beaches and wildlife sanctuaries.

Passage of the Delaware Coastal Zone Act led to my appointment in 1973 as chairman of the Council on Environmental Quality (CEQ). The CEQ had been created when President Richard Nixon signed into law the National Environmental Policy Act (NEPA) of 1969. It defined U.S. environmental policy as follows: "The Congress, recognizing the profound impact of man's activity on the interrelations of all components of the natural environment, and the critical importance of restoring and maintaining environmental quality to the overall welfare and development of man, declares that it is the continuing policy of the federal government to create and maintain conditions under which man and nature can exist in productive harmony" [sec. 101 (42 U.S.C. § 4331)]. We environmentalists call this potent statement America's Magna Carta for the environment.

This act established the CEQ to advise the president and Congress on the national implementation of the policy and to administer the critical environmental impact statement (EIS) process, which required all federal agencies to prepare statements describing the environmental impact of their proposed actions. The agencies strongly doubted the EIS process, but over the course of three years the CEQ managed to convince them of its value. Through EIS, seriously harmful projects were avoided and more environmentally friendly alternatives were developed. Subsequently, the states, some cities, and some other countries adopted the EIS process.

Russell Train, the first chairman of the CEQ, recruited Lee M. Talbot to the organization, for which he became a principal advisor. (Talbot's life and career are detailed thoroughly in "Lee Talbot and the Environmental Movement," below.) Talbot was quickly assigned to help the United Nations plan its first Conference on the Human Environment in Stockholm. Falling back on his earlier, extensive work with conservation and government leaders in Africa and Southeast Asia, Talbot helped gain their support for the UN conference. During this time he also became deeply involved in whaling policy, serving in several capacities as a U.S. delegate to the International Whaling Commission, drafting position papers and strategies, and helping to put the issue of whales on the agenda for the Stockholm conference, actions which led to the United Nations' call for a moratorium on the taking of whales.

Talbot proposed, developed, and negotiated U.S. policy on an endangered species convention, and he ensured that this convention made the agenda of the Stockholm conference, where it was adopted as international policy. Talbot visited countries throughout the world to gain support for the convention. He then served as chief negotiator at a conference in the United States where the convention was agreed upon, and in 1973 he was a major force in Congress's passage of the Endangered Species Act. Talbot also negotiated U.S. policy on the World Heritage Trust Convention, which was endorsed as international policy at the Stockholm conference. (Later, he served as deputy head of the U.S. delegation to a negotiating conference in Paris, where the convention was agreed upon.) Along with Russell Train, who served as head of the U.S. delegation at Stockholm, Talbot made a major impact on the conference. This same conference established the UN Environment Programme (UNEP), to be headquartered in Nairobi, Kenya—an action that truly brought all the nations of the world into the environmental movement. Protecting the environment is a global problem. It requires a global solution.

I succeeded Train as chairman of the CEQ in 1973. By this time, Congress had established the Environmental Protection Agency, the Clean Air Act, the Clean Water Act, the Endangered Species Act, and the Land and Resources Fund. State and local governments had also established environmental protection measures. A number of new, national nongovernmental organizations, including the Natural Resources Defense Council, the Environmental Defense Fund, and the Environmental Law Institute, had joined the existing Sierra Club Legal Defense Fund, offering impressive legal and scientific talents to ensure that the new environmental laws and regulations were effectively implemented. They assisted the Sierra Club, the National Audubon Society, the National Wildlife Federation, the Wilderness Society, and other nongovernmental organizations in building increasingly effective citizen advocacy and environmental educational programs.

Over subsequent years, federal and state courts blessed the new laws and regulations. More and more students graduated from universities and colleges with degrees in such fields as environmental science, environmental law, and ecology. Elementary and secondary schools introduced more youths to the wonders of nature and educated them about the threat posed to the environment by human activities. Members of the news media (e.g., Walter Cronkite) effectively portrayed the positive nature of the environmental movement. Congress, under bipartisan control, remained strongly supportive of the environment. When Jimmy Carter, arguably our best environmental president, came into office, American environmentalism was strong, and growing stronger.

When Ronald Reagan became president, though, this began to change. Business—long the principal culprit in degrading the environment—had been fighting environmental regulations with increasing intensity. Many business leaders and lobbyists falsely claimed that stricter environmental regulations would cause the loss of hundreds of thousands of jobs and significantly increase inflation. Before becoming president, Reagan had been employed for several years by General Electric; in this job he extolled the merits of free enterprise and denounced the threat to it by government regulations. And as president, Reagan verified his lack of environmental credentials by immediately appointing James Watt secretary of the interior and Ann Gorsuch administrator of the Environmental Protection Agency. Their track records showed them to be highly dedicated to fighting the environmental movement. Appropriately, the public rebelled against Watt's and Gorsuch's antienvironmental actions and forced them out of office.

But life on Earth is now at a critical juncture. The cumulative impact of ever more people using ever more resources is seriously degrading the foundations of life—the air, water, croplands, grasslands, forests, and fisheries.

Though definite worldwide progress is being made toward reducing many very negative impacts on the environment, every gain from these efforts will be wiped out unless human population growth can be stabilized at an acceptable level. Overpopulation threatens the quality of life everywhere. In the twentieth century the world's population tripled and consumption actually increased twentyfold, devastating much of the environment in the process (Haupt and Kane 2004). This huge growth occurred primarily in affluent countries, for example the United States, whose people use far more resources than do people in poor countries. Now, at the start of the twenty-first century, the consumption problem is being compounded as many poor countries, for example highly populated China, rapidly climb the ladder to higher affluence, using greater resources per capita.

It is urgent that we, the people of the world, mount a major effort to forgo the environmentally destructive development that has characterized much of our past and replace it with environmentally sustainable development. This will mean exchanging the practice of providing more resources for more people with the practice of increasing the quality of life for a stable population. To reach this goal, we must call for a worldwide effort to develop technology that will permit us to meet human needs while using fewer resources.

We must also make the important choice to go down the "soft energy" path. Soft energy entails using energy more efficiently and developing renewable forms of energy (e.g., wind and solar energy). This is clearly a win-win situation: It will take several decades to make the transition to a soft-energy future, but doing so will provide us with our second-best method, after stabilizing population, of resolving the global predicament.

We will also need business's help if we are to save our environment. Some manufacturers have already voluntarily set technical goals of producing zero waste and zero pollutants and of using zero green sites. That kind of commitment on the part of business is crucial.

And we will need to consider our choices regarding technology and the impacts of research and development more carefully in the future. It may appear heretical for someone who has spent much of his life in the research and development field, who looks to that field to solve some of our current problems, to accuse it of being a source of our current environmental predicament. We must recognize, however, that in addition to its many environmentally safe contributions, the research and development field has given us nuclear weapons, hazardous chemicals, a depleted ozone layer, damaged health, acid rain, contaminated water, urban smog, massive oil spills, harmful pesticides, and mountains of waste.

The challenge now before us is to thoroughly analyze the long-term impact of new technology, weighing its potential benefits against its negative effects, before we put it to use. In the mid-1970s the U.S. Congress established the Office of Technology Assessment (OTA), which became a highly credible institution in the judgment of the scientific community. (I served as its director from 1977 to 1979.) The OTA was unfortunately abolished by a regressive Congress in 1995. The American people should force Congress to reestablish this institution.

There are clear signals that people worldwide are developing the ability to choose wisely among alternative futures. A technology assessment event of major historical significance was the world community's banning in the 1980s of the use of chlorofluorocarbons that attack the earth's protective ozone layer. And the world community is currently carrying out a technology assessment of the impact of greenhouse gases on the atmosphere—the so-called global warming effect, which could be devastating to agriculture, biological diversity, and human coastal settlements.

By the time the UN Conference on Environment and Development, also known as the Earth Summit, was held at Rio de Janeiro, Brazil, in 1992, global warming had become a global priority. In December 1995 the UN Intergovernmental Panel on Climate Change, composed of 2,500 scientists, reported that global climate change was actually occurring and that it was caused, at least in part, by human activities. Two years later, in Kyoto, Japan, representatives from 160 countries (under the auspices of the United Nations and with strong leadership by the United States)—essentially the whole world community—produced the Kyoto Protocol, which called for countries to lower their greenhouse gas emissions. In 2001 the UN Intergovernmental Panel on Climate Change underscored the need for such a protocol, reporting that global warming would be much more severe than predicted six years earlier and would produce dire results. The panel's report cited "new and stronger evidence that most of the observed warming of the last 50 years is attributable to human activities," primarily the burning of fossil fuels such as oil (Watson and the Core Writing Team 2001).

However, six months later at a UN conference on global warming, the United States was the only one of 179 nations to vote against implementing the Kyoto Protocol. And the Bush administration continues to demonstrate its antienvironmentalism. It is currently pushing the production of fossil fuels even as the news media reports on the growing evidence of the devastating potential of global warming.

Since conservative Republican politicians have become a strong force in Washington, their leaders, in concert with automobile, coal, oil, and utility company supporters, have staged a war on environmentalism. When George W. Bush became president, he became commander of that force, assisted by a team of proven antienvironmentalists: Vice President Dick Cheney, Secretary of Energy Spencer Abraham, Attorney General John Ashcroft, and Secretary of the Interior Gale Norton. Together they have nullified proenvironment programs enacted in other administrations and encouraged numerous antienvironmental measures. Why do they do this? Are they ignorant about the growing and well-established environmental threats to all life? Or are they politically motivated, catering to self-serving supporters in business and the religious Right who are opposed to environmental regulations? Whatever its objective, the Bush administration's irresponsible actions have seriously threatened the quality of life on Earth for future generations.

Differences of opinion are normal and can be constructive. But some businesses, particularly the aforementioned automobile, oil, coal and utility industries, have been fighting environmental regulations vigorously and blindly, building ever more potent and expensive advertising and lobbying programs to deliberately deceive the American people. The Bush administration has marched in lockstep with these businesses on environmental issues, compiling one of the worst environmental records in history. Among other things, this administration has opened our national parks, refuges, forests, and wilderness areas to exploitation; adopted an antienvironmental energy plan defined by the oil, coal, gas, and nuclear industries; relaxed provisions of the Clean Air Act and the Clean Water Act at the behest of industry; used the temporary fallout from the manipulation of the California energy market by Enron and others to declare a (nonexisting) national energy crisis, thereby justifying more drilling for oil; effectively turned the powerful Office of Information and Regulatory Affairs over to business; and taken steps to nullify NEPA. Additionally, at the behest of the religious Right, the administration has cut off funds for family planning programs—the best proven route to reducing population growth—in developing countries. (Officially, they do this to prevent abortions. In reality, though, their action may be leading to hundreds of thousands of additional abortions by desperate women striving to avoid unwanted pregnancies.)

Both houses of Congress currently are conspiring with the White House on antienvironmental legislation. The U.S. government has launched an all-out, devastating attack on the laws and regulations that protect our environment. Now is the time for environmentalists to stand up and be counted, or stand guilty of silent complicity in the destruction of our environment.

The United States, the world's biggest despoiler of the environment during most of the past thirty years, was once at the forefront of key global environmental issues. Until relatively recently, we provided the leadership and much of the funding to help alleviate the world's environmental problems. The world is crying out for us to resume that leadership and rejoin the many other countries that have maintained their efforts to save the environment.

I am optimistic that we can still save the earth, but this optimism is dependent on our country playing a strong role in the effort. I believe we will. The power of the culture of environmentalism, growing in today's world, will force the United States to reverse its recent course of action. If you listen carefully, you can hear future generations pleading for us to do so. The fate of the earth rests on how we Americans choose our leadership.

The world community knows how to cope with its most critical environmental issues, such as growth in human population, degradation through development and use of fossil fuels and nuclear energy, and loss of biodiversity through habitat destruction. Although headway is being made on each of these fronts, progress is plagued by the resistance of some businesses and religious groups, and by the failure of governments to provide necessary resources and protective regulations. The challenge for today's students is to help break down those barriers—bringing hope that future generations will at last learn how to live as a sustainable world community.

LEE TALBOT AND THE ENVIRONMENTAL MOVEMENT

My thirty-year collaboration and friendship with Lee Talbot began when I was appointed in 1973 as chairman of the CEQ. A forty-two-year-old with a Ph.D. in geography and ecology from the University of California, Berkeley, Talbot had already acquired a stellar international reputation in wildlife research and other conservation issues while working in sixty countries, funded first by IUCN and later by the National Academy of Sciences and the National Geographic Society. He was, needless to say, a great asset to the CEQ.

Talbot's dedication to conservation was inspired by his family. His father spent forty-four years as a range and wildlife ecologist for the U.S. Forest Service. He worked with Aldo Leopold to establish the United States' first wilderness area, and he served as a professor of forestry at the University of California, Berkeley. Talbot's mother was an anthropologist and biologist who worked as field assistant to her father, C. Hart Merriam, a biologist, naturalist, anthropologist, and pioneer ecologist, and perhaps most important the founder and first director of the agency now known as the U.S. Fish and Wildlife Service. He was also the cofounder of the American Ornithologists Union and the National Geographic Society.

The broad scientific interests that Talbot was exposed to in his family were likely what led him, as a student, to propose to the Berkeley faculty members that they establish an interdisciplinary, interdepartmental Ph.D. program in ecology. The program would entail study of geography, forestry, range management, wildlife management, zoology, botany, soil science, and agricultural economics. The faculty accepted the proposal and made Talbot the program's guinea pig. He loved it, and in 1963 he was awarded an interdisciplinary Ph.D. in geography and ecology. His multifaceted training may well explain why he has been so successful in dealing with various "big pictures"—many interconnected disciplines are needed in decision making.

Interestingly, Aldo Leopold, Talbot's father's colleague, had warned of the shortcomings of teaching narrow disciplines: "All of the sciences and arts are taught as if they were separate. They are separate only in the classroom. Step out on the campus and they are immediately fused" [Leopold (1932) 1991, 302]. Convinced of the wisdom of this observation, I have been trying for years to convince universities to establish a separate College of Integrated Studies to turn out professional generalists. My proposals have been met with great enthusiasm by a minority of faculty members, but the majority, especially heads of departments, have turned down the idea. The day of enlightenment will come.

The breadth of Talbot's approach to problem solving was enhanced by his good fortune in marrying a talented young biologist, Martha "Marty" Hayne, after only six weeks of courtship. They spent their honeymoon living amid glorious wildlife in a tent in the then-roadless Serengeti-Mara plains in East Africa; they continued to live and work together in East Africa for the first six years of their marriage. Their partnership led not only to greater understanding of the life cycles of African wildlife, but also showed the people of Africa the way to capitalize economically and spiritually on their wildlife treasure by saving it for future generations of wildlife enthusiasts to enjoy. Their research also led to an NBC documentary on African wildlife titled *Man, Beast and the Land* (1967). It received the year's highest viewership ratings for a documentary and won the Council on International Non-Theatrical Events (CINE) Golden Eagle Award.

After their long research safari in East Africa, the Talbots, pencils and research journals at the ready, traveled and worked in many other parts of the world. They backpacked from the African, Asian, and Pacific tropics to the American and European Arctic; dived in the Indian and Pacific oceans, the Gulf of Thailand, and the Caribbean, Java, Philippine, Red, and South China seas; whitewater rafted or canoed on four continents; and climbed in North America, Africa, Europe, and Asia.

Lee Talbot not only has had a superb lifetime collaborator, then, but has used the whole world as his laboratory. He has gotten to know its land, water and air, and its plant and animal life; he has also gotten to know its people, from the most disadvantaged to heads of governments and captains of industry. He has learned firsthand of the devastating impact of human activities on the global environment, of successes in curbing that impact, and of opportunities for humans to save the earth. Talbot has put his research and exploration, his publications, his teaching, his consulting, his advocacy, and his international leadership toward furthering a higher quality of all life on Earth.

So you see what interdisciplinary training, collaboration, and a laboratory without intellectual or geographic borders can do for a person. Talbot's contributions to the world only increased after his initial extensive focus on Africa.

In 1964–1965 he conducted the South East Asia Project for IUCN. It involved surveying the remote areas of eleven Asian countries to assess land use and conservation potential. Talbot then presented his recommendations for safeguarding and managing selected areas to heads of state and key ministers, and he worked with staff from several of the nations to begin implementation of his proposals.

In 1965 he organized and chaired the Bangkok Conference on Conservation of Renewable Natural Resources in South East Asia, which provided the foundation for many conservation programs—a first for the region.

In 1966 Talbot became the Smithsonian Institution's field representative for international affairs in ecology and conservation and initiated the institution's first ecosystem-oriented programs both nationally and internationally. At this time, growing concern over man's attack on the environment had triggered a flurry of activities among governments and the United Nations to do something about it.

As I mentioned earlier, Russell Train, the first chairman of the CEQ, recruited Talbot to be his principal advisor, the position in which he helped plan the 1972 Conference on the Human Environment at Stockholm. After the Stockholm conference he worked six more years for the CEQ, the first three with me. I look back on those three years fondly. He took me under his wing as we traveled together to international meetings in Africa, Asia, and Europe, introducing me to conservation institutions and leaders, educating me about key conservation issues, and thrilling me with stories about the spectacular wildlife and habitats with which he had worked. I observed firsthand the tremendous administration and friendship that Talbot had earned during his years of work with the world's leading conservationists and government leaders.

In his last six years at the CEQ, Talbot continued his string of significant accomplishments, contributing to projects such as the Marine Mammal Protection Act, a national predator control policy, management of nongame wildlife, control of off-road vehicles and clear-cutting, and ecological research. He helped to negotiate U.S.-U.S.S.R. and U.S.-Japan environmental agreements. As chief scientist and director of international affairs for the CEQ, he traveled extensively, meeting frequently with leading scientific and other officials of many countries. He served as the key liaison between the CEQ and national nongovernmental environmental organizations.

In 1978 Talbot left the CEQ and moved to Europe, where he simultaneously served as conservation director and special scientific advisor for World Wildlife Fund International in Switzerland and as senior scientific advisor on conservation and natural resources for the International Council of Scientific Unions in Paris.

After two years, Talbot was selected for the prestigious position of director general (essentially the CEO) of Swiss-based IUCN. It was the world environmental community's most representative body. About sixty national governments and numerous individual governmental agencies from other countries had officially signed on as voting members, as had over one hundred nongovernmental environmental organizations. They were represented by high-level, well-informed conservationists. The IUCN agenda covered the whole gamut of environmental issues, and staff members were assigned to work on IUCN projects around the world. No one could have been better trained or more accomplished to head this organization than Lee Talbot.

When Talbot took on the director general position, he knew that IUCN was bankrupt and about to be dissolved by the Swiss government. Talbot saved the organization through tough management and budgetary decisions and effective fundraising. He paid off the deficit and expanded IUCN's assured budget several-fold while developing and managing projects in over a hundred countries. Moreover, he initiated the Conservation for Development Centre to help implement the policy and research recommendations of IUCN, and he created the World Conservation Monitoring Centre, a set of linked, computerized databases and conservation monitoring facilities.

Talbot's experience at IUCN reinforced his growing conviction that the total world funding for environmental issues—by the United Nations, governments, and nongovernmental organizations combined (about $157 million in 1980)—was only a small fraction of what was needed. And it was only 15% of the billions spent by member countries of the Organisation for Economic Co-operation and Development for furthering development in third world countries. Much of that development, such as dam building and logging, was having a devastating impact on the environment, not to mention damaging indigenous peoples' quality of life (Talbot 1987). Thus Talbot decided that he could have a greater influence on saving the environment by working on the development establishment rather than on the environmental establishment. He appreciated the importance of development, but the world had to learn how to carry it out in an environmentally sustainable way.

In 1983, then, Talbot left IUCN and established his own consulting firm: Lee Talbot Associates International, Advisors on Environment and Development. He focused his efforts on the policies and procedures of the World Bank, which has the greatest leverage over the world's development expenditures in developing countries. He gradually expanded his efforts to regional banks (i.e., the Inter-American, Asian, and African Development Banks).

Talbot resumed his earlier approach of working in the most remote areas of Asia, Africa, and Latin America in order to evaluate possible environmental impacts of proposed projects and to advise the banks and governments involved on how to avoid or ameliorate them. To his well-established techniques of exploring by foot, four-wheel drive, or water, he added the use of helicopters, satellite imaging, and global positioning. By 2001, Talbot had raised the total number of countries where he had worked to 129. He repeatedly visited the People's Republic of China, the Kingdom of Lesotho, Mauritius, Zambia, and Nepal, and during the past five years he has made many trips to Laos and Uganda. His work has led to the development of environmental action plans that have in several cases been adopted by the country involved and funded by the World Bank.

Talbot's work has also stopped a dam and irrigation project planned by the Asian Development Bank that would have destroyed the Chitwan National Park in Nepal. It has led the prime minister of Laos to ban all logging in a critical area and implement recommendations for reducing illegal logging. It has resulted in a biodiversity action plan for China that was accepted and launched by its highest government body in 1994. An especially significant accomplishment has been Talbot's development, with François Falloux of the World Bank, of national environmental action plans (NEAPs; see Falloux and Talbot 1993). NEAPs are now required for all countries that receive "most favorable" financial treatment by the World Bank.

Over the years Talbot's consultancy clients have included governments and governmental agencies from Africa, North America, Asia, Australia, and Europe; UN organizations (e.g., the World Health Organization, the UN Environment Programme, the UN Educational, Scientific and Cultural Organization, and the UN Security Force); national academies of sciences; universities; nongovernmental organizations; and private industry.

Lee Talbot's career beautifully illustrates some of the major advances that have been made on behalf of the global environment over the past fifty years. Tens of millions of individuals have become environmentalists, educational institutions have trained ever more students in environmental fields, and the membership numbers and clout of nongovernmental environmental organizations have grown markedly. Governments have created more and more laws and environmental regulations. The United Nations and affiliated world organizations have greatly expanded their roles in protecting the earth through world conferences and other activities. More businesses have joined the cause, recognizing that economic development must be environmentally sustainable, and the faith community is becoming increasingly involved in saving "God's creation."

A culture of environmentalism now permeates much of the world. I believe it will grow into a major corridor of power, making the twenty-first century the Age of Environmentalism and bringing the people of the world closer together.

Lee Talbot has contributed significantly toward building this culture of environmentalism. Through his more than 270 scientific and technical publications— including seventeen books and monographs (some in several languages), as well as numerous popular articles and speeches—Talbot has served as an effective educator. He has received numerous national and international awards for scholarly achievement, and he has been recognized worldwide as a leader in the development and application of scientific knowledge to environmental problems and international development.

In 1994 he decided to accept the opportunity to become Professor of Environmental Science, International Affairs and Public Policy at the Department of Environmental Science and Policy at George Mason University, continuing his consulting on a part-time basis. Talbot emits great enthusiasm and self-satisfaction in this opportunity to help interest young people in his lifelong field. Imagine the great privilege and satisfaction his students receive in learning from the master.

With a good appreciation and understanding of environmentalism, Talbot's students, today's students, should be motivated to bring their individual consciences to bear on saving the earth. Consider some of the many opportunities they have for action: they can select leaders who are dedicated environmentalists, increase the power of nongovernmental environmental organizations, further the involvement of the faith community, influence business and government agencies, work with international organizations, teach, consult, do research. Each of these activities is vital to helping the world community cease its environmentally destructive pursuits and move toward environmentally sustainable development, as delineated most recently at the World Summit on Sustainable Development held in Johannesburg, South Africa, in 2002.

We need more Lee Talbots.

References

Carson, R. 1962. *Silent spring.* Boston: Houghton Mifflin.

Falloux, F., and L. M. Talbot. 1993. *Crisis and opportunity: Environment and development in Africa.* London: Earthscan.

Haupt, A., and T. Kane. 2004. *Population handbook.* 5th ed. Washington, DC: Population Reference Bureau.

Leopold, A. [1932] 1991. *The river of the mother of God.* Ed. S. L. Flader and J. B. Callicott. Repr. Madison: University of Wisconsin Press.

Leopold, A. 1949. *A Sand County almanac, and sketches here and there.* New York: Oxford University Press.

National Environmental Policy Act of 1969. Public Law 91–190. *U.S. Statutes at Large* 83:852, codified at *U.S. Code* 42, § 4321–4347.

Talbot, L. M. 1987. *Agenda for sustainable development: A framework for actions which international development institutions can take to incorporate environmental considerations in their operations.* Washington, DC: World Resources Institute.

Watson, R. T., and the Core Writing Team, eds. 2001. *Climate change 2001: Synthesis report.* Third assessment report of the Intergovernmental Panel on Climate Change (complete). Cambridge: Cambridge University Press.

2

International Environmental Policy: Some Recollections and Reflections

Russell E. Train

In the early 1960s, I was a judge on the U.S. Tax Court. I had been on two hunting safaris with my wife Aileen in the mid-1950s in East Africa; those experiences left me with a deep desire to help save Africa's extraordinary wildlife. So early in 1961, a small group of friends—Nick Arundel, Jim Bugg, Kermit Roosevelt, and Maurice Stans—met with me in my Tax Court chambers, and we began organizing a new foundation devoted to conservation in Africa. This foundation would be known first as the African Wildlife Leadership Foundation, later and more simply as the African Wildlife Federation.

The late '50s and early '60s were a time of political ferment in Africa. Most of the old European colonies achieved independence during that period. In the British East African countries, the countries with which I was most familiar, there were established national parks and game departments, often directed by retired British or Indian army officers or retired colonial administrators. For example, I think of Mervyn Cowie and Ian Grimwood in Kenya, Rennie Bere in Uganda, and John Owen and Bruce Kinloch in (then) Tanganyika. These men and their senior staffs—all European—were highly motivated. They had little or no professional training in national park or wildlife management, but they made up for their lack of training with their passionate commitment to wildlife and its conservation.

The British had trained essentially no Africans to be park or wildlife managers. The highest level reached by Africans was usually that of game scout on a bicycle, carrying an old Enfield rifle. Even to relatively inexperienced conservationists such as my colleagues and I, it was apparent that Africans would soon be taking over the national parks and game departments of East Africa, and it was equally apparent that almost none of them would have any professional qualifications for the job.

Consequently, at that first meeting of our new conservation foundation, we decided that the most immediate need was to train Africans to manage their own wildlife resources. This decision was rooted in the conviction that those same wildlife resources, properly managed, constituted a potentially tremendous economic asset to the countries involved, primarily through tourism. One member of our group, Nick Arundel, contacted Lee M. Talbot—then in East Africa where he and his wife, Marty, had spent five to six years—to get his ideas on the subject. I think it fair to say that we were in full agreement that our first priority would be to develop conservation leadership in the countries of East Africa. Hence the formal name selected for our group: the African Wildlife Leadership Foundation (AWLF). I became president and chairman of the board, positions I would hold until 1969.

Later in 1961, I went to East Africa on AWLF business and attended the Arusha Conference in Tanganyika. This conference on the conservation of African wildlife had been initiated by the International Union for the Conservation of Nature and Natural Resources (IUCN, now the World Conservation Union) and was jointly sponsored by IUCN, the Food and Agriculture Organization of the United Nations (FAO), the UN Educational, Scientific and Cultural Organization (UNESCO), and the Commission on Technical Cooperation in Africa (CCTA). Lee and Marty Talbot were the principal organizers of the conference.

Lee Talbot had already been a key factor in maintaining the integrity of the great Serengeti National Park and the neighboring Mara Reserve in Kenya. During the late 1950s, the newly independent Tanzanian government seriously considered breaking up the Serengeti and turning large pieces of it over to native settlement, ranching, and agriculture. The ecological work that Talbot had already done in the area and later ecological surveys, which he played a major role in initiating, demonstrated without a shadow of doubt the totally destructive effect such a fragmentation of the ecosystem would have had on the great wildlife migration patterns that are the glory and the fascination of the Serengeti.

It was also at that same Arusha Conference that Talbot helped bring together the chief game wardens of most of the participating countries. They discussed the problem of wildlife poaching, which was critical in most African areas, and they reached a vitally important conclusion: that the biggest cause of commercial-scale poaching was the demand created by the international trade in wildlife and wildlife products, a situation perhaps epitomized by the ivory-carving industry in Japan. Two years later, Talbot used that conclusion in a proposal to IUCN for international action to regulate the wildlife trade. He presented the proposal at the 1963 IUCN General Assembly in Nairobi, Kenya, which I attended as well.

The World Wildlife Fund was also launched in 1961, and its U.S. division was established in December of that year. I had the honor of being one of the five founding trustees.

As the AWLF got under way, it was responsible for several key education initiatives. We provided the initial funding for the College of African Wildlife Management, which was established at Mweka (near Moshi) in northern Tanzania. The college provided a two-year nondegree practical course in wildlife and parks management with heavy emphasis on fieldwork, the only such program in all of English-speaking Africa. (It now provides a wide array of wildlife management programs, ranging from short courses and certificate training to postgraduate training. Only recently was a similar institution established in South Africa.) The AWLF likewise provided the initial funding—in this case thanks to the generosity of Laurance S. Rockefeller—for a wildlife and parks conservation school for French-speaking Africans at Garoua, Cameroon. In addition, the AWLF awarded a small number of students from Kenya and Tanzania scholarships that allowed them to take full degree courses in wildlife management at American universities.

After a couple of years in Southeast Asia, Lee and Marty Talbot returned to the United States, and in 1966 Lee Talbot went to work at the Smithsonian Institution. I, meanwhile, had resigned my Tax Court judgeship in 1965 to become president of the Conservation Foundation, moving the AWLF into the same offices on Connecticut Avenue. In 1969, I went back into government in the Nixon administration as undersecretary of the interior. In 1970, I became the first chairman of the Council on Environmental Quality (CEQ). And in January 1970, just as the CEQ was getting under way, I persuaded Lee Talbot to become one of the first members of the CEQ staff as our senior scientist. (He would be of great assistance to the CEQ on both domestic issues, for example, public land matters such as predator control, range management, and clear-cutting, and international issues.) Thus began an extraordinarily fruitful period of environmental policy development and achievement. Few remember this now, but President Richard Nixon was determined to have a strong environmental record, and I believe that what the CEQ was able to accomplish during that period of 1970–1973, with bipartisan support from Congress, was by far a more productive period of environmental policy achievement than before or since.

We at the CEQ were uniquely positioned to orchestrate policy development in a strategic way. For example, in 1970, President Nixon directed the CEQ (at its own suggestion) to do a study of ocean dumping—the discharge of shore-generated wastes into the seas. The CEQ undertook this study under Talbot's direction, involving some eighty participants from various agencies. A comprehensive report was submitted to the president in October 1970. He publicly released the report, and he called for both domestic legislation and an international convention to deal with the problem. The CEQ also ensured that the ocean dumping proposal was on the agenda for the June 1972 UN Conference on the Human Environment, held at Stockholm, Sweden. I headed the U.S. delegation to the Stockholm conference, and Talbot acted as my special assistant. He had a remarkable network of friends among the various national delegations and was an enormous help in moving our agenda along. Our call for an international ocean dumping convention was endorsed unanimously by the conference; in November 1972, a conference in London, chaired by Martin Holdgate, wrote and agreed to the international convention. The completion of the international convention lit a fire under Congress and led to its final approval of our domestic ocean-dumping legislation.

Another example of the CEQ's impact on international policy development can be seen in the approval of the World Heritage Trust Convention. In 1965, long before the formation of the CEQ, I had participated in a White House Conference on International Cooperation and, together with Joseph Fisher, then president of Resources for the Future, had put forward the World Heritage Trust concept. The Johnson administration did not follow up on the idea. So in 1971, when the CEQ was putting together President Nixon's annual environmental message to Congress, we included the World Heritage Trust as one of his proposals. Subsequently, in the UN preparatory meetings leading up to the Stockholm conference (meetings at which Talbot cochaired U.S. participation along with Assistant Secretary of the Interior Nathaniel Reed), the trust program was put on the conference agenda. At the conference, the proposal was endorsed unanimously. The World Heritage Trust Convention was formally approved in Paris on November 16, 1972, the one hundredth anniversary of the establishment of Yellowstone National Park.

In November 2002, I was privileged to participate in the thirtieth anniversary celebration of the World Heritage program, held in Venice, Italy. Sadly, the United States has taken little, and at times no, interest in the program and has never provided it with more than token support. The notion that there are sites around the world, both natural and cultural, of such unique value that they truly belong to the heritage of all people is one that can help unite rather than divide us. We badly need such programs.

Talbot was especially involved in another CEQ project, the proposal for a moratorium on the commercial killing of whales. He took the lead at the CEQ in putting the proposal together. President Nixon espoused the idea and publicly called for the moratorium. We got the issue placed on the 1972 Stockholm conference agenda, where, like the World Heritage Trust proposal, it was endorsed, albeit with two abstentions. Later that month, Talbot and I attended the annual meeting of the International Whaling Commission (IWC) in London. I attended as President Nixon's "personal representative" to demonstrate his strong interest and involvement in the moratorium issue. We got a majority of the vote, but unfortunately the IWC rules required a two-thirds majority, so it was not until several years later that the United States prevailed in the passing of the commercial whaling moratorium. I am glad to say that the United States has stuck successfully to this position over the years despite continuing Japanese opposition. The World Wildlife Fund stays actively engaged in the issue.

Yet another U.S. agenda item (and CEQ project) at the Stockholm conference was a convention dealing with the international wildlife trade, an idea that had its roots in the 1961 Arusha Conference and in Talbot's subsequent proposal to IUCN in 1963. The proposal for this convention was endorsed at Stockholm, and a conference on the convention was held in Washington, DC, in January 1973. Chris Herter chaired the conference, I headed the U.S. delegation and, again, Talbot was my principal assistant. The conference adopted the Convention on International Trade in Endangered Species of Wild Fauna and Flora, commonly known as CITES. The United States' strongest ally in the negotiation of that convention was Kenya—an important factor because the United States was a wildlife and wildlife-product consumer nation while Kenya was a prominent producer. The head of the Kenyan delegation was Perez Olindo, the director of the Kenya Wildlife Service. Olindo had been the AWLF's first university student and graduate in the United States about ten years earlier, and he and I had a very close association. The CITES convention required each country to have both a scientific and a management authority for dealing with endangered wildlife, and that was the spark that enabled us to develop, and Congress to pass, the Endangered Species Act.

I left the CEQ in September 1973 to join the Environmental Protection Agency as its second administrator. But I recall the early 1970s as a period of truly remarkable achievement on the environment front, thanks to a combination of presidential leadership; bipartisan support in Congress; strong international cooperation; active involvement by private, nonprofit environmental organizations; and a deeply concerned public. Our environmental problems today, such as global warming, are no less critical than the problems of the 1970s. If anything, they are even more urgent. Admittedly, our problems have tended to become more complex and more contentious over time—all the more reason there is a tremendous need today for strong U.S. leadership at home and abroad and, most important, for presidential leadership. We need that above all.

3

The *Global 2000 Report* and Its Aftermath

James Gustave Speth

It is easy to long for the clarity of the early days of the modern environmental movement, when the problems could be seen and smelled and the villains were obvious.
Those insults led me, fresh out of law school in 1969, and others to help found the
Natural Resources Defense Council (NRDC). In dozens of lawsuits, we attacked air
and water pollution, clear-cutting, strip-mining, stream channelization, and breeder
reactors and generally partook in a great moment in American history when extraordinary progress was made in cleaning up and protecting the environment here at
home. The 1970s were a wonderful decade for our American environment, beginning with the enactment of the National Environmental Policy Act under President
Richard M. Nixon and culminating in President Jimmy Carter's protection of
Alaskan lands. Efforts were bipartisan; Democrats like Senator Edmund Muskie
joined Republicans like Senator Howard Baker to compile an unmatched record
of tough environmental legislation. Republican environmental leaders like Russell
Train and William Ruckelshaus got the new environmental cleanup effort off to
an excellent start. Democrats like Douglas Costle carried this effort forward. As a
result, air and water pollution have been much reduced in the United States, we got
the lead out of gasoline, and we have built a remarkable system of parks, wilderness,
and other protected areas. Many lakes and rivers have been revived, and forest cover
in the United States has increased, not decreased. These successes have underscored
that the United States can protect its environment without wrecking its economy.

After working through the 1970s exclusively on domestic environmental issues,
my realization that even larger challenges were massing on the global front came
as a rude awakening. I was serving then as chair of President Carter's Council on
Environmental Quality (CEQ), and thanks to the foresight of Lee M. Talbot,
the CEQ's chief scientist, we were busily preparing the *Global 2000 Report to the
President*. (Talbot had suggested that we ask the president to ask us to prepare a
reconnaissance of the future, circa 2000, and in President Carter's first environmental message to Congress, he did so.) The *Global 2000* effort brought home to me and
others that there was a new agenda of global environmental challenges more threatening and difficult than the predominantly domestic concerns that motivated us in
founding the NRDC. Climate change, devastation of ocean fisheries, deforestation

in the tropics, loss of species, land deterioration, and other unwanted processes were occurring on a frightening scale and at a rapid pace. I remember thinking that we were building a fool's paradise here in America by concentrating on local environmental concerns while ignoring these global-scale ones.

It was because of the *Global 2000 Report* that global-scale environmental challenges moved into American politics. We presented the trends that might unfold between 1980 and 2000 in population and environment if societies continued their business-as-usual approach. Already referred to by some critics as "Bad News Jimmy," President Carter showed courage in supporting this big dollop of gloom and doom in an election year.

From today's perspective, we can look back and see what actually happened. Unfortunately, many of our projections proved correct, at least approximately. *Global 2000* projected that population would grow from 4.5 billion to 6.3 billion by 2000. The actual number was 6.1 billion, so we were more or less on target. The report projected that deforestation in the tropics would occur at rates in excess of an acre a second, and for twenty years, an acre a second is what has happened. It projected that 15–20% of all species could be extinct by 2000, mostly due to tropical deforestation (CEQ and U.S. Department of State 1980, I:1–3, 37). Biologists Stuart Pimm and Peter Raven (2000) have estimated conservatively that there are about seven million species of plants and animals. Two-thirds of these species are in the tropics, largely in the forests. Pimm and Raven have also estimated that about half the tropical forests have been lost and, with them, that about 15% of tropical forest species have already been doomed. So there is evidence that our species loss estimate was perhaps high but not far off the mark.

Global 2000 projected that about six million hectares a year of drylands, an area about the size of Maine, would be rendered nearly barren by the various processes we describe as desertification (CEQ and U.S. Department of State 1980, I:2–3). And that continues to be a decent estimate today.

We predicted: "Rising CO_2 concentrations are of concern because of their potential for causing a warming of the earth.... If the projected rates of increase in fossil fuel combustion ... were to continue, the doubling of the CO_2 content of the atmosphere could be expected after the middle of the next century ... The result could be significant alterations of precipitation patterns around the world, and a 2 degree to 3 degree Celsius rise in temperatures in the middle latitudes of the earth" (CEQ and U.S. Department of State 1980, I:37). A quarter century later, this description still falls neatly within the range of current estimates.

I present these numbers not to pat our *Global 2000* team on the back. Some projections, like those on the prices of food and minerals, *Global 2000* got wrong, and the report had many shortcomings. But on most of the big issues of population and environment, the report correctly indicated trends and predicted consequences. Other reports—from the UN Environment Programme, the Worldwatch Institute, the National Academy of Sciences, and elsewhere—were saying much the same around this time. In short, the basics about emerging global-scale environmental concerns were known a quarter century ago. Political leaders then and since have been on notice that there was a new environmental agenda—more global, more threatening, and more difficult than the predominantly domestic agenda that spurred the environmental awakening of the late 1960s and the first Earth Day in 1970.

Global 2000 also called attention to the important ramifications of environmental decline for human security and social stability, noting that environmental threats "are inextricably linked to some of the most perplexing and persistent problems in the world—poverty, injustice and social conflict." It concluded that "vigorous, determined new initiatives are needed if worsening poverty and human suffering, environmental degradation, and international tensions and conflicts are to be prevented" (CEQ and U.S. Department of State 1980, I:4). The 1980 report of the International Commission on International Development Issues was prescient in its plea for attention to these linkages: "War is often thought of in terms of military conflict, or even annihilation. But there is a growing awareness that an equal danger might be chaos—as a result of mass hunger, economic disaster, environmental catastrophes, and terrorism, so we should not think only of reducing the traditional threats to peace, but also of the need for change from chaos to order" (I:3).

President Carter first addressed global-scale environmental issues in February 1980 during the Second Environmental Decade Celebration in the East Room of the White House, noting that they were "long-term threats which just a few years ago were not even considered." He concluded on an optimistic note—"the last decade has demonstrated that we can buck the trends"—and shortly thereafter he requested that I, along with Secretary of State Edmund Muskie, prepare a plan of action to do just that (Speth 1983, 2). In January 1981 we issued our report, *Global Future: Time to Act,* a 198-page agenda of what the federal government could do to address the challenges identified in *Global 2000* (CEQ and U.S. Department of State 1981). By this time, of course, President Carter had lost the election, and our report became merely more fuel for the antienvironmental pyre of the early Reagan years.

Looking back over the past two decades, it cannot be said that my generation did nothing in response to *Global 2000* and similar alerts. Progress has been made on some fronts. There are outstanding success stories, but rarely are they on a scale commensurate with the problems. For the most part, we have analyzed, debated, discussed, and negotiated these issues endlessly. My generation is a generation, I fear, of great talkers, overly fond of conferences. On action, we have fallen far short. As a result, with the notable exception of international efforts to protect the stratospheric ozone layer, the threatening global trends highlighted a quarter century ago continue to this day.

With more than two decades of dilatoriness behind us, it is now an understatement to say that we are running out of time. For such crucial issues as deforestation, climate change, and loss of biodiversity, we have already run out of time: appropriate responses are long overdue.

No president since Carter has given priority to global-scale environmental challenges. The failure has been truly bipartisan. These issues, more than most, require true political leadership, which we have not yet had. But they also require changes that are far more sweeping and difficult than voting in a new slate of political leaders, as useful as that may sometimes be.

ACKNOWLEDGMENT: This chapter has been adapted with permission from my book *Red Sky at Morning* (New Haven, CT: Yale University Press, 2004), ix–xi, 6–9.

References

Council on Environmental Quality and U.S. Department of State. 1980. *The global 2000 report to the president: Entering the twenty-first century.* 3 vols. Washington, DC: Government Printing Office.

Council on Environmental Quality and U.S. Department of State. 1981. *Global future: A time to act.* Washington, DC: Government Printing Office.

Independent Commission on International Development Issues. 1980. *North-south, a programme for survival: Report of the Independent Commission on International Development Issues.* Cambridge, MA: MIT Press.

Pimm, S. L., and P. H. Raven. 2000. Extinction by numbers. *Nature* 403(6772): 843–845.

Speth, J. G. 1983. A new institute for world resources. *American Oxonian* 70(1): 2.

4

Sustainable Conservation: Can It Be Done?

Gene E. Likens

We still talk in terms of conquest. We still haven't become mature enough to think of ourselves as only a tiny part of a vast and incredible universe. Man's attitude toward nature is today critically important simply because we have now acquired a fateful power to alter and destroy nature. But man is a part of nature and his war against nature is inevitably a war against himself. The rains have become an instrument to bring down from the atmosphere the deadly products of atomic explosions. Water, which is probably our most important natural resource, is now used and re-used with incredible recklessness. Now, I truly believe that we in this generation must come to terms with nature, and I think we're challenged as mankind has never been challenged before to prove our mastery, not of nature, but of ourselves.

Man has long talked somewhat arrogantly about the conquest of nature, now he has the power to achieve his boast. It is our misfortune—it may well be our final tragedy—that this power has not been tempered with wisdom, but has been marked by irresponsibility; that there is all too little awareness that man is part of nature, and that the price of conquest may well be the destruction of man himself. Instead of always trying to impose our will on Nature we should sometimes be quiet and listen to what she has to tell us. Only then, might we see the madness of our feverish pace and learn humility and wisdom.

—Rachel Carson, *Of Man and the Stream of Time*

SUSTAINABLE CONSERVATION AND A NEW PARADIGM
FOR ECOLOGY AND CONSERVATION

Conservation and protection of ecosystems and landscapes are seriously jeopardized in the largely destabilized world of the early twenty-first century. Threats come from several fronts, including: (1) war and terrorism, which seriously threaten and degrade

Table 4.1. Some major challenges facing conservation/restoration/protection of ecosystem services (modified from Likens 2001b).

A. Minimize effects of land fragmentation and loss of arable land; adaptively manage land-use changes

B. Maintain adequate quantity and quality of water, including minimizing
    Salinization of inland waters
    Toxic algal blooms and pathogens
    Impacts of dams on flowing waters
    Wasteful use of water

C. Manage ecosystem effects of alien species invasions including agents of infectious disease

D. Establish and maintain ecological reserves (land, freshwater, marine)

---

natural and human-dominated ecosystems and the ecosystem services they provide; (2) increased globalization of the marketplace, which markedly accelerates the invasion of alien species and spread of infectious disease; (3) the accelerating rate of suburban sprawl and fragmentation of landscapes and shorelines; and (4) the deficiency of leadership on these issues from what is currently the most powerful country in the world—the United States (table 4.1).

Future conservation efforts cannot be predicated on the basis of merely putting a fence around an area, walking away, and assuming that the area will remain unchanged and protected in the future (see Ervin 2003; Likens 1995; Parish, Braun, and Unnasch, 2003). War, terrorism, air and water pollutants (including those in groundwater), humans, and global climate change do not respect such "fences." At the very least, diligence is necessary. Most important, systems thinking, adaptive management toward realistic goals, and realization and incorporation of the existence of change will be required to make progress toward sustainable conservation.

There has been a view within the conservation movement that acquiring land or marine areas is an end in itself; that is, that such land or marine areas will be "preserved" once acquired. Obviously, acquiring such areas is a critical first step for conservation and protection. But only nonliving items, like pickles, can be preserved. Ecosystems are composed of living components and thus will change with time through normal ecological and evolutionary processes, including succession, disease, alien invasions, mutations, immigration, emigration, and local extinctions. Therefore, clear goals of conservation for a site must be established early. Innovative adaptive management must then be used toward attaining and/or protecting these goals. The recent major decline of African gorillas due to an epidemic of Ebola hemorrhagic fever is a highly visible example of why fences alone may not protect wild living resources (Kaiser 2003).

An important aspect of successful adaptive management is the potential to learn from experimental manipulations, particularly large-scale manipulations such as watersheds (Likens 1998, 2001a). Experimentation is a very powerful tool in science, especially in ecosystem science (Likens 1985). Unfortunately, appropriate sites for such large-scale, experimental manipulations are relatively rare. I have argued previously that portions of parks and other protected areas could be legitimately used for such large-scale experimental manipulations; the results of the experiments would be critical for the adaptive management of the larger portions of those parks and for other protected areas, thus facilitating and guiding their very protection (Likens 1998). Such experimentation should be done carefully and judiciously, but it could in the long run provide the basis of knowledge for more sustained conservation.

Global climate change represents a particularly serious threat to both conserved and threatened areas. The lack of leadership by the current U.S. administration in this regard is of great concern relative to the future of conservation. For example, how can conservation be achieved, let alone sustained, without reducing the threat of global climate change? I would argue that it cannot.

Increasingly, we seem to take the short view about natural resources—often out of greed and self-interest—rather than the long view held by past political leaders such as Thomas Jefferson and Theodore Roosevelt, a view which benefited future, as well as present, generations. The long view that we do take is dispiriting, thanks to the ever-increasing size of the human population and that population's capacity for "human-accelerated environmental change" (Likens 1991, 1994). Loss of productive farmland in the United States (two acres lost every minute in the United States; prime land converted 30% faster, proportionally, than nonprime, rural land between 1992 and 1997; American Farmland Trust 2002) is a good example of how current prices and priorities are insufficient for protecting and conserving an extremely vital resource. The market value of this resource for development and other nonfarm uses currently is much greater than its market value for food production. The critical question, however, is this: For how long can this finite resource be consumed? The answer seems bleak, given the limitations of the so-called green revolution (see, e.g., Evenson and Gollin 2003), including the increased need for pesticides and their environmental side effects; the environmental and health impacts of intensive agriculture; the increasing "needs" of an ever-expanding human population (see, e.g., Likens 2001b); and the fact that while farmland is being lost, our marine resources are being overharvested (see, e.g., Baum et al. 2003; Ellis 2003; Pauly and Maclean 2003). Will short-term changes in societal behavior or genetic engineering give us a temporary reprieve (Pauly et al. 2003)? What will happen next?

This is a pessimistic vision for the future, and I don't enjoy sharing it. Despite many recent calls to "protect land health" and to "live within our limits" (e.g., Dombeck, Wood, and Williams, 2003), though, I don't believe we can turn this pessimistic view around without commitment from the political leadership of the United States. From 1970 to 1973, the United States was the world leader in environmental concern and protection. During this time, the National Environmental Policy Act and the Clean Water and Clean Air Acts were passed, and the Environmental Protection Agency was established; the United States also led international agreements on ocean dumping, endangered species, and the World Heritage Trust, providing models that were adopted throughout the world. And how does the United States fare today as a role model for conservation and environmental protection? It is no longer a leader. Its secrets of disproportionate wealth, disproportionate utilization of resources, uneven consumption of energy, and other inequalities can no longer be kept from the rest of the world, particularly the poor and disadvantaged.

TWO BRIEF EXAMPLES

Ecology of Infectious Disease

All disease has an ecological component. Lyme disease is present in all conterminous U.S. states; in some areas it has reached epidemic proportions. The ecology of this disease is complicated and has been the subject of much study. The infectious agent is a spirochete bacterium (*Borrelia burgdorferi*) transferred by the bite of a black-legged tick (*Ixodes scapularis*). Following a blood meal, commonly from a white-tailed deer (*Odocoileus virginianus*), the adult tick falls to the ground and lays eggs. During the following summer, the eggs hatch into larvae. These larvae obtain blood meals from mice, such as the white-footed mouse (*Peromyscus leucopus*), from which they may obtain the spirochete bacterium. The larvae transform into nymphs a year or so later, and they pass the spirochete to humans or other mammals when they attach for a blood meal. In general, the risk of Lyme disease in humans is correlated with mast years of acorns in oak forests with a lag of two years (Jones et al. 1998; Ostfeld 1997; Ostfeld and Jones 1999; Ostfeld, Jones, and Wolff, 1996). The disease also correlates with increased fragmentation of forest landscapes, which tends to foster population increases in both white-tailed deer and white-footed mice, both major reservoirs for the spirochete bacterium (Allan, Keesing, and Ostfeld, 2003; Line 2003). In one study, forest patches less than one hectare in size had three times as many ticks and seven times as many infected ticks as did larger patches of forest in the mid-Hudson Valley area of New York (Allan, Keesing, and Ostfeld, 2003). This is exciting research, particularly because the results clearly show a direct connection between biodiversity (species richness), human health, and land use.

Global Marketplace and Invasion of Alien Species

Alien species may have positive, negative, or neutral impacts on ecosystem function or temporal change in ecosystems. In some cases, the ecological impact may be quite serious (see Likens 1998, 2001), and the problems may be exacerbated by a strong connection to the world trade network and increased global travel by humans (e.g., Bright 1999).

Alien forest pests in the United States (e.g., Asian long-horned beetles attacking maple; hemlock wooly adelgids attacking eastern hemlock; gypsy moths; "sudden oak death" attacking oaks; and beech bark disease attacking American beech) certainly do not respect boundaries. They represent serious threats to conservation goals.

A NEW PARADIGM FOR ECOLOGY AND CONSERVATION

S. T. A. Pickett, R. S. Ostfeld, M. Shachak, and I (1997) have summarized a framework for the scientific practice of conservation focused on biodiversity, ecosystem function, and heterogeneity. It features a new and more appropriate paradigm for ecology and conservation: the "flux of nature" paradigm, which replaces the older, more static "balance of nature" paradigm (Pickett, Parker, and Fiedler, 1992). Our book *The Ecological Basis of Conservation* (1997) details the features of this framework. According to Lee M. Talbot, "The new paradigm in ecology makes most previous conservation theory obsolete and in the process shows why previous theory has had limited success" (Talbot 1997). In the new paradigm, "an ecosystem is characterized by uncertainty rather than predictability, and that it must be described as probabilistic rather than deterministic" (Pickett et al. 1997). Accordingly, this paradigm offers scientists "both a unique opportunity and a heavy responsibility to create an effective link between ecology and conservation policy. At the least, the scientist's role can involve communicating appropriate information to decision makers, managers, and the public. Such communication must be scientifically valid and should represent broad scientific consensus, particularly in view of the prevalence of disinformation [about] environmental issues" (Pickett et al. 1997).

In the future, it is likely that conservation may increasingly be focused on restoration and reclamation. If so, these approaches cannot be based on steady-state assumptions. They must also, by necessity, be ecosystem-based, as Talbot established almost fifty years ago.

Visionaries like Talbot, as well as Aldo Leopold, John Muir, and Rachel Carson, have tried to put us on the right track to sustainable conservation, but as Will Rogers said, "Even if you are on the right track, you'll get run over if you just sit there." So where do we go from here in efforts to conserve wild living resources, and how can we sustain and nurture the protection of large, unfragmented areas (ecosystems) on our planet? How do we make this need compelling to the majority and to the powerful?

In my opinion, we have a very long and difficult path to reach sustainability in conservation. However, models for sustainability do exist. Aldo Leopold described the challenges decades ago: "A land ethic, then, reflects the existence of an ecological conscience, and this in turn reflects a conviction of individual responsibility for the health of the land. Health is the capacity of the land for self-renewal. Conservation is our effort to understand and preserve this capacity" (Leopold 1949, 221). Rachel Carson echoed and extended Leopold's ideas: "Your generation must come to terms with the environment. Your generation must face realities instead of taking refuge in ignorance and evasion of truth. Yours is a grave and a sobering responsibility, but it is also a shining opportunity. You go out into a world where mankind is challenged, as it has never been challenged before, to prove its maturity and its mastery—not of nature, but of itself. Therein lies our hope and our destiny. 'In today already walks tomorrow'" (1962).

Given the enormous importance of protecting and sustaining large, unfragmented natural areas, upon which our lives and those of other organisms depend, handwringing, inaction, shortsightedness, and piecemeal approaches cannot be tolerated. We need to find convincing ways to show the ecological importance of wild living resources to the public, to demonstrate their connections to our life support, and then to change the public's short-term and sometimes greedy view into a long-term, more holistic (ecosystem) view about these vital resources' value. This challenge is not new, and there is a large gap between the rhetoric and the values and policies that are necessary to effect change (e.g., Orr 2004). Why aren't we getting this message across to the public (Myers 2003)? The clarion call to action has been issued, but given the current accelerated rate of fragmentation and destruction of landscapes, probably never before has the need to resolve this challenge been more urgent.

CONCLUDING NOTE: INNOVATIONS IN ECOSYSTEM SCIENCE AND MANAGEMENT

In closing, I wish to describe a few of the contributions that Lee Talbot has made toward the goals of sustainable conservation and ecosystem management. His career has shown that one person can indeed make a difference.

Serengeti-Mara Savanna Ecosystem Research Program

Talbot first defined and analyzed the Serengeti-Mara savanna ecosystem (Talbot 1963b). In so doing, he originated one of the first truly long-term, large-scale, ecosystem-oriented research programs. Talbot's achievements through this program include the following:

- First definitive analysis and description of mammalian migrations, including causal factors, routes, and timing (Talbot 1961; Talbot and Talbot 1963)
- Definition and description of the different and complementary food habits (and nutritional requirements) of the nearly thirty principal herbivores (Talbot 1962, 1963a, 1968; Talbot and Talbot 1963)
- First survey of the soils of the Serengeti-Mara savanna, including analysis and definition of the interrelationships between climate, parent material, soils, vegetation, fires, and herbivores, as well as demonstration of how soil factors affected the distribution of grassland types and their use by wild animals (Anderson and Talbot 1965)
- Identification of the nitrogen recycling mechanism, allowing drought-adapted ungulates to thrive on forage, which would not support temperate-zone species (Talbot and Talbot 1963)
- Analysis of resource partitioning by the grassland plants and by the herbivores that use them, and demonstration of the dynamic ecological interactions (Talbot 1962, 1964, 1968)

Ecosystem Science and Management

- Field ecological research and surveys to identify ecosystems that are biodiversity-rich or otherwise significant and require protection (Talbot 1960; Talbot and Talbot 1964, 1966a)
- Ecological research and surveys on endangered species (often the first such efforts for a given species) determining location, status, trends, threats, and ecological requirements of the species involved, for example, Arabian oryx (Middle East), Asiatic lion (South and Southwest Asia), Sumatran rhinoceros (Southeast Asia), and Sumatran tiger (Sumatra) (Talbot 1957, 1960; Talbot and Talbot 1964, 1966a, 1966b)

- Research on the management of wild living species (free-ranging aquatic and terrestrial plants and animals), with particular reference to the impact of management on the target species' ecosystems (Botkin et al. 1995; Talbot 1972b, 1975a, 1975b, 1976, 1978)
- Research on the meat production potential of mixed species of wild herbivores based on the complementary roles in the ecosystem, especially their diets, nutritional requirements, and seasonal movements (Talbot 1963a, 1966)
- Research on the role of fire in tropical savanna ecosystems, primarily in Africa and Asia (Talbot 1963b, 1972a, 1980; Talbot and Kesel 1975)
- Research on the role of humans in the genesis, spread, and maintenance of ecosystems that had often been considered "natural" (Talbot 1969, 1974, 1989; Talbot and Kesel 1975)
- On behalf of the Marine Mammal Commission, production of an international research project on management of wild living resources (this entailed working with over 400 scientists and managers from thirty-three countries, and it produced the finding that virtually all stocks of wild living resources subject to commercial harvesting were or were being depleted (Talbot 1996)
- Organization of a series of meetings and international workshops—rooted in the growing concern about stock depletion of wild living resources—in the 1970s and 1990s, aimed at bringing an ecosystem approach to the management of wild living resources (each produced a set of principles to be published, used, and updated; see table 4.2)
- Whale research: For example, as a member of the Scientific Advisory Committee of the International Whaling Commission and as a deputy to the U.S. delegate to the commission, Talbot analyzed existing research on whales and whaling, recognized for the first time that the existing methods were fundamentally in error, recognized the uncertainties, and initiated an ecosystem approach to these problems
- Major contributions to the initial drafting and passage of the Marine Mammal Protection Act of 1972, the first legislation in the world that placed maintenance of the health of the ecosystem as the overriding policy (Anonymous 2002; Talbot 2003)
- Service as senior scientific advisor on conservation and natural resources to the International Council of Scientific Unions, Paris

Table 4.2. Principles for the conservation of wild living resources (modified from Holt and Talbot 1978; Mangel and Talbot et al. 1996).

A. The ecosystem must be maintained

B. Management decisions should include safety factors

C. Wasteful use of other resources should be avoided

D. Survey or monitoring analysis and assessment should precede planned use and accompany actual use

E. Evaluate uncertainty and use adaptive management

F. Incorporate socioeconomic factors

References

Allan, B. F., F. Keesing, and R. S. Ostfeld. 2003. Effect of forest fragmentation on Lyme disease risk. *Conservation Biology* 17(1): 267–272.

American Farmland Trust. 2002. *Farming on the edge.* Washington, DC: American Farmland Trust.

Anderson, G. D., and L. M. Talbot. 1965. Soil factors affecting the distribution of the grassland types and their utilization by wild animals on the Serengeti plains, Tanganyika. *Journal of Ecology* 53:33–56.

Anonymous. 2002. Passage of the Marine Mammal Protection Act of 1972. *MMPA Bulletin* 22:1–2, 14–15.

Baum, K. A., R. A. Myers, D. G. Kehler, B. Worm, S. J. Harley, and P. A. Doherty. 2003. Collapse and conservation of shark populations in the Northwest Atlantic. *Science* 299(5605): 389–392.

Botkin, D. B., K. Cummins, T. Dunne, H. Regier, M. Soble, L. Talbot, and L. Simpson. 1995. *Status and future of anadromous fish of western Oregon and northern California: Findings and options.* CSE Research Report. Santa Barbara, CA: Center for the Study of the Environment.

Bright, C. 1999. Invasive species: Pathogens of globalization. *Foreign Policy* 119 (Fall): 50–64.

Carson, R. 1962. *Of man and the stream of time.* Commencement address, Scripps College, Claremont, CA, June 12.

Dombeck, M. P., C. A. Wood, and J. E. Williams. 2003. *From conquest to conservation: Our public lands legacy.* Washington, DC: Island Press.

Ellis, R. 2003. *The empty ocean.* Washington, DC: Island Press.

Ervin, J. 2003. Protected area assessments in perspective. *BioScience* 53(9): 819–822.

Evenson, R. E., and D. Gollin. 2003. Assessing the impact of the green revolution: 1960 to 2000. *Science* 300(5620): 758–762.

Holt, S., and L. M. Talbot. 1978. *New principles for the conservation of wild living resources.* Wildlife Monographs, no. 59. Bethesda, MD: Wildlife Society.

Jones, C. G., R. S. Ostfeld, M. P. Richard, E. M. Schauber, and J. O. Wolff. 1998. Chain reactions linking acorns, gypsy moth outbreaks, and Lyme-disease risk. *Science* 279(5353): 1023–1026.

Kaiser, J. 2003. Ebola, hunting push ape populations to the brink. *Science* 300(5617): 232.

Leopold, A. 1949. *A Sand County almanac, and sketches here and there.* New York: Oxford University Press.

Likens, G. E. 1985. An experimental approach for the study of ecosystems. *Journal of Ecology* 73(2): 381–396.

Likens, G. E. 1991. Human-accelerated environmental change. *BioScience* 41(3): 130.

Likens, G. E. 1994. Human-accelerated environmental change: An ecologist's view. Australia Prize Winner presentation, Murdoch University, Perth.

Likens, G. E. 1995. Sustained ecological research and the protection of ecosystems. In *Ecosystem monitoring and protected areas: Proceedings of the Second International Conference on Science and the Management of Protected Areas,* ed. T. B. Herman, S. Bondrup-Nielsen, J. H. Martin Willison, and N. W. P. Munro, 13–21. Wolfville, NS: Science and Management of Protected Areas Association.

Likens, G. E. 1998. Limitations to intellectual progress in ecosystem science. In *Successes, limitations, and frontiers in ecosystem science,* ed. M. L. Pace and P. M. Groffman, 247–271. New York: Springer.

Likens, G. E. 2001a. Biogeochemistry, the watershed approach: Some uses and limitations. *Marine and Freshwater Research* 52(1): 5–12.

Likens, G. E. 2001b. Ecosystems: Energetics and biogeochemistry. In *A new century of biology*, ed. W. J. Kress and G. Barrett, 53–88. Washington, DC: Smithsonian Institution Press.

Line, L. 2003. "3 BR, forest vu" may have added feature: Lyme disease risk. *New York Times*. April 8.

Mangel, M., L. M. Talbot, G. K. Meffe, M. Tundi Agardy, D. L. Alverson, J. Barlow, D. B. Botkin, G. Budowski, T. Clark, J. Cooke, et al. 1996. Principles for the conservation of wild living resources. *Ecological Applications* 6(2): 338–362.

Myers, N. 2003. Conservation of biodiversity: How are we doing? *The Environmentalist* 23(1): 9–15.

Orr, D. W. 2004. *The last refuge: Patriotism, politics and the environment in an age of terror.* Washington, DC: Island Press.

Ostfeld, R. S. 1997. The ecology of Lyme-disease risk. *American Scientist* 85 (July–August): 338–346.

Ostfeld, R. S., and C. G. Jones. 1999. Peril in the understory. *Audubon* (July–August): 74–82.

Ostfeld, R. S., C. G. Jones, and J. O. Wolff. 1996. Of mice and mast: Ecological connections in eastern deciduous forests. *BioScience* 46(5): 323–330.

Parish, J. D., D. P. Braun, and R. S. Unnasch. 2003. Are we conserving what we say we are? Measuring ecological integrity within protected areas. *BioScience* 53(9): 851–860.

Pauly, D., J. Adler, E. Bennett, V. Christensen, P. Tyedmers, and R. Watson. 2003. The future for fisheries. *Science* 302(5649): 1359–1361.

Pauly, D., and J. Maclean. 2003. *In a perfect ocean: The state of fisheries and ecosystems in the North Atlantic Ocean.* Washington, DC: Island Press.

Pickett, S. T. A., R. S. Ostfeld, M. Shachak, and G. E. Likens, eds. 1997. *The ecological basis of conservation: Heterogeneity, ecosystems, and biodiversity.* New York: Chapman and Hall.

Pickett, S. T. A., V. T. Parker, and P. L. Fiedler. 1992. The new paradigm in ecology: Implications for conservation biology above the species level. In *Conservation biology: The theory and practice of nature conservation, preservation, and management*, ed. P. L. Fiedler and S. K. Jain, 65–88. New York: Chapman and Hall.

Talbot, L. M. 1957. The lions of Gir: Wildlife management problems of Asia. *Transactions of the North American Wildlife Conference* 22:570–579.

Talbot, L. M. 1960. A look at threatened species. *Oryx* (4–5): 153–293.

Talbot, L. M. 1961. Preliminary observations on the population dynamics of the wildebeest in Narok District, Kenya. *East African Agricultural and Forestry Journal* 27(2): 108–116.

Talbot, L. M. 1962. Food preferences of some East African wild ungulates. *East African Agricultural and Forestry Journal* 27(3): 131–138.

Talbot, L. M. 1963a. Comparison of the efficiency of wild animals and domestic livestock in utilization of East African rangelands. In *Conservation of nature and natural resources in modern African states*, 328–335. IUCN New Series, no. 1. Morges, Switzerland: International Union for the Conservation of Nature and Natural Resources.

Talbot, L. M. 1963b. Ecology of the Serengeti-Mara savanna of Kenya and Tanzania, East Africa. 2 vols. Interdisciplinary Ph.D. diss., University of California, Berkeley.

Talbot, L. M. 1964. The biological productivity of the tropical savanna ecosystem. In *The ecology of man in the tropical environment*, 88–97. IUCN New Series, no. 4. Morges, Switzerland: International Union for the Conservation of Nature and Natural Resources.

Talbot, L. M. 1966. *Wild animals as a source of food.* Fish and Wildlife Service, Special Scientific Report: Wildlife, no. 98. Washington, DC: U.S. Department of the Interior. (Orig. pub. 1964.)

Talbot, L. M. 1968. The herbivore complex: Recent research and its implications to North Africa. In *Proceedings of the technical meeting on the West Mediterranean, North Africa and Sahara,* 340–359. London: International Biological Programme.

Talbot, L. M. 1969. The wail of Kashmir: Man's impact on the land. In *Population, evolution, and birth control: A collage of controversial ideas,* ed. G. Hardin, 50–51. 2nd ed. San Francisco: Freeman.

Talbot, L. M. 1972a. Grasslands of the world. In *Britannica yearbook of science and the future,* 142–147. Chicago: Encyclopaedia Britannica.

Talbot, L. M. 1972b. Predator control. *Transactions of the North American Wildlife and Natural Resources Conference* 37:395–399.

Talbot, L. M. 1974. Lebanon: Land use and vegetation, and how to make a desert. In *Readings in wildlife conservation,* ed. J. A. Bailey, W. Elder and T. D. McKinney, 382–386. Washington, DC: Wildlife Society.

Talbot, L. M. 1975a. Maximum sustainable yield: An obsolete management concept. *Transactions of the North American Wildlife and Natural Resources Conference* 40:91–96.

Talbot, L. M. 1975b. The need for an ecosystems approach to management of living marine resources. In *Systems thinking and the quality of life,* ed. W. Phillips and S. Thorson, 457–461. Washington, DC: Society for General Systems Research.

Talbot, L. M. 1976. Needed: A revolution in our management of the living resources of the seas; The need to revise the scientific basis for management of living marine resources. In *Nature and human nature,* ed. W. R. Burch, 3–14. Yale University School of Forestry and Environmental Studies, Bulletin Series, no. 90. New Haven, CT: Yale University.

Talbot, L. M. 1978. The role of predators in ecosystem management. In *The breakdown and restoration of ecosystems,* ed. M. W. Holdgate and M. J. Woodman, 307–319. NATO Conference Series I: Ecology, vol. 3. New York: Plenum Press.

Talbot, L. M. 1980. El ecosistema de las praderas. *Informa, Fundacion Vida Silvestre Argentina* (7/8): 24–27.

Talbot, L. M. 1989. Man's role in managing the global environment. In *Changing the global environment: Perspectives on human involvement,* ed. D. B. Botkin, M. F. Caswell, J. E. Estes, and A. A. Orio, 17–33. Boston: Academic Press.

Talbot, L. M. 1996. *Living resource conservation: An international overview.* Washington, DC: Marine Mammal Commission.

Talbot, L. M. 1997. The linkages between ecology and conservation. In *The ecological basis of conservation: Heterogeneity, ecosystems, and biodiversity,* ed. S. T. A. Pickett, R. S. Ostfeld, M. Shachak, and G. E. Likens, 368–378. New York: Chapman and Hall.

Talbot, L. M. 2003. Does public policy reflect environmental ethics? If so, how does it happen? *University of California Davis Law Review* 37(1): 269–280.

Talbot, L. M., and R. H. Kesel. 1975. The tropical savanna ecosystem. *Geoscience and Man* 10: 15–26.

Talbot, L. M., and M. H. Talbot. 1963. *The wildebeest in western Masailand, East Africa.* Wildlife Monographs, no. 12. Washington, DC: Wildlife Society.

Talbot, L. M., and M. H. Talbot. 1964. *Renewable natural resources in the Philippines: Status, problems, and recommendations.* Washington, DC: International Union for the Conservation of Nature and Natural Resources/World Wildlife Fund.

Talbot, L. M., and M. H. Talbot. 1966a. *Conservation of the Hong Kong countryside.* Hong Kong: government printer.

Talbot, L. M., and M. H. Talbot. 1966b. The Tamarau (*Bubalus mindorensis* (Heude)): Observations and recommendations. *Mammalia* 30(1): 1–12.

PART II

BIODIVERSITY, WILDLIFE ECOLOGY, AND MANAGEMENT

Larry L. Rockwood

In part II, the authors grapple with the practical issues of wildlife management. The old model for management was essentially as follows: (1) identify those landscapes critical to the survival of certain species we wish to preserve, (2) move the people out, (3) fence the landscapes off, (4) keep the people out, and (5) restore those landscapes to their "natural" state. But these authors instruct us that such an approach is ineffective, impractical, naive, and perhaps even immoral.

Five themes are common to the chapters. First, humans are part of these ecosystems and landscapes. They must be accommodated, even as we attempt to carry out conservation policies. Second, local community support and involvement is crucial to successful conservation. Third, at the same time, pressure and financial support from nongovernmental organizations and other watchdog groups is often crucial to ensure that conservation goals are not subverted by corruption or lack of resources. Fourth, conservation must move from simple preservation of individual species to the maintenance of those ecosystems upon which all species depend. And fifth, the tension between conservation and development is real, but both goals can be achieved.

In chapter 5, Robert Goodland describes the approaches now taken by the World Bank to conserve biodiversity in developing countries. He immediately points out that conservation of biodiversity also means conserving habitats and ecosystems that are critical to sustaining those environmental services on which the poor, more than the wealthy, depend. Therefore, the presumed conflict between conservation and humans is false.

Goodland notes that the World Bank, in the past, supported projects that caused major environmental degradation. In the late 1980s, though, Lee Talbot persuaded the World Bank to establish rules such that "maintenance of the ecosystem, not just the conservation of individual species," became its main objective. The policy of the World Bank is now also not to "significantly convert" or "degrade" critical natural habitats. But, as Goodland demonstrates, critical natural habitat, significant conversion, and degradation are all terms that must be interpreted.

As the former chief environmental advisor to the World Bank, and as one of the first individuals hired by the group to oversee the environmental effects of the projects it was funding, Goodland has a unique perspective on the tensions between presumed economic benefits and potential environmental catastrophes. He is able to illustrate his points with many examples of projects that the World Bank has modified based on environmental principles. As Goodland shows, the group must be flexible, but its policies are now based on putting "the onus on developers . . . [to produce designs such that] the people and their environments...will be clearly better off if the development is permitted." Furthermore, Goodland states that the group will not even consider development projects within critical natural habitats unless several criteria are met, including best practice design.

Goodland concludes that the World Bank must implement and strengthen its social and environmental policies, grant exceptions only when projects meet a stringent set of criteria that unambiguously demonstrates that benefits outweigh losses, and demand insurance or escrow bonds in such cases. Finally, the World Bank can often promote both biodiversity and social progress through the rehabilitation of degraded areas.

In chapter 6, Kenton R. Miller and Charles Victor Barber stipulate that while about 12% of the world's land surface area (about 17 million km^2) has been officially "protected," the challenge of the twenty-first century is to make these protected areas into something more than areas on a map. They must be more than "paper parks," more than fenced space available only to the elite or to foreign tourists. The ecosystem services provided by protected areas—timber, fodder, biodiversity protection, and clean water, for example—must be recognized by both national and local communities.

Miller and Barber argue that protected areas benefit the poor through watershed protection, through small-scale forestry projects that provide employment, and through protection from both flood and drought. Although conservation once had a reputation as a concern of the rich and well connected, a new model, poverty reduction through conservation, has emerged. As stressed by others in this part, "fortress conservation" doesn't work in the midst of poor and/or dense populations in Asia, Africa, and Latin America. And, as Goodland emphasized, conservation of natural habitats can coexist with development, if done properly.

A related topic is the "empowerment" of local populations. The support and involvement of local communities in conservation issues has been proposed as an exciting new model. The concept is that if local poor people can benefit, perhaps even earning cash, from protecting forests or wildlife, conservation goals will be achieved. Yet as Peter J. Balint and Judith Mashinya (2006) have described, projects held up as positive models—such as Zimbabwe's Communal Areas Management Program for Indigenous Resources (CAMPFIRE)—can devolve completely, as oversight by outside agencies is withdrawn and/or local leadership becomes corrupt. Indeed, Miller and Barber conclude that local communities are not always better managers than central governments. However, they likely have a better chance as contrasted with control by those in remote capitals. Miller and Barber also point out that change, whether it is global climate, institutional, or socioeconomic, is inevitable, and that we must be planning for the future, not protecting the past.

In chapter 7, John Seidensticker begins by stating that one of the great problems in conservation biology is what he terms *ecological amnesia*, the disappearance in the public mind of what environments were like even a few decades ago. Once species and natural environments are lost, mental adjustments are made in the minds of local people; what is "natural" now is not what was natural twenty or fifty years ago. Can people in the eastern United States even imagine an environment in which mountain lions, wolves, and elk roamed the woods? Though we have seen historical references in movies, do Americans truly have a concept of the western United States populated with millions of bison? No, and thus, as Seidensticker points out, "the baseline for what is 'natural' shifts from one generation to the next." Given the growth of human populations and the associated decline of nature, is it possible to restore ecosystems to a natural state? When human populations are very dense, is ecological restoration even possible?

In Asia conservation and restoration, if they are to take place at all, must be done in concert with the human populations that have always been part of the landscape. Seidensticker describes two case studies in this chapter: (1) Asian lions and the Gir Forest (India) and (2) rhinos, tigers, and the Chitwan Valley (Nepal). In each case a baseline was established, to which we can now refer over fifty years later. In the case of the Gir Forest, Lee Talbot visited the region in 1955 and described the land-use patterns and the use of the forest by livestock and wild ungulates, as well as the status of the lions. Talbot found that the number of grazing livestock was unsustainable and was displacing the wild ungulates. But he also recognized that people and their animals had been part of the ecosystem for a long time. There was no way to restore this forest to some kind of "natural" state without people. His recommendations, which are detailed by Seidensticker, therefore took into account the local people, their livestock, the vegetation, and the wild grazing ungulates, in addition to the lions. Talbot's recommendations have been followed and the number of lions has increased. Nevertheless, as Seidensticker indicates, what is wild and natural is not static, but is always changing. As the Gir Forest ecosystem faces new challenges from both inside and outside its boundaries, managers must adapt. Seidensticker believes that "top-down, command-and-control" management is inflexible and that local empowerment is crucial.

The case of the Royal Chitwan National Park parallels that of the Gir Forest. Due to the presence of malaria, and therefore low human population density, much of the Chitwan Valley was still teeming with wildlife in 1950. This area had been an exclusive hunting reserve of one of Nepal's leading families; it was home to elephants, rhinos, tigers, bears, and crocodiles. Lee Talbot visited Chitwan in 1960, just as the region was radically changing due to a sudden massive influx of people (made possible by the introduction of DDT in 1954 to control malaria). As the human population tripled in the 1950s, wildlife was being decimated. Wild elephants, wild water buffalo, and swamp deer were driven to extinction. As settlers streamed in, the forests and grasslands were rapidly converted to agriculture. Rhino and tiger populations were declining precipitously. As he did in the Gir Forest, Talbot recommended basic studies of the ecology of this valley, and Seidensticker was part of the team that undertook these studies. Once basic scientific information was obtained, researchers had to decide how to reconcile the preservation of an ecosystem retaining large mammals, including large carnivores, with twenty-five million people with a per capita gross domestic product of $1,100 living nearby. The mere establishment of a national park was clearly not enough; poaching of wildlife and grazing by domestic livestock within the park continued through the 1980s.

As Seidensticker describes it, one of the answers was the establishment of buffer zones around the park combined with a community-based forestry/guardianship program. Community forests have been planted as buffers around Chitwan. The concept is to provide fodder and wood, returning profits to the people from both the community forests and those tourists who visit to view rhinos and tigers from the secure perch of elephants. Whereas this approach has been a failure in Zimbabwe, it appears to have been a success in Chitwan.

Seidensticker is optimistic about Chitwan's future: "Talbot taught that sound science was the basis for enlightened, effective conservation. His recommendations were founded in the notion that a deep knowledge of both human attitudes and workings of the ecosystem was a way to move conservation away from uninformed confrontation toward some workable solutions."

Finally, in chapter 8, James G. Teer discusses seven trends and issues in global conservation, namely (1) the emergence of ecology as a science, (2) the proliferation and influence of nongovernmental organizations, (3) sustainable use, (4) the transfer of ownership of wildlife to the private sector, (5) game farming and ranching, (6) development of ecotourism and nonconsumptive uses of wildlife, and (7) the strategy of saving what is useful to human society.

Teer emphasizes that although we face many challenges in global conservation, not all the news about its current state of global conservation is bad. Some wildlife populations have recovered to nuisance levels (e.g., white-tailed deer, elk, turkeys). Because of advances in ecology and related disciplines, conservation has become based more on science and less on opinion. And technology (e.g., computer modeling, advanced methods of tracking animals, use of satellites and geographic information systems) has improved our methods of ecosystem management.

Teer also finds the proliferation of nongovernmental organizations, and what he calls "citizen conservationists," to be a positive development. No longer can the proverbial faceless bureaucrat in a city make unilateral decisions about conservation policy. Many organizations as well as the private sector must participate, and regional or local corruption must be eliminated.

Echoing many others in this part, Teer is an advocate of sustainable use and the principle that human needs must be accommodated in conservation plans. To summarize, in his words, what each of these authors has emphasized: "Conservation efforts will succeed to the extent that we mobilize efforts to serve the human condition as well as the species and their habitats."

Reference

Balint, P. J., and J. Mashinya. 2006. The decline of a model community-based conservation project: Governance, capacity and devolution in Mahenye, Zimbabwe. *Geoforum* 37(5): 805–815.

5

Conservation of Sensitive Biodiversity Areas

Robert Goodland

In this chapter I outline the global priority to conserve most of what is left of the world's biodiversity. I also discuss the important role Lee M. Talbot has had in biodiversity conservation, especially successful because of Talbot's ability to influence the World Bank—a major financier of both the loss and the conservation of biodiversity in developing countries.

DEFINING PROBLEMS AND POLICY

The primary reason for protecting sensitive areas is to conserve the biodiversity on which society depends, to help prevent extinctions and other losses and to sustain the environmental services on which the poor, more than the wealthy, rely. Sadly, in the past, development projects have often caused major environmental degradation. This is an especially great problem because much development takes place inside unique or sensitive ecosystems. Rules for more environmentally friendly development are now being clarified worldwide; for example, many (44 of 630) UN World Heritage sites are threatened by development projects. When Russell Train chaired the U.S. Council on Environmental Quality (CEQ), he and Lee Talbot, the CEQ's director, led the campaign to adopt the World Heritage system. As will be seen, Talbot shifted international focus away from both individual charismatic megafauna and single species conservation to habitat conservation.

Development of International Environmental Policy

Talbot has long experience in the political arena, including leading the CEQ and later serving as the director general of the World Conservation Union (IUCN, formerly the International Union for the Conservation of Nature and Natural Resources). He has played key roles in fostering the National Environmental Policy Act (NEPA) of 1969 and in drafting the UN Convention on International Trade in Endangered Species of Wild Fauna and Flora (CITES) and the UN Wetlands Convention, as well as the U.S. Endangered Species Act and the Marine Mammal Protection Act. The Marine Mammal Protection Act was a pathbreaker for two reasons: Talbot introduced the precautionary principle into law for the first time, and he established that maintenance of the ecosystem, not just the conservation of individual species, is our paramount objective. In the 1980s, based on these achievements and on years of awareness-raising, Talbot began to help the World Bank (WB) devise its own biodiversity rules. He helped persuade the WB to hew to two overriding principles. They are detailed by the WB's new rules, stating that from henceforth:

1. "The Bank does not support projects that, in the Bank's opinion, involve the significant conversion or degradation of critical natural habitats" (WB 2004, para. 4).
2. "The Bank does not support projects involving the significant conversion of natural habitats" (WB 2004, para. 5).

The new principles and rules indicate that the WB wants to conserve such sites and rehabilitate them if previously degraded. The project area of influence includes the airsheds and watersheds of the project itself (e.g., exploration, prospecting, extraction, processing, beneficiation, transport, disposal, decommissioning, and restoration), including access roads, pipelines, transmission lines, ports, and terminals, through coastal zones, estuaries, and offshore areas.

The default principle is clear: no significant conversion of such habitat will be supported by the WB, to the extent it abides by its own biodiversity policies. As species extinctions intensify and as protected and other habitats are threatened, the trend by the WB is to strengthen such prudential policies. Taken in conjunction with the WB's precautionary and preventive approaches and the trend to enforce rather than bend these policies, this should be strict guidance for all developers.

Weaknesses of Environmental Policy

The key difference between critical and other natural habitats for development planners is that there are more provisions for exceptions to locating projects in natural habitats, and fewer exceptions for siting projects in critical natural habitats. This policy is weak in that the definitions of the terms "habitat" and "critical habitat" leave flexibility in interpretation to the WB staff. Moreover, the policy says that no "significant conversion or degradation" will be supported in such areas, but it does not define "significant." The policy thus implies that a development project designed to ensure a very small impact that does not "significantly convert" or "degrade" that habitat may be supported by the WB even inside protected areas. The key question remains unanswered: Are there indeed some development projects with impacts so slight that the policies would permit them inside critical natural habitats? The WB staff does not have a list of such cases.

Of course, the devil is in the details. The major scope for judgment and flexibility provided by the policies may not always be exercised by the most knowledgeable WB environmental professionals. And the policies' provision of so much room for interpretation and judgment tends to hamper their consistent and rigorous implementation; the definitions of critical natural habitats, other (noncritical) natural habitats, significant conversion, and degradation all are open to interpretation. The WB tacitly admits subjectivity to the extent that it relies on its own opinion. This internal opinion making may not be transparent. Often the critical first interpretation will be by the proponents' own environmental assessment team, and hence it may not always be fully objective.

The natural habitats policy of the WB contains further scope for exceptions and interpretation, allowing "wiggle room" when "there are no feasible alternatives for the project and its siting, and comprehensive analysis demonstrates that overall benefits from the project substantially outweigh the costs." The no-feasible-alternatives clause is addressed by the "analysis of alternatives" part of each project's required environmental assessment, which has become critical in exercising such judgments. The comprehensive analysis clause, in which the overall benefits from the project ostensibly outweigh the costs, is not addressed as rigorously as the analysis of alternatives. Quantification of social and environmental costs and benefits is problematic, and it is rarely undertaken systematically. In addition, it would be subjective to balance a clear, quantitative economic benefit (e.g., one ton of oil exported) with a qualitative impact (e.g., seepage of oily process brines and drilling muds into the nearby forest creek).

If a project will cause significant conversion or degradation of natural habitat, the WB's requirement for mitigation is similarly subjective—there is little guidance on what mitigation will be found acceptable. Granted, minimizing habitat loss is better than not minimizing such loss, but if a series of projects all successfully minimize habitat loss, a country could end up with practically no habitat. The WB's clauses exemplifying acceptable mitigation ("strategic habitat retention . . . [and] . . . protecting an ecologically similar area") while begging the question "How much is enough?" do offer an opportunity for offsets and for the net benefit approach discussed later in this chapter.

In addition, not all policies are rigorously and consistently enforced by the WB staff due to human limitations. The WB's track record on biodiversity conservation is mixed, and loss of habitat is not always avoided in WB-supported projects. The group's Operations Evaluation Department noted in 2001 that it has failed to mainstream environmental concerns; its main role remains development rather than conservation (WB Independent Evaluation Group 2006). Although the development goal has recently been environmentally improved by adding the goal of sustainable development, the WB remains unclear about what sustainability should be (WB 2003).

The other reason for the WB's policy flexibility is to avoid any risk of the chilling effect of banning development inside protected areas or in natural habitats in general, instead of merely banning significant conversion or degradation of protected habitat. At the WB, there are conflicting goals of environmental sustainability and traditionally conceptualized development. Similarly, governments sometimes promote risky development rather than conservation. If development were to be totally banned inside protected areas, governments might begin to be loath to permit creation of protected areas. This constraint can be avoided by surveying areas proposed for conservation before they are officially created and added to national conservation systems. The WB's policy depends on such surveys in order to determine if a project will significantly convert or degrade an area, but is rarely so interpreted. In view of the irreversibility of extinctions, if the area in question is exceptionally valuable, this should take precedence over a damaging project, as argued by Nigel Dudley and Sue Stolton (2002).

What Policy Means for Development

The WB's policy offers the opportunity to negotiate the default presumption that conversion projects will not be supported in such areas. This puts the onus on developers to come up with best practice designs, demonstrating that the people and their environments (e.g., habitats, biodiversity) will be clearly better off if the development is permitted and that the people agree in advance.

Developers have little flexibility in negotiating gray areas between critical and noncritical habitats (see case study, "Mitsubishi versus the Gray Whale"). But through excellent project design it is theoretically possible to support a claim that conversion of habitat will not be significant, or that the conversion will not result in degradation. An example of such a case might be a small underground mine with no ore treatment and no access roads promoting unplanned settlement. In practice, actual cases are difficult to find.

Lee Talbot pragmatically noted (1985, 2003) that in addition to excellent project design, there is scope for thinking outside the envelope in financing net benefit off-sets (see below). If a project has been designed to impose a very small impact and if the developer is willing to pay to improve net conservation, trade-offs are possible, as

---

Case Study: Mitsubishi versus the Gray Whale

The Mitsubishi Corporation was involved in an expensive six-year fight to build the world's largest (116 km²) salt mine. This mine would have impacted the UN World Heritage site of Laguna San Ignacio, Baja California Sur, in Mexico; Laguna San Ignacio is a UN biosphere reserve and has long been an official protected area under Mexican law. The project proposed to pump 6,000 gallons per minute of lagoon water into 21,000 acres of evaporation ponds linked to a mile-long conveyor pier. Mitsubishi's 1994 environmental assessment described the site as "wastelands, with little biodiversity and no known productive use." This contrasts with the fact of the affected people and the environmental view that it is the last undeveloped nursing ground for the gray whale (*Eschrichtius robustus*), which has a major role in the web of life. (Mitsubishi submitted its proposal for the salt mine one month after the gray whale was removed from the endangered species list.) The protected area supports a rich array of endangered and threatened species, such as dolphins, marine turtles, golden eagles, pronghorn antelope, and mountain lions. The mangroves of Laguna San Ignacio are the start of a food chain that supports the important community fisheries of the area, also a migratory bird refuge. In March 2000, Mitsubishi abandoned its proposal after five years of bitter international struggles.

long as stakeholders are fully informed (and as long as they agree that biodiversity will be better conserved with, rather than without, the trade-offs). Suppose a proponent approaches the WB showing that there is an especially profitable mineral or fossil fuel source inside a sensitive area, and suppose this proponent is seeking partnerships in order to design and finance measures to ensure that the people and the environment will be markedly better off afterward. The WB is likely to be less doctrinaire with this proponent. The absence of a successful track record makes this risky. But the financing of two huge new national parks as offsets in the Chad-Cameroon oil pipeline case, and Talbot's initiatives in securing $30 million for conservation of a 3,000 km^2 watershed to protect Laos's Nam Theun hydropower reservoir, are recent examples of what it is possible to get the WB to finance for conservation (see also chapter 19 in this volume).

National systems of protected areas are not static. They need to adapt in order to keep up with changing realities and to meet national standards of conservation of representative samples of the nation's biota. While critical natural habitats should remain conserved, there is scope for offsets in the case of noncritical natural habitats.

The lessons for proponents are:

1. Contract with the most reliable and independent environmental and social assessment team at the beginning of the project. A weak environmental assessment will lead to greater costs in the future.
2. Recognize that any proposal in or near any protected area will be held to higher standards than elsewhere. Such projects will demand meticulous designs to prevent significant impacts, and they will have to be accompanied by clearly beneficial net offsets. In some cases it will not be possible to find a substitute protected area (this occurred in the "Mitsubishi versus the Gray Whale" case presented in this chapter, in which the gray whale migrates 5,000 km to calve at Laguna San Ignacio in Mexico).
3. Know that the mandated analysis of alternatives has to indicate whether substitutable areas are available. (This is also illustrated in this chapter's case study—there are many sites for salt mining where marine salinity, solar evaporation, and transport factors are propitious and could be lucrative with few or no environmental impacts.)
4. If you opine that you must site an industry inside a critical habitat, you must propose major net benefits or offsets.

DEFINING SOLUTIONS: INTEGRATING CONSERVATION AND DEVELOPMENT

Flexibility in the WB's Policy

The WB's policy is designed to foster sustainable development, of which conservation of the environment is an essential part. As previously discussed, though, interpretations of the policy can be subjective at times. The reality is that much valuable biodiversity remains unprotected and at risk while some protected areas have lost most of their biodiversity. In addition, at the time of protection, the existence of valuable ores or fossil fuels may have been unknown. Operationally, because most projects involve natural habitat conversion, the WB may be persuaded to support a project if it is meticulously designed to avoid major impacts (e.g., directional drilling from outside the area in question) and if the environment and biodiversity will be clearly better off (see below) in a net benefit sense—of subtracting the area to be converted by development from offsets or improved conservation elsewhere Thus proponents can site low-impact projects inside natural habitats, but only if they finance protection of an equivalent area in perpetuity. This can be a win-win opportunity both for industry and for conservation.

More and Better Conservation: The Net Benefit Approach

Experience shows that by creating new parks and by improving existing parks, we can contribute significantly to long-term biodiversity conservation. In order to achieve this, we must make irreproachable social and environmental assessments if there is to be a project inside a protected area. The WB wants these assessments to become more independent and rigorous than in the past. The environmental assessments must show how the project can be designed to prevent major impacts, particularly irreversible ones. The assessments will also show how the development project will achieve net benefit for the community, the nation, and the environment. "Net benefit" means minimizing negative impacts and contributing positively to conservation—being clearly better off with the project. One effective way of ensuring net benefit is by means of sustainably financed offsets or set-asides. These also have to be better than what may be converted. Here, "better" means informed agreement of stakeholders that the proposed offset is (1) more extensive in area than the site to be developed, (2) greater in environmental value, (3) less disturbed and/or damaged, (4) likely to result in more biodiversity, (5) possessed of greater environmental service value, and (6) under a more secure level of protection (e.g., such as by financing in perpetuity).

Essentially, net benefit and offsets are pragmatic approaches seeking to integrate economic development into sustainable development and conservation. The key is the degree of substitutability of the site in question. Critical natural habitat is not substitutable; noncritical habitat may be. Most sites in the 118 countries protected under the UN World Heritage List, for example, are critical because they are not substitutable; they are conserved for geographic or geological features that often support endemic biodiversity. Of course, there are many more critical natural habitats than UN World Heritage sites. The latest UN World Heritage site, East Rennell Island in the Solomon Islands, is a good example of nonsubstitutability. It is the world's largest raised atoll and has the largest freshwater lake in the Pacific; both features support vast numbers of endemic species.

Improved benefit-sharing greatly enhances conservation to the extent that affected people are better off and do not have to resort to predatory hunting, fishing, overcollecting of forest products such as medicinals, logging, and cultivating unsuitable lands. In mining projects, the environmental assessment will ascertain if metalliferous species restricted to the ore body are present and in need of conservation, by means of an offset if those species occur elsewhere. The impacted people will know where else such species can be found and are usually the most effective people to conserve offsets in perpetuity. These offsets can and should benefit impacted people.

There are few rules of thumb here. Each project is different (e.g., directional drilling vs. mountaintop removal), each area is biologically different (many areas can be offset, given goodwill), and the conservation needs of each area also are different (some countries already have a network of protected areas; others may have some parks marked on a map, but little meaningful conservation). Technologies are improving gradually. For example, the best practice goal of closed systems—zero emissions into air or water, therefore operating with no significant impacts—is being achieved in the pharmaceutical industries. In another example, leading manufacturers of industrial carpets have now ceased to use landfills for worn carpets: all components are recycled. The track record of development in sensitive ecosystems is quite poor, though, and the offset, set-aside, and net benefit approaches are quite new, so the WB has to act with great caution. Permits should be given only where there is full agreement of all stakeholders, including the local population, that offsets and other mitigations are unambiguously better than the alternative. The track record of developers must improve markedly. The time for minor, incremental improvements is over.

Each case should be tailored to specifics. In some cases one conservation priority may be the captive breeding of a species on which local communities depend, such as those exporting a decorative palm or fern. The propagation of pollinators of such species may not be known. Likewise, aquatic resources on which communities depend (and which are often harmed by river or ocean pollution from development projects) may require special precautions. And if an endangered species may be further threatened by a project, its captive breeding for subsequent wider reintroduction may be useful.

One example of captive breeding involves the Kihansi spray toad, *Nectophrynoides asperginis,* which lives only in the fine mist created by the cascading waters of the Kihansi Falls in the southern Udzungwa Mountains of Tanzania. The toad has been threatened with extinction by diversion of the water for a 180-MW hydropower project. Although artificial spray systems have been installed to preserve its habitat, a captive breeding program may be the toad's only hope for survival. The original WB environmental assessment was unacceptable, and the Norwegian Agency for Development Cooperation (Norad) had to finance a new environmental assessment in order to reduce the damage. Now the toad is breeding well in several zoos.

Captive breeding can be a useful tool in limited cases. But it cannot mitigate environmental degradation, nor should it be applied only to a small number of charismatic (or notorious) species. For every known endangered species listed from an area, there are inevitably many more undiscovered species. That is why conservation of a large tract of environmentally equivalent ecosystem adjacent to a project is almost always the main mitigation. Captive breeding can be a useful, minor add-on.

Definitions of "No-Go" Areas

The WB's definition of critical natural habitats includes the following (presented on IUCN's Red List of Threatened Species [2007]):

1. Protected areas (e.g., UN World Heritage sites, UN biosphere reserves, Natura 2000 sites in Europe, UN Wetland Convention sites)
2. Areas meeting IUCN's categories I–VI, and marine categories I–V (e.g., fishing or fish breeding reserves)
3. Proposed protected areas (e.g., as designated in ecoregion action proposals, regional assessments, or land use plans)
4. Areas recognized as protected
5. Areas maintaining conditions vital for protected areas (e.g., watersheds, buffer zones)
6. Areas on supplementary lists or as determined by the WB's environmental staff
7. Areas highly suitable for biodiversity conservation (meaning that areas in which biodiversity is unknown need to be assessed before they can be categorized)
8. Areas critical for rare, vulnerable, migratory, or endangered species (e.g., redlist@ssc-uk.org)

Additionally, areas used by vulnerable ethnic minorities (WB 2007) are, if anything, even more of a no-go than the items listed under the natural habitats policy (Colchester et al. 2003). The concerns of potentially affected people are paramount and need to be addressed before, or at the same time as, the above criteria.

Offsets

Offsets are areas that can be conserved by developers, either adjacent to a project or elsewhere, such that the environment is unambiguously better off with the project. Offsets were codified for air pollution following the United States' 1970 Clean Air Act Extension, whereby a firm can expand or build a new plant if it reduces emissions at existing plants by more than the amount to be produced at the new facility. Each U.S. state mandates the ratio of pollution reduced to pollution added (this can be 10:1). Firms must also provide their new plants with control technology to achieve the lowest achievable emission rate, regardless of cost. If the firm reduces its pollution more than required, it may bank the credit or sell it as a marketable permit to another firm.

Environmental offsets follow this pattern, though they are not yet as widely mandated in developing nations. Offsets typically substitute one area for another and are strongly supported by the WB, which may even consider assisting with the financing. For environmental offsets, the "best technology" mandated in the air pollution case becomes the best practice design. The ratio of conserved habitat to converted habitat is best practice if it also follows the 10:1 ratio of the Clean Air Act Extension.

Depending on the size of habitat to be converted, a 1:1 ratio probably would not function adequately to conserve biodiversity and thus would not be acceptable. There is a minimum critical size of habitat below which fires spreading from adjacent agricultural lands, or winds drying the edges of the habitat, render the area unsustainable. This size varies with ecosystem and local conditions; for example, 10 km^2 may be on the large size for an actual mine, whereas a 10 km^2 conserved area may not be viable. It is often better to add the offset area to an existing conservation unit rather than creating a new or small one. So the range for offset ratios is significantly more than 1:1. The best practice ratio is 10:1. While there are no hard-and-fast rules, the offset ratio needs to be an unambiguous net benefit in order for developers to be granted the exception to the rules that they seek. In the case of Laos' Nam Theun hydropower project, Talbot and Thayer Scudder helped conserve about 4,000 km^2 of habitat as an offset for a 400 km^2 reservoir (Scudder, Talbot, and Whitmore, 2001). Usually, the ratio will be part of the net benefit package of offsets, damage prevention and mitigation, and sustainable financing. The other part may include a foundation financing national biodiversity priorities or a fund receiving a small fraction of profits or linked to volume of wastes or area affected.

Given the many variables and the need for flexibility, we must look to examples to see the range of solutions available. (Specifics can be provided by environmental assessments.) In 2000, during construction of a Chad-Cameroon oil pipeline, routing was adjusted to avoid all protected areas. Exxon conserved habitat by financing two new national parks supporting fairly intact forest, many times the area lost to the pipeline right-of-way and by the access roads, camps, wells, staging areas, etc. The presidentially decreed national parks exceed 4,000 km^2 of essentially intact ecosystem, while the converted areas total less than 100 km^2, mainly following existing roads (hence already disturbed). The offset ratio is 40:1 in this case. In another case, that of Uganda's Bujagali hydropower project, Talbot, Jason Clay, and William Jobin (2001) persuaded the government to create major conservation units (at Kalagala) on both banks of the Nile downstream of the small run-of-river reservoir. Both of these examples are recent, so their long-term impacts are still unknown.

Talbot pointed out that the most effective way to achieve net benefit is often through strengthening the national protected area system by allocating a fraction of profits to an endowment in perpetuity. Similarly, allocating a fraction of the profits to conservation may also be developers' preferred way to ensure a net benefit offset. This follows the approach mandated by the governments of China, Brazil, and others of allocating small fractions (1–2% each) of total project costs to social and environmental priorities.

If Offsets Are So Effective, Why Are There So Many No-Go Areas?

Despite the flexibility of offsets and the net benefit approach, the fact remains that there are some areas in which risky development should not take place:

1. *Indigenous lands:* The main category where offsets may not be possible includes land used by indigenous or vulnerable communities that cannot be successfully resettled elsewhere, even if there is more extensive and productive land available.

2. *Lands for which there are no substitutes available:* The second category of explicitly no-go areas concerns situations in which no similar habitats are available to be conserved as offsets by the developer. If the ore or well is in tropical wet forest, there may not be enough undisturbed forest nearby to substitute meaningful offsets for the converted forest. If the much rarer tropical dry forest is involved, there will be even less scope for offsets. If the slated ecosystem is already very restricted, such as a

mountaintop cloud forest in a desert region, there will be little or no opportunity for offsets. (Here again, technology may eventually be able to exploit ore from beneath, leaving the cloud forest intact on top.)

3. *"Mines are small, impacts are extensive"*: The area of a mine or well itself usually is modest, but its area of impact may be orders of magnitude greater. For example, Papua New Guinea's Porgera Open Pit gold mine covers less than a dozen km². However, the mine pit produces 40,000 mt³ per day of tailings that are released into rivers connected with Lake Murray, Papua New Guinea's largest freshwater lake, located 140 km downstream. Under WB policies, the 140 km stretch of river would be declared an impact zone. Since people commonly walk 10 km to fish (less to cultivate their riverbank gardens), any heavy metals or fish declines might affect (2 × 1400) 2800 km², depending on assumptions. As the mine is over 3,000 m in altitude, substitute ecosystems are likely to be available, but there is little one can do for riparian ecosystems and for the seven different ethnic communities downstream, unless disposal into rivers is banned. Access roads or rights of way have similar effects; they can be rerouted away from villages and habitat to reduce impacts, but tailings and roads often impact the people and environment more than a mine or development site itself.

Projects inside Protected Areas

The case commonly arises that the government creates a protected area that already contains some concessions inside it. The concessions often have not yet been exercised, although sometimes they are being exploited at the time of creation of the protected area. In this special case, the reasons for creation of the protected area should guide subsequent development. For creation of a new protected area means the government's priorities have evolved, shifting from converting natural habitat to conservation. No new concessions should be granted; where the government once granted conversion permits, it now wants a protected area. Of course, the government should not unilaterally void previously granted concessions without compensation, and it can refuse to renew exploration or other permits, or accelerate their lapse.

One example of a project that would be inside a protected area, should it happen, involves the substantial gold deposit in Ireland's holy mountain, Croagh Patrick. (This perfectly conical mountain is where Ireland's patron saint, Patrick, reportedly fasted during all forty-four days of Lent in 441 A.D.). Ireland's minister for energy denied permits to mine Croagh Patrick following widespread public opposition to the mining proposal in the 1980s. Still, not all sacred mountaintops are sacrosanct. A second example of a project that would be inside a protected area, and in fact is inside a protected area, involves gold mining on the sacred Amungme mountaintop in Papua New Guinea. Amungme is being mined by Freeport McMoRan Copper & Gold, Inc., an Australian mining corporation. Freeport dumps 120,000 tons of waste into the Aikwa River every day.

Yet another example of a project inside a protected area involves Arizona's Mount Graham. The Vatican was given permission to build a telescope on Mount Graham, despite the fact that it is a San Carlos Apache sacred site and an endangered species habitat, despite vigorous protests since the 1970s, and despite the presence of many better alternative locations for the telescope. As of 2005, the U.S. government has reconsidered decisions to overturn the Endangered Species Act and the Religious Freedom Act in this case.

As opposed to these examples, it would be preferable, when drafting national legislation to conserve an area, to include provisions to compensate concession holders. Clauses accelerating the lapse rate could be strengthened so that if concessions are not activated within a certain time they may be withdrawn. If the concessions are old and unlikely ever to be taken up, the compensation can be nominal. If the concessions are being lucratively exploited, the legislation should gradually mandate redesign to reduce impacts to the minimum compatible with the purposes for which the protected areas were enacted, along with retraining for the employees to sustainable jobs elsewhere. There is also the useful possibility of compensating the concessionaire with a new concession well outside the new protected area. This buyback mechanism has worked well in some countries. Specifically, "buyback" means that the concessionaire's permits are bought for cash; this happened in May 2002 when the U.S. government bought back $120 million of oil and gas leases from the Collier family for land inside three protected areas in the Everglades. Sometimes concessions inside sensitive areas can be exchanged for future drilling credits elsewhere. The WB's policy would not permit it to support any risky development inside such sensitive areas, even if there was a prior legal concession.

Treason is a matter of dates; so, too, with conservation. If a development permit was granted before official recognition of the conserved area, it is unfortunately legal. If the permit was granted after gazettement, it is illegal and inadvisable. If gazettement occurs around and just outside a legal development concession, it is a gray area and a recipe for problems. Possibly the best account of an oil concession that was grandfathered into a national park involves a Basic Resources Corporation operation in Guatemala's Petén province (Bowles et al. 1999). In 1985 the corporation received and exploited an oil concession in an area that was later, in 1990, included in a national park. Problems arose because the concessionaire was granted a second, much larger concession two years after the gazettement of the national park. The WB was internationally criticized for misinterpreting its own natural habitat and other policies when it approved a loan to Basic Resources Corporation in 1994 and a second loan in 1996. Eventually the International Finance Corporation (IFC), the private-sector arm of the WB, pressured the proponent into supporting the national park, but the proponent prepaid the loans rather than continue to work with the IFC. Prepayment in order to reduce scrutiny has, on occasion, been resorted to elsewhere. Such was the case with Freeport's previously described Amungme mining project in Papua New Guinea. Freeport unilaterally canceled its WB contract as soon as it heard that a WB environmental auditor was scheduled to inspect the project in 1999. The use of prepayment to nullify the WB's social and environmental responsibilities must be prohibited. Freeport's commitment to remediation seems to have started shortly thereafter (Shuey 1999).

Where there is concern, governments may define conservation boundaries more from the location of extractable deposits rather than from social and environmental criteria, such as in the Lorentz National Park World Heritage site in Indonesia and in Conoco's Warim concession in Papua New Guinea. While not perfect, conservation has to exploit the second best on occasion.

A different example involves the UN World Heritage Kakadu National Park in Australia, a Rio Tinto Corporation uranium mining site since the mid-1990s. Rio Tinto overrode protests by the indigenous peoples and many other groups, proceeding to mine uranium from inside the national park. Whenever one ore body was depleted, Rio Tinto began extracting a new ore body; protests broke out repeatedly. After more than a decade of protests, all major mining corporations finally agreed not to mine inside Kakadu National Park, and in 2003 they began returning 50,000 tons of mined uranium ore back down the mining shafts before withdrawing from the park.

CONCLUSION

Although biodiversity conservation goals are modest, they are hard to secure. The area needed for conservation in each ecosystem type is said to be about 15% (Pimm et al. 2001). Does this mean that we conservation professionals really acquiesce to the destruction of 85% of all natural habitats? But achievement of the 15% goal, if inadequate for some conservation priorities in the future, would leave us with a much better situation than the one today. Is it not odd that the UN Biodiversity Convention does not prioritize prevention of extinctions? Is it not sad that conservation of biodiversity receives less treatment than patents and intellectual property rights in the Biodiversity Convention?

For the future, conservation professionals must first raise their sights to reach and then exceed the 15% goal, and they must begin restoration of degraded areas. Second, we must follow Lee Talbot's lead and persuade developers to finance net benefit arrangements so that our environment will be unambiguously better off with the projects developed. Third, we need to convince institutions to implement the rules already on their books and to strengthen such rules commensurate with scientific evidence. Fourth, we need to hold CEOs accountable to mandatory codes of conduct, combined with insurance and performance bonds, backed up by litigation for failure to respect prevailing laws.

Developers are increasingly willing to work closely with stakeholders, affected people, and social and environmental specialists to design projects so that their social and environmental impacts are minimal from the outset. Developers must guarantee that people and the environment will be markedly better off in the net benefit sense. If so, there is a major opportunity for mutual benefits between developers and environmental conservation. Over the years, as natural habitats are becoming even rarer than they are today, the need for offsets is increasing commensurately.

In closing, we can summarize the WB's policy options for protected area development as follows:

1. The WB should consistently implement its social and environmental policies, specifically those on involuntary resettlement, indigenous peoples, natural habitats, and environmental assessment. The WB should cease weakening its policies; on the contrary, these policies need to be strengthened as world environments deteriorate. The WB must accept full responsibility for the conscientious implementation of its policies.

2. The default presumption is that conversion or degradation from development projects is not permitted in or near natural habitats (WB definition), nor in or near protected areas (IUCN definition).

3. After assessing all alternatives, if proponents strongly seek an exception, they must acknowledge that the risks and liabilities will be greater, that scrutiny will be more intense, both by financiers and by civil society, and that less than best practice will be unacceptable. Best practice project design (very low impacts) hence becomes important.

4. The WB may agree to a one-off exception in certain cases, but only if the project unambiguously: (a) meets more stringent standards of decision criteria, especially social and environmental assessment, poverty reduction, and sustainable development; (b) designs out or fully mitigates significant impacts beforehand; and (c) finances net benefits and offsets in perpetuity, effectively substituting for the noncritical habitat to be lost to the project.

5. The WB should support governments in their clarification of protected areas and previous concessions. The WB should support those countries that still permit developers inside protected areas to update their policies and to phase out such concessions.

6. The WB can conserve much biodiversity by supporting rehabilitation of old, abandoned, and risky development areas before they damage more people or natural resources.

7. Insurance or escrowed performance bonds, managed by an independent board, must be mandatory in all appropriate cases (Goodland 2003).

References

Bowles, I. A., A. B. Rosenfeld, C. F. Kormos, J. D. Nations, C. C. S. Reining, and T. T. Ankersen. 1999. The environmental impacts of International Finance Corporation lending and proposals for reform: A case study of conservation and oil development in the Guatemalan Petén. *Environmental Law* 29(1): 103–128.

Colchester, M., A. L. Tamayo, R. Rovillos, and E. Caruso, eds. 2003. *Extracting promises: Indigenous peoples, extractive industries and the World Bank.* Baguio City, Philippines: Tebtebba, Indigenous Peoples' International Centre for Policy Research and Education; Moreton-in-the-Marsh, UK: Forest Peoples Programme.

Dudley, N., and S. Stolton. 2002. *To dig or not to dig? Criteria for determining the suitability or acceptability of mineral exploration, extraction and transport from ecological and social perspectives.* Gland, Switzerland: World Wildlife Fund.

IUCN—The World Conservation Union. 2007. Red list of threatened species (search site, with categories, for entire list). http://www.iucnredlist.org/search/search-basic.

Pimm, S. L., M. Ayres, A. Balmford, G. Branch, K. Brandon, T. Brooks, R. Bustamante, R. Costanza, R. Cowling, L. M. Curran, et al. 2001. Can we defy nature's end? *Science* 293(5538): 2207–2208.

Scudder, T., L. M. Talbot, and T. C. Whitmore. 2001. *Environmental and social issues, Nam Theun 2 Hydro Project: V.* Vientiane, Laos: Government of the Lao Peoples Democratic Republic.

Shuey, S. A. 1999. Freeport's commitment to remediation. *Engineering and Mining Journal* 200(12): 34–37.

Talbot, L. M. 1985. *Helping developing countries help themselves: Toward a congressional agenda for improved resource and environmental management in the third world.* Washington, DC: World Resources Institute.

Talbot, L. M. 2003. Does public policy reflect environmental ethics? If so, how does it happen? *University of California Davis Law Review* 37(1): 269–280.

Talbot, L. M., J. Clay, and W. Jobin. 2001. *Environmental and social considerations in the Bujagali Hydro, Uganda: VIII.* Arlington, VA: AES.

World Bank. 2003. *World development report 2003: Sustainable development in a dynamic world; Transforming institutions, growth, and quality of life.* Washington, DC: World Bank.

World Bank. 2004. Operational policy statement (OP) 4.04, revised ("natural habitats").

World Bank. 2007. Operational policy statement (OP) 4.01, operational directive 4.20.

World Bank Independent Evaluation Group. 2006. Assessing the effectiveness of World Bank assistance for the environment.

6

Protected Areas: Science, Policy, and Management to Meet the Challenges of Global Change in the Twenty-First Century

Kenton R. Miller & Charles Victor Barber

The past half century has been witness to a fundamental transformation in how humanity defines and orders the use of land and its natural resources across the planet's landscape. Officially designated "protected areas" have become one of the earth's major land uses, encompassing slightly more than 12% (17,125,893 km^2) of the planet's land surface, an area larger than the United States and nearly six times the size of India. This is a truly new and dramatic departure from the past: in 1952, protected areas accounted for a mere 1.25 million km^2 (Mulongoy and Chape 2004).

This achievement represents the outcome of more than a century of work by visionary, highly motivated individuals, communities, and governments exerting a rare sense of commitment to stewardship of our planet's natural wealth and wonders, the future of humanity's well-being, and the sustainability of life on Earth. In the words of the 2003 Durban Accord, "We celebrate one of the greatest collective land-use commitments in the history of humankind—a worldwide system of some 100,000 protected areas and tripling of the world's protected areas over the last 20 years" (IUCN—The World Conservation Union [IUCN] 2005, 220).[1] This achievement was reinforced in 2004 when the parties to the UN Convention on Biological Diversity (CBD) adopted a Programme of Work on Protected Areas, an ambitious document that embodies in international law, for the first time, a comprehensive commitment to strengthen and expand protected areas across the globe (Dudley et al. 2005; Secretariat of the Convention on Biological Diversity 2004).

Turning this commitment into effective conservation on the ground, however, is one of the great challenges of the twenty-first century: the achievement celebrated at Durban is a fragile one. As a major global land use, protected areas are no longer simply isolated, minor set-asides of a pretty park here, a breathtaking geological feature there. Without a doubt, they will continue to be valued for their aesthetic features, their potential as recreational and tourist attractions, and for the preservation of historic and cultural sites. Rather, as we look to the world of the twenty-first century, protected areas are now enmeshed in complicated and conflicted struggles over fundamental human values, politics, economics, and culture—in a world that is profoundly changed from that of a mere half century ago and continues to change in demographic, economic, and biophysical ways at a dizzying pace.

Of all the issues facing people and governments at this time, why place the issues surrounding national parks and protected areas front and center on their political agenda? Because the security of this great legacy is now threatened by processes of change occurring on local, national, and global scales. In the words of the recently completed Millennium Ecosystem Assessment: "Human actions are fundamentally, and to a significant extent irreversibly, changing the diversity of life on Earth, and most of these changes represent a loss of biodiversity. Changes in important components of biological diversity have been more rapid in the past 50 years than at any time in human history. Projections and scenarios indicate that these rates will continue, or accelerate, in the future" (Millennium Ecosystem Assessment 2005, 2).

Biodiversity loss cannot be stemmed by protected areas alone, but protected areas can and must play a central role. For that to happen, however, we need to think about protected areas—and manage them—in ways considerably different from the past.

Specifically, we first need to expand our vision of what protected areas contribute to human well-being. We must recognize in our thinking about policies and actions that protected areas are fundamental to maintaining vital ecosystem services— such as our supplies of clean water—upon which human well-being depends. We must acknowledge and act on the fact that a significant proportion of the planet's biodiversity—the very diversity of life on Earth—lies within existing protected areas, or deserves the protection of new ones. And we must accept, as a core societal value, the central spiritual and cultural importance of wild nature for all cultures and peoples, both traditional and modern. Furthermore, in some regions (such as in Africa), the majority of biodiversity may exist outside defined protected areas, across landscapes that have long been settled and cultivated. In such instances, strategies for biodiversity conservation will need to incorporate protected areas within mosaics of diverse land uses that include farms and managed forests, human settlements, roads, and other limited infrastructure. To the extent possible, such approaches to bioregional planning can maintain biodiversity within a context of development through community-based management (Miller 1996). Building on that expanded vision of protected areas as a central pillar of human well-being and our relationship to this planet and its living wonders, we will need to take practical action in five areas:

1. Reinventing and broadening the theory and practice of protected area governance and management
2. Strengthening the management capacity of protected areas at all levels, from the nation-state to the local community, reflecting broadened visions of protected area functions and governance
3. Planning and managing protected areas adaptively, for a world that has changed dramatically over the past few decades and will continue to change rapidly in the future
4. Finding innovative ways to build synergies between protected areas and development and poverty alleviation priorities

5. Developing stronger and more innovative ways to increase and sustain financing for protected areas

PROTECTED AREAS: THEIR VALUE AND ROLE

What are protected areas? According to IUCN, a protected area is "an area of land and/or sea especially dedicated to the protection and maintenance of biological diversity, and of natural and associated cultural resources, and managed through legal or other effective means" (1994, 7).

In the past such areas were generally called parks—national and otherwise—and their central purpose was considered to be to conserve wild nature and outstanding natural features, as well as to foster public education and recreation in the natural environment. The term "protected area" was only introduced into the general discourse in the 1960s in order to recognize those reserves managed for nature conservation purposes while under the jurisdiction of other than central governments. In some cases protected areas permit limited and controlled extraction of natural resources, including the gathering of wildlife and plant materials, along with limited and regulated timber harvesting, hunting, and fishing. The internationally recognized protected area categorization scheme produced by IUCN (1994) now recognizes a range of categories, from strictly protected areas to cultivated and other productive landscapes that nonetheless serve important conservation objectives.

Protected areas are important because they maintain and protect large, representative samples of the variety of life on Earth. They provide security for the myriad values of biodiversity, including the supply of clean and abundant freshwater and air, and many other products and services. Access to recreation opportunities, to places important for cultural identity and spiritual renewal, and to high-quality scenery are paramount to the quality and meaning of life in all cultures. They also help keep the environment resilient against natural and man-made disasters, like hurricanes, flooding, and drought. Furthermore, they provide options for the future, including wild genetic resources for the development of new and improved foods and medicines.

Protected areas can also help address the issues of poverty if they are managed to supply rural communities with water, food products, energy resources, medicines from the wild, and employment opportunities. Similarly, watershed restoration and farming and forestry projects can protect towns and farms against flood and drought. They can also offer work for the seasonally unemployed when crops such as coffee, various fruits, and sugarcane need little direct management. Furthermore, protected areas can strengthen the resolve and cultural identity of communities facing poverty by supplying materials from the wild traditionally used for subsistence and ceremonial purposes, and by maintaining sites of historical and spiritual significance.

In addition to these values, many places are held in high regard by indigenous peoples for cultural, spiritual, and historic reasons. In many parts of the world, such sites have already been incorporated into parks and protected areas to provide them with added protection from looting, trespassing, and disrespectful activities. On the other hand, in some cases, the activities permitted in parks (e.g., through the presence and passage of hikers who leave a light footprint in wilderness areas) may create conflict with indigenous values and perspectives.[2] In other cases, protected areas have restricted indigenous communities' long-standing access to their own customary territories and resources, giving rise to serious conflicts that threaten the viability of both indigenous lifestyles and the very biodiversity that protected areas are designed to conserve.

Protected areas—and, specifically, the biodiversity that they protect—are directly and indirectly related to the future well-being of humankind, although humanity's actions have often not reflected this reality: "Conservation of biodiversity is essential as a source of particular biological resources, to maintain different ecosystem services, to maintain the resilience of ecosystems, and to provide options for the future. These benefits that biodiversity provides to people have not been well reflected in decision-making and resource management, and thus the current rate of loss of biodiversity is higher than what it would be had these benefits been taken into account" (Millennium Ecosystem Assessment 2005, 7).

SCIENCE, DEVELOPMENT, AND EMPOWERMENT: THE CONTESTED TERRAIN OF BIODIVERSITY

As protected areas have expanded and diversified across the rapidly changing planet—and the human economies and societies that increasingly dominate it—so too have people's visions of what they are, what they provide (or should provide), and how they should be governed and managed. Conflicts between protected areas and other resource uses have also multiplied, as might be expected with the rapid expansion of a relatively new land use regime to such a vast area of the globe. As the last great repository of the planet's wild biological diversity, a key source of numerous ecosystem services, and home to many of the planet's traditional and indigenous communities, protected areas have become a template for broad debates about the relationships among nature, humanity, science, development, policy, and culture.

As biodiversity has become more prominent on international and national agendas—and has become the major rationale for expanding and maintaining protected areas—three distinct approaches have developed, with their own visions of what biodiversity means, what priorities should be for saving it, and who should do what to implement those priorities. These three poles can be called "science," "development," and "empowerment." The tension between these views is sometimes creative, sometimes not. Finding ways to reconcile and build synergies among these three approaches is a major challenge facing those at all levels who seek to maintain and expand protected areas, particularly in the tropical developing countries, where most biodiversity is located and where its loss is most rapid.

Many people can agree, on a general level, that robust approaches to protected areas in the developing world must be based on (1) good science, while simultaneously addressing issues of (2) development and poverty alleviation and (3) imbalances of power over decision making and resource allocation. Emphasis on one or another element of this triangle yields profound differences in views on what should be conserved, for what purposes, how scarce funds should be allocated, what methods should be used, and who should have ultimate authority and control over protected areas.

Science

The discipline of conservation biology has grown rapidly over the past two decades, strengthening practical efforts to conserve biodiversity through development of both basic and applied conservation science methods and information. Advances in conservation biology inform the "science" approach to biodiversity conservation, most prominently through various priority-setting exercises at the global and regional levels—for example, biodiversity hotspots, priority ecoregions, key biodiversity areas, endemic bird areas, the global gap analysis, etc.—and the elaboration of sophisticated methodologies for systematic conservation planning and protected area management effectiveness evaluation at the protected area system and site levels (see Barber 2004a; Gordon, Franco, and Tyrrell, 2005; Groves 2003; Hockings, Stolton, and Dudley, 2000; and Rodrigues et al. 2003).

Championed most prominently by the big nongovernmental organizations (NGOs) concerned with international conservation, this school of thought takes a rationalist approach, grounded in the biological sciences, on the assumption that there is an objective scientific reality to the world's biodiversity (and threats to it) that can be discovered and mapped, and that priorities for protected area action and funding can be set accordingly. It is—compared to the other two schools of thought—an essentially biocentric perspective that employs a relatively narrow definition of biodiversity bounded by what is known to science about the diversity of genes, species, and habitats, and ways of measuring them.

The "science" approach is undeniably a key foundation for any scientifically defensible strategy to reduce biodiversity loss by means of protected areas. It is criticized by some, however, as overly reductionist, blind to the economic realities of poverty, dismissive of the concerns of local and indigenous communities, and unduly self-serving for the NGOs that most vocally promote (and fund-raise for) it (Brechin et al. 2004; Chapin 2004; Dowie 2005; Khare and Bray 2004; Zerner 2000).

Development

These days, most conservationists would agree that for protected areas to be sustainable they must be designed and implemented to ensure equitable sharing of costs and benefits, particularly with the poor, on both moral and practical grounds. Indeed, ascendance of the UN Millennium Development Goals—which include the environment in passing but essentially focus on alleviating hunger and poverty—as the primary touchstone for international development assistance agencies and efforts has put conservation advocates on the defensive, recasting their efforts as development-related and calling for balance that does not ignore biodiversity in development aid priorities (see, e.g., Lapham and Livermore 2003).

Some conservation and development practitioners, however, go further, arguing for a new model of pro-poor conservation. This model seeks to redefine the very goal of biodiversity conservation, moving further along the continuum from conservation through poverty reduction (poverty reduction as a necessary means for conservation) toward poverty reduction through conservation (conservation primarily defined as a means to reduce poverty) (Roe and Elliott 2003).

Proponents frequently justify this approach with practical arguments that "exclusionary fortress" approaches to protected area conservation don't work in the midst of dense and poor local populations. But it is, in the end, a morally based approach rooted in the philosophical view that human well-being—and eradication of the inequalities that cripple it—is the greatest moral good, trumping the well-being, or even the existence, of other species. The "development" school of thought therefore often focuses on win-win strategies and situations where poverty alleviation and protected areas can go hand-in-hand.

Critics of this approach point out, however, that (1) win-lose strategies, involving difficult trade-offs, are more common in areas where biodiversity is being lost most rapidly; (2) the "development" approach plays fast and loose with the hard-and-fast biological imperatives governing the survival of species and natural habitats; and (3) it is unrealistic and unfair to expect the last remaining bits of relatively wild nature to bear the burden of alleviating the world's poverty.

Empowerment

The "empowerment" school of thought grows out of the deconstructionist school of political ecology (derived from postmodern currents in literary criticism), which holds, essentially, that there is no one true version of history or current reality, just competing narratives which are constructed in the service of one interest group or another, with political and economic elites and their allies constructing the dominant narrative and imposing it on those marginalized from the levers of political and economic power.[3]

Indigenous and local communities and their advocates have, therefore, put forth a counternarrative with both moral and practical elements. In moral terms, the rights of indigenous and other resource-dependent communities to their long-standing territories and their purported accomplishments as natural resource stewards are strong moral arguments in favor of increasing local control over biodiversity resources and decisions.

In instrumental terms, concurring with the "development" school of thought, they argue that biodiversity conservation in developing countries is a practical impossibility in the face of determined resistance from those who live in and around high-priority conservation areas. This counternarrative has achieved considerable success on the international stage (more so, some would say, than it has on the ground in many countries), and community-based conservation, participation of indigenous and local communities, and similar concepts are now firmly rooted in the international protected area dialogue.[4] These are positive developments in many ways, as we argue below. Less positive is a growing trend to cast protected areas and those who support them as a malignant force intent on dispossessing indigenous and local communities in the service of an abstract and idealized vision of wilderness (Chapin 2004; Dowie 2005).

While most concede that stronger emphasis on the rights, capacities, and participation of indigenous and local communities is essential for effective protected area efforts in many places, critics sound a number of caveats, arguing that community-based management is not a panacea for a number of reasons, mainly because local communities vary greatly in their cohesiveness and in the quality of their environmental stewardship. The forces of global change (discussed below) have, in many cases, undermined formerly sustainable traditional resource management practices. Moreover, the sum of a local community's conservation objectives will not inevitably encompass all national and global conservation and protected area objectives. Some governments also complain that advocates for a much broader human rights agenda are utilizing international biodiversity forums to push claims that are beyond the scope and competence of the biodiversity and protected area arena.

Increasingly, protected area debates are more about who should make decisions through what structures and processes than they are about what needs to be done in substantive scientific and technical terms. A reasonably broad consensus has developed about what aspects of biodiversity are threatened and about how to alleviate those threats, in the technical sense, although the underlying science and available management tools are continually being refined and improved.

The core debates are about the governance structures—and the distribution of power over those structures—that make and implement decisions about protected areas. These debates animate protected area politics from the local level—where the relative merits of government agency versus local community control over protected areas are debated—to the global level, where diplomats are presently contesting governance and power arrangements over proposed marine protected areas on the high seas.[5]

These tensions between science, development, and empowerment and the competing visions of protected areas that they champion are both caused and exacerbated by the accelerating forces of global change, discussed below.

Global Change

Our world has changed beyond the recognition of those who lived just a century ago.[6] We now live in a world of shifting climates, sea level rise, swelling human populations, invasions of alien species, shrinking and fragmented habitats, and the myriad processes and pressures of industrialization and ever-expanding globalized commerce and communication. To protected area managers, charged with maintaining sites either in the state our ancestors found them or under controlled sustainable use practices, these myriad and often malignant forces of change may appear well beyond the scope of local or even national capacities to respond. The challenges are indeed daunting, but managers must adapt and respond to these new realities, or protected areas as we know them today will soon cease to exist.

Changing the way that we do business is no easier for protected area managers than it is for anyone else. We cannot, however, fulfill our duties as stewards of the earth's last natural ecosystems if we plan and manage for a world of the past, a world that no longer exists. To manage for a changing future, we must better understand the multifaceted nature of the global changes that are upon us, including its socioeconomic, biophysical, and institutional dimensions.

Socioeconomic Change

"Global change" typically conjures images of climate change, biodiversity loss, and the like. But these biophysical changes are driven by equally momentous socioeconomic changes. From 1900 to 2000, the global population grew from 1.2 billion to 6 billion—a fivefold increase unprecedented in human history—and is expected to stabilize at nearly 9 billion by 2050. Humanity was largely a rural species until recently, but by 2030 60% will live in cities, mostly in coastal areas, although the absolute number of rural people will continue to increase. Most population growth will occur in the high-biodiversity developing countries of the tropics.

The global economy has grown at an even more stunning rate: global gross domestic product grew from $17 trillion in 1950 to over $107 trillion in 2000. However, the fruits of this global economic boom have not been equally shared. Today, fully 78% of the world's people may be considered poor, while 11% are middle income and another 11% rich. 1.1 billion people live on less than one dollar per day.

Humanity now appropriates some 40% of the earth's net primary productivity (solar energy captured by photosynthesis). Economic sectors with the most direct impacts on protected areas include agriculture, livestock raising, fisheries, the wood products industry, the trade in wild products, the generation of energy, and the appropriation and alteration of freshwater.

Biophysical Change

People are fundamentally transforming the earth as a result of this expanding human footprint. For example, climate change, driven by anthropogenic greenhouse gas emissions, is already upon us. CO_2 concentrations are now about 30–35% higher than the natural background of the past 10,000 years. The global average surface temperature increased over the twentieth century by about 0.6°C and is expected to increase by a further 1.4–5.8°C by 2100, an increase greater than any during the past 10,000 years. As a result, ice and snow cover is receding, sea levels are rising, and weather patterns are changing. Impacts are greater at higher latitudes and vary from region to region, but all regions of the planet can expect to be affected. Species ranges are shifting relative to the elevation of their habitats, and along latitudes; sea surface temperatures are rising, and natural resource-based production systems such as agriculture and forestry are expected to experience significant impacts, although the precise nature and distribution of future changes are difficult to specify.

Habitat conversion and fragmentation are increasingly altering the basic context for conservation efforts. Only one-fifth of original forest cover remains in relatively large, undisturbed tracts, grasslands have been extensively converted to agriculture and pasture, and half the world's wetlands have been lost during the twentieth century. Remaining natural habitats are increasingly fragmented into smaller and smaller patches, with growing negative effects on species abundance and distribution, and on the provision of ecosystem services from natural systems.

Alteration of hydrological cycles is greatly diminishing the quantity and quality of the world's freshwater. Driving forces include impoundment (by dams) and other alterations of river flow, conversion of wetlands, pollution, and skyrocketing human demand for freshwater. Resulting changes in natural water flows, alteration of sedimentation processes, and water quality degradation negatively affect the biodiversity and ecosystem functions of water systems worldwide.

Alien species have spread dramatically as a result of the increased mobility and trade of people, goods, and species across the planet, and are now recognized as one of the greatest threats to the stability and diversity of ecosystems, second only to habitat loss. Small islands and freshwater ecosystems are particularly at risk.

Biodiversity loss is occurring as fast today as at any time since the dinosaurs died out some sixty-five million years ago, and the current extinction rate is thought to be at least 1,000 times higher than the rates typical throughout Earth's history. Some 20,000 species are known to be threatened with extinction, although the actual number may be considerably higher. Key drivers of biodiversity loss include habitat loss and fragmentation; introduction of alien species; overexploitation of wild species; pollution; climate change; and the shrinking spectrum of species used in industrial agriculture and the corresponding loss of agricultural, forest, and livestock genetic diversity.

Institutional Change

Taken together, these unprecedented forces of socioeconomic and biophysical change present protected areas with sobering and critical challenges. The third dimension of global change, however—institutional change—provides reason for hope, as well as new strategies, tools, and partners for action.

In our view, institutional global change encompasses new global norms of conduct, new forms of knowledge generation and communication, and new governance arrangements, all very relevant for protected areas and their managers.

The Emergence of Global Norms of Conduct   All human societies have norms—rules and expectations concerning how people and their institutions should behave in a given situation. For the first time, however, global norms that transcend particular countries and cultures are emerging. These include norms concerning universal human rights and equality; democracy, accountability, and the rule of law; and global cooperation, including stewardship of the global environment. Like all norms, these new global norms are only variably respected or enforced. What is important, however, is that they are increasingly accepted as norms that should apply to and guide the behavior of everyone, everywhere—a truly revolutionary concept in the broad sweep of human history.

Globalization of Communications, Knowledge, and Culture   The way that humanity communicates—and thereby shares and transforms knowledge and culture—has changed beyond recognition in just the past fifty years. Telecommunications, television, and the Internet—virtually unknown to most people not long ago—are now ubiquitous, knitting the world's consciousness and culture together into a truly global village. Increasingly, we all have access to the same impossibly vast and diverse fund of human knowledge and experience, but we are at the same time drawn into an increasingly common global culture.

New Institutions and Forms of Governance   These new global norms—and the communications web that spreads and reinforces them—have catalyzed the development of new kinds of institutions and forms of governance important for the future of protected areas. These include new global environmental institutions and agreements, as well as a proliferation of NGOs at local, national, and global levels, forming a new and potent third force alongside government and the private sector.

With respect to the governance of protected areas, trends toward decentralization and devolution are particularly striking. Where protected areas in the past were largely seen as the purview of national (or sometimes state or provincial) governments, this view is rapidly changing. Across the globe, many governments are increasingly decentralizing natural resource management authority and responsibility to local government units, at the same time transferring many such functions to local and indigenous communities, NGOs, and private landowners. A wide variety of innovative protected area arrangements have resulted, including diverse comanagement arrangements, community conserved areas, indigenous protected areas, private land trusts, and the like. (For examples, see Barber 2004b; Borrini-Feyerabend, Kothari, and Oviedo, 2004; Borrini-Feyerabend et al. 2004; Jaireth and Smyth 2003; and Miller et al. 1997.)

Local governments and local communities are not inevitably better managers or stewards—guidance and oversight from central government is still essential. But these new, decentralized protected area governance arrangements are often better able to incorporate local rights and interests, and mobilize local resources and commitment for protection, than are regimes dictated and controlled by remote capitals.

A VISION FOR THE FUTURE

If the protected areas of the future are to serve humanity and nature in a rapidly changing and stressful world, they will need to shift beyond being "nature's playgrounds" to becoming seen and treated as part and parcel of the human life-support system—not simply places to enjoy, but very much places that are necessary for human survival.

To serve that end, protected area policies of the twenty-first century will need to be formulated around three central goals.

1. *Provide the full range of nature's goods and services (ecosystem services):* Society is generally aware of the importance of managing watersheds carefully if we are to maintain the quantity and quality of freshwater supplies. Similarly, people are increasingly cognizant of the need to be good stewards of forests and oceans to maintain supplies of timber, fish, and other goods essential to human well-being. It is likewise widely appreciated that greenways can help maintain air quality and provide recreational opportunities in and around urban centers. Perhaps less well understood is the fact that to maintain each of these services of nature, specific managerial policies, strategies, and practices need to be established, implemented, and adapted as conditions change.

2. *Secure Earth's diversity of life (biological diversity):* Biodiversity is so complex that it is not possible to measure directly. We can, however, measure and monitor the status and trends of better-known groups of species (e.g., mammals, birds) and of the particular ecosystems and habitats of which they are a part. Thus, for example, forests, wetlands, coastal zones, and coral reefs are places of high diversity; by conserving large wild examples of these places and ensuring connectivity among them, the bulk of Earth's biodiversity can be secured.

3. *Integrate a sense of the sacredness of nature into the values of contemporary societies (natural and cultural heritage):* Many indigenous and modern communities have long held dear particular sites and natural features that demonstrate outstanding natural beauty or scientific value, or that are linked to historical events or spiritual beliefs and practices. At the same time, however, managers are being asked to justify the existence of protected areas on purely economic (monetary) terms. The challenge to traditional and modern societies is to strengthen their appreciation of nature and its wild places, to ensure that they hold an increasingly important place in the human heart and spirit in ways that are often as important as the goods and services that we can obtain from them.

Meeting these three central goals for the future will require action today in five key areas.

Expand and Strengthen Protected Area Systems to Meet the Biodiversity Crisis

While protected areas are widely recognized as the cornerstone of efforts to slow biodiversity loss, most protected areas were not designed explicitly for that purpose. New insights and methodologies from the discipline of conservation biology now provide us with the tools to remedy this situation. Recent work such as the global gap analysis (Rodrigues et al. 2003), the global amphibian assessment (IUCN, Conservation International, and NatureServe, 2006), and the mapping of key biodiversity areas (Eken, Bennun, and Boyd, 2004)—and corresponding work at the national or ecosystem level—provides an increasingly accurate map of what most urgently needs to be conserved in order to slow biodiversity loss. New tools for systematic conservation planning at the system and site levels provide the methods to implement these priorities as protected area networks are expanded and strengthened (See, e.g., Groves 2003; Margules and Pressey 2000). To be effective, however, all such methods need to balance two sets of considerations.

First, ecological considerations—such as species richness and endemism—are the essential basis for designing protected area systems, because a "system" designed in the absence of clear scientific goals and systematic scientific methods is not a system at all, but rather an ad hoc assemblage of protected areas.

Second, and equally important, this conceptual priority map needs to be filtered through a set of socioeconomic and political considerations, including threats, opportunities, available resources, and the relative balance of costs and benefits. Without factoring in these considerations, even the most elegant conceptual map of ecologically set priorities—whether at the global, national, or site level—will remain a mere piece of paper.

Reinvent and Strengthen Protected Area Governance

Responding to socioeconomic and political considerations, however, is more than a conceptual exercise for protected area planners and managers. Rather, we need to reinvent and expand protected area governance so that it may better respond to changing societal demands, to the challenges posed by the need to conserve bio-diversity across the landscapes and seascapes where people live and work (Borrini-Feyerabend 2004; Jaireth and Smyth 2003).

First, more attention must be paid to broadening the spectrum of governance models and mechanisms beyond the centralized, state-managed parks that currently dominate protected area practice. Second, more effective and diverse protected area governance requires participatory decision-making and management processes that incorporate and respond to the interests of a broader range of stakeholders—particularly the indigenous and local communities living in and around protected areas. Third, these new models and methods for governance and participation need to ensure that both the costs and benefits of protected areas are shared equitably. A greater shift to community-based management (CBM) is a central element of these transitions.

This type of management is vitally important for a number of reasons. Contrary to popular images, there are very few places where wild biodiversity exists in isolation from human communities and activities; many "natural" ecosystems have in fact been shaped by anthropogenic disturbance; considerable local knowledge about biodiversity and its management has developed over human history, and is an important resource for management in many places; indigenous and local communities in many parts of the world directly depend on ecosystem goods and services, and so have considerable incentives to conserve them—if they can reap an equitable share of the ensuing benefits; and modern-day conservation authorities are hard-pressed to cope with the costs and logistics of management on their own, requiring the help of local communities to succeed.

By the same token, CBM is not a panacea for protected area management. Local communities vary greatly in the cohesiveness and quality of their environmental stewardship. The forces of global change have, in many cases, undermined formerly sustainable traditional resource management practices. Moreover, the sum of a local community's conservation objectives will not inevitably encompass all national and global conservation objectives. The challenges facing protected areas in the twenty-first century require a diversity of governance models and approaches, ranging from community-managed initiatives (with a substantial focus on poverty alleviation in managed landscapes) to state-led efforts to conserve relatively unpopulated, undisturbed large tracts of natural habitat.

Build Climate Change Resilience and Adaptation into Protected Area Systems

Climate change is already having impacts on many habitats and ecosystems, and these impacts are projected to intensify in coming decades (Hannah and Lovejoy 2003). Climate change therefore has enormous implications for protected area design and management. Given uncertainties about the intensity of climate change and the specific nature of its impacts on different species and habitats, we need to manage adaptively and proactively, continuously monitoring progress to adjust management and, in some cases, taking a more active hand in shaping ecological processes.

Climate change adds a crucial element to the review of existing protected areas, since the biodiversity features conserved in a particular area may move beyond its boundaries in a few decades. Thus the analysis needs to ask not only which biodiversity features a protected area currently encompasses, but also which of those features will no longer lie within its boundaries in the future—and which new features may move into it.

The same issues apply to the selection of potential new areas. Rules that factor in potential species range shifts, precipitation and fire-regime shifts, and the like need to be incorporated into site-selection algorithms and the decision-making processes they support. Potential new protected areas should be selected to conserve the future location and needs of priority biodiversity features, not their present location, and they need to be chosen with regard to their contributions to climate change resilience and adaptation, not just their contributions to biodiversity per se.

Climate change factors will also affect the setting of priorities for action on the ground. Consideration of likely shifts in species in habitat ranges, for example, may alter the balance of priority given to expanding the size of new sites versus establishing new sites. Alternatively, climate change may decrease the future integrity and utility of some existing sites and increase the priority that should be given to new sites. In virtually all cases, the importance of connectivity to allow movement of species and habitats is likely to grow.

Finally, conserving examples of habitat types that are particularly resilient to climate change should now be a key ecological criterion for protected area system design and site selection.

Strengthen Synergies between Protected Areas and Poverty Alleviation

There are both moral and practical reasons for protected area policy makers and managers to move in the direction of more active "pro-poor" conservation (Adams et al. 2004; Roe and Elliott 2003). Morally, it is difficult to justify conservation for its own sake in places where local people are living in misery. Practically, protected areas cannot long survive the pressures of impoverished adjacent populations. In addition, the international aid community has, with the adoption of the UN Millennium Development Goals in 2000, largely united around an agenda that puts reduction of poverty, hunger, and disease at the center of the international development agenda. Given the dependence of protected areas in many developing countries on continued international aid, protected area policy makers ignore the antipoverty agenda at their own financial peril.

Nevertheless, protected area policy makers and managers should treat the call for pro-poor conservation with some caution. Some protected areas—such as those with significant wildlife- or dive-tourism potential—may indeed be able to serve a significant poverty reduction purpose while still meeting their biodiversity conservation objectives. In other cases, however, a protected area may not have significant commercial potential, due to lack of marketable ecotourism attractions, isolation, or the presence of especially sensitive or endangered species and habitats. Ultimately, protected areas should not be pushed into a situation where their whole rationale for existence is dependent on their ability to reduce poverty in surrounding human communities.

Establish Mechanisms to Sustainably Finance Protected Area Sites and Systems

In the end, protected areas, no matter how well designed, cannot be effectively managed without sufficient financing that is sustainable over time. There has been extensive analytical work on conservation financing mechanisms, and a number of innovative examples are being implemented around the world.[7] But protected area managers in most countries already face a funding shortfall for their existing sites, and efforts to expand and strengthen protected area systems will require a redoubling of efforts to secure the financial means to do so. To develop a sustainable financial base for protected areas, decision makers can:

- Design a sustainable financing plan for each individual site and for the entire system and assign personnel to implement it
- Apply methodologies to calculate realistic costs of protected area systems that include all necessary expenditure items, including minimum salaries, infrastructure, equipment, operation and maintenance, outreach, and education
- Develop mechanisms to complement core budgetary funds with other financing sources such as solicitation of grants from donor agencies and individuals
- Expand the use of "debt-for-nature" swaps, which reduce developing country debt in exchange for local-currency financial commitments to conservation
- Establish payment for ecosystem services systems
- Endow conservation trust funds dedicated to protected areas
- Earmark user fees, taxes, and other charges for protected areas
- Establish ecotourism and other local sustainable development activities that may benefit local communities and protected area management alike

CONCLUSION

Achieving effective conservation of biodiversity on the ground is one of society's greatest challenges in the twenty-first century. Protected areas already provide a major tool to this end, covering a significant portion of Earth's terrestrial surface. However, to secure these sites in the face of the biophysical, social, economic, and institutional changes already taking place at accelerating rates, current approaches to protected area management must strengthen their reliance on science. Protected areas must contribute to poverty alleviation. And new governance methods will need to promote cooperative planning and management, as well as the equitable sharing of costs and benefits.

Existing goals for protected area management will need to be strengthened. Their emphasis should find a new balance that explicitly seeks the protection of the diversity of life on Earth and secures and provides the full range of ecosystem services.

Most significantly, goals for the twenty-first century must seek to incorporate into modern and traditional cultures a commitment to protect and respect the sacredness of nature and wild places. The conservation of nature and natural resources must become a standard of behavior and performance for governments, communities, and individuals if sustainability is to be achieved in our children's future.

Notes

1. The Durban Accord was the major political statement emerging from the Fifth IUCN World Parks Congress in Durban, South Africa, September 2003. The Congress attracted more than 3,000 protected area professionals and policy makers, and it was the largest such gathering to date.

2. In Australia, both Ayers Rock (Uluru) National Park and Kakadu National Park are comanaged by the aboriginal community and the government of the Commonwealth. Together they control access to sacred sites such as cave paintings.

3. For extensive examples of the political ecology approach to conservation issues, see Brechin et al. (2004) and Zerner (2000).

4. For an extensive exposition of this school of thought, see Borrini-Feyerabend et al. (2004). For a review and critique, see Barber (2004b).

5. The issue of marine protected areas on the high seas was being vigorously debated during 2006 in both the Convention on Biological Diversity and a special working group on the topic established by the UN General Assembly.

6. This section has been adapted from Barber, Miller, and Boness (2004).

7. For extensive analysis, cases, references, and links on conservation finance, see the work of the Conservation Finance Alliance consortium at http://www.conservationfinance. org.

References

Adams, W. M., R. Aveling, D. Brockington, B. Dickson, J. Elliot, J. Hutton, D. Roe, et al. 2004. Biodiversity conservation and the eradication of poverty. *Science* 306(5699): 1146–1149.

Barber, C. V. 2004a. Designing protected area systems for a changing world. In *Securing protected areas in the face of global change: Issues and strategies,* ed. C. V. Barber, K. R. Miller, and M. Boness, 41–94. Gland, Switzerland: IUCN.

Barber, C. V. 2004b. Parks and people in a world of changes: Governance, participation and equity. In *Securing protected areas in the face of global change: Issues and strategies,* ed. C. V. Barber, K. R. Miller, and M. Boness, 97–134. Gland, Switzerland: IUCN.

Barber, C. V., K. R. Miller, and M. Boness, eds. 2004. *Securing protected areas in the face of global change: Issues and strategies.* Gland, Switzerland: IUCN.

Borrini-Feyerabend, G. 2004. Governance of protected areas, participation and equity. In *Biodiversity issues for consideration in the planning, establishment and management of protected area sites and networks,* 100–105. Convention on Biological Diversity Technical Series, no. 15. Montreal: Convention on Biological Diversity.

Borrini-Feyerabend, G., A. Kothari, and G. Oviedo. 2004. *Indigenous and local communities and protected areas: Towards equity and enhanced conservation: Guidance on policy and practice for co-managed protected areas and community conserved areas.* Best Practice Protected Area Guidelines Series, no. 11. Gland, Switzerland: IUCN.

Borrini-Feyerabend, G., M. Pimbert, M. T. Farvar, A. Kothari, and Y. Renard. 2004. *Sharing power: Learning-by-doing in co-management of natural resources throughout the world.* Tehran, Iran: International Institute for Environment and Development.

Brechin, S. R., P. R. Wilshusen, C. L. Fortwangler, and P. C. West, eds. 2004. *Contested nature: Promoting international biodiversity with social justice in the twenty-first century.* Albany: State University of New York Press.

Chapin, M. 2004. A challenge to conservationists. *World Watch Magazine* 17(6): 17–31.

Dowie, M. 2005. Conservation refugees: When protecting nature means kicking people out. *Orion* (November/December): 16–26.

Dudley, N., K. J. Mulongoy, S. Cohen, S. Stolton, C. V. Barber, and S. B. Gidda. 2005. *Towards effective protected area systems: An action guide to implement the Convention on Biological Diversity Programme of Work on Protected Areas.* Convention on Biological Diversity Technical Series, no. 18. Montreal: Secretariat of the Convention on Biological Diversity.

Eken, G., L. Bennun, and C. Boyd. 2004. Protected areas design and systems planning: Key requirements for successful planning, site selection and establishment of protected areas. In *Biodiversity issues for consideration in the planning, establishment and management of protected area sites and networks,* 37–44. Convention on Biological Diversity Technical Series, no. 15. Montreal: Secretariat of the Convention on Biological Diversity.

Gordon, E. A., O. E. Franco, and M. L. Tyrrell. 2005. *Protecting biodiversity: A guide to criteria used by global conservation organizations.* Yale School of Forestry and Environmental Studies Reports, no. 6. New Haven, CT: Yale School of Forestry and Environmental Studies.

Groves, C. R. 2003. *Drafting a conservation blueprint: A practitioner's guide to planning for biodiversity.* Washington, DC: Island Press.

Hannah, L., and T. Lovejoy, eds. 2003. *Climate change and biodiversity: Synergistic impacts.* Advances in Applied Biodiversity Science, no. 4. Washington, DC: Center for Applied Biodiversity Science, Conservation International.

Hockings, M., S. Stolton, and N. Dudley. 2000. *Evaluating effectiveness: A framework for assessing the management of protected areas.* Best Practice Protected Area Guidelines Series, no. 6. Gland, Switzerland: IUCN.

IUCN—The World Conservation Union. 1994. *Guidelines for protected area management categories.* Gland, Switzerland: IUCN.

IUCN—The World Conservation Union. 2005. *Benefits beyond boundaries: Proceedings of the fifth IUCN World Parks Congress: Durban, South Africa 8–17 September 2003.* Gland, Switzerland: IUCN.

IUCN—The World Conservation Union, Conservation International, and NatureServe. 2006. Global amphibian assessment. http://www.globalamphibians.org.

Jaireth, H., and D. Smyth. 2003. *Innovative governance: Indigenous peoples, local communities and protected areas.* New Delhi, India: Ane Books.

Khare, A., and D. B. Bray. 2004. *Study of critical new forest conservation issues in the global south.* New York: Ford Foundation.

Lapham, N. P., and R. J. Livermore. 2003. *Striking a balance: Ensuring conservation's place on the international biodiversity assistance agenda.* Washington, DC: Conservation International.

Margules, C. R., and R. L. Pressey. 2000. Systematic conservation planning. *Nature* 405(6783): 243–253.

Millennium Ecosystem Assessment. 2005. *Ecosystems and human well-being: Biodiversity synthesis.* Washington, DC: World Resources Institute.

Miller, K. R. 1996. *Balancing the scales: Managing biodiversity at the bioregional level.* Washington, DC: World Resources Institute.

Miller, K. R., J. A. McNeely, E. Salim, and M. Miranda. 1997. *Decentralization and the capacity to manage biodiversity.* Washington, DC: World Resources Institute.

Mulongoy, J., and S. Chape. 2004. *Protected areas and biodiversity: An overview of key issues.* Cambridge: UN Environment Programme World Conservation Monitoring Centre/ Secretariat of the Convention on Biological Diversity.

Rodrigues, A. S. L., S. J. Andelman, M. I. Bakarr, L. Boitani, T. M. Brooks, R. M. Cowling, et al. 2003. *Global gap analysis: Towards a representative network of protected areas.* Advances in Applied Biodiversity Science, no. 5. Washington, DC: Conservation International.

Roe, D., and J. Elliott. 2003. Pro-poor conservation: The elusive win-win for conservation and poverty reduction? Paper presented at the Fifth IUCN World Parks Congress, Durban, South Africa, September 8–17.

Secretariat of the Convention on Biological Diversity. 2004. *Programme of work on protected areas.* Montreal: Secretariat of the Convention on Biological Diversity. http://www.cbd.int/doc/publications/pa-text-en.pdf.

Zerner, C., ed. 2000. *People, plants, and justice: The politics of nature conservation.* New York: Columbia University Press.

7

Ecological and Intellectual Baselines: Saving Lions, Tigers, and Rhinos in Asia

John Seidensticker

ECOLOGICAL AMNESIA, SHIFTING BASELINES, AND CONSERVATION PRACTICE

Ecological amnesia—forgetting the environmental conditions of the past—is arguably the greatest problem faced by conservationists. We lack the most basic history of most of our landscapes, but we need this history to provide a frame of reference for measuring ecological change. When landscapes are degraded, and species and natural goods and services are lost, people living in those landscapes make mental adjustments. A father can tell his son how it once was, but the son will never really comprehend what he hasn't experienced himself. The baseline for what is "natural" shifts from one generation to the next. Geographer Clarence Glacken (1967) emphasized that this has been a fundamental pattern in the history of people as agents of ecological change.

Whether implicitly or explicitly stated, the ecological amnesia issue underlies and dominates much of our conservation thinking and becomes an assumption in planning and executing recovery plans for endangered species. The idea that it should be possible to restore ecosystems to their baseline or "natural" states is now widely recognized as naive (Knowlton et al. 2003). Achieving consensus about past events in the social and ecological systems involved, and their interplay, is usually unattainable. Human population growth and environmental impacts seem irreversible: a trajectory of recovery is unlikely to mirror exactly the trajectory of decline.

But we must act anyway to craft effective solutions for endangered species recovery, simultaneously meeting proximate and ultimate conservation challenges. Proximate challenges include providing security for species and the dynamic genetic and ecological processes that characterize and sustain at-risk species and ecosystems. We have to base our actions on our best understanding of the ecological and behavioral needs of the species and characterize the naturalness of anthropogenic alterations to their ecosystems by judging (1) degree of change; (2) degree of sustained control, including both the amount of cultural energy required to sustain the current state and the effects that persist without inputs of cultural energy, such as dams, introduced species, and severe pollution; (3) spatial extent of change; and (4) abruptness of change (Angermeier 2000; Mace, Balmford, and Ginsberg, 1998). Ultimate challenges to restoration are presented by cultural history, as well as by valuation, management, and policy systems. These must be identified and incorporated into our restoration efforts (Clark, Curlee, and Reading, 1996).

We can move forward in an always uncertain world by setting biologically defensible recovery goals and managing for survival rather than engaging in hospice conservation in which we dutifully record events as we watch species slip away (Tear et al. 1993). The conservation of large carnivores—lions (*Panthera leo*) and tigers (*P. tigris*)—and of other demanding species, such as the greater Asian one-horned rhinoceros (*Rhinoceros unicornis*), in the human-dominated landscapes of Asia requires great commitment on the parts of conservation biologists, activists, land managers, and political leaders. It also requires a good bit of tolerance on the part of people who live and work in the places where these splendid great animals still live, and where we would like them to live in the future (Noss 1996). What I have always found thrilling about working on conservation issues in Asia is that this commitment is as important an assumption in endangered species recovery as the issue of ecological amnesia. From a North American's perspective, with our limited understanding of our own environmental history and our myths of a nature untouched, we look at an Asian situation and the prospects look bleak. We see little hope that restoration is possible. I think we are overwhelmed by the heavy human footprint; we don't usually see people as part of the solution to ecological restoration even though people dominate landscapes. An Asian, on the other hand, would point out that we have always lived here. People are a fact of life. Change happens. Now let's get on with what we are trying to do and ask: "What useful ideas and tools do you bring to help us keep our wonderful wildlife?"

Establishing baselines is one such idea and a tool that is essential in endangered species recovery. Lee M. Talbot's ecological reconnaissance of South Asia in 1955 established baselines that we can use to measure both the ecological and intellectual changes in the landscapes he visited and the way we think about the conservation of those landscapes and species he profiled (see Talbot 1960). In this chapter, I examine two areas where Talbot established baselines in 1955: the Gir Forest in India and the Chitwan Valley in Nepal. A half century later, much has changed in the social, economic, and natural landscape-conservation matrix that supports the large, charismatic, but extinction-prone species that Talbot championed. Using these baselines, and from my experience in these human-dominated landscapes over that last thirty years, I see a future for wildlife on the Indian Subcontinent, rather than a continual deterioration and decline. Without Talbot's baselines, established a half century ago, we would be challenged to gauge how much ground has been gained and how we can use the past to inform the future.

ASIAN LIONS AND THE GIR FOREST

The number of Asian lions (*P. leo persicus*) and the extent of their distribution greatly diminished during the *shikar* (the practice and the culture of the hunt) and the British

Raj in India. Lions were greatly valued as hunting trophies and demand exceeded supply. Historically, lions were widespread throughout northern India. For example, fifty lions were reported to have been killed by a single British officer in the vicinity of Delhi between 1857 and 1859. By the 1880s, lions could be found only in the Kathiawar (also called the Saurashtra) Peninsula in the western Indian state of Gujarat (Pocock 1939; Rangarajan 1996). The imperturbable I. R. Pocock put it this way: "In all parts of the world occupied by Europeans where lions occur, the disappearance of lions is merely a question of time" (1939, 219).

Contrasts rush to meet you as you fly from Mumbai to Rajkot in Gujarat. You glide over the rippling expanses of the Arabian Sea, tranquil with scattered fishing boats. On making landfall at the southern tip of the Kathiawar Peninsula, the land beyond the coastal fringe looks heavily used by a dense rural population. After this first belt of farms and villages, an island of dissected forested hills rises starkly some 40 km from the coast, and then disappears behind you, almost before you can focus on the details. Again, the rural human activities are all you can see. The forested islands in the landscape are the volcanic Gir Hills that rise to 600 m above sea level. The forest connections have been severed between the Gir Hills and the Girnar Range to the north and the Mythiala Range to the east, where a few lions could be found in 1955.

In the 1870s, the forest, valuable for its teak (*Tectona grandis*) and *Acacia* woodland and scrubland, covered an area of more than 3,000 km². Today the Gir Forest ecosystem is about 1,900 km²–1,154 km² in the Gir Wildlife Sanctuary, 259 km² in Gir National Park, and 470 km² in Gujarat Forest Department lands. It is the only remaining extensive forest area in the entire peninsula of some 60,000 km². The climate features hot, wet summers and dry, cool winters. Summer monsoon rains water this landscape, but the amount of rain that falls varies greatly from year to year. When the rains fail, local residents traditionally descend on the forest in search of fodder for their cattle and buffalo. The deep major *nalas* (river courses) in the forest have permanent water and there is permanent subsurface water along some of the tributary streams, and these sources support livestock and wildlife alike during drought years. If you are looking for lions, this is where you are likely to encounter them (Chellam and Johnsingh 1993; Seidensticker 1985).

By 1884 Asian lions were limited to the vicinity of the Gir and at most a few tens of individuals remained; some observers placed this number at less than a dozen, but others estimated there were about one hundred at the time. They have been protected since 1900, but some trophy hunting by permit was allowed until the middle of the twentieth century. Problem lions—man- or livestock-eaters—were killed as necessary.

How to count large carnivores has been a contentious issue among Indian wildlife managers. Some rely on counting pugmarks (tracks) or on counting the individual lions coming to buffalo baits to arrive at total counts of lions and tigers. Wildlife scientists counsel using a sampling-based approach to estimate density with confidence limits and population trends (Jhala et al. 1999; Karanth et al. 2003). As a result, we don't know with statistical confidence the course of the lion recovery from the lows of the 1880s to the present numbers, although counts were made periodically (see Chellam and Johnsingh 1993 for a summary.) In 1995, Jhala et al., using vibrissae spot pattern for individual identification and conducting a "mark-recapture" census, estimated there were 201 ± 23 lions greater than eighteen months of age in the Gir National Park and Wildlife Sanctuary (1999). Concurrently, in 1995, the Gujarat Forest Department censused lions using baits over three days and arrived at a total count of 204 lions. Now, a decade later, the entire lion population is thought to number about 300 (Johnsingh et al. 2004), and a new, statistically rigorous census is planned. Lions are very good dispersers. Even in the human-dominated landscape, lions in small groups (ten to fifteen individuals) have moved to the Girnar and Mityala hills and along the coast near Rajula and Kodar, 10–30 km from the Gir protected areas (Johnsingh et al. 2004). By this measure, the lion population in the Gir has been recovering over the past century. In spite of this, it comes as some surprise that one of our most eloquent writers of natural history, David Quammen, in his tome *Monster of God: The Man-Eating Predator in the Jungles of History and the Mind* (2003), suggests that the Asian lion will be lost in the next 150 years or so. Quammen's gloomy predictions are based on dated UN human population density predictions that show a continual climb, while recent predictions show that the growth of the human population increase is slowing (Flannery 2003). The relationship between human population density and threat intensity is a complex one (Cardillo et al. 2004; McKinney 2001). Most important, it is not that people *are* that matters most—it is what people *do.*

Based on his 1955 visit, Talbot made many useful descriptive observations on ecology and land use in the Gir Forest, and on the lions (see 1960, 238). He also included recommendations for the recovery of the Gir Forest and its lions. These recommendations for a future for the Gir Forest established our intellectual and ecological baselines for this ecosystem.

In Talbot's view, the overwhelming number of grazing livestock was unsustainable and was displacing wild ungulates. For wildlife in the Gir to thrive, these livestock pressures had to be reduced. Talbot recognized the political and cultural issues around buffalo and cattle and endorsed a scientific approach to understanding the ecology of the vegetation, the wildlife, and the domestic livestock as a foundation for taking management actions that would transcend such cultural and political barriers and serve as a foundation for sustainable land-use and wildlife conservation policies.

Talbot also recommended that the Gir Forest be given park or sanctuary status. But he recognized that the people who lived by grazing their animals in the Gir had been part of that system for a considerable time and that it was naive to think you could restore the ecosystem to any unpeopled, pristine, "baseline" state. He assumed that grazing animals would remain, and that, through administrative processes, their impacts could be reduced to sustainable levels based on our understanding of the natural variation and succession patterns in this ecosystem.

Talbot observed that enlightened and effective protection had resulted in the increase of the Asian lion population. He proposed that lion conservation actions be informed by scientifically derived information on the number and behavioral ecology of lions. He recognized that difficulties might arise from inbreeding in a population that had undergone such a significant bottleneck. Surely this was one of the first instances of concern about what today we consider a principal risk in small populations (Frankham, Ballou, and Briscoe, 2004). Talbot also saw that there was no additional extensive forest where lions could live that could be linked to the Gir. He proposed establishing additional populations away from the Gir to buffer such risks.

The government of Gujarat, the central government of India, the Bombay Natural History Society, and, more recently, the Wildlife Institute of India have mounted a vigorous response to the conservation challenges facing the Gir and the lions that Talbot articulated a half century ago. They have been assisted by a wide array of international supporters, including IUCN—The World Conservation Union (formerly the International Union for the Conservation of Nature and Natural Resources), the World Wildlife Fund, the Smithsonian Institution, Flora and Fauna International, and the U.S. Fish and Wildlife Service, among others.

In recent years, there has been a concerted effort to establish a statistically rigorous methodology to monitor the trend in lion numbers; this has been supported by the Gujarat Forest Department, the responsible management authority (Y. Jhala, pers. comm., 2002; Jhala et al. 1999). The lions in the Gir appear to have recovered to the point where they are a viable and self-sustaining component of their ecosystem, and we can agree that 200–300 lions are far more secure than their mere tens of counterparts a century ago. But this population still faces considerable intrinsic risks from demographic stochasticity and genetic deterioration and extrinsic risks from pathogens and poachers, floods, fires, droughts, cyclones, and earthquakes, a changing vegetation mosaic, and the interactions of all these factors.

We know now that the Asian lion population did lose nearly all of its genetic variability and, perhaps as a result, that there is an extremely high incidence of structurally abnormal spermatozoa in male lions both in the wild and in the captive population (Wildt et al. 1987). An effort to establish a second population away from the Gir in the Chanddraprabha Wildlife Sanctuary in Uttar Pradesh was made in 1957, but this failed because of inadequate preparation and the small size of the sanctuary (Negi 1965). Periodic cyclones batter the Gir Forest (Oza 1983), and the deaths of about half the lions in the Serengeti and Masai Mara in East Africa due to canine distemper, likely contracted through village dogs (Roelke-Parker et al. 1996), alerted Indian wildlife management authorities and wildlife scientists to the fact that catastrophes actually do befall small populations. This has made more urgent the evaluation and planning that have been under way for some time for a second lion reintroduction in the Sitamata or Darrah-Jawaharsagar wildlife sanctuaries in Rajasthan or the Kuno Wildlife Sanctuary in Madhya Pradesh (Johnsingh, Chellam, and Sharma, 1998; Johnsingh et al. 2004).

The 345 km² Kuno Wildlife Sanctuary lies within a 3,300 km² block of forest. Twenty-four villages with 1,547 families were moved from the sanctuary to new homes outside. About US$3.4 million was invested by the government of India in readying the sanctuary for lions through 2005, and ungulate biomass in the sanctuary recovered sufficiently to support five to eight lions (Johnsingh, Goyal, and Qureshi, 2007).

We know much more about the lion's ecological and behavioral needs than we did half a century ago. Indian and foreign-national ecologists have undertaken studies over the last thirty years to bolster our knowledge of lion life history and behavioral ecology. Principal among the efforts are those of Paul Joslin (1973), S. P. Sinha (1987), and Ravi Chellam (1993), and the next generation of Indian scientists is continuing the work. Such information can be used to establish the criteria by which the lion's ecological and behavioral needs are met when management actions are undertaken (Chellam and Johnsingh 1993). Our knowledge of the ecology of the dry, deciduous Gir Forest has been usefully articulated in contributions by Toby Hodd (1970), Stephen Berwick (1974, 1976), Sinha (1987), Jamal Khan (1993), and H. S. Sharma and A. J. T. Johnsingh (1996).

Lions need meat and lots of it. When Berwick (1974) measured the biomass of wild ungulates available to lions in 1970 in the Gir, excluding livestock, the amount was quite low, less than 400 kg/km² (summarized in Eisenberg and Seidensticker 1976). From the mid-1970s until today, forest officials have made a concerted effort to eliminate cattle and buffalo in the core national park area and reduce their numbers in surrounding wildlife sanctuary area of the reserve. In 1989, Khan et al. (as reported in 1996) found that the numbers of axis deer (*Axis axis*), sambar (*Cervus unicolor*), chowsingha (*Tetraceus quadricornus*), and chinkara (*Gazella gazella*) had increased; axis deer numbers had increased by a huge 1,320%. Nilgai (*Boselophus tragocamelus*) numbers had decreased by about 30%, however, probably as a result of increased shrubs and tree cover; nilgai prefer open habitat. The grazing axis deer were the clear beneficiaries of the removal of cattle and water buffalo. Wild ungulate biomass available to lions had increased to 3,290 kg/km² in the national park and 1,900 kg/km² in the sanctuary. There remained, however, nearly 9,000 kg of livestock/km² in the West Sanctuary and about 75% of this amount in the East Sanctuary. The lions have responded to this change by increasing the number of wild ungulates in their diet. Wild ungulates made up 21% of the lions' diet in the early 1970s (Joslin 1973), 52% of their diet in the early 1980s (Sinha 1987), and 74% by the late 1980s (Chellam 1993).

Talbot (1960) recommended that the responsible authorities start to reduce domestic livestock in the forest. And they have done so. Before 1972, there were an estimated 17,000 cattle and buffalo resident in the Gir and an additional 25,000–50,000 that would be herded into the Gir during the dry season. In 2004, there were about 13,000 livestock in the Gir protected areas, and when transient livestock foraged in the periphery of the protected area, the number increased to 20,000 (A. J. T. Johnsingh, pers. comm., 2004). In 1972, the Gujarat Forest Department resettled 308 of 845 *Maldhari* families outside the protected areas. Nonetheless, the area remains under intense pressure from livestock—both those permitted and those that illegally use the forest (Johnsingh, Chellam, and Sharma, 1998). The Forest Department provides protection, periodic censuses of the wildlife populations, fire and water-hole management, compensation payments for lion-killed livestock and attacks on people, management of the *Maldhari* graziers, and maintenance of building and roads. Numerous water holes have been created in the last fifty years. The history of the management of the Gir Forest is contained in the forest working plans (cf. Joshi and Karamchandani 1976).

The relationship between the local graziers, the *Maldharis*, their livestock, and the lions is complex. When wild ungulate numbers were lower and lions were killing mostly *Maldhari* livestock, the devout Hindu *Maldharis* would drive lions from their kills by throwing rocks at them, holding them off long enough for members of the untouchable caste to come and skin the cow or buffalo before lions and other scavengers again descended on the carcass (Berwick 1976). Lions were habituated to this routine and lion attacks on humans were few (Seidensticker and Lumpkin 2004). With a decreased interaction, lion attacks on humans have markedly increased (Saberwal et al. 1994), and these attacks are even higher in drought years when outside livestock illegally inundate the sanctuary.

Quammen's (2003) pessimism on the future of lions in India is based largely on his reading of UN human population density estimates and the heavy human footprint on the landscape. I suggest, though, that if Quammen had started with an examination of the baselines, both ecological and intellectual, that Talbot established, instead of with his own anecdotal experiences fifty years later, he might have come to see that in Asia what is wild and natural is a very elusive concept. What is natural is always changing. The last generation of conservation leaders left us with more lions than the proceeding one. In the Gir, conservation leaders have accomplished, and are continuing to accomplish, a great deal. Today we know the number of lions with statistical confidence. We know that they have increased significantly over the last 120 years. We know that the lions living in the Gir have virtually no genetic variation, a point of great concern, and this is prompting a serious effort to establish a second lion population away from the area. More interventions may have to be planned and implemented, much as has been done for the Florida panther (*Puma concolor coryi*) to increase genetic variation (see Beier et al. 2003). We know a great deal about the behavioral ecology of the lions. Detailed work has been undertaken on the wild ungulate–domestic livestock–vegetation interaction. Livestock numbers have been reduced; wild prey numbers have greatly increased. Lion-human conflicts have been investigated, and ways to mitigate these are under continual discussion.

It seems to me that for the Gir to prosper, conservationists must begin to see the Gir as the greater Gir Forest ecosystem as defined by the lions. This is also the view of Chellam and Johnsingh (1993), Singh (1997), and Johnsingh, Chellam, and Sharma (1998). The next major conservation intervention is to expand our point of reference from the Gir with its present 1,900 km^2 to about 2,500 km^2 by reconnecting it to nearby hills and coastal forests with forested corridors. An integrated approach toward conservation in the greater Gir Forest ecosystem was initiated in 1996 and is ongoing (Mukherjee and Borad 2004). To sustain the Gir and its wildlife, a concerted effort needs to be made to promote local guardianship of the lions and the Gir ecosystem by the *Maldharis,* as is being done in the Chitwan Valley with the local *Tharu* people and throughout the entire Terai Arc Landscape for tigers and rhinos (see below).

Management of the Gir Forest ecosystem is a moving target, and management should maintain ecosystem resilience in the face of change from both outside and inside the ecosystem boundaries. (Ecological resilience is the amount of change a system can undergo and still remain within the same state or domain of attraction, capable of self-organization, and adapt to changing conditions.) Top-down, command-and-control management tends to become inflexible and lose sight of original goals and focus, while seeking control itself (Holling and Meffe 1996). The future of the Gir ecosystem rests, it seems to me, in an old Asian tradition of making the transition from despair to that of local empowerment. Isn't this why India is a democracy today? We acknowledge the crisis. Now we can transition to informed conservation interventions. I believe that local conservation leadership will take these next steps to maintain the greater Gir Forest ecosystem with its lions intact. We will never secure a future for wild lions in our lifetime or any lifetime. It requires that we build on the foundations of earlier efforts and be prepared to adjust our approach to meet emerging new challenges in this ever-changing world. This is what the practice of conservation is.

RHINOS, TIGERS, AND THE CHITWAN VALLEY

The Royal Chitwan National Park in the Nepal lowlands holds a spectacle of big mammals that is as close as we will ever come to experiencing what it was like when so many big animals—elephants, rhinoceroses, tigers, bears, crocodiles, and people—lived together at the end of Pleistocene. Today, almost everywhere, at least one if not more of these players (humans excepted) has exited the scene. Between 1846 and 1950, the Chitwan Valley was the exclusive hunting reserve of the Rana family, Nepal's hereditary prime ministers. A British representative to Nepal, E. A. Smythies, chronicler of the *shikar,* described the Chitwan Valley as "one of the most beautiful places in the world," "saturated with tigers," with "rhinos . . . a positive nuisance" (1942, 80, 83). Smythies recounts that during the 1933 winter hunt, forty-one tigers, fourteen rhinos, and two leopards were killed. Five years later, over a three-month period, with Lord Lilithgow, viceroy of India, participating, 120 tigers, 38 rhinos, and 27 leopards were shot.

When you approach the Himalayas from the Ganga Plain, the foothills rise as a dim outline in the heavy haze before you, but before you reach the base of the hills, the ground steadily rises, formed by alluvial sands, silts, and gravels that have accumulated. This is the northernmost geomorphic element of the Ganga Plain, the Piedmont zone; it was created by the coalescing of discreet alluvial fans from the many rivers and streams that descend the mountains, forming a 10–15 km wide belt, or fan surface, that is divisible into the gravelly *bhabar* (which developed from coarse sediments brought down from the main Himalaya and deposited against the foothills) and more distal sandy *terai* areas. Water is scarce in the *bhabar* due to the porous soils and streams disappear beneath the gravel. Many streams run above ground in the *bhabar* only during the monsoon, and the streams reappear only when they reach the finer sediments carried further and form the *terai* (Shukla and Bora 2003). The flat *terai* was once covered in forest with extensive tall grass areas and riverine forests along the watercourses (Stainton 1972). The Chitwan Valley floodplain (also called the Rapti Valley because the Rapti River flows east to west through the valley ~120–200 m above sea level) lies in "inner *terai*," or *dun*—the intermediate zone between the outermost Himalayan range, the Mahabharat (2,700 m), and the outer foothills, the uplift ridge system called the Churia Hills (up to 1,400 m), the eastern extension of the Shivalik Hills in India. Until about fifty years ago, the well-watered, malaria-ridden Chitwan Valley had largely been left behind by the rest of the world, with only the indigenous *Tharus,* resistant to malaria (Modiano et al. 1991), living in scattered villages. Gentle people, the *Tharus* farmed and did not hunt, and they snared small game only for subsistence. This valley and others like it along the base of the Himalayas also formed a political buffer zone. Its uncut forests, wide rivers, vast stretches of giant grasses, and dangerous wild animals—including mosquitoes carrying malaria—discouraged invasions into Nepal from the south, and kept other Nepalese from moving in as well. The widespread use of DDT to suppress mosquitoes, beginning in 1954 when this chemical was still hailed as a miracle weapon against insect disease vectors, changed this dynamic (Rose 1971).

Talbot's visit to the Chitwan Valley came at a tipping point in its environmental history: the trajectory of the valley ecosystem was being determined not by natural forces, but by the weight of a staggering accumulation of people. Talbot (1960, 119) describes how difficult it was even to reach this valley, much less to gain any accurate information about animal life there. In spite of this remoteness, the destruction of the valley's fabled wildlife was well under way, with reports of seventy-two rhinos poached in 1954 alone. Talbot also visited other rhino areas in northern India and reported on their status, conservation contexts, and the land-use matrices in which these were embedded. Up to that time the model of equilibrium/climax succession without human influences underlay almost all conservation thinking. Talbot was shifting the assumptions of that model to the state-and-transition model that we now recognize depicts ecosystem structure, function, and response to disturbance/perturbations, including those resulting from human activities (Westoby, Walker, and Noy-Meir, 1989).

By presenting the different rhino conservation areas in their contexts, Talbot was identifying different ecosystems' states and scales and the mix of natural and human activities that could support rhinos in different landscapes. This was an essential step in envisioning a possible future for the Chitwan Valley and the entire *terai* where people were simply not going to go away.

Once the Chitwan Valley was declared "malaria free," settlers from the hills, who were not naturally malaria resistant, streamed into the valley to take advantage of its rich soils and more benign climate. The human population mushroomed, tripling during the 1950s to about 100,000 by 1960. Wild elephants *Elephas maximus,* wild water buffalo *Bubalus bubalis,* and swamp deer *C. duvaucelii* were extirpated. The numbers of gaur *Bos frontalis* and rhinos were much reduced. Tigers were rare by 1960. Also by 1960, 65% of the valley's 2,600 km² forests and grasslands had been converted to agricultural land. The Chitwan Rhinoceros Reserve, an 800 km² former hunting reserve that was declared a protected area to save Nepal's only and vanishing population of the greater one-horned rhinoceros, was inundated with settlements. Hard-won steps toward a future for Chitwan's spectacular wildlife began in the mid-1960s, even as the human population in the valley continued to climb. In 1964, the Nepal government formed a Land Settlement Commission to look into the land settlement chaos in the valley. As a result, 22,000 people were removed and settled elsewhere. To protect the Chitwan Valley's remaining value as wildlife habitat, the government established the country's first national park there in 1973. Recognizing that there had to be people-free core areas, park officials focused on strict protection as their first order of business, and soon the number of rhinos and tigers began to increase. More land was added to the park in 1977, bringing the total area to nearly

1,000 km². An additional 750 km² was designated as buffer zone to cushion human impacts on the edge of the park and to provide local people places to find wild food, fodder for their livestock, wood for cooking fires, and thatch grass for roofing their homes (summarized from Mishra and Jeffries 1991).

In 1972, Kirti Man Tamang and I traveled to Nepal, and, with Hemanta Mishra, we founded the Smithsonian-Nepal Tiger Ecology Project (SNTEP); this eventually became the Smithsonian-Nepal Terai Ecology Project when the research emphasis shifted from the tiger and its prey to the rhino and its habitat as reported by Eric Dinerstein (2003). We knew that we needed a good scientific understanding of how tigers and other wildlife were responding the changes in their landscape, what the limits of their tolerances were, and how wildlife and people could fit together. We realized that conservation was not a project to finish and be done with. Conservation is a process of continually adapting and targeting responses to measured changes, new threats, and new opportunities as they emerge in this dynamic system. The key to long-term success in this process is inspired local leadership.

As he did in the Gir, Talbot recommended that studies to acquire basic scientific understanding of the species and processes that underpin the valley's ecology be undertaken. The results of these efforts are described in many dissertations, papers, monographs, and books (e.g., Dhungal and O'Gara 1991; Joshi 1996; Maskey 1989; McDougal 1977; Mishra 1982; Seidensticker 1976a, 1976b; Smith 1984, 1993; Sunquist 1981; Sunquist and Sunquist 1988; Tamang 1982; see also Seidensticker, Christie, and Jackson, 1999, on tigers and Dinerstein 2003 on rhinos).

The life histories of many of the large animals and the dynamics of this spectacular ecosystem were systematically investigated (see references in the preceding paragraph) in the nearly two decades that the SNTEP was in operation and by other workers who followed. We learned, for example, that the biologically rich Chitwan Valley river floodplains in the national park supported an ungulate biomass that rivaled that of the Rift Valley parks in Africa (Seidensticker 1976b). The ungulate assemblage included rhinos, axis deer, hog deer (*C. procinus*), sambar, barking deer (*Muntiacus muntjak*), wild swine (*Suf scrofa*), gaur, and elephants, and each had its own wide or narrow habitat affinities. The valley is a pulse-perturbation system with cool season fires and monsoon flooding and is in a state of constant renewal; hence it derives its great productivity. After the fires burn the rank grass on the floodplain, new grass shoots appear almost immediately because of the moist, sandy soils and high water table. After the fires burn through most of the forested areas on the subcontinent, any new grass growth is dependent on the monsoon rains; the interval between the fires and new grass growth can be months, greatly reducing the productivity for grazing ungulates in these forests. Within the Chitwan system, the areas that support the greatest rhino densities are the earliest successional habitats where the giant grass (>4 m tall) *Saccharum spontaneum* grows. Within the *S. spontaneum* grass areas are grazing lawns, characterized as short swards of mixed grasses, including *Imperata cylindrical,* maintained through intense grazing by rhinos and other large ungulates (Dinerstein 2003; Karki, Jhala, and Khanna, 2000). The *S. spontaneum* and grazing lawn areas are critical rhino habitat and are maintained by frequent monsoon flooding. Rhinos can do well in degraded scrub as long as it is next to these *S. spontaneum* grass areas/grazing lawn areas. Not all tall grasses are equally appealing to rhinos. Most of the tall grass in the park is *Narenga porphya-coma,* which rhinos shun except the first few weeks of new growth. Rhinos are not dependent on proximity to croplands, but they are restricted to living close to rivers because of their foraging preferences and thermoregulatory needs (Dinerstein 2003). We know now that this valley's ungulate biomass supports a greater density of tigers than nearly anywhere in the tiger's range (Karanth and Nichols 1998, Karanth et al. 2004). Despite their size and power, tigers are relatively poor dispersers. They don't usually cross open agricultural areas, and this greatly influences the metapopulation dynamics of Nepal's (and northern India's) tiger populations (Wikramanayake et al. 2004).

By the mid-1980s, Chitwan was essentially an island tucked between the great sea of humanity that lived to the south in the state of Bihar, India, and the Nepalese farmers separating the park from first towering ranges of the outer Himalayas. Rhinos had been translocated to other parks in the western Nepal *terai* beginning in 1986 after numbers climbed above 400 from a low of about 100 animals remaining in the 1960s. Chitwan is a spectacularly rich biological jewel, but an island none-theless. The sustainability of the system depends on meeting both biological and political criteria for survival. Biologically adequate core blocks of habitat, protected as national park and equivalent reserves, need to be reconnected to other core areas along the base of the Himalayas. Human-induced mortality had to be reduced to what the species at risk could sustain through their own life history adaptations. And this required the tolerance and support of those people who were living next to these habitat cores and corridors (Dinerstein 2003).

In 1991, Talbot returned to the Chitwan Valley for the first time since his 1955 visit because of concerns raised both by the Asian Development Bank (ADB) Office of Environment and by some national and international nongovernmental organiza-tions (NGOs) concerning the adverse environmental impacts of the planned East Rapti Irrigation Project (pers. comm., 2004). He described his 1991 encounter with the Chitwan Valley to me: "When I had walked down the Rapti Valley in 1955 there had been no human settlements or agriculture, and there were forests as far as one could see. In 1991 the only forest remaining was the green island of the Chitwan; people and farms were crowded everywhere else from the tops of the denuded mountains to the river banks" (pers. comm., 2004). The US$40 million East Rapti Irrigation Project, approved by the ADB in 1987, was to irrigate 9,500 ha of rice cul-tivation on the north side of the Rapti across from the national park. Water would be diverted from the Rapti by a weir built within the southern border of the park. Using hydrological data from the project's own figures, Talbot concluded that the project was not viable because there simply was not enough water in the river in the dry sea-son to provide project requirements. Additionally, of the proposed 9,500 ha, 8,000 ha were already irrigated by a series of village schemes. The project would essentially dry the areas adjacent to the Rapti in the dry season. The hydrologic system and processes supporting the park's grassland and floodplain forests, so essential for the rhino and other wildlife, would have been irreversibly altered. Talbot presented his conclusions; the project was not built.

In addition to deflecting this wrongheaded irrigation scheme, Talbot proposed some positive actions that would benefit people and wildlife alike: provide (1) flood protection on the north bank of the river and (2) agricultural and rural assistance including agricultural forestry, community health, family planning, and provision of electricity. He also proposed a buffer zone along the north bank with trees and thatch grass based on the preliminary results from some popular pilot plots planted by the National (formerly the King Mahendra) Trust for Nature Conservation.

When the Royal Chitwan National Park was established, buffer zone management was not very sophisticated; as the human population grew, these lands were soon degraded and overused in a classic example of the well-known tragedy of the commons. People had to turn back to the park, illegally, to meet their basic needs. To relieve some of the pressure, villagers were permitted to cut grass in the park for two weeks each year just to keep roofs over their heads. For these poor people, the park did not create wealth—it sequestered it from them. Furthermore, tourism to the park was creating wealth for outsiders and the state. Conventional wisdom has it that a well-established tourist industry based around a national park will have a trickle-down effect in the local economy. Surveys showed, however, that only a small fraction of the money being generated each year by Chitwan was recycled into the local economy (Sharma 1991). Resentment grew, and when tiger and rhino poaching emerged as a renewed threat in the 1980s, the local people were little interested in cooperating with authorities to stop it. The park and the people living in the valley were on a collision course. If Chitwan's wildlife was to survive, there would have to be a dramatic shift in the way conservation was being pursued there (Bookbinder et al. 1998).

After plummeting in the early 1990s because of heavy poaching, the number of tigers and rhinos in Chitwan rebounded as a result of innovative new conservation initiatives. The big question: How do you enable tigers and rhinos to persist in Nepal, where annual per capita gross domestic product is about $1,100 (this figure for the United States is $33,900), the human population is growing at more than 2% per year, and about twenty-five million people live in a country the size of Arkansas, which has a population one-tenth that size? In many other places, this combination of poverty and rapid population growth is proving lethal for large carnivores and has eliminated rhinos, which require the most productive floodplain habitats to live.

When I arrived in Chitwan, our scientific interest was the big mammals living in the park. From the outset we were keenly aware that we could not live here without the help and support of the local people. Now we know that the survival of the park's big mammals also depends on local support. Led by the National Trust for Nature Conservation (a Nepal-based NGO that began as the Smithsonian-Nepal program), a dialogue was initiated with local farmers to learn how to get this support and how to improve the human-wildlife relationship in zones between their fields and the park. This zone was full of domestic cows and goats, and tigers were killing them. In return, people were putting poison into the carcasses, killing tigers. Arun Rijal put it this way: "This is what we had to solve after the video tape stops if we wanted to save our animals" (pers. comm., 2000). He was referring to the videotapes that environmental educators were showing local villagers, videos telling them why they should appreciate Chitwan's natural treasures. The fact was that the reason people tolerated rhinos in their rice fields and tigers killing their cattle, and even their loved ones, was that they couldn't stop them. Villagers were taking the brunt of wildlife damage and seeing no return from the increasing numbers of dangerous animals in the park.

The experiment to change all this began in the local community of Bagmara, located on the north bank of the Rapti River, across from the national park. Key legislation was adopted in 1993 that sanctioned the organization of user group committees to manage buffer-zone forests. This innovative conservation effort in Bagmara started with a small $10,000 award from the U.S. Agency for International Development (USAID) that was used to establish a native tree nursery in 1995. Shankar Choudhury, a ranger for the National Trust for Nature Conservation, donated the land for the nursery after he heard about community forestry initiatives elsewhere. This sparked further efforts; during the next year a 32-ha plot of government-owned buffer zone was fenced and planted to provide for the timber and firewood needs of the villagers. A stand of native rosewood (*Dalbergia sissoo*), which takes about twenty years to mature, was planted in the zone—an investment in the future that will yield a lucrative reward. In the following years, more plantations were established and nonplantation areas were fenced off to allow natural regeneration and to discourage livestock grazing. Stall feeding of livestock was initiated. A wildlife observation tower was built to rent out to local hotels so their guests could spend the night. Villagers began to offer nature walks and elephant rides through the forest, where both rhinos and tigers have reestablished as permanent residents living outside the national park boundaries (summarized in Dinerstein 2003; Seidensticker 2002).

The community forestry/guardianship program keeps growing. Its needs have been modest: fence and fence posts, seeds and saplings, people to help. And the response has been tremendous. The program's essential concept has been to return the profits from the community endeavors to create natural wealth for improving community wealth—supporting schools, health clinics, and job training—rather than letting the income become individual wealth. In Chitwan, tigers and rhinos are thereby becoming stars in efforts to improve the lives of local people, and not only for the income generated. Women once walked two to three hours a day to sneak into the park to collect firewood and fodder. Inside the park, they risked being caught by the army or attacked by a tiger or startled rhino, and then they made the hours-long return trip carrying 40-pound bundles of wood on their heads. This crushing daily trek, coupled with all the other chores considered women's work, left little time for anything else. The local community forests changed all that. Thus Bagmara is the birthplace of what has come to be known as the guardianship program. This effort by the National Trust for Nature Conservation and their partners gives local people a reason to care about the park and its wildlife: they improve their own lives when, with little investment, large, potentially dangerous mammals can be helped to successfully coexist with the nearly 300,000 people living in thirty-six villages adjacent to the park.

By 2000 seeing a greater one-horned rhino was commonplace in Chitwan. Wild elephants are a recent addition, after a forty-plus-year absence from the area. In 1973, when the park was established, Chitwan hosted 800 visitors; in 1999, nearly 100,000 came, more than half of them from India and from within Nepal. This influx has changed Chitwan and its surrounds. People who want to explore pristine wildernesses where the only sounds are made by calling birds, roaring tigers, trumpeting elephants, and their own footsteps find this tragic. But without these changes, without an increasing human presence, paradoxically, tigers and rhinos and elephants might be going or gone instead of gaining ground. The tourists are a necessary part of the Chitwan tale of progress (Seidensticker 2002). They may be disruptive at times, but "better to put up with a few bloody obnoxious tourists than lose all this spectacular wildlife," as Hemanta Mishra puts it (pers. comm., 2000). Tourists brought village user committees in the buffer zones about US$350,000 annually as part of a 50% share of park revenues (Dinerstein et al. 2007)

The Maoist insurgency that rapidly expanded in the early 2000s had a devastating effect on Nepal's tourist economy, which is the nation's third largest source of foreign exchange. In 2005, Nepal recorded 275,000 foreign visitors, down from 500,000 in 1999. In Chitwan, tourists declined to 25,000 in 2005. A 2005 census could not account for 192 of the 544 rhinos previously counted in the park. Guard posts—once widely dispersed through the park—were burned or abandoned as the army consolidated its forces and discontinued patrolling beyond the parameter of its own posts. This power vacuum made it easy to trap deer and other animals in the park (Baral and Heinen 2006). In my visit to Chitwan in 2005, I found the deer population in the park much reduced compared to previous visits in 2000 (Seidensticker and Lumpkin 2006) and to my own work there in 1974. Charles McDougal, who had been monitoring a 100 km² lowland, grassland area of the park since 1972, did not find a reduced number of tigers during the Maoist insurgency period (pers. comm., 2005). In my subsequent visits I found general agreement among NGO and park officials that the bottom-up benefits coming from parks to local communities were a significant factor in the very survival of the parks during the insurgency.

With the basic needs for human-wildlife coexistence model in place—with local guardianship supporting the core parks and sustainable forestry supporting the connecting corridors—the idea that you can connect and sustain a necklace of parks and corridors at the base of the Himalayas along the 1,000 km from Chitwan in Nepal to Corbett National Park in India has become a conservation vision that is a force in itself. In the local conservation vernacular, this is known as the TAL–Terai Arc Landscape. This 49,000 km² conservation corridor project connects twelve wildlife reserves and national parks (Seidensticker and Lumpkin 2006). Nepal has embraced this vision as a government policy, and various government ministries act to facilitate international donors and NGOs in support of this vision. NGO investment in the Nepal TAL between 2000 and 2002 was about US$25/km², one-tenth of what Nepal's government invested (Dinerstein et al. 2007). I am confident that India will follow suit because so many younger conservation leaders there understand how important this vision is to sustaining Indian forests and wildlife and to bettering the lives of people who live next to wildlife. The details are being worked out all along the *terai* each day (Dinerstein 2003; Johnsingh et al. 2004).

The challenge in designing and restoring sustainable conservation landscapes that benefit both people and wildlife is ensuring that they (1) matter for people and (2) meet the needs of the wildlife species living in the areas. Stephen Kellert noted: "Support for endangered species conservation will emerge when people believe this effort enhances the prospects for a more materially, emotionally, and spiritually worthwhile life for themselves, their families, and communities. This may not constitute a particularly easy task, but it may be an unavoidable one" (1996, 78). The vision of a conservation landscape extending from Corbett National Park to Chitwan and the understanding of the conservation engineering it will take to implement the vision are in place. But you have to meet political criteria, and there are those who continue to voice the bitterness of impotence and decline. The question we now face is: Will this vision of a sustainable conservation landscape, with its large animals extant, transcend the chaos of a recent Nepal Maoist insurgency (Satyal Pravat 2004; Vigne and Martin 2004), the waves of migrants that involve a quarter of Nepal's population (Baral and Heinen 2006), and/or proposed major resource extractions to supply India's and Nepal's growing economies? These are no doubt formidable challenges. The Maoists joined in the process of writing a new constitution for Nepal in 2007, and there is hope that with stability we will see recovery in Chitwan and other protected areas in the TAL. Because this conservation vision is sustained with bottom-up, human-oriented processes, though, and because of the life-support services the Piedmont zone provides to the economy and well-being of the entire Ganga Plain, local conservation leaders predict that it will succeed (C. Gurung et al., pers. comm., 2004), and they are working to make it so.

A half century after Talbot first entered the Chitwan Valley, bore witness to a landscape in decline, and established the ecological and intellectual baselines for a sustainable conservation landscape recovery, we can step back and see that there is a vision and a way to recover and sustain this natural wildlife spectacle for the future and for the people in this human-dominated landscape.

CONSERVATION VISION: CONCLUDING THOUGHTS

The lessons from the conservation efforts of the half century since Talbot's 1955 visits to these two landscapes can perhaps best be summarized as follows: Lee Talbot has a vision, a right vision for the ecosystems, the people, and the splendid species he champions. According to the late Neil Postman (1996), "schooling" boils down to two problems. The first problem is really an engineering problem. It is the assessment of change, dynamics, and cause and effect. To understand is to "know what causes provoke what effects, by what means, at what rate" (Tufte 1997, 9). It is the problem of the means by which we become learned. The other challenge we face in schooling is metaphysical. For schooling to make sense, the young, their parents, and their teachers must have a vision. I believe this paradigm of seeing schooling in terms of its engineering side and its metaphysical vision helps us to think about this elusive concept we call conservation. Talbot taught that sound science was the basis for enlightened, effective conservation. His recommendations were founded in the notion that a deep knowledge of both human attitudes and the workings of the ecosystem was a way to move conservation away from uninformed confrontation toward some workable solutions. On the metaphysical side of this equation, Postman emphasized the importance and the power of our narratives, our stories, and our images to inform our vision. In conservation, our stories and narratives start with the environmental history of a landscape and our accurate image of what a landscape is now. We then use these to inform our vision of what a conservation landscape can be, and this vision becomes our target in conservation practice.

Author's Note: When I first read Lee Talbot's "A Look at Threatened Species" (1960) after my first trip to India in 1972, it wasn't the details of his observations that first impressed me and then stayed with me. It was his vision. The landscapes and marvelous species he described were threatened, but the vision he so vividly left with me was this: With informed conservation interventions, these species and landscapes can be sustained. I believe that Lee Talbot looks at the world a little differently than most. He saw threatened species as symbols for recovery rather than symbols of ecosystems in decline, even in human-dominated landscapes. The honored Indian conservation biologist K. Ullas Karanth puts it this way: a combination of "vision, persistence, thinking at the right social and spatial scales, and constructive dialogue is the key to the tiger's future" (Seidensticker 2001, 4). To this we can add lions and rhinos, the Gir, Chitwan, and everywhere we work with threatened ecosystems and species. Karanth and Talbot and so many other practicing conservationists are committed to bringing sound science to conservation; more important, though, we recognize the importance of the right vision, informed by the past, grounded in the present. The tiger, the lion, the rhino, and the landscapes where they live will be lost without this vision, no matter how good the science.

On a late afternoon in May 1976, I walked the 5 km from my office at the National Zoo, where I was then a visiting scientist, to the offices of the Council on Environmental Quality, located near the White House. The World Wildlife Fund had asked me to go to Indonesia to see what we could do to save the last Javan tigers. Lee Talbot ushered me into his quiet, unpretentious office and spent the next hour sharing with me his experiences and observations on the ecology of Java and Bali. He suggested who I should see, and he slyly infused me with some ideas on how to think about past and future conservation in Indonesia. I remember his warning: "Practically every plant at the level you are when walking in a Javan tropical wet forest has thorns." He was as right about the thorns as he was about how to think about conservation in these Javan human-dominated landscapes. My time in Indonesia was enriched by those few minutes of introduction by Lee Talbot because his observations and insights were so all-encompassing and yet to the point. Over the years I have continued to suggest that he use his 1955 trip as a basis for looking at environment change in Asia, and he may yet. Here I wanted to report on how it is to follow in Lee Talbot's footsteps from my perspective two decades later. Those footprints are large and long-lasting, yet light on the land. It has been my honor.

ACKNOWLEDGMENTS: Susan Lumpkin has been my partner for a quarter century and my own thinking about conservation is completely commingled with hers, as we have worked together on our books and articles, travel, and life. My visits to South Asia have been supported by the Smithsonian Special Foreign Currency Program, *Smithsonian* magazine, the World Wildlife Fund, Friends of the National Zoo, the Save the Tiger Fund, and the U.S. Fish and Wildlife Service. Paul Joslin introduced me to the Gir in 1984. My experience and time in Chitwan were shaped and supported by my fellow investigators, associates, and administrators in the Smithsonian-Nepal Tiger Ecology Project: Kirti Man Tamang, Hemanta Mishra, Tirtha M. Maskey, Andrew Laurie, Mel Sunquist, Dave Smith, Eric Dinerstein, Chuck McDougal, Ross Simons, David Challinor, and Chris Wemmer, and more recently by Arup Rajouria, Arun Rijal, Shanta and Sarita Jnawali of the National Trust for Nature Conservation, and Chandra Gurung and Mingma Sherpa of the World Wildlife Fund. Ullas Karanth has been an inspiration for the more than two decades I have known him. I first met A. J. T. Johnsingh when he came to work with me at the Smithsonian's National Zoological Park Conservation and Research Center in 1980, and our collaboration has continued as he has devoted himself to mentoring and training the next generation of outstanding Indian conservation scientists at the Wildlife Institute of India, including the foremost expert on lions in India, Ravi Chellam, and on the Gir ecosystem, Jamal Khan.

On September 23, 2006, Mingma Norbu Sherpa, Chandra Gurung, Harka Gurung, Tirtha Man Maskey, Narayan Poudel, Yeshi Choden Lama, Jennifer Headley, Jill Bowling Schlaepfer, and Mathew Preece died in a helicopter crash in the Kanchenjunga Conservation Area, Nepal. I dedicate this essay to them.

References

Angermeier, P. L. 2000. The natural imperative for biological conservation. *Conservation Biology* 14(2): 373–381.

Baral, N., and J. T. Heinen. 2006. The Maoist people's war and conservation in Nepal. *Politics and Life Sciences* 24 (1–2): 2–11.

Beier, P., M. R. Vaughan, M. J. Conroy, and H. Quigley. 2003. *An analysis of scientific literature related to the Florida panther.* Tallahassee: Florida Fish and Wildlife Conservation Commission.

Berwick, S. H. 1974. The community of wild ruminates in the Gir ecosystem, India. Ph.D. diss., Yale University.

Berwick, S. H. 1976. The Gir Forest: An endangered ecosystem. *American Scientist* 64 (January–February): 28–40.

Bookbinder, M. P., E. Dinerstein, A. Rijal, H. Cauley, and A. Rajouria. 1998. Ecotourism's support of biodiversity conservation. *Conservation Biology* 12(6): 1399–1404.

Cardillo, M., A. Purvis, W. Sechrest, J. L. Gittleman, J. Bielby, and G. M. Mace. 2004. Human population density and extinction risk in the world's carnivores. *PLoS Biology* 2(7): 900–914.

Chellam, R. 1993. Ecology of the Asiatic lion (*Panthera leo persica*). Ph.D. diss., Saurashtra University.

Chellam, R., and A. J. T. Johnsingh. 1993. Management of Asiatic lions in the Gir Forest, India. *Symposia of the Zoological Society of London* 65:409–424.

Clark, T. W., A. P. Curlee, and R. P. Reading. 1996. Crafting effective solutions to the large carnivore conservation problem. *Conservation Biology* 10(4): 940–948.

Dhungel, S. K., and B. W. O'Gara. 1991. *Ecology of the hog deer in Royal Chitwan National Park, Nepal.* Wildlife Monographs, no. 119. Bethesda, MD: Wildlife Society.

Dinerstein, E. 2003. *The return of the unicorns: The natural history and conservation of the greater one-horned rhinoceros.* New York: Columbia University Press.

Dinerstein, E., C. Loucks, E. Wikramanayake, J. Ginsberg, E. Sanderson, J. Seidensticker, J. Forrest, G. Bryja, A. Heydlauff, S. Klenzendorf, et al. 2007. The fate of wild tigers. *BioScience* 57(6): 508–514.

Eisenberg, J. F., and J. Seidensticker. 1976. Ungulate biomass in southern Asia: A consideration of biomass estimates for selected habitats. *Biological Conservation* 10(4): 293–308.

Flannery, T. 2003. The lady, or the tiger? Review of *Monster of God: The man-eating predator in the jungles of history and the mind*, by D. Quammen. *The New York Review of Books* (October 9): 13–14.

Frankham, R., J. D. Ballou, and D. A. Briscoe. 2004. *A primer of conservation genetics.* Cambridge: Cambridge University Press.

Glacken, C. J. 1967. *Traces on the Rhodian shore: Nature and culture in western thought from ancient times to the end of the eighteenth century.* Berkeley: University of California Press.

Hodd, K. T. B. 1970. The ecological impact of domestic stock on the Gir Forest. In *Papers and proceedings, 11th Technical Meeting of the International Union for the Conservation of Nature, New Delhi 25–28 November 1969,* vol. 1, 259–265. IUCN Publications New Series, no. 17. Morges, Switzerland: IUCN.

Holling, C. S., and G. K. Meffe. 1996. Command and control and the pathology of natural resource management. *Conservation Biology* 10(2): 328–337.

Jhala, Y. V., Q. Qureshi, V. Bhuva, and L. N. Sharama. 1999. Population estimate of Asiatic lions. *Journal of the Bombay Natural History Society* 96(1): 3–15.

Johnsingh, A. J. T., R. Chellam, and D. Sharma. 1998. Prospects for conservation of the Asiatic lion in India. *Biosphere Conservation* 1(2): 81–89.

Johnsingh, A. J. T., K. Ramesh, Q. Qureshi, A. David, S. P. Goyal, G. S. Rawat, K. Rajapandian, and S. Prasad. 2004. *Conservation status of tiger and associated species in the Terai Arc Landscape, India.* Dehradun: Wildlife Institute of India.

Johnsingh, A. J. T., S. P. Goyal, and Q. Qureshi. 2007. Preparations for the reintroduction of Asiatic lions *Panthera leo persica* into Kuno Wildlife Sanctuary, Madhya Pradesh, India. *Oryx* 41(1): 93–96.

Joshi, A. P. 1996. The home range, feeding habits, and social organization of sloth bears (*Melursus ursinus*) in Royal Chitwan National Park, Nepal. Ph.D. diss., University of Minnesota.

Joshi, R. R., and K. P. Karamchandani. 1976. *Working plan for Gir Forest.* Vols. 1 and 2. Gandhinagar, India: Government of Gujarat.

Joslin, P. 1973. The Asiatic lion: A study of ecology and behavior. Ph.D. diss., University of Edinburgh.

Karanth, K. U., and J. D. Nichols. 1998. Estimating tiger densities in India from camera trap data using photographic captures and recaptures. *Ecology* 79:2852–2862.

Karanth, K. U., J. D. Nichols, N. S. Kumar, W. A. Link, and J. E. Hines. 2004. Tigers and their prey: Predicting carnivore densities from prey abundance. *Proceedings of the National Academy of Sciences* 101(14): 4854–4858.

Karanth, K. U., J. D. Nichols, J. Seidensticker, E. Dinerstein, J. L. D. Smith, C. McDougal, A. J. T. Johnsingh, R. S. Chundawat, and V. Thapar. 2003. Science deficiency in conservation practice: The monitoring of tiger populations in India. *Animal Conservation* 6(2): 141–146.

Karki, J. B., Y. V. Jhala, and P. P. Khanna. 2000. Grazing lawns in Terai grasslands, Royal Bardia National Park, Nepal. *Biotropica* 32(3): 423–429.

Kellert, S. R. 1996. *The value of life: Biological diversity and human society.* Washington, DC: Island Press.

Khan, J. A. 1993.Ungulate-habitat relationships in Gir Forest ecosystem and its management implications. Ph.D. diss., Aligarh Muslim University.

Khan, J. A., R. Chellam, W. A. Rodgers, and A. J. T. Johnsingh. 1996. Ungulate densities and biomass in the tropical dry deciduous forests of Gir, Gujarat, India. *Journal of Tropical Ecology* 12(1): 149–162.

Knowlton, N., E. Sala, J. Jackson, and S. Mesnick. 2003. *Marine biodiversity: Using the past to inform the present.* San Diego: University of California, Scripps Institute of Oceanography, Center for Marine Conservation. http://cmbc.ucsd.edu/content/1/images/report_2003.pdf.

Mace, G. M., A. Balmford, and J. R. Ginsberg, eds. 1998. *Conservation in a changing world.* Cambridge: Cambridge University Press.

Maskey, T. H. 1989. Movement and survival of captive-reared gharial (*Gavialis gangeticus*) in the Narayani River. Ph.D. diss., University of Florida.

McDougal, C. 1977. *The face of the tiger.* London: Rivington Books.

McKinney, M. L. 2001. Role of human population size in raising bird and mammal threat among nations. *Animal Conservation* 4(1): 45–57.

Mishra, H. R. 1982. Ecology and behaviour of chital (*Axis axis*) in the Royal Chitwan National Park, Nepal, with comparative studies of hog deer (*Axis porcinus*), sambar (*Cervus unicolor*), and barking deer (*Muntiacus muntjak*). Ph.D. diss., University of Edinburgh.

Mishra, H. R., and M. Jeffries. 1991. *Royal Chitwan National Park: Wildlife heritage of Nepal.* Seattle, WA: The Mountaineers.

Modiano, G., G. Morpurgo, L. Terrenato, A. Novelletto, A. Di Rienzo, B. Colombo, M. Purpura, M. Mariani, S. Santachiara-Benerecetti, A. Brega, et al. 1991. Protection against malaria morbidity: Near-fixation of the α-thalassemia gene in a Nepalese population. *American Journal of Human Genetics* 48(2):390–397.

Mukherjee, A., and C. K. Borad. 2004. Integrated approach towards conservation of Gir National Park: The last refuge of Asiatic lions, India. *Biodiversity and Conservation* 13(11): 2164–2182.

Negi, S. S. 1965. Transplanting of Indian lion in Uttar Pradesh state. *Cheetal* 12(1): 98–101.

Noss, R. F. 1996. Conservation or convenience? *Conservation Biology* 10(4): 921–922.

Oza, G. M. 1983. Deteriorating habitat and prospects of the Asiatic lion. *Environmental Conservation* 10(4): 349–352.

Pocock, R. I. 1939. *The fauna of British India, including Ceylon and Burma.* London: Taylor and Francis.

Postman, N. 1996. *The end of education.* New York: Vintage Books.

Quammen, D. 2003. *Monster of god: The man-eating predator in the jungles of history and the mind.* New York: W. W. Norton.

Rangarajan, M. 1996. *Fencing the forest: Conservation and ecological change in India's central provinces 1860–1914.* Delhi, India: Oxford University Press.

Roelke-Parker, M. E., L. Munson, C. Packer, R. Kock, S. Cleaveland, M. Carpenter, S. J. O'Brien, A. Pospischil, R. Hofmann-Lehmann, H. Lutz, et al. 1996. A canine distemper virus epidemic in Serengeti lions (*Panthera leo*). *Nature* 379(6564): 441–445.

Rose, L. E. 1971. *Nepal: Strategy for survival.* Berkeley: University of California Press.

Saberwal, V. K., J. P. Gibbs, R. Chellam, and A. J. T. Johnsingh. 1994. Lion-human conflict in the Gir Forest, India. *Conservation Biology* 8(2): 501–507.

Satyal Pravat, P. 2004. *Country profile report: Forestry sector in Nepal.* Cambridge, UK: Forests Monitor.

Seidensticker, J. 1976a. On the ecological separation between tigers and leopards. *Biotropica* 8(4): 225–234.

Seidensticker, J. 1976b. Ungulate populations in Chitawan Valley, Nepal. *Biological Conservation* 10(3): 183–210.

Seidensticker, J. 1985. Asian lions: A report from the field. *ZooGoer* 14(6): 11–16.

Seidensticker, J. 2001. Elements of tiger conservation. In *Tigers in the 21st century, saving the tiger: Assessing our success,* ed. J. R. Ginsberg, 1–7. New York: Wildlife Conservation Society.

Seidensticker, J. 2002. Tiger tracks. *Smithsonian* 32(10): 62–69.

Seidensticker, J., S. Christie, and P. Jackson, eds. 1999. *Riding the tiger: Tiger conservation in human-dominated landscapes.* Cambridge: Cambridge University Press.

Seidensticker, J., and S. Lumpkin. 2004. *Cats: Smithsonian answer book.* Washington, DC: Smithsonian Books.

Seidensticker, J., and S. Lumpkin. 2006. Building an arc. *Smithsonian* 37(4): 56–63.

Sharma, H. S., and A. J. T. Johnsingh. 1996. *Impact of management practices on lion and ungulate habitats in Gir protected area.* Dehradun: Wildlife Institute of India.

Sharma, U. R. 1991. Park-people interactions in Royal Chitwan National Park, Nepal. Ph.D. diss., University of Arizona.

Shukla, U. K., and D. S. Bora. 2003. Geomorphology and sedimentology of Piedmont zone, Ganga Plain, India. *Current Science* 84(8): 1034–1040.

Singh, H. S. 1997. Population dynamics, group size and natural dispersal of Asiatic lion. *Journal of the Bombay Natural History Society* 94(1): 65–70.

Sinha, S. P. 1987. The ecology of wildlife with special reference to Gir lion in Gir Lion Sanctuary and National Park. Ph.D. diss., Saurashtra University.

Smith, J. L. D. 1984. Dispersal, communication, and conservation strategies for the tiger (*Panthera tigris*) in Royal Chitwan National Park. Ph.D. diss., University of Minnesota.

Smith, J. L. D. 1993. The role of dispersal in structuring the Chitwan tiger population. *Behaviour* 123(3–4): 165–195.

Smythies, E. A. 1942. *Big game shooting in Nepal.* Calcutta: Thacker, Spink.

Stainton, J. D. A. 1972. *Forests of Nepal.* London: John Murray.

Sunquist, F., and M. Sunquist. 1988. *Tiger moon.* Chicago: University of Chicago Press.

Sunquist, M. E. 1981. *The social organization of tigers (Panthera tigris) in Royal Chitawan National Park, Nepal.* Smithsonian Contributions to Zoology, no. 336. Washington, DC: Smithsonian Institution Press.

Talbot, L. M. 1960. A look at threatened species. *Oryx* (4–5): 153–293.

Tamang, K. M. 1982. The status of the tiger *Panthera tigris* and its impact on principal prey populations in Royal Chitwan National Park, Nepal. Ph.D. diss., Michigan State University.

Tear, T. H., J. M. Scott, P. H. Hayward, and B. Griffith. 1993. Status and prospects for success of the Endangered Species Act: A look at recovery plans. *Science* 262(5136): 976–977.

Tufte, E. R. 1997. *Visual explanations: Images and quantities, evidence and narrative.* Cheshire, CT: Graphics Press.

Vigne, L., and E. Martin. 2004. Nepal's new strategies to protect its rhinos. *Oryx* 38(3): 252–253.

Westoby, M., B. Walker, and I. Noy-Meir. 1989. Opportunistic management for rangelands not at equilibrium. *Journal of Range Management* 42(4): 266–274.

Wikramanayake, E., M. McKnight, E. Dinerstein, A. Joshi, B. Gurung, and D. Smith. 2004. Designing a conservation landscape for tigers in human-dominated environments. *Conservation Biology* 18(3): 839–844.

Wildt, D. E., M. Bush, K. L. Goodrowe, C. Packer, A. E. Pusey, J. L. Brown, P. Joslin, and S. J. O'Brien. 1987. Reproductive and genetic consequences of founding isolated lion populations. *Nature* 329(6137): 328–331.

8

Observations on Trends and Issues in Global Conservation

James G. Teer

In this chapter, I will review some of the important trends and issues in conservation that have evolved over the past few decades. I have selected seven issues to discuss, specifically:

1. The emergence of ecological science and technology
2. The proliferation of nongovernmental organizations (NGOs) and citizen conservation groups
3. The strategy of sustainable use and development in conservation
4. Property rights and the transfer of ownership of wildlife to the private sector
5. The development of game farming and ranching as an industry
6. The development of nonconsumptive uses of wildlife and ecotourism
7. The strategy of saving what is useful to society

Conservation problems are difficult to solve. The complexity and dynamics of the issues can be vexing; the origins of the issues are social, economic, and cultural. They often play out in conflicts over resource and land use in developing regions of the world.

Many advances and successes in wildlife conservation have occurred in the nineteenth and twentieth centuries. Although lists of threatened and endangered species have continued to grow (with few species taken off the lists), some species have recovered locally, at least, and are being used by people for their own welfare. In North America, white-tailed deer, elk, wild turkeys, mountain lions, and many furbearer populations have recovered, some even to nuisance levels. And in Africa, many species—such as antelope, elephants, rhinoceroses, and some species of large cats—are now recovering after a drop-off in pressure for their products.

The timbre and tone of conservation of wildlife have passed through overuse, profligacy, apathy, and, in some cases, unreasoned stridency. Now, hopefully, conservation has entered a period of science and reasoned judgment. This evolution has not been as straightforward or as sure as my words suggest; nonetheless, major advances have been made. Wildlife resources reflect progress. We are far ahead of where we were just two or three decades ago.

In the twenty-first century we operate in a world with increasing human numbers, increasing civil unrest, and environmental degradations that span the world, reaching its most remote regions and cultures. The human population is increasing in areas where resources for both conservation and human needs are the greatest. Human numbers are the apex, and the specter, that impact all that we are or can do. As many conservationists have warned, we are at the crossroads of saving the natural world.

MAJOR TRENDS AND ISSUES

The Emergence of Ecological Science and Technology

The emergence of ecological science and its acceptance by the general public have led to an appreciation of the linkages among species within an ecosystem. We may not know all the answers, but we can now ask the right questions, and we are better equipped to address them.

Ecology has become the fundamental science for the conservation and management of wildlife and its habitats. Since its emergence, conservation has become much more scientific and global, and results are more predictable and quantifiable. Technologies, as well as data management and analysis, have made offered us unbelievable progress toward solving problems. In the past the average field biologist used a pencil, a pair of binoculars, and a notebook to record observations. Today the conservation biologist uses an arsenal of tools including computers for population and ecosystem modeling, electronics for monitoring animals, delivery systems for pharmaceuticals, and policy development and conflict resolution. These tools have made the understanding and management of nature less difficult and more transparent.

The Proliferation of NGOs and Citizen Conservationists

Management and use of wildlife resources can no longer be parochial in design and scope. A conservation action in one area influences actions in other regions. The adage "act locally, think globally" is truly germane.

National governments, once the sole authorities and administrators of wildlife resources in their jurisdictions, are now joined in decision making by hundreds of NGOs and by their citizenries. As a result, conservation issues are being examined and evaluated for their far-reaching impacts by an army of citizen conservation groups, many of which are international in charter and scope. Conservation issues have also been made more transparent and less regulatory by government agencies, for the private sector has demanded a voice in conservation decisions and accordingly has become more participatory in conservation affairs. Very few conservation decisions are made by governments or user groups without being validated and accepted by society. Hearings and testimonies usually accompany any major proposal for change in policies that affect nature.

The World Conservation Union (also known as IUCN, formerly the International Union for the Conservation of Nature and Natural Resources), the World Wildlife Fund (WWF), the Convention on International Trade in Endangered Species of

Wild Fauna and Flora (CITES), the Convention on Biological Diversity (CBD), the UN Environmental Programme (UNEP), and the Food and Agricultural Organization of the United Nations (FAO) are just a few of the organizations dealing with global issues.

The Strategy of Sustainable Use

Most conservation issues and trends affect people and their livelihoods. Human needs are inextricably linked with conservation plans. We generally subscribe to the axiom that conservation must first serve the people if it is to succeed.

The adoption of sustainable use by the conservation community has been the most important strategy of the twentieth and twenty-first centuries. Many people and cultures must live with or close to wildlife. The Brundtland report (formally the report of the World Commission on Environment and Development) and the little book it produced, *Our Common Future* (1987), has had an enormous, if tardy, influence on the policies of conservation science and uses of natural resources.

Sustainable use, however, has had a tough time being adopted by the global conservation community. Heated, interminable debate over the issue took place in at least four general assemblies before IUCN formally accepted the Sustainable Use Initiative in 1995. Sustainable use is not wholly embedded in the policies of many international conservation organizations, nor is it entirely safe in the organizations that have accepted it. But it has to work. It makes sense. It is the policy for the future.

Property Rights and Devolution of Ownership of Wildlife to the Private Sector

In most of the world, ownership of wildlife resources has its basis in English and Roman common law. These laws (which some believe are archaic) vested resource ownership in the people; the state, however, held the resources in trust and administered them on the people's behalf. Free-ranging wildlife could not be owned by the private sector.

We now see a trend in which resource ownership is being devolved to private landowners, communities, NGOs, and other private entities. Such privatization is taking place throughout the world. This trend toward private ownership is the result of commercialization and changes in the viability of conventional forms of land uses. As agricultural crops and animal agriculture became economically marginal in many parts of the world, landowners have searched for ways to diversify their income. Consequently, property rights of landowners began to be expressed and extended to other resources as never before. For example, fencing to confine large mammals is now practiced throughout major grassland and savanna systems.

Through one device or another, usually legislation, wildlife is becoming the property of the private sector, though in some cases partnership arrangements between the state and the private sector remain. Many nations in western Europe have had vested ownership in the private sector for centuries, and it appears that the rest of the world is following the European system. Ownership in southern African nations is being transferred to private landowners who claim it. The debate concerning ownership and the long-term efficacy of private ownership rages in North America and elsewhere around the globe.

The values of life and the markets that attend them exert great influence on functional ecosystems and the conservation of wildlife and its habitat. Community-based conservation projects, transboundary parks and conservancies, treaties and compacts between nations over migratory species, cooperative ventures in controlling trade, and memberships in international agencies are working for cooperative conservation of wildlife and for those that use it.

The long-term results of these trends will be interesting. The question still to be answered: "Who best is able to manage the natural world: the government, or the people who own the land?" Less than 6% of the land surface of the planet is protected by parks and reserves. About 50% of these lands are truly protected, while the other 50% are protected only on paper. Protected areas are absolutely necessary for conservation; however, biodiversity will be saved, if at all, on private property.

Game Farming, Ranching, and Nonnative Species

The farming and ranching of large mammals, birds, and nonnative species are relatively new industries throughout the world. These industries are huge, and they are growing, especially in southern Africa, North America, and Australia. South Africa, for example, contains thousands of game ranches and farms. Most landowners there have de-stocked livestock and fenced their pastures, and they now own the wildlife on those pastures. New Zealand, Australia, Canada, and the United States are other regions with growing confined populations of native and nonnative large mammals.

The potential impacts of these "agricultural enterprises" on free-ranging wildlife and its habitats are serious. Spread of diseases and pathogens, competition by nonnative species with native species for resources, generic dilutions and changes in gene pools from interbreeding, and disruptions of ecosystem integrity (by favoring one element over another) are serious risks. Because of these connections, the conservation community has a huge stake in the management of confined wildlife and in the movement and release of nonnative species.

The discovery and the spread of chronic wasting disease in cervids in the United States and Canada have been connected to the game farming industry, but the methods of transfer of the disease are not fully known. If the disease continues to spread to free-ranging populations of cervids, the recreational hunting industry could be devastated.

According to IUCN (2006), the transfer and release of nonnative species constitute one of the five most important problems confronting ecosystem integrity and conservation of wildlife. In Texas we have literally made the natural world an outdoor zoo. More than 125 species of large mammals have been "tried" (released) on large cattle ranches. Six of these have become successful as free-ranging, unhusbanded populations; curiously, five of the six are natives of the Indian subcontinent. The Nilgai antelope, or blue bull, now numbers over 30,000 in three or four counties of southern Texas. Their presence has impacted native species, livestock, and the natural systems on these large cattle ranches. Whether they can be controlled, much less removed, is highly problematic.

The Development of Nonconsumptive Uses
of Wildlife and Ecotourism

Emphases in conservation have expanded in scope from consumptive to nonconsumptive uses and from consideration of game species only to all life, large and small. Nature study and outdoor activities of one kind or another (e.g., bird-watching, camping, and hiking) have transformed many government conservation agencies and have spurred the creation of citizen conservation organizations. Legislation protecting scarce and sensitive beings and their habitats has come largely from societal interests in nature. "Saving the last of the best and the life in it" is the theme of a growing segment of cultures and societies throughout the world.

Unfortunately, ecotourism has been neither as dependable nor as profitable as many nations had hoped. The trend in user numbers nonetheless favors the nonconsumptive segment of nature enthusiasts, and conservation interests now must consider all creatures, including nongame, in conservation plans and budgets.

Recreational hunting still remains the largest source of income and foreign exchange for many developing nations. Funds from licenses, concession fees, and safaris often fuel conservation efforts. If recreational hunting profits were taken away, budgets for conservation would often be at the very bottom of expenditures by government agencies. Thus antihunting groups, in a sense, defeat their own cause through their positions on sustainable use (hunting) of wildlife resources.

The Strategy of Saving What Is Useful to Society

If there is a single axiom in conservation policy, it has to be that people on this globe will save what they can use. Yes, people in developing nations may be, and often are, reckless in their uses of wildlife resources. This is appropriate, because survival is their first need. Nations struggle for sovereignty, social acceptance, equality among nations, economic stability, and political freedom. People want jobs, food, places to live, health care, schools, and the other everyday necessities and amenities of life that we in the more affluent world take for granted. Conservationists must recognize these needs.

CONCLUDING OBSERVATIONS

The major difference in the practice of conservation today as compared to yesterday is an enlightened and concerned world. The world community understands the need for conservation of living resources and realizes that losing them threatens its own future. Most nations have now institutionalized conservation into their national plans and budgets. Admittedly, many have insufficient funds to support their needs, but most are valiant in their efforts.

We will continue to have divided opinions on some trends and issues in conservation. We should remember, however, what history tells us: that conservation efforts will succeed to the extent that we mobilize efforts to serve the human condition as well as the species and their habitats.

References

IUCN—The World Conservation Union. 2006. Red list of threatened species (site presenting facts about threatened species). http://www.iucn.org/themes/ssc/redlist2006/threatened_species_facts.htm.

World Commission on Environment and Development. 1987. *Our common future*. Oxford: Oxford University Press.

PART III

FOREST AND RANGE ECOLOGY AND MANAGEMENT

Ronald E. Stewart

The authors of part III review the development of public policy and management of range-lands and forests. Specifically, Frederic H. Wagner discusses rangeland management in the United States, while Walter J. Lusigi focuses on the same in Africa; Jack Ward Thomas and James A. Burchfield focus on federal forest management in the United States. Policies can be helpful or harmful, as is especially highlighted for African rangelands. Public policy and management practice have been affected in recent years by a growing environmental awareness and by the development of environmental advocacy groups.

In chapter 9, Frederic Wagner details the development of the livestock industry and grazing management and policy in the United States. Livestock were introduced into the southwestern United States by the Spanish in the sixteenth and seventeenth centuries. Following acquisition of the Southwest by the United States, the number of cattle and sheep increased, peaking in the mid- to late 1900s. In the western United States, these increases were on public lands; in the 1900s, though, as scientific management concepts were developed, the number of animals grazed on national forest lands and on Bureau of Land Management lands decreased. Estimates suggest that today around 70% of U.S. cattle are grazing on private lands in the West.

The ecologist Frederick Clements first put the impacts of grazing into an ecological context in the early 1900s. A. W. Sampson, a student of Clements, produced the first range-management text, *Range and Pasture Management,* in 1923. Perhaps the most influential text on grazing management was *Range Management,* published in 1943 by L. A. Stoddart and A. D. Smith. This book gave a strong ecological foundation to range management and became known as the "range-management bible."

Wagner describes the climatic and vegetative conditions in the four major beef-producing regions in the United States: the Southeast, the Great Plains, the Intermountain West, and the California grasslands. He also discusses grazing ecology and grazing systems, addressing the effects on plant community structure (plant succession and the development of holistic management), the effects on plant community production, and the multiple life-form vegetation concept of the Intermountain West. One consequence of the ecological conditions and human history in the Intermountain West is that much of its vegetation has undergone change that is essentially irreversible, and the Great Plains practice of removing livestock to allow the vegetation to return to pregrazing condition will not work there.

Wagner concludes with a major discussion of changing rangeland policies as affected by shifting public values. Range-management policies in the United States are dominated by federal legislation and executive actions applied to that half of the eleven western states in federal ownership, as well as by broader environmental legislation applicable to public rangelands. The discussion is divided into two time periods: the agrarian era between 1900 and the 1960s and the environmental era from the 1960s to the present. The development of strong environmental advocacy groups that grew out of the environmental movement of the 1960s and 1970s continues to bring about changes in grazing policy in the United States.

In chapter 10, Walter Lusigi identifies seven key policy failures that have contributed to deterioration of rangelands in English-speaking East Africa. These are (1) land tenure policy issues, (2) anti-pastoral colonial policies, (3) wildlife management policies, (4) institutional policy issues, (5) poverty alleviation policies, (6) rangeland-specific fiscal policy issues, and (7) policies on diversification of rangeland economies.

Lusigi calls for the adoption of innovative policies that recognize the "complexity of linkages between pastoral people and their environment." National parks and game reserves must be accepted by local rangeland people to be successful, suggesting that policies that contribute to the present threat of the wildlife resources must be changed. Planning must be based on an appropriate consideration of cultural, political, and socioeconomic factors, in addition to ecological factors. Conservation objectives must be balanced against short- and long-term local human needs, and the conservation program implemented in a manner that is acceptable to the local African population. Pastoral people continue to feel insecure because "they view wildlife conservation as part of the reason why they have been denied tenure of their land and development of their economies."

If user rights in crop production are acquired by taking up cultivation, there is an incentive to exhaust the soil and move on. Laws should be implemented to protect the rights of farmers who are prepared to undertake longer-term investment in the land. At the local level, investment priorities are often targeted to development of important community infrastructure. The main problem in defining local responsibilities for investments is matching local development priorities in terms of the use of national resources with local initiatives and funds. Such investments must be internally sustainable and compatible with the natural resource base.

With respect to African rangelands, notes Lusigi, there has been a political and economic "history of opportunistic exploitation, impelling ever-increasing degrees of human hardship and ecological damage that in some cases might be irreparable." He concludes that these truths must be addressed to restore the balance between consumption and production, thus ensuring the future for millions of rangeland people of Africa.

In chapter 11, Jack Ward Thomas and James Burchfield discuss the development of the concept of sustainability as applied to the management of the national forests in the United States. While a number of definitions of sustainability have been used, they suggest that those who deal with natural resources might be most comfortable with the definition proposed by the World Commission on Environment and Development in 1987: "Sustainability is the ability of present generations to meet their own needs while ensuring that future generations are able to meet their own" (267–268). Thomas and Burchfield trace the development and application of sustainability policy regarding the national forests to the concept (incorporated into the Organic Act of 1897) that national forests would be established to provide a continuous supply of timber, thereby implying sustainability.

Key national forest policies were established under the leadership of two key conservationists, President Theodore Roosevelt and Gifford Pinchot, the first chief of the Forest Service and the first American trained as a forester. As important figures in the Progressive Era, they strongly believed in the application of science to management. Thomas and Burchfield indicate that this also meant that forest management was strongly influenced by a utilitarian outlook. Early direction provided to the Forest Service by Secretary of Agriculture James Wilson, in a letter actually written by Pinchot, "emphasized two points: (1) conservative use to assure the 'greatest good to the greatest number in the long run' and (2) [a form of] 'sustainable management' using such descriptive words and phrases as 'permanent good of the whole people,' 'permanence of resources,' 'continued prosperity,' and 'permanent and accessible supply of wood, water, and forage.'"

As a professional forester, Pinchot sought to bring scientific management to the natural resources of the United States. The Forest Service established a natural resource research organization beginning in 1908, when forestry expert Raphael Zon recommended to Pinchot that the agency "establish forest experiment stations in the national forests to carry on 'experiments and studies leading to a full and exact knowledge of American silviculture, to the most economic utilization of the products of the forest, and to a fuller appreciation of the indirect benefits of the forest.'" Today, the Forest Service has one of the largest natural resource research organizations in the world. While the research organization retains an independence from the land management organization, scientists and land managers now work closely together to solve the complex problems of ecologically based sustainable management.

From its beginning, the Forest Service's emphasis was on the production of goods and services from the national forests for the American people. This utilitarian view continued to guide the agency until the 1990s. However, social pressures and changing public values—as expressed in legislation such as the Endangered Species Act of 1973 and the National Forest Management Act of 1976—brought about significant changes in agency management priorities and practices. The changes were accelerated by actions taken by federal courts in the Pacific Northwest regarding timber harvesting and the status of the northern spotted owl. An injunction on timber harvesting on federal lands in Oregon, Washington, and northern California within the natural range of the owl resulted in several scientific assessments and the development of management alternatives by a team of Forest Service and university scientists. This effort, according to Thomas and Burchfield, laid the groundwork for a more intimate association of scientists in the land management planning process and an ecologically based approach to natural resource management.

Today, sustainable management of the full range of ecological and economic values of the national forests guides both policy and resource management. The changing policy and practice have also led to an emphasis on what Thomas and Burchfield call "wild science." Agency focus is on maintenance of biodiversity and the science to support that management direction. Science is now more intimately incorporated into the mandated multiple-use planning processes at the national forest level. This has resulted in the need for synthesis of scientific information and the active engagement of scientists in the planning process. As the authors demonstrate, these changes were not easy, nor were they implemented smoothly. The result is a more cautious, adaptive approach to natural resource management.

Reference

World Commission on Environment and Development. 1987. *Our common future.* Oxford: Oxford University Press.

9

Half Century of American Range Ecology and Management: A Retrospective

Frederic H. Wagner

To review developments in range ecology, livestock grazing systems, and rangeland policy in the United States over the past fifty years is to review the major developments in and evolution of these fields. The livestock industry in the United States actually arose much earlier, with the first sheep and cattle brought into what is now Texas, New Mexico, Arizona, and California by the Spaniards in the sixteenth and seventeenth centuries. Limited husbandry initially developed around Spanish missions and was picked up by some Native American tribes; there is some unpublished evidence that sheep ranching vastly increased in the upper Rio Grande valley of northern New Mexico in the late 1700s and early 1800s, again in connection with Spanish settlements (T. W. Box, pers. comm., 2004).

But the industry, particularly cattle ranching, did not burgeon in the West as a whole until after Texas became an independent republic in 1836 and the Treaty of Guadalupe Hidalgo ceded a major portion of northwestern Mexico to the United States in 1848. Some growth developed in the 1850s and 1860s to supply the California gold rush. It slowed, though, until the end of the Civil War upheaval and pacification of Indian hostilities in the late 1800s.

At that point industry growth surged, one estimate placing thirty to forty million cattle in the seventeen western states in 1884 (Wildeman and Brock 2000). New Mexico and Arizona are estimated to have had over five million sheep at that time. The prevailing view in range-management circles is that livestock numbers peaked in the American West in the 1880s and 1890s and that the major range impacts occurred at that time.

However, my own compilations of livestock numbers reported by the U.S. Department of Agriculture Statistical Reporting Service (SRS) for the eleven western states (Wagner 1978) and California (Wagner 1989) show beef cattle numbers rising steadily from the 1800s through at least the 1980s. The numbers of cattle grazed on western national forests and U.S. Bureau of Land Management lands did decline during the 1900s (Wagner 1978, 2003), but still remain at substantial, if declining, numbers. Debra L. Donahue (1999) points out that 70% of cattle in the eleven western states are reared on private land which, if my SRS numbers are correct, could suggest that the twentieth century growth in cattle numbers occurred largely on private land. This question needs to be resolved.

The SRS statistics do indeed show sheep numbers peaking in the West shortly before or after 1900. They remained high until the end of World War II, then declined steadily through the twentieth century. Today the U.S. sheep industry has virtually disappeared, driven out of existence by lack of economic viability associated with low demand for lamb in American cuisine and replacement of wool by synthetic fibers in a great variety of fabrics (Wagner 1988). Hence this review focuses largely on the ecology, management, and policies of cattle grazing.

Concerns for range impacts were first expressed shortly before the turn of the twentieth century, along with the reports of high livestock numbers. Their impacts were first viewed in an ecological context as the plant ecological writings of Frederick E. Clements began to appear. Assigning an 1899 date, Clements credits J. G. Smith with giving "the first clear recognition of grazing as a fundamental field of investigation" (1928, 225).

The early 1900s witnessed a large number of grazing studies in the southwestern United States, the western national forests, and the central North American grasslands. It was also a period during which a number of large, federal research areas (under the Department of Agriculture's Forest Service and Agricultural Research Service) were established to study vegetation and soil responses to different patterns of grazing: the Santa Rita Experimental Range in southern Arizona in 1905, followed by the Desert Experimental Range in western Utah, the Jornada Range Reserve in southern New Mexico, the Central Plains Experimental Range in eastern Colorado, and the Fort Keogh Livestock and Range Research Laboratory in eastern Montana.

Clements's student, A. W. Sampson, produced the first range-management text, *Range and Pasture Management,* in 1923. Sampson's book was strongly oriented toward pasture management and livestock husbandry. After a twenty-year hiatus in such texts, L. A. Stoddart and A. D. Smith published one with strong ecological underpinnings in 1943—*Range Management,* which came to be known as the "range-management bible." Published in second and third editions in 1955 and 1975, the "bible" was followed by a stream of other texts and symposia. With publication of the first volume of the *Journal of Range Management* in 1948, and with the convening of the First International Rangeland Congress in 1978, range ecology as a science, and range management as a profession, were well under way by the mid-twentieth century.

Policy decisions affecting American range management have largely been addressed to that half of the eleven western states in federal land, as well as to broader environmental legislation applicable to public land. The majority of these decisions have been made since 1960. Thus rangeland policy history, like the development of range ecology and management, primarily spans the latter half of the twentieth century. Consequently, this chapter's retrospective focuses on that latter half, although in some cases it revisits the earlier, historic roots of the half century's development.

Livestock are produced in several regions of the United States that differ climatically, physiographically, and ecologically, and consequently have different livestock grazing systems. They also vary in land ownership and consequent histories of policy action. The treatment that follows is subdivided according these contingencies.

MAJOR U.S. BEEF-PRODUCING REGIONS

The Southeast

With mean annual precipitation ranging from 1,000 mm upward in the southeastern United States, primary production of herbaceous vegetation is high by North American standards. But it has generally been considered to be of low nutritional value for livestock. As recently as 1967, a U.S. Soil Conservation Service inventory placed 99% of privately owned rangelands in the United States in the seventeen western states (Pendleton 1978). The entire Southeast—from Louisiana and Arkansas eastward to the Atlantic coast, a thin coastal strip extending northward to New Jersey, and all of Missouri, Iowa, and Minnesota—was calculated to contain only 1% of U.S. private rangelands.

Yet by 1999, Debra Donahue was reporting that 81% of all beef cattle in the United States were produced on private lands in the Southeast. Florida, with 1.4 million cattle in 1976, produced as many animals as Utah, Nevada, Arizona, and Washington combined. Florida is today second only to Texas in beef production. Virginia produced twice as many beef cattle in 1976 as Nevada, Arizona, and Utah combined.

This explosion of beef production in the Southeast in recent decades has been made possible by region-wide fertilization of natural vegetation with chemical fertilizers and droppings from the recently expanding poultry industry; by development of improved pastures; and by supplementation with forage crops from cultivated land. The fertilization has particularly enhanced the nutritional value of the winter vegetation, which had traditionally been considered to have the lowest nutritional quality.

Thus it is ironic that the expansive fields of range ecology and management in the United States, the extensive literature documenting them, and decades of public policy action have all developed in the context of a region that now produces only 17% of the nation's beef. There is as yet no significant livestock-herbivory literature for the Southeast in the mainline range-management literature. In a roughly 2,000-page report produced in 1984 by the fourteen-person Committee on Developing Strategies for Rangeland Management, impaneled by the National Research Council/National Academy of Sciences, none of the eighty chapters addressed matters in the Southeast. Only two of the committee members were at universities, or posted with public agencies, outside the West, and neither of these was located in the Southeast. With the available material focused almost entirely on the West, this chapter continues that emphasis.

The Great Plains

W. K. Lauenroth et al. (1994) consider the U.S. Great Plains to encompass the states of Texas, Oklahoma, Kansas, Nebraska, both Dakotas, the eastern two-thirds of Montana and Wyoming, and the eastern halves of Colorado and New Mexico. The Great Plains portion of the latter four states, and hence the entire region, is bounded on the west by the Rocky Mountain system.

Annual precipitation in the region grades from about 1,000 mm on the east to about 300 mm on the west, with heaviest occurrence in April, May, and June in the north, and in July, August, and September, the monsoon season, in the south. Mean annual temperatures decline from 20°C in the south to 4°C in the north.

Cropland, occupying 31.3% of the entire region, is most extensive in the east and declines to the west. The area in rangeland is the complement of this trend, and it occupies 57.6% of the entire region. Combined with 5.1% in pasture, Lauenroth et al. (1994) generalize that 63% of the Great Plains are grazed by livestock.

Physiognomically, the Great Plains vegetation subdivides into four grassland types, coinciding roughly with the east-west precipitation and north-south temperature gradients. A north-south strip of tallgrass prairie bounds the region on the east, as does a shortgrass steppe along the west against the Rocky Mountains. Natural vegetation of the intervening zone is southern mixed prairie from Texas northward through Kansas, and a northern mixed prairie from Nebraska northward through the Dakotas, Wyoming, and Montana. North-south isopleths of annual above-ground net primary production (ANPP) coincide with the east-west precipitation gradient, ranging from 600 g/m² on the east to 150 g/m² on the west.

The region, and hence the livestock-producing rangelands, is 90% private land. Donahue (1999) reports that 17% of American livestock production occurs in the seventeen western states, with well over half in the Great Plains.

The Intermountain West

That portion of the lower forty-eight states between the Rocky Mountains on the east and the Sierra Nevada–Cascade chains on the west is the most diverse region of the United States climatically, physiographically, and ecologically. Positioned in the rain shadows of these two mountain systems, the region has the most arid climates in the nation. But that aridity is not uniform across the region because of its many mountain ranges—the state of Nevada alone has 120—and the close correlation between mean annual precipitation and elevation (MacMahon and Wagner 1985). The latter vary from sea level and below in southwestern Arizona and southeastern California to 4,400 m atop the highest mountains. Consequently mean annual precipitation varies from 50 mm to 1,300 mm on the highest mountaintops.

The region is commonly considered the desert portion of the United States, based on the precipitation levels and vegetation types of the lower elevations whose area exceeds that of the mountain ranges. The deserts, in turn, are subdivided into the southern or "hot" deserts situated largely below 37° N in southern Nevada and southeastern California, and below 35° N in Arizona, New Mexico, and the Trans-Pecos of western Texas. Elevations generally range from below sea level to 1,300 m. Extensive portions of western Texas, and southern New Mexico and Arizona, originally a grassland type termed "desert grassland" (Lofton et al. 2000), have been invaded by woody species and are now considered desert. Except for higher elevations of the larger mountain ranges, year-round grazing is possible in many areas.

The northern or "cool" deserts extend northward from the hot deserts to the Columbia Plateau of eastern Oregon and southeastern Washington. With elevations mostly above 1,280 m, the northern deserts are generally higher than the southern. The shrub-steppe of these elevations occupies roughly 60% of the region. The remaining 40% of the region is cloaked with woody vegetation, occurring on average at elevations above about 1,850 m (West and Young 2000).

Mean annual precipitation of the cool deserts generally ranges between 100 mm and 400 mm—the 229 mm average for the state of Nevada is perhaps a metaphor for the region (Wagner 2003). And given the higher latitudes and elevations, 60% of the annual precipitation typically falls as snow (West 1983). The extreme winters at the high elevations foreclose their use in winter; hence livestock are moved seasonally to the higher elevations in summer to utilize forage production at those levels, and then returned to the lower elevations for grazing in winter.

Three unique characteristics of the entire Intermountain region shape the character of rangeland management and livestock husbandry. One is the physiognomic diversity of the vegetation. The lower, "desert" elevations support a diversity of vegetative life-forms: annual and perennial grasses and forbs, succulents, shrubs, and small trees. These change at progressively higher elevations to lower slope grasslands, "pigmy" conifers, well-developed conifer forests, and alpine tundra. This diversity complicates the ecological effects of grazing on the vegetation, and hence management strategies.

A second characteristic is the physiographic diversity itself. This makes possible the seasonal shifting of livestock grazing described above.

A third unique characteristic of the region is the prevalence of publicly owned land. Half of the region's area is in federal ownership, including 86% of Nevada and approximately two-thirds each of Utah and Idaho. An additional 5–6% of the region is state-owned. Some 89% of the federal land is managed by two agencies: The Bureau of Land Management (BLM) in the Department of the Interior, and the Forest Service in the Department of Agriculture. The BLM, managing 60% of the two agencies' total area, oversees primarily the lower elevations. The Forest Service is responsible for the national forests on the higher elevations of the mountain ranges.

A common ranching operation centers on home ranches with limited (e.g., hundreds) acres of private land. Cattle are transported to high-elevation national forest or BLM lands for grazing between spring and fall. They may winter on low-elevation BLM lands and/or on the home ranch where they are fed supplemental forage. And while Donahue (1999) points out that only 22% of beef producers in the eleven western states graze animals on public land, most of that land supports grazing. It is this livestock grazing on public lands that has elicited most of the range-oriented policy actions during the last half century.

California Grasslands

California west of the Sierra Madre cordillera is obviously part of the mountainous western United States, and it contains a large amount of federal land (37% of the state). But I am separating its grasslands from the remainder in the United States because of its climatic differences and unique vegetation changes. Precipitation levels range from semiarid in the south to subhumid in the north, and a major climatic difference is the moderating effect on temperatures of the maritime influences. Its grasslands, once composed largely of native, perennial grasses, are now dominated by exotic, annual grasses. The type is commonly referred to as the California annual grasslands.

Based on a number of early sources, Laura F. Huenneke (1989) generalizes that the area of California grasslands prior to European arrival was approximately eight million hectares, an area equivalent to 37% of the states of Utah and Idaho. She further concludes that the grasslands were located largely in the California Central Valley.

Huenneke estimates that the area of the contemporary grasslands is roughly comparable with that of the pre-European period. But the Central Valley is now largely cultivated, and many of the grasslands today occupy terrain formerly covered with other vegetation—*Atriplex* scrub, chaparral, riparian and coastal types, and oak woodlands—that has been altered by clearing, burning, and grazing.

The ecological causes of the conversion from native grasses to exotic annuals have received a great deal of study, and livestock grazing has been invoked as one likely major cause. I have shown more than a twofold increase in beef-cattle numbers (cattle in feedlots excluded) in the latter half of the twentieth century (Wagner 1989), and suggested that the native perennial grasses did not coevolve with large numbers of grazing ungulates, which left them vulnerable to introduced livestock. The 2.5 million cattle in the state in the late 1980s exceeded cattle numbers in the Intermountain portions of any of the western states.

GRAZING ECOLOGY AND ASSOCIATED GRAZING SYSTEMS

Grasslands

Effects on Plant-Community Structure

Successional Models   As mentioned above, the consideration of range management in a plant-ecological context began with the studies of Frederick Clements (1928). He emphasized vegetational changes, termed succession, that take place over time on bare areas when freed from disturbance. Vegetation changes from a predominance of annual, aggressive, largely herbaceous species, through stages of perennial forms, to self-sustaining, largely perennial plant communities termed climax states. Disturbance, including grazing, can push vegetation back down the scale, the degree of change depending on the intensity of disturbance. When disturbance is removed, the vegetation moves back up the scale.

E. J. Dyksterhuis (1946, 1949), working in the grasslands of northern Texas, adapted the Clementsian paradigm into a formal range-management conceptual model. His model was based on the premise that climax species were most palatable to livestock and that palatability declined progressively down the successional scale. He also theorized that when cattle were placed in a previously ungrazed, climax vegetation, they first grazed the climax species. These were the first to decrease, thus they were termed "decreasers." They were replaced by somewhat less palatable "increaser" species. If grazing pressure was intense and/or continuous, the animals would turn to the increaser species, reducing their abundance and opening the plant community to low-palatability, early-successional forms termed "invaders."

How far down the scale the system is moved is considered to be a function of the intensity and duration of grazing. And the assumption is generally present for grasslands that easing or release from grazing allows the system to move back toward, or to, the climax (decreaser) stage.

The basic ecological process mediating the interaction between grazing and the vegetation composition is considered to be competition among the plant species. The climax or decreaser stage is assumed to be composed of the most competitive species, which exclude the increaser and invader forms. Grazing the climax forms reduces their competitive advantage and allows intrusion of the lower-successional species. Removal of grazing restores the competitive prowess of the decreasers, which then resume their dominance in the community.

This conceptual scheme has been widely adopted by the range-management profession. And a set of criteria has been developed to characterize range condition according to four classes—excellent, good, fair, or poor—depending on the proportions of plant species in each of Dyksterhuis's three categories. These concepts prevail in much of range management to the present (Milchunas, Sala, and Lauenroth, 1988).

If the management objective for a given rangeland area is to derive maximum sustainable beef production (I will discuss other objectives below), the operator's challenge as viewed in the Dyksterhuis paradigm is to ascertain the maximum number of animals that will alter the vegetation to some intermediate seral stage that is maximally and sustainably productive over time. Grazing too few animals may reduce beef production below the system's potential. Grazing too many may degrade the vegetation and soils, ultimately reducing the area's productivity.

The process of arriving at a framework for such maximization is extremely complex. It will vary for different areas with different climates and ecologies. It may depend on the lengths and timing of grazing and rest periods that enable the vegetation to recover. It may depend on the proper seasons for grazing—vegetations are often vulnerable to damage when grazed during the growing season. Vegetations tend to be more sensitive during drought periods and stocking rates may need to be reduced.

As a result of this complexity, there have been a large number of experimental grazing studies, and numerous grazing systems are in operation. As examples, Rodney K. Heitschmidt and C. A. Taylor Jr. (1991) show (1) deferred rotation involving four pastures with alternating eight- to ten-and-a-half-month grazing interspersed with rest periods; (2) three-pasture rest-rotation systems with two- to ten-month grazing periods and two- to twelve-month rest periods; (3) high-intensity, low-frequency eight-pasture systems with three two-week grazing periods per year alternated with three-and-a-half-month rest periods; and (4) short-duration, eight-pasture systems with eight one-week grazing periods alternating with eight six-week rest periods per year.

Despite this range of alternatives, Heitschmidt and Taylor comment that "the claim that sustainable rates of stocking in [short duration] systems are well above those under continuous year-long grazing environments" is not supported by the evidence (1991, 176). Furthermore, "no grazing system has been shown to be universally superior to any other in terms of its ability to enhance livestock production" (177). Rex Pieper and Heitschmidt (1988) aver that rotational grazing systems are generally no more beneficial than moderate, continuous grazing.

Lauenroth et al. (1994) interpret the extensive results of grazing studies in an evolutionary context, pointing out that the Great Plains vegetation has coevolved over at least the past 10,000 years with large numbers of grazing ungulates. They suggest that the relatively modest response of short- and mid-grass communities to grazing reflect millennia of adaptation to this use. Richard N. Mack and John N. Thompson (1982) pose a similar argument by pointing out the morphological characteristics of Great Plains perennial grasses (rhyzomatous growth form) that enable them to spread vegetatively and make them less dependent on flowering and seeding for reproduction. Moreover, some Great Plains species form cleistogenes, seeds produced in underground leaf folds that are invulnerable to grazing. The implication is that these morphological characteristics coevolved with heavy herbivory, and Lauenroth et al. infer that cattle are ecological equivalents of native, grazing ungulates.

Lauenroth et al. (1994) conclude that the primary goal of a management plan in the Great Plains is the sustainability of resource production. Cultivated agriculture is the second most extensive land use in the region, and global population growth is likely to exert pressures—and provide financial incentives—to expand crop production. But the authors point out that cultivation increases breakdown of soil organic matter and loss of nutrients and exposes soils to erosion. Consequently, the most critical management decision, with regard to long-term effects on the ecosystems of the Great Plains, is the conversion of a grassland to cropland. Grazing should be the predominant use of the land for long-term sustainability (Lauenroth et al. 1994).

Holistic Resource Management　　Allan Savory (1988) spent his early professional years as a wildlife biologist in what was then known as Rhodesia. He observed a common tendency for wild ungulates to aggregate in large herds as a protection against predators. As a result, he inferred the following grazing pattern: the animals grazed an area intensively as a result of their numbers and then promptly moved on. The result was virtually complete utilization of the vegetation in a localized area, after which the animals' departure allowed vegetative regrowth. The concentration of animals also scarified the soil through hoof action and provided an environment for seedling germination and growth. The utilization of all the plant species gave none a competitive advantage, and community composition therefore did not change.

When Savory began examining American rangelands, he considered all of them to be mismanaged. Vegetation, in his observations, was patchy. Some patches were heavily grazed; others had remnants of tall, drying grasses that were too coarse to be eaten, and that, if grazed earlier, could have contributed to beef weight gains. Primary production was being wasted. He attributed this pattern to the scattered distribution of cattle with casual feeding and movement over time and space. They tend to consume the palatable species and leave the less palatable forms that, freed of competition, increase in abundance.

Savory advocates a cattle-grazing system similar to the wild-ungulate patterns he saw in Africa, and commonly termed short-duration, high-intensity use. A rangeland area is divided into a number of pastures. A large number of cattle are introduced into a pasture and grazed intensively for a short period of time, but long enough to graze off all the vegetation. They are then moved to the next pasture for the same complete use, and then moved again, and so on. The vegetation is assumed to regrow after each grazing period.

The Savory system appeals to many American ranchers who traditionally have restricted animal numbers under the constraints of light to moderate grazing implicit in the succession models. This new system provides a rationale for increasing stocking rates and, hopefully, greater beef production and profits than under the succession models. Savory has established an office in New Mexico. He lectures and provides advice for ranchers throughout the West. N. E. West has commented that Savory has had more impact on livestock husbandry in the United States in the past quarter century than the entire range-management professional field (pers. comm., 2004).

But professional ecologists in the United States are generally critical of the Savory grazing model. There is no substantial body of scientific evidence attesting to the virtues of the approach—in fact, Savory scoffs at any scientific evaluation of his system, commenting that is it not possible. Heitschmidt and Taylor comment: "We know of no data in support of this hypothesis...This generalization does not support...the claim that sustainable rates of stocking in SD [short duration] type systems are well above those under continuous grazing in year-long grazing environments...the major benefit derived from the employment of a grazing system is directly related to the forced use of moderate rates of stocking" (1991, 161).

However, a number of my colleagues in the range profession (e.g., T. W. Box, pers. comm., 2004) comment that the major contribution of Savory's system is that it forces ranchers to examine and more critically evaluate the details of their operations, including careful scrutiny and monitoring of their vegetation. Moreover, Savory's system is termed "holistic resource management." He exhorts ranchers to reflect deeply on their personal, social, and economic goals and to fashion their ranching operations in the context of these goals.

Effects on Plant-Community Production

Investigating the effects of grazing by large herds of wild ungulates on the primary production of grasslands in Serengeti National Park, Tanzania, Samuel J. McNaughton (1979a, 1979b) measured ANPP associated with different rates of forage utilization, and in exclosures protected from grazing. McNaughton's measurements showed maximum ANPP at intermediate grazing intensities, which he termed optimization.

He hypothesized that the pattern reflected a coevolved mutualism between grazers and vegetation that resulted in both maximum production of the vegetation and maximum forage for, and maximum secondary production by, the grazers. The implication follows that, up to a certain point, grazing is beneficial to plants—a process termed overcompensation, in which grazed plants produce more herbage than the same plants would produce if ungrazed.

A number of studies have investigated whether grazing optimization occurs in North American vegetation, and some authors have inferred this. For example, Douglas A. Frank (Frank, Kuns, and Guido, 2002; Frank and McNaughton 1993) has inferred grazing optimization in wild-ungulate effects on grassland vegetation in Yellowstone National Park. However, I have questioned whether the conclusion is supported by the evidence (Wagner 2006).

The subject has become quite controversial, eliciting critical reviews (cf. Belsky 1986; Patten 1993) and an entire symposium on the subject published in volume 3, issue 1, of *Ecological Applications*. The reality of overcompensation in North American vegetation is largely rejected by North American range ecologists, first because the concept of grazing as a beneficial effect on grasses contradicts the prevailing paradigm that defoliation has a negative influence on the plants' physiology and ecology (Caldwell et al. 1981), and second because the mass of the evidence does not support it. Daniel G. Milchunas and W. K. Lauenroth (1993) reviewed 236 grazing studies and found evidence of overcompensation in only 17% of the cases. Brian J. Wilsey, James S. Coleman, and Samuel McNaughton (1997) observed that the established cases were from outside the Western Hemisphere.

David D. Briske and Rodney Heitschmidt conclude that "it does not appear to be a significant ecological process operating on a regular basis in grassland systems" (1991, 19). And Heitschmidt and Taylor conclude that "it is...a well-researched, refuted hypothesis...relative to the general effects of grazing on plant growth in multi-species arid, and semi-arid rangelands" (1991, 176). However, T. W. Box (pers. comm., 2004) suggests that it may occur in moist, southern vegetations with long growing seasons, such as the Gulf of Mexico coastal prairie.

Lauenroth et al. (1994), generalizing the effects on grasslands across the west-east precipitation gradient in the Great Plains, commented that ANPP usually declines with grazing except in the high-precipitation, eastern tallgrass prairie. Here, ANPP often increases with some grazing, not because of any overcompensating response by the plants' physiology, but because grazing allows more light penetration into the community and reduces dense litter formation that physically restrains plant growth.

Multiple Life-Form Vegetation

The evolutionary, ecological, and policy context for livestock grazing in the Intermountain West is quite different from that in the Great Plains. A growing body of evidence is now converging on the view that wild ungulates occurred in low numbers in the region prior to European contact (for syntheses, cf. Keigley and Wagner 1998; Wagner 2006). While a number of authors have been writing in this vein, Charles E. Kay (1990, 1994, 2002; Kay et al. 2000) has most comprehensively and aggressively explored the question by quantifying the archaeological and historical evidence for both the United States and Canada. His paradigm is one of low ungulate numbers dating back to the last glacial retreat, and aboriginal hunting as a major contributor to the low numbers.

Kay's paradigm resonates with the prevailing view among Intermountain range ecologists that wild ungulates were not numerous in pre-Columbian times. Hence the Intermountain vegetation had not coevolved with ungulate herbivory, and it was therefore sensitive to, and damaged by, the introduction of European livestock (Donahue 1999; Lofton et al. 2000). This paradigm is also supported by M. M. Caldwell's physiological investigations of Intermountain grasses, which showed them to be sensitive to defoliation (Caldwell et al. 1981), and by Mack and Thompson's (1982) observation that Great Basin perennial grasses, unlike Great Plains species,

are primarily erect, tussock forms that depend on flowering and seeding for reproduction and therefore are vulnerable to grazing. Those authors also noted the scarcity
of dung beetle species in the Intermountain region.

Ecologically, the vegetation of the region is comprised of numerous life-forms:
grasses and forbs, both perennial and annual; cactaceous and liliaceous succulents of
varying statures; shrubs and trees of varying sizes. The combinations vary with elevation, grading from xeric types in the arid, low altitudes, often through grasslands
on the foothills, to varying forest types at higher levels. As a result, the competitive
interactions determining community structures are much more complex than those
in grasslands with their relatively uniform physiognomies.

Prehistorically, fire, commonly set by native peoples, and the relative scarcity of
ungulates maintained perennial grasslands and reduced woody species (cf. Lofton
et al. 2000). But the arrival of Europeans and their actions have sent the vegetation
in varying directions, depending on the site and factors brought to bear, including
introduction of livestock, increase of native ungulates afforded by protective laws,
and suppression of fires.

The nature of the ungulates plays a role. Grazers (e.g., bison, cattle) may impair
the competitive ability of herbaceous species and shift vegetative composition to
woody and succulent forms. Browsers (e.g., deer, pronghorn) produce alternate
trends, reducing woody species and favoring herbaceous.

Thus, over much of the West, original introduction of cattle reduced perennial
grasses and shifted much of the vegetation toward a dominance of woody species: mesquite (*Prosopis juliflora*), creosote bush (*Larrea tridentata*), and sagebrush
(*Artemisia* spp.) at the lower elevations; junipers (*Juniperus* spp.), pinyon (*Pinus* spp.),
and others on the foothills (cf. Archer 1994; Tausch 1999). S. Archer (1994) points
out that a number of factors have contributed to the change, but considers livestock
grazing to have been the overriding influence. It functions both to alter the competitive interactions within the plant community and to remove the fine fuels produced
by the herbaceous vegetation that carries fires and inhibits growth of woody species.

In recent decades this trajectory has been reversed in many areas. On foothill
ranges of the Great Basin where cattle have been removed, and where wintering
deer populations have increased under protective laws, the deer have browsed out
the shrubs and perennial grasses have increased, virtually converting the sites to
grassland.

Season of use plays a role. Research at the Department of Agriculture's Dubois,
Idaho, research station has shown that when grazed in summer, sheep feed primarily
on herbaceous vegetation and shift community structure to a dominance of sagebrush
(Laycock 1967). When grazing is deferred to fall, and herbaceous vegetation becomes
dormant and of low quality, the sheep browse sagebrush more intensively, reduce its
presence in the vegetation, and shift vegetation structure toward grassland.

Yet another complicating factor is the inadvertent introduction of nonnative
annual species. Cheatgrass (*Bromus tectorum*) in the Great Basin is perhaps the most
notorious. The intact, perennial vegetation can thwart its growth. But disturbance
such as grazing, especially if it eliminates the perennial grasses, opens the vegetation
to cheatgrass invasion. It now grows as an understory in sagebrush stands, and as a
winter annual that dries by the end of spring and provides fine fuel for fires, human or

lightning set. Sagebrush is highly sensitive to fire and is eliminated by a cheatgrass-fueled fire. At that point, cheatgrass thickens into a carpet providing annual fuel that burns periodically, facilitates the spread of fire to adjacent areas, and blocks recovery of native vegetation. The result is permanent vegetation change. P. A. Knapp (1996) reported that 20% of Nevada sagebrush steppe has been converted to cheatgrass.

Thus much Intermountain vegetation has undergone change that is essentially irreversible, and the Great Plains paradigm of removing livestock to allow the vegetation to return to pregrazing condition does not apply. W. H. Kruse and R. Jemison comment that "many traditional livestock grazing systems simply do not fit" (2000, 32). And the changes have prompted development of new descriptive models termed "state and transition models" (cf. Westoby, Walker, and Noy-Meir, 1989).

The changes impose added operational costs on ranchers and public agencies. Forage production may be reduced and management costs may involve mechanical, herbicidal, and fire control of brush; reseeding; and control of escaped, introduced forage species as well as unpalatable, invasive species. Given all of this ecological complexity—including ungulate type, season-of-use considerations, low- and high-elevation ranges, and private versus public lands—the design of grazing systems becomes exceedingly complex. With vegetation that did not coevolve with heavy ungulate use, it is not clear that grazing can be a sustainable form of land use in the Intermountain West as described above for the Great Plains. These uncertainties and the marked changes in vegetation have elicited policy pressures from a variety of advocacy groups, and will be described next.

CHANGING PUBLIC RANGELAND POLICIES DRIVEN BY SHIFTING SOCIAL VALUES

Perspective

As noted at the beginning of this chapter, range-management policies have largely consisted of (1) federal legislation and executive actions applied to that half of the eleven western states in federal ownership and (2) broader environmental legislation applicable to public rangelands. Private ranchers pay grazing fees to the federal government for grazing their livestock on these lands for portions of the year, especially on national forest and BLM lands and to a lesser degree on some National Park Service units, national wildlife refuges, and others. Federal rangeland policies are set to address this use and are not directed at private land, although some environmental legislation (e.g., the Endangered Species Act) does apply to the latter. Some federal legislation has been directed to private-land ranchers in the form of attachments to the annual farm bills to provide drought relief, assistance in erosion control, reseeding, etc. But what I am here terming rangeland policies are those directed to public land use in the West.

These policies have emerged and changed over time as values of the American public attached to public lands in particular, and environmental issues more generally, have changed and been translated into political action. For the purposes of this brief review, I divide the period from the early 1900s to the present into three overlapping eras of public values and associated policy actions. Donahue (1999) provides an excellent and much more detailed review of this entire subject.

The Agrarian Era: 1900–1960

As the eleven western states achieved statehood (all before 1900 except for New Mexico and Arizona in 1912), the federal government retained ownership of what had been territories, but strove to divest itself of the land through homestead and land grants to the private sector. Of each thirty-six square miles, three or four were granted to each of the states as state trust lands, established to support education. The government further appropriated some of the federal land into national forests, national parks, federal wildlife refuges, and military reservations. But there was no demand for the remainder because of its aridity and lack of agricultural potential. As late as 1930, President Herbert Hoover offered this remainder to the states, and all declined to accept it. Hence it remained in federal ownership and was termed the unappropriated public domain (UPD).

In the late 1800s and the first decade of the twentieth century, the UPD and national forests, along with some of the other federal lands, were widely grazed by privately owned livestock without any regulation or controls. But with the explosion of livestock numbers around the turn of the century, concerns began to be expressed over the degradation of the public ranges. The nation was strongly agricultural at that point in its history, and the major public values attached to the public lands were utilitarian: mining, timber production, and, in the case of rangelands, continued production of livestock products, soil conservation, and protection of water quality. The realization grew that it was necessary to impose policy constraints on public land grazing to preserve those values.

In 1905, the Forest Service began charging ranchers a fee for livestock use in the national forests. By 1910, the agency began controlling grazing in the forests. After two failed attempts, Congress passed the Taylor Grazing Act of 1934, which divided the public ranges into districts, and established a Division of Grazing, or Grazing Service, in the Department of the Interior to administer livestock use in the grazing districts of the UPD. The Grazing Service was now given responsibility for prescribing livestock numbers on the UPD districts, charging fees, and forming advisory councils in each state to oversee the Grazing Service's operations.

Immediately after passage of the Taylor Grazing Act, Congress assigned the Forest Service the task of conducting a survey of range conditions on public lands. The result, in 1936, was a landmark 600-page volume, *The Western Range,* which concluded that the western ranges were seriously degraded. In 1948, the Grazing Service was combined with the General Land Office, an agency that had been responsible for divesting the public lands, to form the Bureau of Land Management (again, the BLM).

Despite these actions, there was widespread concern that the condition of the public rangelands was continuing to decline. Six surveys, conducted at four- to eleven-year intervals between 1947 and 1994, all concluded that the ranges were in dire straits. A BLM review in 1975 classified its lands as 17% good or excellent, 50% fair, 28% poor, and 5% bad. The evidence was growing that both the Forest Service and the BLM were too much under the influence of the livestock industry, and lacked the power to force the significant reductions in livestock numbers required to halt range decline and permit recovery.

The Environmental Era: 1960s to Present

The roots of the conservation and environmental movement in the United States extend before 1900: establishment of the first national parks and forest reserves (later to become the national forests), the presidency of Theodore Roosevelt in the early 1900s, and the early writings of Aldo Leopold in the 1930s. But it did not expand into a major social movement that pressed for policy action until the 1960s.

By that time, the American population had changed from primarily rural to an urban society; with the change, the values society attached to the environment and to public lands shifted and diversified. After World War II, the population became more mobile, and large numbers of tourists traveled in the West to admire the scenic amenities of the region. As Box commented, "In the past two decades the demand for products—food, fiber, and fuel—from ranges has decreased. The desire for services and amenities has increased" (2003, 34).

New advocacy groups developed and began to make their voices heard in the public-land political process: environmental organizations; hunting, fishing, wilderness, and other outdoor groups; wild-horse advocacy groups, etc. These organizations demanded not only that their values be given equal consideration alongside livestock management in policy setting, but that they be given a voice in decision making. They concluded that livestock grazing by privately owned animals was degrading rangelands owned by the entire public and was competing with other valued resources.

The landmark environmental legislation that legitimized public participation in environmental policy making was the National Environmental Policy Act (NEPA) of 1969. This act required federal agencies to conduct environmental impact analyses of likely consequences of proposed actions affecting public assets. The agencies were then required to seek public comment on these consequences. NEPA was followed four years later by the Endangered Species Act (ESA) of 1973, which actually restrained actions on both public and private land that might threaten the survival and habitat of threatened and endangered plant and animal species.

NEPA and the ESA were followed in rapid succession by a number of policy actions in the 1970s that addressed public rangelands. The Forest Service had already been ordered by the Multiple-Use Sustained-Yield Act of 1960 to broaden its management commitment to all the resources on the national forests. The agency was further instructed by the Forest and Rangeland Renewable Resources Planning Act of 1974 to engage in long-term planning for the management of each of the national forests, and by the National Forest Management Act of 1976 to include representatives of all concerned public interests in the planning process.

In 1974, the Natural Resources Defense Council (NRDC)—a private legal organization that advocates legal action on environmental issues—brought suit against Secretary of the Interior Rogers Morton, demanding that the BLM produce environmental impact statements (EIS), under the auspices of NEPA, on all of its grazing districts before issuing grazing permits to ranchers. The NRDC prevailed in the action. And in 1976, Congress passed the Federal Land Policy and Management Act, instructing the BLM to devote equivalent management attention to all the resources on its lands (much like the instructions to the Forest Service in the Multiple-Use Sustained-Yield Act). Other legislation followed.

All of this action gained momentum through the late 1900s, with some organizations demanding removal of livestock from public lands. Automobile bumper stickers declared "Cattle Free by '93." Agency actions were stymied in the courts with lawsuits brought by the contending interests.

The most detailed and scholarly advocacy in this vein was a 1999 book by Debra Donahue titled *The Western Range Revisited: Removing Livestock from Public Lands to Conserve Native Biodiversity*. Donahue, a University of Wyoming professor of law at the time of the writing, had two degrees in wildlife biology, a J.D., and work experience in three federal agencies and the National Wildlife Federation. In her book, she analyzes in detail the history of livestock rearing and policy in the West. She reviews the decades of surveys on range conditions and concludes that the western public rangelands are degraded, threatening the survival of a number of threatened and endangered plant and animal species and the diversity of the biota as a whole.

Donahue further notes that by 1977 only 18% of all U.S. livestock were produced in the eleven western states, and only 2% of all livestock in the United States are produced on public lands. Moreover, only 22% of western ranchers graze their animals on public lands, and the financial straits of the ranching community are so difficult that members of many ranching families seek employment outside the ranches in order to assist the family economies. Given this minor contribution of public lands to national livestock production, the limited contribution of grazing on public lands to the ranching economy, and Donahue's assessment of the negative effects of livestock grazing, she advocates termination of all livestock use on public land where annual precipitation is less than 12 inches (305 mm).

Contemporary Second Thoughts

The western U.S. livestock industry has long operated on a thin profit margin. As commented above, the sheep industry has virtually disappeared because of high operating costs, low U.S. demand for lamb, and competition from synthetic fibers for wool. Cattle ranching is similarly hard pressed, with J. Winder (1999) citing typical net return on assets at only 2–3%. A number of surveys have shown that many ranchers stay in the business primarily because they value the lifestyle (Starrs 1998).

Coincident with this economic plight, the western United States is experiencing the fastest population growth of any region in the nation. Most of the growth is in urban areas. With that growth, land values are rising, and suburban sprawl leapfrogs out into what have been ranching areas. Ranchland is purchased by real estate developers, and by affluent individuals who build upscale, rural retreats on large tracts of land. Ranchers now see their properties as newly appreciated assets that could be sold to ease their economic plight.

In response to these negative pressures and positive incentives, the western ranching industry is declining. Peter Decker (1998) describes a decline in the number of ranches in Ouray County, Colorado, from seventy to forty between the 1970s and the present. The number of ranchers using BLM lands in the West shrank by a third between the 1950s and 2000 (Wagner and Baldwin 2003). Some economists predict the end of public-land ranching within fifty years.

These trends are now being viewed with alarm by a number of groups, including environmental and outdoor advocacy organizations. The realization is growing that ranching is a bulwark against the pell-mell suburban sprawl and real-estate development that intrude on the largely unpopulated, wilderness character of the region. That character is a major component of the quality of life for both rural and urban residents, and it is a major attraction for the tourist influx that contributes to the region's economy.

As a result, a number of policy initiatives are being developed to strengthen the economic condition of the ranching community and help perpetuate the industry. One is a tax provision termed "conservation easements." In return for placing his or her land in a permanent easement, including all subsequent owners, a rancher is given a charitable income-tax deduction equivalent to the difference between the appraised agricultural value of the land and what he or she could realize by selling the land to a developer (Decker 1998). The arrangement also has the advantage of keeping the land value low, thereby reducing the estate taxes which the rancher's heirs must pay upon his or her demise.

While the easements provide clear economic advantages, they do not supply the continuing cash flow that ranchers need to supplement their thin profit margins. A number of organizations are purchasing "development rights" from ranchers: paying them cash for the difference between the agricultural value of the land and what a developer would pay to acquire it. Some of these are private groups of local residents—Philip Huffman (2000) describes such a Colorado Cattlemen's Agricultural Land Trust. Some are governmental—Decker (1998) discusses Great Outdoors Colorado, a state agency that receives its funding from the tax proceeds of the state lottery. Sherman Janke (pers. comm., 2000) states that the Montana legislature has appropriated funds to buy development rights in that state.

These measures are designed to improve the financial status of ranchers by raising the value of their private-land assets as one means of perpetuating the industry. But they do not address the negative pressures imposed by advocacy groups that oppose livestock use on the public lands, and the resulting legal logjams that inhibit policy action by the public agencies.

However, these groups also now recognize the risk of development in western, open lands. The risk is not in development on the public lands; rather, it is in residential and commercial development on private lands scattered among the public lands. The advocacy groups now recognize that perpetuation of much of the livestock industry depends on continuing use of public lands by many ranchers. Donahue (1999), as noted above, stated that 22% of cattle ranchers in the West are using the public lands. But the percentage is higher in a number of states: 49% in Nevada; 35% each in Utah, Arizona, and New Mexico; and 37% in California.

This recognition is now promoting a willingness among the contending groups to compromise in establishing livestock grazing systems that are ecologically sustainable, yet economically feasible, for the ranchers. The groups come together in voluntary coalitions of environmental and outdoor organizations, ranchers, and public agencies that work together to design mutually agreed-upon land-use plans. Participants actually contribute time and effort by fencing land, improving streams, modifying vegetation, and carrying out other physical actions that help ranchers improve their holdings.

One such group in western New Mexico calls itself "The Radical Center" (Taylor 2004). The Quivira Coalition, a three-person group in Santa Fe, is promoting other, similar arrangements. Still others are the Malpai Borderlands Group in southeastern Arizona and the Diablo Trust in northern Arizona. Hundreds of such groups are forming in the West. All are voluntary and locally developed. None is directed or authorized from above by government agencies.

Whether or not these economic and collaborative stewardships and decision-making projects will succeed in preserving ranching in the West—while at the same time improving the overall condition of western rangelands—remains to be seen. But they do appear to be moving the issue of range management on public lands beyond conflict, controversy, and litigation.

References

Archer, S. 1994. Woody plant encroachment into southwestern grasslands and savannas: Rate patterns, and proximate causes. In *Ecological implications of livestock herbivory in the West,* ed. M. Vavra, W. A. Laycock, and R. D. Pieper, 13–68. Denver, CO: Society for Range Management.

Belsky, A. J. 1986. Does herbivory benefit plants? A review of the evidence. *American Naturalist* 127(6): 870–892.

Box, T. W. 2003. The last 25 years: Changes and reflections. *Rangelands* 25(6): 31–35.

Briske, D. D., and R. K. Heitschmidt. 1991. An ecological perspective. In *Grazing management: An ecological perspective,* ed. R. K. Heitschmidt and J. W. Stuth, 11–26. Portland, OR: Timber Press.

Caldwell, M. M., J. H. Richards, D. A. Johnson, R. S. Nowak, and R. S. Dzurec. 1981. Coping with herbivory: Photosynthetic capacity and resource allocation in two semi-arid *Agropyron* bunchgrasses. *Oecologia* 50(1): 14–24.

Clements, F. E. 1928. *Plant succession and indicators: A definitive edition of plant succession and plant indicators.* New York: H. W. Wilson.

Decker, P. C. 1998. *Old fences, new neighbors.* Tucson: University of Arizona Press.

Donahue, D. L. 1999. *The western range revisited: Removing livestock from public lands to conserve native biodiversity.* Norman: University of Oklahoma Press.

Dyksterhuis, E. J. 1946. The vegetation of the Fort Worth prairie. *Ecological Monographs* 16(1): 1–29.

Dyksterhuis, E. J. 1949. Condition and management of range land based on quantitative ecology. *Journal of Range Management* 2(3): 104–115.

Frank, D. A., M. M. Kuns, and D. R. Guido. 2002. Consumer control of grassland plant production. *Ecology* 83(3): 602–606.

Frank, D. A., and S. J. McNaughton. 1993. Evidence for the promotion of aboveground grassland production by native large herbivores in Yellowstone National Park. *Oecologia* 96(2): 157–161.

Heitschmidt, R. K., and C. A. Taylor Jr. 1991. Livestock production. In *Grazing management: An ecological perspective,* ed. R. K. Heitschmidt and J. W. Stuth, 161–177. Portland, OR: Timber Press.

Huenneke, L. F. 1989. Distribution and regional patterns of California grasslands. In *Grassland structure and function: California annual grassland,* ed. L. F. Huenneke and H. A. Mooney, 1–12. Dordrecht, Netherlands: Kluwer.

Huffman, P. 2000. Caretakers of the land: An interview with rancher Lynne Sherrod. *Orion Afield* 4(3): 18–21.

Kay, C. E. 1990. Yellowstone's northern elk herd: A critical evaluation of the "natural regulation" paradigm. Ph.D. diss., Utah State University.

Kay, C. E. 1994. Aboriginal overkill: The role of Native Americans in structuring western ecosystems. *Human Nature* 5(4): 359–398.

Kay, C. E. 2002. Are ecosystems structured from the top-down or bottom-up? A new look at an old debate. In *Wilderness and political ecology: Aboriginal influences and the original state of nature,* ed. C. E. Kay and R. T. Simmons, 215–237. Salt Lake City: University of Utah Press.

Kay, C. E., B. Patton, and C. A. White. 2000. Historical wildlife observations in the Canadian Rockies: Implications for ecological integrity. *Canadian Field-Naturalist* 114(4): 561–583.

Keigley, R. B., and F. H. Wagner. 1998. What is "natural"? Yellowstone elk population: A case study. *Integrated Biology* 1(4): 133–148.

Knapp, P. A. 1996. Cheatgrass (*Bromus tectorum*) dominance in the Great Basin Desert: History, persistence, and influences to human activities. *Global Environmental Change* 6(1): 37–52.

Kruse, W. H., and R. Jemison. 2000. Grazing systems of the Southwest. In *Livestock management in the American Southwest: Ecology, society, and economics,* ed. R. Jemison and C. Raish, 27–52. Amsterdam: Elsevier.

Lauenroth, W. K., D. G. Milchunas, J. L. Dodd, R. H. Hart, R. K. Heitschmidt, and L. R. Rittenhouse. 1994. Effect of grazing on ecosystems of the Great Plains. In *Ecological implications of livestock herbivory in the West,* ed. M. Vavra, W. A. Laycock, and R. D. Pieper, 69–100. Denver, CO: Society for Range Management.

Laycock, W. A. 1967. How heavy grazing and protection affect sagebrush-grass ranges. *Journal of Range Management* 20(4): 206–213.

Lofton, S. R., C. E. Bock, J. H. Bock, and S. L. Brantley. 2000. Desert grasslands. In *Livestock management in the American Southwest: Ecology, society, and economics,* ed. R. Jemison and C. Raish, 53–96. Amsterdam: Elsevier.

Mack, R. N., and J. N. Thompson. 1982. Evolution in steppe with few large, hooved mammals. *American Naturalist* 119(6): 757–773.

MacMahon, J. A., and F. H. Wagner. 1985. The Mojave, Sonoran and Chihuahuan deserts of North America. In *Ecosystems of the world 12: Hot deserts and arid shrublands,* ed. M. Evenari, I. Noy-Meir, and D. W. Goodall, vol. A, 105–202. Amsterdam: Elsevier.

McNaughton, S. J. 1979a. Grassland herbivore dynamics. In *Serengeti: Dynamics of an ecosystem,* ed. A. R. E. Sinclair and M. Norton-Griffiths, 46–81. Chicago: University of Chicago Press.

McNaughton, S. J. 1979b. Grazing as an optimization process: Grass-ungulate relationships in the Serengeti. *American Naturalist* 113(5): 691–703.

Milchunas, D. G., and W. K. Lauenroth. 1993. Quantitative effects of grazing on vegetation and soils over a global range of environments. *Ecological Monographs* 63(4): 327–366.

Milchunas, D. G., O. E. Sala, and W. K. Lauenroth. 1988. A generalized model of the effects of grazing by large herbivores on grassland community structure. *American Naturalist* 132(1): 87–106.

Patten, D. T. 1993. Herbivore optimization and overcompensation: Does native herbivory on western rangelands support these theories? *Ecological Applications* 3(1): 35–36.

Pendleton, D. G. 1978. Nonfederal rangelands of the United States: A decade of change, 1967–1977. In *Proceedings of the First International Rangeland Congress, Denver, Colorado, USA, August 14–18, 1978*, ed. D. N. Hyder, 485–487. Denver, CO: Society for Rangeland Management.

Pieper, R. D., and R. K. Heitschmidt. 1988. Is short-duration grazing the answer? *Journal of Soil and Water Conservation* 43(2): 133–137.

Savory, A. 1988. *Holistic resource management.* Washington, DC: Island Press.

Starrs, P. F. 1998. *Let the cowboy ride: Cattle ranching in the American West.* Baltimore, MD: Johns Hopkins University Press.

Tausch, R. J. 1999. Historic pinyon and juniper woodland development. In *Proceedings: Ecology and management of pinyon-juniper communities within the interior West*, 12–19. Rocky Mountain Research Station Proceedings, no. RMRS-P-9. Fort Collins, CO: U.S. Department of Agriculture, Forest Service.

Taylor, M. 2004. Notes from the radical center. *Forest Magazine* 6 (Fall): 39–43.

U.S. Forest Service. 1936. *The Western range: Letter from the secretary of agriculture transmitting in response to Senate resolution no. 289; a report on the Western range—a great but neglected natural resource.* Washington, DC: Government Printing Office.

Wagner, F. H. 1978. Livestock grazing and the livestock industry. In *Wildlife and America*, ed. H. P. Brokaw, 121–145. Washington, DC: Government Printing Office.

Wagner, F. H. 1989. Grazers, past and present. In *Grassland structure and function: California annual grassland*, ed. L. F. Huenneke and H. A. Mooney, 151–162. Dordrecht, Netherlands: Kluwer.

Wagner, F. H. 2003. Natural ecosystems III: The Great Basin. In *Preparing for a changing climate: The potential consequences of climate variability and change*, ed. F. H. Wagner, 207–239. Logan: Utah State University.

Wagner, F. H. 2006. *Yellowstone's destabilized ecosystem: Elk effects, science, and policy conflict.* New York: Oxford University Press.

Wagner, F. H., and C. K. Baldwin. 2003. Cultivated agriculture and ranching. In *Preparing for a changing climate: The potential consequences of climate variability and change*, ed. F. H. Wagner, 113–139. Logan: Utah State University.

West, N. E. 1983. Overview of North American temperate deserts and semi-deserts. In *Ecosystems of the world 5: Temperate deserts and semi-deserts*, ed. N. E. West, 321–330. Amsterdam: Elsevier.

West, N. E., and J. A. Young. 2000. Intermountain valleys and lower mountain slopes. In *North American terrestrial vegetation*, ed. M. G. Barbour and W. D. Billings, 2nd ed., 256–284. New York: Cambridge University Press.

Westoby, M., B. Walker, and I. Noy-Meir. 1989. Opportunistic management for rangelands not at equilibrium. *Journal of Range Management* 42(4): 266–274.

Wildeman, G., and J. H. Brock. 2000. Grazing in the Southwest: History of land use and grazing since 1540. In *Livestock management in the American Southwest: Ecology, society, and economics*, ed. R. Jemison and C. Raish, 1–25. Amsterdam: Elsevier.

Wilsey, B. J., J. S. Coleman, and S. J. McNaughton. 1997. Effects of elevated CO_2 and defoliation on grasses: A comparative ecosystem approach. *Ecological Applications* 7(3): 844–853.

Winder, J. 1999. Resolving resource conflicts: A bigger pie. *Proceedings of the Annual Meeting of the American Association for the Advancement of Science.*

10

Policy Failures in African Rangeland Development

Walter J. Lusigi

The ultimate objective of most development activities is improvement of human welfare. If ecological considerations are not adequately taken into account, the objective may not be realized.
> —Lee M. Talbot, "Ecological Aspects of Aid Programs in East Africa"

The persistent suffering of the people who live in the rangelands of Africa has been blamed chiefly on the deterioration of rangeland resources caused by natural disasters such as droughts and floods. Well-intended interventions—in rangeland development, water development, livestock marketing, education and health infrastructure development, irrigated agriculture, small-scale rural industries, fisheries, and tourism—have had mixed, largely negative results that have led to human suffering. I suggest that these negative trends have stemmed from public policies that have failed to acknowledge or address the complexity of the human/environment balance in African rangelands. Where there exists a human population, there are also bound to be complex ethnic, biological, and social influences, which, unless they are understood and incorporated into development policies, will lead to unpredictable outcomes for development efforts. The policies put in place for the development of African rangelands have primarily been outgrowths of experiences in development elsewhere that failed to consider the needs of the people whose livelihoods were immediately affected. It is unfortunate but true that despite over a half century of mixed results from such interventions, development experts have continued to assume uniformity of populations in both high-potential land areas and rangelands.

This chapter focuses on the linkages between public policy and rangeland utilization in Africa. In order to avoid overgeneralization I will use examples from English-speaking East Africa, with which I am most familiar. But with few exceptions these examples also apply to rangelands across Africa. Since rangeland degradation is the result of a long-term process, my chapter will in part take into account the history of rangeland resource use. This will involve an analysis which, if overly critical of policy makers, is not presented as an argument from ecologists; on the contrary,

I present it only as a constructive critique of the policy-making profession from within. The conservation classic "What Kind of Animal Is an Ecologist?" (1957) is Lee Talbot's unparalleled reflection on key ecological and developmental issues witnessed through his travels, which started at an early age and have extended through over fifty countries around the world. He concluded that "the health of the natural resources base of other countries of the world has become an important thread in the American ecological web" (1957, 6). This conclusion helped shape the foreign development policy of the United States concerning natural resources management in third world countries. Such attention to policy development is critical. For a common source of rangeland degradation is not only policy failure but a complete lack of sound policies. Poverty, for example, is an acknowledged cause of rangeland degradation, but poverty is itself caused by failure of governments to institute policies that lead to creation of employment opportunities and equitable distribution of national wealth. Likewise, "market failure" is really the failure of governments to create well-functioning markets. In this chapter, accordingly, "policy failure" will also be understood to mean lack of policies.

The economists who dominate the public policy arena have traditionally given more attention to market failure than to other linkages between public policy interventions and rangeland degradation. The bias of this chapter toward aspects of public policy that directly affect the biological basis for sustainable rangeland management is therefore deliberate. I will be looking specifically at policy issues related to land tenure, agricultural development, marketing of rangeland products, poverty reduction, population growth, infrastructure, and administration of rangeland areas.

THE CURRENT STATUS OF AFRICAN RANGELANDS

Land is an important resource on which Africa's economies are based. According to a recent environmental status report by the UN Environment Programme/New Partnership for Africa's Development (2003), agriculture contributes about 40% of regional gross domestic product (GDP) and provides livelihoods to about 60% of the population. In addition to growing subsistence crops for a large portion of Africa's population, there are increasing demands on land to produce cash crops for export, facilitating economic growth. Africa contains the world's largest expanse of dry lands, covering roughly two billion hectares of the continent, or 65% of Africa's total land area. One-third of this is hyperarid desert, while the remaining two-thirds consist of arid, semiarid, and dry subhumid areas—home to about two-thirds of the continent's population. Approximately 22% of the total land area is under forest, 43% is characterized as extreme desert, and 57% is vulnerable to desertification. Only 21% is suitable for cultivation. Although reliable data are lacking, it is estimated that some 500 million hectares of land in Africa have been affected by soil degradation since 1950, including as much as 65% of agricultural land. Approximately 50% of land degradation in Africa is from overgrazing, 24% from activities related to crop production, 14% from vegetation removal, and another 13% from overexploitation of the land. Wind and water erosion is extensive in many parts of Africa; about 25% of the land is prone to water erosion, about 22% to wind erosion. Information regarding rates of soil loss in Africa is varied and country specific, with estimates ranging from 900 t/km² per year to 7,000 t/km² per year. Likewise, studies of the economic impacts of soil loss are localized and varied, but these impacts are estimated to reach up to 9% of GDP.

LAND TENURE POLICY ISSUES

The failure of independent African nation-states to rectify the boundaries established during colonial rule ("divide and rule") is arguably the most fundamental cause of rangeland degradation in Africa. For when the present-day nation-states of Africa were being colonized, it was Europeans' deliberate policy to establish boundaries that divided major tribes, leaving them with no coherent power to challenge colonial authority. This affected rangeland people more than sedentary tribes because they controlled large expanses of territory across which they moved with their livestock. Pastoralism based on nomadism is a biological necessity for survival on low-potential rangelands affected by uneven distribution of rainfall and frequent droughts. The establishment of state boundaries across these territories, and the enforcement of regulations that limited nomadic movement, disrupted the basic fabrics and functions of rangeland societies. This can be said to be the real beginning of the deterioration of the rangelands. The establishment of state boundaries was accompanied by the declaration of land to be crown land (state land) and by introduction of new forms of land ownership patterns.

In Kenya, for example, orders and rules were formulated to implement settlement schemes. The Crown Land ordinances of 1902 and 1915 were used to lease land to European settlers for periods of 99 to 999 years and to sell land on easy terms to European settlers. The Kenya Order in Council of 1920 gave all land the legal status of crown land. Subsequent orders delimited and affirmed the boundaries of the African land reserves (later, in the 1940s, called native land units). The white highlands, also known as "scheduled areas," were created. Pastoral territory was reduced due to forced reduction in movement across state boundaries, as well as annexation of land for commercial settlement. This resulted in increased human and livestock population pressure, ultimately leading to serious problems including overgrazing and soil erosion on rangelands. Weather fluctuations resulting in increased droughts have further compounded human suffering in these pastoral areas; the droughts have led to inherent food deficits, and famine relief has become an almost permanent feature of the rangelands.

Continued political instability across Africa, caused in part by insecurity of land tenure, seems to throw into question the viability of the modern African states as nations. It would seem prudent at this time to revisit the issue of land tenure and ownership as a broader issue of policy both at the national and regional levels. Changing regional conglomerations of countries like those called the East African community into single federal states might be one type of solution, but the issue will not go away on its own or by doing nothing.

ANTI-PASTORAL COLONIAL POLICIES

The political programs of colonial and postcolonial governments were very much directed toward one feature of pastoral systems—their mobility (Niamir-Fuller 1999). Colonial officials brought with them a cultural bias against mobile people, who were viewed as primitive, shiftless, and immoral. European ignorance of the nature of pastoral mobility resulted in it being seen as antithetical to good land husbandry because there were very few visible ties to the land. The "nomad" would destroy the land with his livestock and then move toward "greener pastures." Mobile populations were also less easy for colonial governments to administer, so the governments often instituted measures to limit the mobility of their subjects and their subjects' animals. In eastern Africa, this was done through the establishment of grazing schemes, grazing blocks, and group ranches as good models of proper land use and livestock management; the governments' hope was that pastoralists would be convinced of the necessity to reduce their livestock numbers, to overcome overgrazing, and to reduce soil erosion to manageable levels. Once settled, a rural population would be medically treated, schooled, and taxed. Other forms of settlement developed from the famine relief centers created to provide food for people after they lost their livestock to droughts. Such situations were sometimes exploited to introduce unplanned settlements around Christian mission famine relief centers and nonviable irrigation schemes and fishing villages around rivers and lakes. The cultural and political biases leading to failed policies against nomadic pastoralism have in many cases been continued in today's postcolonial, independent Africa. One independent African government even insisted on changing the pastoral dressing tradition of loose-fitting garments adapted to the hot, dry climate to traditional western-style tight-fitting shirts and trousers.

Many pastoral people still feel very insecure about their present position. They see continuing encroachment into their remaining territory by agricultural people, and their continuing exploitation by the sedentary middlemen, as part of the reason secure tenure of their properties continues to be denied. Innovative policies that recognize the complexity of linkages between pastoral people and their environment will go a long way in relieving the current situation. In the words of Talbot when discussing the Masai situation, "Clearly new approaches to deal with the demographic pressures at work in Maasailand must be undertaken within a system framework that comprises the ecological system of the rangelands themselves, within which nomadic pastoralism was developed, and the demographic system operating in Kenya generally and in Maasailand in particular. In short, they must recognize the intimate relationship between the demographic and the environmental factors and their effects on the capacity of these lands to support a growing and varied population" (1986, 451).

WILDLIFE CONSERVATION AS A PUBLIC LAND USE POLICY

Africa is proud of the fact that 10% of its territory is now set aside in national parks or other protected areas for wildlife conservation. Most of these parks are found on rangelands, which support a broad array of migratory wildlife and pastoral live-stock that have coexisted here for centuries. The introduction of the parks, which are specifically designated for wildlife conservation alone, has introduced an ele-ment of instability into rangeland communities; the practice of allocating land for a single purpose such as national parks is completely foreign to the African cultures on which it has been imposed. Furthermore, the process of setting aside these national parks was done largely by decree instead of through negotiated compromise, alienat-ing societies that had hitherto coexisted peacefully with wildlife. Apart from taking large tracts of these societies' grazing land, wildlife conservation laws, which pro-hibited subsistence hunting, caused a large portion of the male population of these societies to be put in prison for what came to be called poaching. All of these factors have contributed to a hardening of local attitudes toward wildlife conservation as it is currently implemented, and national parks are not yet fully accepted in their new environments. Although the independent African governments have been quick to quote lucrative tourist revenues as a justification for national parks, tourist revenues often go into the pockets of corrupt government officials, which does little to console pastoral people. They see wildlife use their remaining grazing lands every day, but they cannot take their livestock into national parks even during the dry season. The situation is even made worse when wildlife damages, without any compensation, the few crops the pastoral societies have introduced to survive.

Wildlife conservation is still an indispensable part of the African culture and landscape. It can also contribute substantively to the ailing African economies in terms of tourist revenues. But national parks must be accepted by local range-land people, and for this to happen they must contribute to those people's welfare. Accordingly, current policies that have contributed to the present threat of the wild-life resources must be changed, and planning must be based on a proper evaluation of cultural, political, and socioeconomic (as well as ecological) factors; conservation must be balanced against local human population needs in both the short term and the long term (Lusigi 1975). Above all, the resulting conservation program must be acceptable to the local African population. Pastoral people continue to feel uneasy about their present position, as they view wildlife conservation as part of the reason why they have been denied tenure of their land and development of their economies. This need for a more accommodating policy is more urgent today than ever before in light of increased populations and accompanying poverty. The policy must include possibilities of organized game harvesting to control wildlife populations outside protected areas and reduce human wildlife conflicts. As stated by Talbot, "from an ecological standpoint, wildlife makes more productive use of parts of Maasailand than domestic livestock and the economic potential of wildlife harvest is worthy of serious consideration" (1971, 45).

INSTITUTIONAL POLICY ISSUES

As part of their nation-building efforts, central governments have taken over the ownership of natural resources (e.g., land, trees, and water) that were formerly under local community control. Usufructuary rights are granted to individual cultivators, and traditional management practices continue to prevail in many rangeland areas, though they are denied legitimacy in the event of a conflict of interest. Central governments have not been able to establish any effective alternative management systems; different users therefore tend to have open access to common properties for grazing, water, and fuelwood purposes. The government has difficulty protecting forest reserves against encroachment. Influential people close to the decision-making powers are frequently able to obtain rights to and encroach upon the common lands.

The traditional society has undergone other fundamental changes. Leaders have been co-opted by the state for administration, tax collection, and law enforcement and may now be more aligned with national powers (and the associated individual benefits) than with local interests. Recent developments have shown that distrust for the pastoral way of life now runs deep—not only among people practicing agriculture, but also among the pastoralists themselves. In Kenya, for example, where a pastoralist head of state and his vice president have been in power for the last thirty years, life for pastoral people has gotten worse. The head of state is listed among the wealthiest people in the world; however, the district that he represented in the parliament for over forty years is still the hardest hit by famine every year. As a result, local motivation and initiatives have been sharply reduced and replaced by a dependency on central government. The virtue of decentralization is being extolled by many governments, but so far, with very few exceptions, there has been little real action in this direction. The fear of reinforcing tribal divisions may act as a deterrent. The distribution of wealth (e.g., livestock) seems to have become more skewed, and this will tend to further erode local cohesion. Given the low population densities, large distances, and poor road conditions, government services are relatively weak and, in pastoral situations, difficult to deliver. Research has mainly emphasized high-potential areas while providing little guidance for dry land development.

Public interventions in water supply have proven very difficult to sustain in the absence of effective policies to ensure functional mechanisms of cost recovery, well-trained technicians for the pastoral areas, and availability of affordable, imported spare parts. Unplanned water development has led to serious cases of rangeland deterioration around watering sites. The IUCN—World Conservation Union Sahel studies program reports some 4,500 development projects in the Sahel supported by 180 different donors (Lusigi and Nekby 1991, 8). In spite of such impressive numbers, though, successes are relatively few, scattered, and difficult to generalize because of uncoordinated or nonexistent policies. Because of war and internal unrest, development efforts still face major setbacks in many countries.

In many cases, institutional frameworks in Africa have not provided sufficient authority and scope to deal with the problem of rangeland development because they are too narrowly focused. To improve development programs, governments should address complexity through integrated land use policies and creation of essential new institutional frameworks. These new institutions must encourage the participation and cooperation of the local people.

POVERTY ALLEVIATION POLICY ISSUES

The steady increase in poverty on African rangelands over the years can be related to four mutually reinforcing trends: (1) high human population growth, (2) poor performance of agriculture, (3) natural resource deterioration, and (4) increasing poverty and food insecurity. Despite the AIDS epidemic, which is now decreasing the population, better health care across the continent has greatly increased infant survival and prolonged life. And there has been little adjustment of family size, as children apparently are still seen as assets in the struggle for survival and as security for old age. Agricultural growth continues to be based on area expansion; intensification through new technology, cash inputs, or adjustments in farming systems is slow. The environmental consequences of population expansion thus tend to be reduction of fallow periods, soil exhaustion, and cultivation of shallow soils and steep slopes, followed by accelerating erosion, overexploitation of forest and range areas around settlements, and consequent denudation and erosion. This results in worsening prospects for future agricultural growth. Moreover, the failure of agriculture to keep pace with population growth—in a situation when the nonfarm sector is only able to absorb a fraction of the added labor force—has resulted in decreased per capita income and, in some areas, chronic food insecurity. A large part of the rangeland population now faces a situation where per capita access to resources in the form of land and livestock is dwindling. The exploitation of more marginal rangelands also implies that the risk of crop failure is enhanced and that food relief is required with increasing frequency. Although current and quantitative information about what is happening in African rangelands is scanty, it is likely that we have only seen the beginning of an accelerating deterioration of income and natural resources. This situation will not be reversed unless sound policies to correct it are put in place. Issues to be confronted must include corruption, transparency, and good governance.

RANGELAND-SPECIFIC FISCAL POLICY ISSUES

Price policy can have a substantial impact on the use of natural resources, but it is hard to generalize about its potential effects. A general price increase will tend to increase the demand for basic factors of production such as land, labor, and capital, the supply of which are relatively fixed in the short term. Although both future and present benefits are enhanced by such a general increase, it may result in land mining if the permanency of the change is questioned or if future returns are heavily discounted by farmers. The open access to rangelands may encourage overexploitation. A policy on agricultural subsidies may similarly result in undesirable resource allocation if inputs are not valued at world market prices. Large-scale mechanized farming may result and encroach on fragile rangelands. A general price increase will also affect export crops more than food crops, since the impact will be related to the degree of market orientation. The consequent change in the production mix may be good or bad from an environmental point of view, depending on the type of export crops that are encouraged. The production of tree crops, such as tea or coffee, or nitrogen-fixing crops, such as groundnuts, if expanded, may contribute to the preservation of natural resources. The government may wish to consider the pros and cons of deviating from pricing principles to promote crops with a positive impact on the environment through an increase in their relative price. If the price of fuelwood is increased in relation to competing products, this will lower demand and encourage plantations. Changing factor prices may similarly have environmental effects. Promoting cost recovery for water may diminish overgrazing in the vicinity of boreholes.

One of the big problems in pastoral livestock production is distress sales in times of drought; these sales result in sharply depressed prices and deteriorating terms of trade with other food items. It may be possible to stabilize prices in localized events, but no way has been found to deal with a countrywide stress situation. The only way to limit distress sales will be through the promotion of a more commercial type of pastoralism, under which the breeding herd is restricted to what can be sustained under "normal" conditions and marketing of nonbreeding stock is a continuous process. This will require both incentives (e.g., availability of consumer goods) and efficient marketing arrangements, as well as alternative and attractive means of keeping wealth (e.g., banking). In other words, banked assets can be utilized to survive during emergencies and subsequently used to rebuild the breeding herd.

Apart from the devolution of ownership of common land to local groups referred to earlier, other land tenure constraints exist. If user rights in crop production are acquired simply by taking up cultivation, there is an incentive to exhaust the soil and move on. This is particularly the case when the fallow in shifting cultivation reverts to cultivated legume fallow or conservation structures. Rules need to be introduced to protect the rights of farmers who are prepared to undertake such investment. The existence of such rules would be a powerful conservation incentive.

POLICIES ON DIVERSIFICATION OF RANGELAND ECONOMIES

In many rangeland areas a substantial part of family income comes from nonfarm activities. Some of this income constitutes remittances by family members who have left the area permanently or seasonally. But another part has its origins in trading and crafts, of which we need a much better understanding before we can identify constraints and opportunities. The investment priorities of local communities frequently refer to improvement in infrastructure (water supply and roads) and social services (education and health), which are of profound importance both for production and living conditions, and which also may promote specialization, mobility, and migration. The big problem in defining the local responsibilities for such investments will be to match local development priorities in terms of use of national resources with local initiatives and funds. Thus, in order to ensure the sustainability of whatever facilities are created, arrangements for cost recovery will be needed to ensure maintenance and operation (e.g., water supply); even so, the task may be complicated by the lack of skills and imported spare parts. Recent studies indicate that there may be opportunities for local communities to engage in wildlife management and tourism. Those who choose to establish such activities may encounter problems in maintaining control over their area of operations similar to problems faced in management of rangelands for livestock production. Additionally, they may have to deal with poaching and the migration of wildlife.

CONCLUSION

A condition of extreme ecological fragility extends over very large portions of the African rangelands. Scientifically, nothing is more important than land use and management calculated to conserve the limited elements of productivity. Nothing must be done to frustrate the natural ecosystem processes of growth and decay. The productivity of any ecosystem, or the health of any habitat, is greatest only when entire biogeochemical cycles proceed without interruption, and to the limit of whatever capacity local nutrients or climatic circumstances can create. In these rangelands, there has been a political and economic history of opportunistic exploitation, impelling ever-increasing degrees of human hardship and ecological damage that in some cases might be irreparable. It is important that all concerned return to the drawing board and realistically confront these bitter truths in order to reverse the trend. Only by doing so will we restore the balance between consumption and production and ensure the future of the millions of rangeland people of Africa.

References

Lusigi, W. J. 1975. *Planning human activities on protected natural ecological systems.* Munich: Technical University of Munich.

Lusigi, W. J., and B. A. Nekby. 1991. *Dryland management in sub-Saharan Africa: The search for sustainable development options.* AFTEN Working Paper, Environmental Division, Africa Technical Department. Washington, DC: World Bank.

Niamir-Fuller, M., ed. 1999. *Managing mobility in African rangelands: Legitimization of transhumance.* Stockholm: Beijer International Institute for Ecological Economics.

Talbot, L. M. 1957. What kind of animal is an ecologist? *California Monthly* 68(2): 26–31.

Talbot, L. M. 1971. Ecological aspects of aid programs in East Africa, with particular reference to rangelands. In *Ecology and the less developed countries,* ed. B. Lundholm, 21–51. Ecological Research Bulletin, no. 13. Stockholm: Swedish Natural Science Research Council.

Talbot, L. M. 1986. Demographic factors in resource depletion and environmental degradation in East African rangeland. *Population and Development Review* 12(3): 441–451.

UN Environment Programme/New Partnership for Africa's Development. 2003. *Environment action plan.* Midrand, South Africa: UNEP.

11

The Role of Science and Scientists in Changing Forest Service Management Relative to Sustainability

Jack Ward Thomas & James A. Burchfield

Since their creation one hundred years ago, the national forests of the United States have been the locus of a series of social conflicts and experiments in the human relationship to nature. Yet throughout it all, a single, powerful concept—sustainability—has driven the evolution of the management of these treasured public lands. It has been and is the basic underlying principle that has guided management over a century of adjustments. As we learn more about the profound ecological and political implications of this noble goal, the target only becomes more urgent.

Merriam-Webster's Collegiate Dictionary defines sustainability as "a method of harvesting or using a resource so that the resource is not depleted or permanently damaged." To those who deal with natural resources, the definition of sustainability that rings most clearly is the one advanced by the World Commission on Environment and Development in 1987: "Sustainability is the ability of present generations to meet their own needs while ensuring that future generations are able to meet their own" (2). Clearly the concept of maintaining sustainability in forests has become more complex and more inclusive of new and confounding factors over the past century (Johnson and Ditz 1997). In our view, the principle has been woven into management practice through the influence and interactions of two tightly entwined factors: science and policy. Each has its own feedback loops—changes in policy emphasis are expressed in laws, regulations, court decisions, and budgets, while science responds to the accumulation of knowledge as more sophisticated empirical measures and greater appreciation of complexity become apparent (Cortner and Moote 1999; Lee 1993). However, both factors are vital contributors to the visionary goal of sustainability, and both struggle to find means to design conscious interactions with natural systems to fulfill human demands.

John Fedkiw, in considering these interacting factors, proposed "a pathway hypothesis" that described the evolutionary processes of science and policy as follows: "We have been on a path toward a more holistic, ecological approach to sustainability since the beginning of professional, science-based resource management in the United States" (2004, 9). That pathway is described as both broad and bounded—as broad as societal demands and as bounded as the acceptability of the consequences of actions allowed and the limits of sustainable outcomes. Both factors have changed dramatically over the past century and can be expected to continue to evolve (MacCleery 2004; Sample 2004).

ORIGINS OF SUSTAINABILITY AS A MANAGEMENT GOAL
FOR THE FOREST SERVICE

The Creative Act of 1891 provided the original authority for the president of the United States, through proclamation, to create forest reserves from the public domain. There was, however, no direction included as to how, and for what purpose, those reserves were to be administered. In 1896, the secretary of the interior requested that the National Academy of Sciences recommend a national policy for management of the forest reserves. The recommendations of the committee assigned to that task were incorporated in the Organic Act of 1897 (U.S. Department of Agriculture, Forest Service 1906). That act restated the president's authority to designate forest reserves and, in addition, specified the criteria for such designation (Cubbage, O'Laughlin, and Bullock, 1993): "No national forest shall be established, except to improve and protect the forest within the boundaries, or for the purpose of securing favorable conditions of water flows, and to furnish a continuous supply of timber for the use and necessities of citizens of the United States."

The words "to furnish a continuous supply of timber" imply that sustainability is a guiding principle for management of the national forests. Clearly, it would be impossible to produce a continuous supply of timber without the activity being sustainable.

Interestingly, this guiding principle was an optimistic wager that a professional cadre of managers and scientists could carry out that task. The profession of forestry did not yet exist in the United States, and the underlying science base to support management activities would have to be quickly developed. What science that did underlie forestry was initially extrapolated from whatever existed in Europe and the empires of European nations such as British India (Pinchot 1947).

Adding complexity to this bold gamble was the harsh political reality. In 1887, the forest reserves were under the jurisdiction of the Department of the Interior and could hardly be characterized as being under rational management, let alone science-based stewardship with sustainable management as a goal. Furthermore, the institutional house was in disorder, with rules and regulations applying to the forest reserve being promulgated by the Department of Agriculture and responsibility for surveying, mapping, and classifying the lands resting in the hands of the U.S. Geological Survey in the Department of the Interior. In reality, there was precious little "management" of the forest reserves for the next eight years. Timber theft and trespass for reason of livestock grazing was rampant. Charges of official corruption and incompetence were common (Cubbage, O'Laughlin, and Bullock, 1993; Pinchot 1947). "The technical and complex problems arising from the necessary use of forest and range soon demanded the introduction of scientific methods and a technically trained force, which could not be provided under the existing system" (U.S. Department of Agriculture, Forest Service 1906, 15–16).

In a bold political act to force a science/policy connection, the forest reserves were transferred from the Department of the Interior to the Department of Agriculture in 1905 to be managed by the newly created U. S. Forest Service as dictated by the Transfer Act of 1905. That act stated that "the permanence of the resources of the reserves is therefore indispensable to continued prosperity, and the policy...for their protection and use will be invariably be guided by this fact, always bearing in mind that the conservative use of these resources in no way conflicts with their permanent value...continued prosperity...is directly dependent upon a permanent and accessible supply of water, wood, and forage." Words like "permanence" and "continued prosperity" emphasized something altogether new—the national forests were to be managed in a sustainable fashion.

Subsequently, Secretary of Agriculture James Wilson issued instructions to the newly appointed chief of the Forest Service, Gifford Pinchot (in a letter actually written by Pinchot), on the management of the national forests. These instructions were, essentially, taken directly from the Organic Act, and they emphasized two points: (1) conservative use to assure the "greatest good to the greatest number in the long run" and (2) what today would be called sustainable management using such descriptive words and phrases as "permanent good of the whole people," "permanence of resources," "continued prosperity," and "permanent and accessible supply of wood, water, and forage" (U.S. Department of Agriculture, Forest Service 1906, 11).

BUILDING A SCIENCE BASE

It was understood that a sound technical background (i.e., science base) would be required for the agency and its budding professional workforce to manage the national forests under these instructions. As stated in the Forest Service's 1906 *Use Book:* "Improvement in the standard of the technical management alone can secure steady and constant increase in returns without depleting the forest. To this end careful investigation is essential...In these and many other ways the basis of knowledge necessary for the best forest work will be laid...Whether the work is done under the supervision of...officers sent out from Washington or under the technical assistants stationed permanently on the reserves, the local officers will in every case assist and cooperate in the work" (18–19).

Yet scientific capabilities in those early days were sorely lacking. Pinchot frankly admitted, forty-two years after the creation of the Forest Service, that the "practical" forestry (i.e., forestry that turned a profit) being put into action in the early years was based on "next to nothing" in terms of scientific knowledge and that the agency was "short on underlying facts" (1947, 307). He also said that two monographs—*The White Pine* and *The Adirondack Spruce*—and "what was known about forestry in general, had equipped us to deal with the immediate problems of forest management on the National Forests...That knowledge, fortified by practical experience and common sense, got us by...On the whole, however, it was too little and too narrow when compared with the vastness of the field and the complexity of the problems...There was no time for lengthy observation or the search for precise forest facts" (307).

The scientists who first populated the agency were scattered and few in number, and Pinchot recognized that these researchers could only gain influence by first playing along.

> In those days the research men were...attached to the Forest Supervisors, to help with technical problems. Their presence was often resented on the ground that they were Eastern tenderfeet (which had nothing to do with the case) but more commonly because of their persistent and sometimes unreasonable habit of asking embarrassing professional questions...we usually had to split the difference between what was best for the forest and what was practical in logging...That research in the Forest Service today [added 1947] gives sound and practical results is due to this constant struggle between what is ideally good and what is practically possible. (Pinchot 1947, 308)

It may be surprising to some that these same words would sound familiar to Forest Service research personnel active today—at least some of the time. In fact, this condition has likely been sustained over the years as the information emerging from research activities steadily increased—both in volume and variety—and was brought to bear in management. Change in management approach is often expensive and almost never easy or comfortable.

Although Pinchot (1947) lamented the struggle to more fully incorporate research into management, research had begun very early in the agency's history. Experiment stations for grazing were formed in 1903, two years before the official establishment of the Forest Service. Work relative to the development of wood products were instituted in 1901 under the Bureau of Forestry in the Department of Agriculture, and then bloomed when these activities, now under Forest Service sponsorship, were moved to Madison, Wisconsin, in 1909.

In May of 1908, Raphael Zon recommended to Chief Pinchot that the Forest Service establish forest experiment stations in the national forests to carry on "experiments and studies leading to a full and exact knowledge of American silviculture, to the most economic utilization of the products of the forest, and to a fuller appreciation of the indirect benefits of the forest" (Pinchot 1947, 309). That summer, the first forest experiment station was established in the Coconino National Forest near Flagstaff, Arizona. Yet it was not until the passage of the McSweeney-McNary Act of 1928 that the forest experiment stations were authorized in law for the limited purposes of cooperative research on reforestation and forest product development.

Forest Service experiment stations were made independent and placed under the direction of Earle Clapp, who was later to become acting chief. Though criticized by some in the national forest system (then and now), the independence of the research division has paid huge dividends. Many of the changes in the course of national forest management, often reluctantly accepted by line officers in the national forest system, resulted from application of research conducted in the independent research division (Pinchot 1947).

The first chief clearly came to appreciate the benefits of an independent research division, saying in 1947 that "the greatest contribution of Forest Research is the spirit it has brought into the handling of natural resources. Under the pressure of executive work, the technical ideas of the forester at times grow dim. It is Forest Research which has kept the sacred flame burning and has helped raise Forestry to the level of the leading scientific professions…The research man must anticipate coming needs. He must of necessity be ahead of his time. That means standing alone, exposed to the ridicule of those who live only day to day" (Pinchot 1947, 310).

That, in our opinion, is as true today as it was then.

IN THE BEGINNING: THE FOCUS ON UTILITARIAN VALUES

From the time of the emergence of forestry as a profession in the United States in the early years of the twentieth century, research efforts related to forestry, whether in the private or governmental sector, were primarily directed at developing knowledge and understanding of natural processes and interactions necessary to manage and exploit forests at a profit. This knowledge included harvesting, followed by the assured replacement of those forests with the species deemed likely to be the most profitable at the next harvest. Therefore, initial research efforts were primarily directed at reforestation (with emphasis on development and planting of "superior" trees), enhanced growth of commercially valuable trees, protection of the forest, insect and disease control, tree harvesting, wood product development, fire control, and grazing (Pinchot 1947).

The underlying assumption was, too often, that "good forestry"—for example, forestry that turned a profit and kept harvested sites in tree cover—was good for everything else, including soil stabilization, retention of watershed values, wildlife and fish, recreation, and aesthetics. By 1979, that assumption had been shattered (Thomas 1979). Furthermore, science information helped establish that timber management had other impacts, eventually undermining the long-held misconception that foresters are fully equipped and qualified to address all aspects of the management of forested ecosystems (Bunnell 1976a, 1976b).

SOCIAL PRESSURES CHALLENGE THE UTILITARIAN MODEL

Through the middle of the twentieth century, forest management policy underwent a series of transformations in response to the expansion of American prosperity, the new accessibility of forests to a growing middle class, and an expanded understanding of the role of forests relative to social goods and services beyond timber. Outdoor recreation, to a large degree associated with demands for hunting and fishing, was building a constituency fueled by a surge in ownership of automobiles and a rapidly expanding road system which was, ironically enough, constructed primarily for purposes of fire protection and potential timber management. This new constituency had few ties to the direct economic benefits of logging practices, but they liked visiting the forest and began to raise questions about the functions performed by forests for wildlife habitat, water quality, and aesthetics. National forests were (and remain) strongholds for wildlife species valued for hunting that had been reduced by well over 90% in the previous 200 years and, in some cases, provided unique settings for species hovering on the edge of extirpation or even extinction. These forests were also the locations of some of the nation's most sought-after natural treasures for recreational use, notwithstanding the competition to corner this market by the National Park Service. It would be hard to argue that the forests could continue to focus attention only on timber production and grazing (Cubbage, O'Laughlin, and Bullock, 1993).

By 1933 the National Park Service was coming into its own and was heavily supported by the Franklin D. Roosevelt administration. A large expansion of the national park system took place in 1933 when a presidential order transferred sixty-three national monuments and military sites from the Forest Service and War Department to the National Park Service. The Forest Service was both unappreciative and resentful of those "raids." By 1964, there were thirty-two national parks, seventy-seven national monuments, and ninety-three other areas under National Park Service management. The Forest Service realized it had to enter the recreation and fish and wildlife management areas or surrender even more lands to the National Park Service and the U.S. Fish and Wildlife Service (Cubbage, O'Laughlin, and Bullock, 1993).

The Forest Service, in spite of its ambitions to be a primary source of timber and other wood products for a growing nation, was effectively relegated in its first fifty years of existence to a minor, essentially local role as a timber supplier. The efforts of private timber producers, who wanted no competition from government timber, had kept appropriations for sale preparation and administration at relatively low levels. The Depression, which began in 1929, kept timber demand at low levels throughout the 1930s. So it wasn't until preparations for World War II began in 1939 that the potential of timber supplies from the national forests received renewed attention. However, the Forest Service was again squeezed out of the timber supply business when the timber industry gobbled up wartime demand with favorable "cost plus 10%" contracts (Cubbage, O'Laughlin, and Bullock, 1993).

THE TIMBER ERA: TOO MUCH TOO FAST?

This situation changed quickly and dramatically with the end of the war in 1945. The "easy pickings" for timber companies from private lands were near depletion and there was a concurrent and soaring demand for wood to supply houses to meet the nation's pent-up demands, which had accrued since the beginning of the Depression sixteen years earlier. In appreciation for military veterans' services during the war, Congress passed an act, the GI Bill, which subsidized a college education and a house loan for the veterans. Both college-level education demands and housing demands boomed as a result (Clary 1986; Hirt 1994).

The timber industry, with its well-established lobbying capabilities, now turned to the national forests as a stopgap supplier of raw wood suitable for conversion to building materials at relatively low costs. The Forest Service enthusiastically and effectively responded to the direction from Congress and the Truman and Eisenhower administrations to their initial applause, along with that of the body politic. The heyday of timber production from the national forests ensued. The Forest Service, however, zealously guarded its flanks relative to incursions by other federal agencies with interests in recreation, fish and wildlife, and livestock grazing (Clary 1986; Hirt 1994).

Oddly, the shift from single and group selection timber harvest regimes, so pushed by the early Forest Service in sharp contrast to the activities of private industry, had evolved from Forest Service research that had been narrowly focused on how to more efficiently grow, harvest, regenerate, and manage stands to produce select timber species for commercial purposes as prescribed by the McSweeney-McNary Act.

Accumulating research results produced by Forest Service scientists and their academic colleagues indicated that, purely in terms of production of timber of desired species, even-aged timber management utilizing clear-cutting was superior to uneven-aged management approaches using single-tree or group-selection harvest. In keeping with the signals from political overseers in the form of budgets, the Forest Service adopted those practices for timber management in most forest ecosystems, ranging from conifer systems in the West to the hardwoods of Appalachia (Clary 1986).

This was a complete reversal from the position that was taken and maintained by the Forest Service in its early years. In fact, during that period, the Forest Service vigorously attacked the clear-cutting being carried out on private lands and even proposed taking over the regulation of forest management and timber harvest from private lands. That effort failed. The position behind it was based, ostensibly, on the imported "science" of silviculture as practiced in Europe (Pinchot 1947). But the broader social interests for forests would force management and research to look beyond the tree stem to a fuller range of forest benefits.

SCIENCE EMERGES AS A DRIVING FORCE IN POLICY

By 1960, the Forest Service had operated under the auspices of the Organic Act of 1897, as amended by the Transfer Act of 1905, without legal challenge and without impingement of other legislation—or any court challenge to their operations—for sixty-three years. In 1960, the Forest Service, to protect itself from encroachments from the National Park Service and the Fish and Wildlife Service upon its land base, pushed for passage of the Multiple-Use Sustained-Yield Act (MUSY). This act authorized activities in the management of fish and wildlife, recreation, and grazing, in addition to timber production and watershed protection (Cubbage, O'Laughlin, and Bullock, 1993). In addition, it provided for the institution of research efforts within the areas of fish and wildlife and recreation.

Around the same time as MUSY's passage, science-based discoveries began to assert themselves more openly in the policy arena. Prior to the 1960s, science and scientists were carefully insulated from forest policy, by consent of both scientists and managers. As Jerry Franklin describes the model that had held sway for the first half of the century: "Direct involvement of scientists with development of policy or with decision makers was strongly discouraged—largely by mutual agreement" (1999, xi).

Scientists didn't want the pressures and potential influences of policy making in their work, and managers didn't believe that scientists would understand the compromises and practicalities of the real world. But science couldn't be constrained as researchers began to understand forests as ecosystems instead of tree stands (or tree farms). Latent public tensions regarding the very purposes of management began to be examined more closely by the science community, coming to a head in 1962 with the publication of Rachel Carson's *Silent Spring*. Scientists would no longer be relegated to the back bench on policy issues related to national forest management. Their time in the sun had come.

THE ENVIRONMENTAL LAWS OF THE 1970s: THE BEGINNING OF THE ENVIRONMENTAL ERA

Policy makers wasted little time reacting in the wake of the two-pronged attack from pent-up public demand and gathering scientific evidence. Preservation of nature, a strong chord in American interests toward nature since Thoreau, found political champions. The Wilderness Act of 1964 not only was a landmark for resource protection, but also provided stimulus for research related to wilderness and backcountry recreation. The Water Resources Planning Act demanded increased attention to watersheds and was reinforced by the Wild and Scenic Rivers Act of 1968. Research efforts in these fields were correspondingly increased by the Forest Service and other entities.

The National Environmental Policy Act of 1969 forced dramatic changes and significantly increased demands on the Forest Service for assessment of effects of proposed management actions, personnel to quickly comprehend what was required, and development of processes for carrying out such assessments. This required a quick buildup in professional expertise in the areas of fish and wildlife biology, ecology, social sciences, recreation, hydrology, and engineering, among others. That, in turn, was mirrored in the expansion of agency's research activities in those fields—in terms of both in-house research and contracted research. "Best guesses" and "professional opinions" would no longer suffice as a basis for management actions.

WILD SCIENCE EMERGES

Franklin divided this overt infusion of science into the natural resource policy debate into two categories (1999, xi). He called research relative to improved implementation of existing methods and policies "domesticated science," a situation where managers are not only highly influential but comfortable. However, research that probed basic assumptions underlying current policies—such as clear-cutting or ecological values of old-growth forests—was called "wild science" and more likely occurred outside a manager's comfort zone. As the Forest Service's research branch expanded in size and breadth, more and more researchers were involved in wild science and that, in turn, would have dramatic impacts on management.

John (it's always been J.J.) Hickey considered the appropriate role of science in natural resources management to be a journey more than a destination: "The purpose of science is not to conquer the land, but to understand the mechanisms of ecosystems and to fit man into the resources he has available on the planet on which he has evolved" (1974, 168). The domesticated science that Franklin described was aimed at conquering the land, turning it into a tame and compliant servant of rather narrow human demands. The application of that domesticated science brought rewards and repetition when it was applied in isolation. However, it was the emergence of the efforts in wild science that made it possible to apply the sum of the science to all the uncertainties and new questions that form the basis of a dynamic relationship with nature.

CONCERNS OVER BIODIVERSITY RETENTION
BECOME A DRIVING FORCE

During the same period, scientists were becoming increasingly aware of the value and the precipitous worldwide (including North American) decline in biodiversity—this would lead to the passage of the Endangered Species Act (ESA) in 1973 (see Noss and Cooperrider 1994 for an excellent review). Old-growth (i.e., late-successional) forests in the Pacific Northwest were a focus of concern, as they were quite obviously in a different state of forest development and were being cut at a rapid pace. Most of the remaining old-growth forests were on federal lands—primarily national forest lands. And research relative to biodiversity and ecosystem function was rapidly increasing due to wild-science research findings by such Forest Service scientists as Jerry Franklin and James Trappe and their associates in the Pacific Northwest. Much of this research and philosophical discussion was captured by E. O. Wilson (e.g., 1985, 1988, and 1992) in later publications.

Social scientists were also beginning to examine the Forest Service as an organization, describing it in cultural terms and admiring its cohesion and focus (Kaufman 1960). In the immediate aftermath of World War II, the agency had proven remarkably efficient in fulfilling the mission defined for it by Congress and a series of presidential administrations. Its overriding purpose, to "get out the cut" to satisfy the demands of an increasingly robust economy had, indeed, been fulfilled. The postwar generation of agency professionals, sometimes referred to as the Marine Corps of the civil service, emerged with jubilant optimism from forestry schools, ready to handle any problem.

They entered the agency's ranks at the lowest levels (ranger districts) where they were acculturated into the "family" or the "outfit." They then advanced through the ranks with others of their brethren to be rewarded for their substantial achievements (Kaufman 1960). Although these Forest Service professionals were almost universally praised for their efforts at the time, the accumulating implications of their actions would be seriously questioned in the future. But questions did arise at the time— many based on the results that were accumulating from the newly expanded efforts of wild science. Nothing would shatter the forester's domestic comfort more than a widely popular initiative to save species from extinction—the ESA. The ESA would eventually prove to be the 800-pound gorilla of forest management, storming across the agency's safe and stable cultural landscape and elevating scientists to positions of authority and impact that they had neither sought nor could effectively handle.

The ESA's stated purpose, "to provide a means whereby the ecosystems upon which endangered species and threatened species may be conserved," seemed simple enough at first. Compliance did not appear to be problematic; after all, foresters had been on the vanguard of conservation for half a century. But the complexity of ecosystem conservation demanded more of an altered response in management than originally anticipated—it implied restraint, nimbleness, and an adaptability that hadn't been part of the standard, comfortable, straightforward, management procedures of the past. Worse, it implied that scientists would have a far different role than simply serving as efficiency experts for a foregone design—they could become the designers. The ESA would come to have dramatic, and largely unforeseen, impacts which would permanently transform the management of the national forests. Suddenly, Forest Service research units dedicated to fish and wildlife were shifting attention away from big game species to species declared, or likely to be declared, threatened or endangered.

THE BACKLASH

It was not only the ESA that was upending the forester's comfortable world. There was also a backlash of public opinion relative to common silvicultural practices centered around clear-cutting. That backlash focused first on two disparate regions: the Bitterroot National Forest in Montana and the Monongahela National Forest in West Virginia (Cubbage, O'Laughlin, and Bullock, 1993).

Increasingly, public concerns were being expressed concerning the consequences of such practices relative to wildlife, fisheries, recreation, and aesthetics. The Forest Service, in its zest to maintain its widely acclaimed rate of timber harvest, had instituted a highly visible silvicultural system (even-aged timber management) with dramatic visual and ecological impacts without comprehensive understanding of the consequences (Cubbage, O'Laughlin, and Bullock, 1993; Hirt 1994; Langston 1995). Application of the concentrated results of tame science relative to enhanced timber yields of desired species led to a dramatic change in management practice that was evolving into a political disaster, despite the fact that results were usually positive in the strict sense of enhanced timber production. This collision with public understanding and support was due to a lack of full appreciation on the part of Forest Service leaders relative to the potential political, ecological, and social consequences of broad-scale intensive timber management on public lands.

SCIENCE AND PLANNING

By the time of the passage of the Forest and Rangeland Renewable Resources Planning Act (RPA) in 1974, it was clear that the fabric of domestic science that had bound the timber-based model of management was frayed beyond repair. The federal courts held in 1975 that the Forest Service, by clear-cutting, was in clear violation of the Organic Act of 1897, which required that timber harvest be limited to "dead, matured, or large-growth" trees and required individual marking of trees to be removed. The sister act to the RPA, the National Forest Management Act of 1976 (NFMA), was essentially developed and pushed by the Forest Service to overcome the results of court actions that had declared clear-cutting illegal and ended the old domestic science era (*West Virginia Division of the Izaak Walton League of America Inc. v. Butz*). Moreover, the RPA had repealed and replaced the McSweeney-McNary Act of 1928 as the authorizing statute for Forest Service research activities, both legitimizing and opening the door for rapid expansion of wild research activities into ecological and social conditions and processes.

The RPA also directed increased attention to land use planning, inviting new areas of science-based modeling and simulations to forecast the impacts of alternative courses of action. More important, the RPA/NFMA changes stipulated that a "committee of scientists" was to be named to help promulgate the regulations that would guide the actual means of implementing the law. This requirement was either a clear signal that Congress's confidence in the management arm of the Forest Service was waning or recognition that scientists and science had something more significant to contribute to management than had been formerly true, or both. Data requirements stimulated by an explosion in land use planning resulting from both the RPA and the NFMA, in turn, produced an increased demand for Forest Service research to supply the essential science base upon which the broadly expanded demands for knowledge could be supported.

Even as policy pushed new forms of science and broader management priorities, the agency's fixation on producing more timber more rapidly and efficiently— routinely reinforced by both Congress and presidential administrations through budget direction—continued. Organizational culture doesn't change overnight, particularly in the face of such signals. Throughout the 1970s and 1980s, this focus would begin to drive a wedge between the agency and a significant proportion of the American people who insisted on greater attention to watershed values, aesthetics, wildlife habitat, and biodiversity (Hirt 1994; Langston 1995).

The combination of new societal concerns, embodied in the environmental laws of the 1970s, and the onslaught of new information flooding in from research scientists would dramatically shift the trajectory within Fedkiw's pathway hypothesis relative to national forest system management (2004).

Forest Service planning regulations emerging from the NFMA contained a phrase that would have significant and unforeseen results: "All species of native vertebrates and desirable nonnative vertebrates will be maintained in viable numbers well-distributed in the planning area." This requirement, far more stringent than any requirement of the ESA, was, over time, to create a sea change in Forest Service research and management.

THE EMERGING DEMAND FOR SYNTHESIS OF APPLICABLE SCIENCE

Scientific activity surrounding species diversity and their distributions had not been absent prior to this landmark regulation; rather, it had been advancing at a steady pace. Observations of species interactions and the emerging patterns of conclusions about species viability challenged simple solutions. An illustrative case emerged from eastern Oregon. In 1973, the Blue Mountains of Oregon and Washington were in the third year of an outbreak of Douglas-fir tussock moth (*Orgyia pseudotsugata*) that had resulted in the defoliation of hundreds of thousands of acres of fir species (Wickman 1992). Managers of the three national forests involved, backed by their supervisors, were set on broad-scale application of DDT to suppress the outbreak.

Public concern over such a massive application of a chlorinated hydrocarbon pesticide in the forest environment was at a high level and had already delayed the operation by a year. Though the technology and efficacy of such treatments were well established, the newest research under way by Forest Service scientists strongly questioned the longer-term effectiveness of such treatments. Furthermore, scientists on the ground believed that the outbreak was in the process of collapse. But the researchers were ignored, and the treatment went ahead as scheduled in spite of public opposition and the ongoing collapse of the population (as a footnote, this was the last such application of chlorinated hydrocarbon-based pesticides in the forest environment in the United States).

In the wake of the Douglas-fir tussock moth outbreak in the Blue Mountains, demands for a large-scale salvage of the trees killed in the outbreak required new information to comply with the requirements of the new Forest Service planning regulations. The response was a synthesis of available information relative to 379 vertebrate species in terms of their association with plant communities and successional stages for the purposes of breeding and feeding. *Wildlife Habitats in Managed Forests: The Blue Mountains of Oregon and Washington* (Thomas 1979) provided an early synthesis of science-based information that was to become a model for science applications in the future. Single-purpose domestic science would no longer be sufficient. Wild science and its synthesis were becoming the standard.

By the late 1980s, additional synthesis projects were under way to determine and quantify the impacts of timber management activities on various species of wildlife, but these summary documents were as much "patchwork" as science (e.g., Thomas and Radtke 1989). It wasn't until events unfolded relative to national forest management in the Pacific Northwest during 1985–2005 that the nexus between science and policy reached the next step of maturity for public land management.

A WRENCHING CHANGE IN SCALE IN ASSESSMENT AND PLANNING

By 1980, a new branch of wildlife biology was emerging; it was referred to as landscape ecology or conservation biology, wherein habitat considerations moved up in scale from stand level to the level of "landscapes." The condition, sizes, and configuration of habitat patches, their arrangements on the landscape, and the connectivity between patches were operative factors in planning and evaluation. Concepts of "coarse filter" approaches relied on the idea that species richness would be preserved if the correct combinations of ecological conditions were maintained. "Fine filter" approaches were also used to deal with individual species or guilds of species. Simultaneous compliance with the requirements of the ESA and the NFMA, related to maintenance of species diversity, produced a demand for the achievement of both objectives (Hunter 1999; Meffe et al. 2002).

ENTER THE SPOTTED OWL: THE HARBINGER OF CHANGE

The federal forest lands west of the Cascades in Washington, Oregon, and northern California were the most productive in the United States relative to timber. Timber harvests from those lands had gradually increased from less than 0.5 billion board feet per year in the late 1940s to some 5 billion board feet per year by 1989. The stands being so aggressively harvested were predominately virgin late-successional coniferous forests, commonly called old-growth forests (Thomas et al. 1990).

The harvest technique being utilized was generally clear-cutting, although some shelter-wood and seed-tree harvest units were applied (in the end, these latter techniques were not much different in terms of either habitat or visual impact). Regeneration of clear-cut units was accomplished primarily through planting of pure stands of genetically "superior" seedlings. These new stands were, by and large, to be managed as plantations for maximal production of high-value timber. But the appearance of these harvest units, peppering the landscape of the Cascades and the Coast ranges, was increasingly distressing to a growing number of citizens. Forest Service research scientists, as well as university cooperators, were routinely publishing research results that were bringing the general application of this silvicultural regime into question—particularly as it related to retention of biodiversity. It was intuitively obvious that the late-successional forests characterized by structural and species diversity were apt to have characteristics of unique and highly evolved ecosystems. If so, a collision with the ESA was inevitable (Norse 1990).

Studies ranged widely, from the below-ground consideration of nitrogen fixation by mycorrhizael fungi and their ecological function to studies of the ecological activities of the late-successional forest canopy and associated species. Studies of individual avian species such as the northern spotted owl (*Strix occidentalis caurina*) and the marbled murrelet (*Brachyramphus marmoratus*) were indicating that these species were likely candidates for listing as threatened or endangered by the Fish and Wildlife Service under the auspices of the ESA (Yaffee 1994). Similar concerns were emerging concerning aquatic species as well, including several species and subspecies of anadromous fish.

INERTIA HAS ITS DAY: RELUCTANCE TO CHANGE

Although as late as 1985 the Forest Service had maintained a reputation within government as a paragon of bureaucratic efficiency (Clarke and McCool 1985), the cracks in the foundation were beginning to show. The Reagan administration had appointed John Crowell, a former attorney for a large timber company, as assistant secretary of agriculture over the Forest Service. He and his successor, George Dunlop, maintained pressure on the Forest Service for an even larger timber program (Clarke and McCool 1996).

"Funds for recreation, planning, research, and technical assistance to state and private forestry were continually targeted for reductions, while appropriation for production oriented programs, such as energy, minerals, road building, and timber, were increased" (Clarke and McCool 1996, 60). Any activities that did not produce results that increased monetary returns were clearly suspect. Wild research efforts were clearly within that group. Luckily, Congress thwarted many of the more draconian cuts proposed by the Reagan administration, and the Forest Service's deputy chief for research, Robert Buckman, and his associates skillfully maneuvered to keep the research division essentially intact.

Even before the ascendancy of the Reagan administration, it was evident that the timber cut levels of 11.5 billion reached in 1965–1970—a departure from sustained-yield levels—could not be sustained. Yet Reagan's appointees continued to demand additional increases in those cut levels, levels that were already producing a growing backlash from segments of the public and, increasingly, from within Forest Service ranks. Elements within the Forest Service resisted these pressures as best they could short of outright insubordination (Yaffee 1994). Shortly before leaving office in January 1985, Crowell sent a memo to the agency chief which instructed, in part, that the chief was to reply, in ten days, with a plan showing a rise in timber sales that would "reach a 20 billion board foot annual level by 2030" (Clarke and McCool 1996, 64). Crowell resigned shortly thereafter, and such a plan was never delivered.

A growing environmental community was increasingly distressed with the consequences of the timber harvest program, which had leveled off at around 10–12 billion board feet per year. A common thread in the complaints was that "Reaganomics was transforming the Forest Service from a multiple-use agency to single-use one" (Clarke and McCool 1996, 64). Some line officers in the national forest system began to push back against the mandated timber harvest levels.

Scientific results continued to challenge a status quo of high harvest levels. Additionally, the social and political momentum of the timber harvest program was proving increasingly difficult to maintain. The so-called easy stuff had been logged. What timber remained was largely on less productive, steeper, or more remote ground, and standards for the construction of roads and more environmentally and aesthetically sensitive harvesting increased costs as the value per unit of the timber declined. Several attempts to adjust timber harvests to accommodate the requirements of the environmental laws and the rapidly accumulating research results were thwarted by political overseers.

SCIENTISTS THRUST INTO THE POLICY ARENA

Late-successional forests continued to be cut at the same rate with concurrent increasing fragmentation of what late-successional forest remained. New scientific evidence was building to the inevitable conclusion that high harvest levels were increasingly outside the bounds of the acceptable pathway to sustainability—socially, ecologically, and legally. By the beginning of the 1990s, "decisions were taken out of the hands of the resource professionals as the agencies lost their scientific credibility with the public and all three branches of the federal government" (Franklin 1999, xi).

A new era ensued in which science and scientists were more routinely involved in natural resource policy development.

The first significant move in this direction came in the last year of the presidency of George H. W. Bush, and the four federal land management and regulatory agencies involved—the Forest Service, the Bureau of Land Management, the Fish and Wildlife Service, and the National Park Service—had no choice but to turn to the science community for assistance. The northern spotted owl had become the poster child for the conflict in the national forests of the Pacific Northwest. It seemed likely that the owl was headed for listing as a threatened species, so the Bush administration commissioned a team of scientists to develop a management strategy for the owl. This team, the Interagency Scientific Committee to Address the Conservation of the Northern Spotted Owl, became known as the ISC (Thomas et al. 1990). Under tight deadlines, the team delivered the requested plan in six months.

The ISC team was composed entirely of scientists, all of whom were biologists—that is, no management personnel were included. Their task was limited, despite their protests, to the development of a single strategy for the management of forested habitats on federal lands of the Pacific Northwest that would balance a high probability of the viability of the northern spotted owl, well distributed within the area of concern. There was no effort to consider economic or social effects or the effect on other species of concern.

The ISC plan was heavily influenced by the emerging field of conservation biology (Soule and Orians 2001). This was to be the first truly regional-scale plan developed by the agencies involved. Regional assessments became more commonplace thereafter and were increasingly considered essential if ecosystem management approaches were to succeed (Johnson et al. 1999). The ISC plan was initially accepted by the agency heads, but then it was rejected by the director of the Bureau of Land Management, Cyrus Jamison (who was alarmed by the significant drop in timber harvest that would result), in favor of the so-called Jamison strategy.

THE STRING RUNS OUT ON THE STATUS QUO

This led to a shutdown, by the federal court, of all timber harvesting operations within the range of the northern spotted owl. The presiding judge, William Dwyer, ordered the government to respond to two questions. First, would withdrawal of the Bureau of Land Management from applying the strategy provided by the ISC compromise that strategy's efficacy? Second, what would be the effect of the long-term viability of other species thought to be associated with late-successional forests on federal lands if the strategy were implemented? Judge Dwyer clearly understood the stated purpose of the ESA: "to provide a means whereby the ecosystems upon which endangered species and threatened species may be conserved" (Thomas et al. 1993, 2).

Yet another team of scientists, the Scientific Assessment Team, was assigned to prepare answers to Judge Dwyer's questions (Thomas et al. 1993). The first question was impossible to answer because, in reality, the "Jamison strategy" existed only as a press release and the team could not attain access to any other potentially related documents. The team then assessed the likelihood of survival in a viable state of 667 species (555 terrestrial species and 112 fish stocks or species). All were considered likely to maintain viability under the ISC strategy.

Secretary of the Interior Manuel Lujan called for the Endangered Species Committee, also known as "the God Squad," to overturn the decision and allow a return to the status quo. (This committee is authorized in the ESA, and it is empowered to allow a species to go to extirpation or, perhaps, extinction if the social/economic costs of preservation are considered too much to bear.) So the ISC team was essentially put on trial by government and timber industry attorneys in a weeklong hearing in Portland, Oregon. In a politically humiliating decision for the administration, the Endangered Species Committee, which was composed almost entirely of the administration's own high-level political appointees, refused to excuse the northern spotted owl from the ISC strategy, or even to criticize the strategy or the ISC. The Bureau of Land Management's director, in agreement with the secretary of the interior (i.e., the administration) continued to refuse to comply with the strategy, and the judicially imposed shutdown of timber harvest on federal lands within the range of the northern spotted owl remained in place. The administration decided to make this situation an issue in the looming presidential election.

In the meantime, Congress, through the auspices of the House Agriculture Committee chaired by Eligio "Kika" de la Garza of Texas, entered the conflict by establishing a committee of four scientists—forest economist K. Norman Johnson of Oregon State University, forest ecologist Jerry Franklin of the University of Washington, forester John Gordon of Yale University, and wildlife biologist Jack Ward Thomas of the Forest Service's research division—to develop an array of alternatives for the consideration of Congress. A broad array of alternatives based on the best science available, and, this time, with assessment of social and economic consequences for each alternative, was delivered to the committees on schedule (Johnson et al. 1991). The so-called Gang of Four report was praised by Congress, but in the end no action was taken. All concerned settled back to wait for the outcome of the presidential election between President George H. W. Bush, H. Ross Perot, and William Clinton.

ENTER ECOSYSTEM MANAGEMENT

In the meantime, Forest Service Chief Dale Robertson issued two position statements that dramatically and conceptually changed the trajectory of the management of the national forests. The first was that the Forest Service would abandon clear-cutting as a timber harvest/regeneration technique, except in rare cases where it was obviously the best treatment. Second, the Forest Service would henceforth utilize ecosystem management as a guiding principal. President Bush reiterated this as national policy at the 1992 UN Conference on Environment and Development (the "Earth Summit") in Rio de Janeiro, Brazil.

The spotted owl debacle was a prominent issue in the 1992 presidential campaign. Bush and Perot condemned the effect of the ESA on the timber industry and promised modifications after the election that would eliminate or significantly diminish adverse effects on the industry and those employed in the industry. Clinton maintained support of the ESA, although he acknowledged that there was a significant problem that required resolution and promised a solution to this after the election (Yaffee 1994). Clinton won Washington, Oregon, and California and the overall election with a plurality of the votes cast.

The essential shutdown of timber harvest from public lands within the range of the northern spotted owl continued until newly elected President Clinton appointed yet another committee of scientists (the Forest Ecosystem Management Assessment Team, or FEMAT) to provide him with options for the management of the federal lands within the range of the northern spotted owl. FEMAT was expressly instructed to utilize an ecosystem management approach and to encompass all the federal lands within the range of the northern spotted owl in the plan (FEMAT 1993).

FEMAT was composed entirely of scientists and support staff. But this time, social scientists were involved in addition to economists, foresters, and biologists, and together they examined consequences of various alternatives on the nation, region, political entities, and communities. The team provided the president with ten options. He selected one of those options; following significant modifications by teams charged with preparation of the Record of Decision and Environmental Impact Statement, the option was instituted in 1994 and was still in place at the time of this writing. It became known as the Northwest Forest Plan, or the "President's Plan."

Unfortunately, late addendums to the plan, following its initial selection by President Clinton, added a fine filter approach dealing with more than 400 species thought to be associated with late-successional forests to what had been designed as a coarse filter plan. It has been argued that the fine filter addendum was necessary if the plan were to receive judicial approval. Whether or not that was true, the admixture of coarse and fine filter approaches made application unwieldy and inordinately expensive to execute, and helped produce a result wherein predicted timber yields fell far below projections and effects on rural timber-dependent communities were greater than expected.

Science, once again, had played a powerful new role in setting timber policy on the federal lands that had been ratified at the highest level. Lawsuits challenging the decision were immediately filed by those on both sides of the conflict—the so-called conflict industry. Judge Dwyer upheld the plan as satisfying applicable law.

The Northwest Forest Plan was the first truly large-scale ecosystem management plan—the result of a long evolutionary process that resulted from the interactions of evolving science, both the cumulative results of research studies and the interactions of laws previously discussed. Which action came first, and which action prompted what reaction, is a tangle of evolving public perceptions, new supporting science, and the laws of the environmental era. It is a classic example of the old dilemma: "Which came first, the chicken or the egg?"

NOW, A STEP TOO FAR

The Northwest Forest Plan dealt only with federal lands of the Cascade Range west to the Pacific Ocean. Obviously, the anadromous fish that were listed as threatened, or considered likely candidates for such listing, were associated with the Columbia

River system, which extended far beyond those confines. In 1995, President Clinton, speaking through Katie McGinty, director of the Council on Environmental Quality, asked the now chief of the Forest Service, Jack Ward Thomas, and the director of the Bureau of Land Management, Michael Dombeck, to extend ecosystem assessment, and ultimately ecosystem management, to federal lands on the "east side" of Oregon and Washington. The two agency heads explained that such assessment and planning (ostensibly leading to land-use planning operations of the two agencies) would need to encompass the entire watershed of the Columbia River if ecosystem management principles were to be appropriately applied—and they noted that the assessment should include private lands as well as public lands. The president agreed.

So, 117 years after John Wesley Powell's 1878 observations that the organization of government assessments and activities related to the management of natural resources should be based on watersheds as opposed to ecologically meaningless political boundaries derived from cadastral surveys, the government began to move in that direction in a serious fashion. However, this time around, agency heads and subheads, who had been precluded from control or even oversight of the ISC, Scientific Assessment Team, Gang of Four, and FEMAT, which were led and primarily executed by scientists, demanded that activities be directed by a collage of regional directors of federal agency stakeholders.

In addition, efforts were made to fully inform and seek advice from political leaders at county and state levels during the process. The mixed leadership produced a situation that was later likened to a square dance without a caller. Elected senators and congressmen, who did not like the outcome of the FEMAT exercise, quickly discerned that no action would be required as long as the effort did not reach completion. So by asking for more and more information, analysis, and review, they ensured that the effort would not be completed until there was a change in administrations.

The decision was then reached that there would be no attempt at a watershed-level plan for federal land management actions—that is, planning would proceed through individual management units (e.g., national forests and grasslands, Bureau of Land Management districts, etc.). It is unlikely, however, that it will be possible to produce a set of land management plans that are coordinated in terms of dealing with threatened or endangered fish species. These plans must be aimed at the preservation of ecosystem function at the appropriate landscape scale—another name perhaps for ecosystem management. That remains to be seen.

PUTTING THE WHEELS UNDER ECOSYSTEM MANAGEMENT

In the midst of this process, for two weeks in December 1995, over 350 natural resource managers, ecologists, economists, sociologists, and administrators gathered near Tucson, Arizona, to begin the process of assembling a knowledge base for the stewardship of the lands and waters in the United States using an ecosystem management approach. The purpose was to put a foundation under the concept of ecosystem management. Sponsors included the Forest Service, the Fish and Wildlife Service, the Bureau of Land Management, the National Park Service, the National Oceanic and Atmospheric Administration, the Geological Survey, the National Biological

Service, and the World Resources Institute. The results were published in three volumes with over 1,787 pages (Sexton et al. 1999)—a truly remarkable synthesis that contained all there was to know at the time relative to ecosystem management. Obviously, techniques and approaches to ecosystem management were being quickly and comprehensively developed (Boyce and Haney 1997; Hunter 1999). "Between 1992 and 1997, eighteen federal agencies adopted many of the core elements of an ecosystem-based approach. Agencies in several states worked to shift their natural resource agencies to adopt an EM perspective. Ecosystem management had boosters in high places, most notably Secretary of the Interior Bruce Babbitt and USDA Forest Chief Jack Ward Thomas" (Yaffee 2002, 89).

By 1996, 169 projects were under way that could be categorized as ecosystem management. All seemed to point to a bright future for ecosystem management. There was also momentum for adoption of the approach across the world (di Castri and Younes 1996).

TWO STEPS FORWARD, ONE STEP BACK

In spite of that progress, by 1999, the term "ecosystem management" had been essentially abandoned as far as agency policy statements were concerned because "in a legal and political system that rewards extreme positions, a policy that emphasizes science-based balancing and integration of interests satisfies no one" (Yaffee 2002, 89). However, ecosystem management persists despite lack of political popularity simply because, given the operative laws of the nation and the science behind ecosystem management approaches, there is simply no going back. To paraphrase Shakespeare, "A rose by any other name would smell the same."

THE EVOLUTION OF POLICIES RELATED TO FIRE: PART AND PARCEL OF ECOSYSTEM MANAGEMENT

Following the dramatic fire year of 1910, the Forest Service adopted the "10 A.M. Fire Policy," which simply stated that the objective was to extinguish any and all wildfires originating from any source reported in the national forest by 10 A.M. of the following day. Suppression of wildfire became on overriding mission on the theory that, until wildfires were under control, the broad scale institution of "scientific forestry" would be impractical, if not impossible. Other land management agencies, both federal and state, adopted that policy (Pyne 2001).

By the mid-1980s, detailed and voluminous research and assessment, spearheaded initially by research personnel at the Tall Timbers Research Station and followed by the National Park Service and the Forest Service, had clearly revealed that wildfire plays a significant role in fire-adapted ecosystems and is essentially impossible to completely control over the long term despite increased sophistication and resources. Policies were dramatically changed to allow some fires to burn "under prescription" (i.e., under specific circumstances), and "controlled burns" became a routine part of forest and range management activities (Pyne 2004).

Clearly, land management activities in ecosystems that evolved with fire must, both in reality and in keeping with principles of ecosystem management, sooner or later come to grips with the role of fire in land management. That accomplishment will likely be the big challenge for public land managers in the first quarter of the second century of the existence of the Forest Service (Pyne 2004).

Thus the trajectory of action within the pathway "toward a more fully holistic approach to resource and ecosystem use and management" was altered yet again (Fedkiw 2004, 7). Coupled with, or more properly included within, ecosystem management, this is the most recent dramatic example of a change in management approach. Intricate interactions of accumulating knowledge from research and experience, along with changes in public attitudes and perceptions, have led to changes in management direction from Congress and presidential administrations that are dramatically altering the management of the national forests. These dramatic changes in attitude and management have been initiated, stimulated, and essentially dictated by the evolving related science and by the influence of scientists over many decades.

SCIENCE AND POLICY: AND THE BEAT GOES ON

As a result of careful consideration of the history of the conflict over the northern spotted owl, Steven Yaffee observed:

> Given a set of policy processes that exist largely to satisfy the perceived needs of a variety of human groups, and manage the conflict between them, science and scientists are anomalies. Scientists often see themselves as purveyors of absolute truth, whose knowledge might be limited by available information, but are unbiased by the kinds of value often expressed in the political process. Political executives...believe that most truths are relative and dependent on the motives of individuals and groups that bring them forward. In their view, the decision making process works via persuasion, influence, and political compromise—activities seen by scientists as illegitimate and unprofessional. As a result, both groups are uncomfortable with the other. (1994, 231–232)

Yet science and scientists have had, and will continue to have, an increasingly important role in defining the pathway to sustainability relative to the management of the national forests in the Forest Service's second century. This was made clear in the first revision of the agency's planning regulations in three decades, conducted in 2004. The new rules established a dynamic process to account for changing forest conditions—emphasizing science and public involvement—that will drive planning activities and, in turn, national forest management. Furthermore, the new rule directs forest managers to take into account the best available science applied at the landscape level, considering ecological, social, and economic sustainability. Seemingly, ecosystem management, albeit by another name, is alive and well, and sustainability remains a guiding concept and principle. Continued evolution of the concept and application of sustainability in land management will likely continue to be driven by an ever-changing understanding derived from scientific inquiry and the synthesis of extant knowledge for application in national forest management.

It is not a comfortable fit—science and management—and never has been. That will likely always be so. But nonetheless, science and scientists have had and will continue to have considerable influence over the means the Forest Service uses to provide goods and services for the American people, in a sustainable fashion, in the second century of its existence.

References

Boyce, M. S., and A. Haney, eds. 1997. *Ecosystem management: Applications for sustainable forest and wildlife resources.* New Haven, CT: Yale University Press.

Bunnell, F. L. 1976a. Forestry-wildlife: Whither the future. *Forestry Chronicle* 52(3): 147–149.

Bunnell, F. L. 1976b. The myth of the omniscient forester. *Forestry Chronicle* 52(3): 150–152.

Carson, R. 1962. *Silent spring.* Boston: Houghton Mifflin.

Clarke, J. N., and D. McCool. 1985. *Staking out the terrain: Power differentials among natural resource management agencies.* Albany: State University of New York Press.

Clarke, J. N., and D. McCool. 1996. *Staking out the terrain: Power and performance among natural resource agencies.* Albany: State University of New York Press.

Clary, D. A. 1986. *Timber and the Forest Service.* Lawrence: University Press of Kansas.

Cortner, H. J., and M. A. Moote. 1999. *The politics of ecosystem management.* Washington, DC: Island Press.

Creative Act of 1891. U.S. Statutes at Large 26:1103, codified at *U.S. Code* 16, § 471.

Cubbage, F. W., J. O'Laughlin, and C. S. Bullock III. 1993. *Forest resource policy.* New York: Wiley.

di Castri, F., and T. Younes. 1996. *Biodiversity, science, and development: Toward a new partnership.* Wallingford, UK: CAB International.

Endangered Species Act of 1973. Public Law 93–205. *U.S. Statutes at Large* 87:884, codified at *U.S. Code* 16, § 1531–1536, 1538–1540.

Fedkiw, J. 2004. *Pathway to sustainability: Defining the bounds on forest management.* Durham, NC: Forest History Society.

Forest and Rangeland Renewable Resources Planning Act of 1974. Public Law 93–378. *U.S. Statutes at Large* 88:476, amended; codified at *U.S. Code* 16, § 1600 (note), 1600–1614.

Forest Ecosystem Management Assessment Team. 1993. *Forest ecosystem management: An ecological, economic, and social assessment.* Portland, OR: U.S. Forest Service.

Franklin, J. F. 1999. Foreword to *Bioregional assessments: Science at the crossroads of management and policy,* ed. K. N. Johnson, F. F. Swanson, M. Herring, and S. Greene, xi–xii. Washington, DC: Island Press.

Hickey, J. J. 1974. Some historical phases in wildlife conservation. *Wildlife Society Bulletin* 2(4): 164–170.

Hirt, P. 1994. *A conspiracy of optimism: Management of the national forests since World War II.* Lincoln: University of Nebraska Press.

Hunter, M. L., Jr., ed. 1999. *Maintaining biodiversity in forest ecosystems.* Cambridge: Cambridge University Press.

Johnson, N., and D. Ditz. 1997. Challenges to sustainability in the U.S. forest sector. In *Futures of sustainability,* ed. R. Dowser, D. Ditz, P. Faeth, N. Johnson, K. Kozloff, and J. J. MacKenzie, 191–280. Washington, DC: Island Press.

Johnson, K. N., J. F. Franklin, J. W. Thomas, and J. Gordon. 1991. *Alternatives for management of late-successional forests of the Pacific Northwest: The scientific panel on late-successional forest ecosystems*. A report to the Agriculture Committee and the Merchant Marine and Marine Fisheries Committees of the U.S. House of Representatives. Washington, DC.

Johnson, K. N., F. Swanson, M. Herring, and S. Greene, eds. 1999. *Bioregional assessments: Science at the crossroads of management and policy*. Washington, DC: Island Press.

Kaufman, H. 1960. *The forest ranger: A study in administrative behavior*. Baltimore, MD: John Hopkins University Press.

Langston, N. 1995. *Forest dreams, forest nightmares: The paradox of old growth in the inland West*. Seattle: University of Washington Press.

Lee, K. N. 1993. *Compass and gyroscope: Integrating science and politics for the environment*. Washington, DC: Island Press.

MacCleery, D. W. 2004. The historical record for the pathway hypothesis. In *Pathway to sustainability: Defining the bounds on forest management*, ed. J. Fedkiw, D. W. MacCleery, and V. A. Sample, 25–48. Durham, NC: Forest History Society.

McSweeney-McNary Act of 1928. U.S. Code 16, § 581, 581a, 581b–581I.

Meffe, G. K., L. A. Nielsen, R. L. Knight, and D. A. Schenborn. 2002. *Ecosystem management: Adaptive, community-based conservation*. Washington, DC: Island Press.

National Environmental Policy Act of 1969. Public Law 91–190. *U.S. Statutes at Large* 83:852; codified at *U.S. Code* 42, § 4321 (note), 4321, 4331–4335, 4341–4346, 4346a–6, 4347.

National Forest Management Act of 1976. Public Law 94–588. *U.S. Statutes at Large* 90:2949, amended; codified at *U.S. Code* 16, § 472a, 476, 476 (note), 500, 513–516, 518, 521b, 528 (note), 576b, 594–592 (note), 1600 (note), 1601 (note), 1600–1602, 1604, 1606, 1608–1614).

Norse, E. A. 1990. *Ancient forests of the Pacific Northwest*. Washington, DC: Island Press.

Noss, R. F., and A. Y. Cooperrider. 1994. *Saving nature's legacy: Protecting and restoring biodiversity*. Washington, DC: Island Press.

Organic Act of 1897. U.S. Statutes at Large 30:11, amended; codified at *U.S. Code* 16, § 473–475, 477–482, 551.

Pinchot, G. 1947. *Breaking new ground*. New York: Harcourt, Brace

Powell, J. W. 1878. *Report on the lands of the arid region of the United States*. 45th Cong., 2nd sess. H. Doc. 73m.

Pyne, S. J. 2001. *Year of the fires: The story of the great fires of 1910*. New York: Penguin Books.

Pyne, S. J. 2004. *Tending fire: Coping with America's wildland fires*. Washington, DC: Island Press.

Sample, V. A. 2004. Sustainability and biodiversity: The challenge of the future. In *Pathway to sustainability: Defining the bounds on forest management*, ed. J. Fedkiw, D. W. MacCleery, and V. A. Sample, 49–64. Durham, NC: Forest History Society.

Sexton, W. T., A. J. Malk, R. C. Szaro, and N. C. Johnson. 1999. *Ecological stewardship: A common reference for ecosystem management*. 3 vols. Oxford: Elsevier.

Soule, M. E., and G. H. Orians. 2001. *Conservation biology: Research priorities for the next decade*. Washington, DC: Island Press.

Thomas, J. W. 1979. *Wildlife habitats in managed forests: The Blue Mountains of Oregon and Washington*. Agriculture Handbook, no. 553. Portland, OR: U.S. Department of Agriculture, Forest Service.

Thomas, J. W., E. D. Forsman, J. B. Lint, E. C. Meslow, B. R. Noon, and J. Verner. 1990. *A conservation strategy for the northern spotted owl*. Report of the Interagency Scientific Committee to Address the Conservation of the Northern Spotted Owl. Portland, OR: U.S. Department of Agriculture, Forest Service.

Thomas, J. W., and R. Radtke. 1989. Effects of timber management practices on forest wildlife management. In *The scientific basis for silvicultural and management decisions in the national forest system,* ed. R. M. Burns, 107–117. General Technical Report, no. W-55. Washington, DC: U.S. Department of Agriculture, Forest Service.

Thomas, J. W., M. G. Raphael, R. G. Anthony, E. D. Forsman, A. G. Gunderson, R. S. Holthausen, B. G. Marcot, G. H. Reeves, J. R. Sedell, and D. M. Solis. 1993. *Viability assessments and management considerations for species associated with late-successional and old-growth forests of the Pacific Northwest.* Portland, OR: U.S. Department of Agriculture, Forest Service.

Transfer Act of 1905. Public Law 58–33. *U.S. Statutes at Large* 33:628, amended; codified at *U.S. Code* 16, § 472, 554.

U.S. Department of Agriculture, Forest Service. 1906. *The use book: Regulations and instructions for the use of the national forest reserves.* Washington, DC: Government Printing Office.

Water Resources Planning Act. Public Law 89–80. *U.S. Statutes at Large* 79:244, amended; codified at *U.S. Code* 42, § 1962m, 1962a, 1962a-1–a-2, 1962b.

West Virginia Division of the Izaak Walton League of America Inc. v. Butz, 367 F. Supp. 422 (1973), 522 F.2d 945 (1975).

Wickman, B. E. 1992. *Forest health in the Blue Mountains: The influence of insects and disease.* General Technical Report, no. PNW-GTR-295. Portland, OR: U.S. Department of Agriculture, Forest Service, Pacific Northwest Experiment Station.

Wild and Scenic Rivers Act of 1968. Public Law 90–542. *U.S. Statutes at Large* 82:906, amended; codified at *U.S. Code* 16, § 1271 (note), 1271–1287.

Wilderness Act of 1964. Public Law 88–577. *U.S. Statutes at Large* 78:890, amended; codified at *U.S. Code* 16, § 1131 (note), 1131–1136.

Wilson, E. O. 1985. The biological diversity crisis. *BioScience* 35:700–706.

Wilson, E. O. 1988. The current state of biological diversity. In *Biodiversity,* ed. E. O. Wilson and F. M. Peter, 3–18. Washington, DC: National Academy Press.

Wilson, E. O. 1992. *The diversity of life.* Cambridge, MA: Belknap Press.

Wilson, E. O., and F. M. Peter, eds. 1988. *Biodiversity.* Washington, DC: National Academy Press.

World Commission on Environment and Development. 1987. *Our common future.* Oxford: Oxford University Press.

Yaffee, S. L. 1994. *The wisdom of the northern spotted owl.* Washington, DC: Island Press.

Yaffee, S. L. 2002. Ecosystem management in policy and practice. In *Ecosystem management: Adaptive community-based conservation,* ed. G. K. Meffe, L. Nielson, R. L. Knight, and D. A. Schenborn, 89–94. Washington, DC: Island Press.

PART IV

MARINE CONSERVATION, ECOLOGY, AND MANAGEMENT

Lee M. Talbot

In part IV the authors examine the status of marine conservation, ecology, and management, analyzing the causes of present problems and examining what is being done to address the situation. The world's marine habitats have been significantly altered; marine living resources have been depleted by overexploitation, pollution, and conversion of estuarine habitats. Until recently, even among many scientists, the conventional wisdom was that the oceans and the marine resources were so vast that human activities could not greatly affect them. There are still many marine scientists and managers who downplay overfishing and who continue to apply and teach the scientifically discredited, simplistic approaches to fisheries management that have led to the depletion of stock after stock throughout the world's oceans.

However, new research has shown that human activities have profoundly impacted many of the world's marine ecosystems. Around 90% of the world's larger predatory fish already have been lost to overfishing, with significant and, in some cases, probably irreversible changes to their ecosystems. Further research is demonstrating that for whales and other forms of exploited marine living resources, the original population levels appear to have been very much higher than previously assumed. This means that the present populations represent severe depletion, even though they are being managed as if they were at healthy levels, with no signs of overexploitation.

In chapter 12, Sidney J. Holt analyzes the idea—fundamental to most marine exploitation to date—that extractive use of "renewable" living resources can be sustained by imposing simple management measures. He first reviews the history of the analyses and models developed to describe the reproductive behavior of marine living resources in relation to exploitation. Many of these were deterministic models; they led to the development of the concept of maximum sustainable yield (MSY), an approach to fisheries management that is still dominant even though it has been discredited scientifically for many years.

The deterministic models, such as MSY, predict reversible changes from one steady-state condition to another. They are based on the idea that the ecosystem is stable and unchanging, and they focus on a single species, ignoring the multiple interactions with the other species in the system. As such, they have little relationship with the real world.

Holt describes the origins of "the power of the otherwise disreputable MSY," showing that it came from the insistence in 1958 by the United States that MSY be accepted as the internationally acceptable reference point for management. This insistence was based on political considerations aimed at protecting the economically valuable U.S. fisheries from foreign exploitation.

Holt also addresses the question of sustainability. For a fundamental problem with the way sustainability is used is the question of the time frame: "How long is sustainable? In perpetuity? For a fish or human generation? Or two or three?" Holt notes that the matter of the time frame is "rarely mentioned explicitly and practically never discussed in depth," yet how to deal with variations through time is a core problem in efforts to manage for sustainability. Natural intrinsic population cycles can be very long (perhaps a century or more), yet the time period considered for a management regime is generally much shorter. Consequently Holt emphasizes that "the notion of sustainability is essentially meaningless except within a specified and appropriate time frame." And he notes the confusion between biological sustainability and economic sustainability. A biologically sustainable harvest might well be too low to be economically sustainable, and conversely, within a fixed time period, an economically sustainable harvest could lead to extermination of the resource.

Holt concludes by discussing problems of the current approaches to multispecies management and ecosystem management. Because of the complexity of the ecosystems and the limitations of knowledge, these approaches also suffer from a simplistic approach. He states that an ecosystem approach can be perverted to justify unsustainable exploitation of some of its parts; for example, proposed overfishing of some large predators in order to "save" the prey, or to regain a "balance." And he determines that "the economic and social forces favoring unsustainability are alive and kicking and looking for loopholes in the scientifically based conservation matrix."

John R. Twiss Jr., Robert J. Hofman, and John E. Reynolds III focus in chapter 13 on the U.S. Marine Mammal Commission, the principles of marine mammal management and conservation associated with it, and the status and conservation issues involved with varying groups of marine mammals. The authors' approach is to show that much of the past several decades' history of marine mammal conservation "is reflected in the background, content, and implementation of, and changes to, the U.S. Marine Mammal Protection Act (MMPA)."

The MMPA is one of a series of U.S. environmental laws passed in the early 1970s. At that time, there were a number of highly publicized incidents of abuses of marine mammals: hundreds of thousands of dolphins were killed in the tuna fishery, the International Whaling Commission failed to stop the decline of the great whales, and baby seals were killed on the Canadian ice floes. The public reaction was strong, and members of Congress said that during this period they received more mail on marine mammal conservation than on any other issue save the Vietnam War. This, and other factors described by Twiss, Hofman, and Reynolds, led to passage of the act.

The MMPA was particularly significant for a number of reasons, including a series of firsts in resource management legislation. For example, it was the first act to mandate an ecosystem approach, the first to establish the precautionary principle (now established in national and international law), and the first to require optimum sustainable populations.

The authors detail the key provisions of the MMPA, including the functions and composition of the Marine Mammal Commission and its Committee of Scientific Advisors, both of which the act established. They then discuss how the optimum sustainable population determinations evolved, and they detail how the ecosystem approach to marine conservation was developed. Several amendments were made to the MMPA; the authors examine their origins, significance, and effects. The Marine Mammal Commission's important role in a number of international marine mammal conservation issues is also considered. For example, the authors write at length about the commission's involvement with the International Whaling Commission and the negotiations leading to the Convention for the Conservation of Antarctic Marine Living Resources.

After providing a broad overview of past and current marine mammal conservation issues and the steps taken to address them, Twiss, Hofman, and Reynolds point out that conservation must be a "dynamic process" that takes into account both socioeconomic and biological-ecological factors. They emphasize that the public, the scientific community, environmental and industry groups, Congress, and the courts all play important interacting roles in marine conservation.

In addition to a number of areas where marine mammal conservation has been effectively achieved, the authors address the issues that have resisted solutions, and the new, previously unrecognized, and sometimes controversial issues that have arisen. One of the key attributes of the Marine Mammal Commission, they note, is its "ability to look forward and attempt to proactively address issues before they reach the crisis stage." Looking ahead, the authors state that the issues judged to be the most pressing in the near future are the "direct and indirect effects of fisheries, environmental contaminants, harmful algal blooms, disease, underwater noise, habitat degradation and destruction, climate change, difficulty identifying optimal management units, and ineffective management strategies."

In chapter 14, Michael L. Weber focuses on the ideologies that underlie and play a major role in determining the policies of government, management, and even science. Weber defines ideology as "the body of doctrines, myths, and beliefs that guides an individual, social movement, institution, class or large group." He considers that it includes a mixture of myth and fact that guides how we view the world and make decisions.

Prior to the 1970s the policies for marine wildlife were dominated by an ideology of abundance. Weber believes that the passage of the MMPA introduced a counterideology, the ideology of scarcity.

The ideology of abundance has a long history for marine species. It reflected a conviction that humans could manipulate nature to provide maximum use of resources, and that unused resources were wasted resources. A major effect of the ideology of abundance was to remove the need for caution. For example, believing in abundance, and thinking that the fishing yield was only limited by the capacity of the fishers, until the late 1980s the U.S. government had as a principal goal the expansion of the U.S. fishing fleet. With this approach, it followed that there was greater danger of underexploiting fish than overexploiting them, so the burden of proof was on those who urged conservation.

Weber describes the factors that led to the passage of the MMPA, with its underlying ideology of scarcity. But he also describes the continuation and hardening of the ideology of abundance in the fishing industry, and the subsequent passage of the Fishery Conservation and Management Act in 1976. This led to a vast increase in the U.S. fishing boat fleet and, by the late 1980s, to the decline, through overfishing, of many fish stocks, which Weber notes emphasized how wrong our earlier views were about the potential impacts of fishing. In reaction, reform efforts reflecting the ideology of scarcity led to passage of the Sustainable Fisheries Act of 1996. The author discusses the developments since then and contends that there are significant challenges ahead before the ideology of scarcity really overcomes the practitioners of the ideology of abundance.

12

Sustainable Use of Wild Marine Living Resources: Notion or Myth?

Sidney J. Holt

The idea that extractive use of "renewable" living resources can be sustained, and even optimized, by imposing simple management measures, such as placing limits on the scale and type of the extractive process, is wishful thinking. It was conceived in late-nineteenth-century Europe and emerged full-blown in the 1930s. It is closely associated with the development of capitalism as a globalizing market economy and with technological innovation, especially the use of fossil fuels to provide cheap and mobile power. I have reviewed this history in some detail elsewhere (Holt 2006) and would draw attention also to the final chapter of the same book (Lavigne et al. 2006). Here I mention some elements of that review as a preliminary to consideration of the derived notion of maximum sustainable yield (MSY) and related "indicators" as desirable objectives in managing extractive uses.

The idea was to modify the 1803 proposition by Thomas Malthus that populations tend to increase exponentially, but that the means of generating such increase grew only linearly; when the exponential and straight lines intersected there would be a catastrophe, even revolution. The modification was to suppose that the exponential rate of increase itself decreases as the population gets bigger over time so that eventually an equilibrium would be attained. Pierre Verhulst (1838, 1845, 1847) put that in mathematical form by supposing that the (initial) exponential rate of increase would be a linearly decreasing function of population size. That leads to the familiar logistic curve of population growth, with an inflection at the point where the population has reached half its final equilibrium size.[1] If it is reasonable to assume this growth process, then extracting catches (harvests) in such a way as to hold the population at the inflection size should provide a maximum continuing yield equal to the slope of the population growth curve at that point.

This theory was applied in the 1930s to the stock(s) of blue whales in the Southern Ocean and later to cod in the North Sea and to yellowfin tuna and sardines in the eastern tropical Pacific, although the former application referred to numbers of animals while the latter three referred to population biomass (Graham 1935, 1939; Hjort, Jahn, and Ottestad, 1933; Schaefer 1954; Schaefer, Sette, and Marr, 1951).[2]

Subsequently, scientists working with fish and whales adopted modified versions of the simple logistic, in which the exponential rate of increase was arbitrarily assumed to vary in a curvilinear manner (expressed as segments of a parabola or other polynomial, but always oriented such that the highest point occurred at zero population size) with population abundance or density—these two quantities have not always been clearly distinguished from each other—so that a maximum continuing yield would be obtained by maintaining the population at somewhat below half of its "natural" size (about 30% in most applications to exploited stocks of fish) or above that level (about 60% in applications to whale stocks made in the 1970s under the New Management Procedure adopted by the International Whaling Commission [IWC] in 1975) (Pella and Tomlinson 1969; Richards 1959).[3]

In parallel with these approaches to fisheries management by application of variants of the logistic curve, another idea was pursued in the late 1930s through the 1960s and beyond, associated primarily with Michael Graham of the English Fisheries Laboratory in Lowestoft, Suffolk. To obtain high continuing catches with minimal fishing effort, the fish should be caught when they had reached an optimal age and weight—that is, when each cohort (generation) had reached a maximum total weight through the contrary processes of natural mortality and growth in individual size. Since it is not practicable to catch in a single sweep an entire cohort of wild fish, the fishing effort and the age/size at which the fish first became liable to capture should be adjusted, together, in order to increase and possibly maximize the total catch from a cohort during its exploitable life. This approach was formally elaborated by R. J. H. Beverton and S. J. Holt ([1957] 2004),[4] illustrated by curves of yield per recruit (Y/F) against fishing effort and, by implication, population size, as well as by three-dimensional illustrations of "eumetric fishing," defined by a proper balance between fishing intensity and size-electivity of fishing gear. Curves of Y/F against fishing mortality rates (presumed to be directly proportional to calibrated indexes of fishing effort) could be asymptotic or peaked, according to the relative values of the parameters of growth in body weight and of natural mortality in the exploited phase of the population.

The focus of these studies was stocks that were thought to be overfished. For the demersal fish commonly exploited in the northeastern Atlantic it was found that currently, or in the recent past, a reduction in fishing effort would provide a higher sustained Y/F and hence generate a more efficient industry. That is, we were dealing with situations to the right of the peaks of domed curves. Any error in such an assessment was most likely to underestimate the benefits of reducing effort, since there might also be an increase in the number of recruits, coming from a larger population of parents. Later studies of some pelagic species suggested that their parameter sets more likely generated asymptotic curves of Y/F against fishing effort. Even in those cases, while decreasing effort would not be likely to lead to higher catches, about the same level of catch could be obtained with considerably less effort.[5]

Beverton and Holt supplemented this analysis by adding a simple function relating the number of recruits to the size of the parent population, thus producing what they called a *self-regenerating population model*. This model could only be evaluated by tedious iteration, and their computing capacity was limited to producing only one trail calculation, for the North Sea haddock. Subsequently, with more computing power available, Beverton was able, just before his death in 1995, to begin to explore more fully the properties of this model (Beverton and Anderson 2002). Most important, these simulations showed the population being exterminated if the fishing effort were to be sustained above a certain level, which would generate a critical or higher value of the fishing mortality rate. This theoretical phenomenon was noted and commented on in 1998 by A. J. Pitcher.

All the above-mentioned single-species models of exploited wild populations have essentially the same basic properties, exemplified through the shapes of their curves of "sustainable yield" against population size or density. These are all peaked, and the locations and amplitudes of the peaks provide the rationale for identification of both MSY and MSY levels of population size (MSYL) as possible management objectives. Other possible objectives have been suggested based on the slopes and amplitudes of these curves at defined locations (Food and Agriculture Organization of the United Nations [FAO] Fisheries Department 1997).[6] Despite many published criticisms of these ideas, most famously by Peter Larkin (1977), they persist in the scientific literature and in political debates about the questions of conservation and management. I shall later examine why that might be.

Other features of the models referred to above are that the curves are smooth, without inflections, and that the highest rate of natural population increase occurs when the population is so small as to be almost extinct. Furthermore, that extinction only occurs if the critically high exploitation rate is maintained. It can reasonably be argued that this latter situation is unrealistic because the exploitation rate would be expected to fall, or exploitation to cease for economic reasons, before biological extinction would occur, a situation that has been called "commercial extinction" in the fisheries literature. However, that condition may not be met in multispecies fisheries in which exploitation of the more vulnerable species continues while economic gain comes mainly from more abundant, if perhaps less valuable, species. One oft-cited example of this phenomenon is the continued exploitation in the Antarctic of the smaller species of baleen whales after the blue whale had been depleted to an uneconomic level; a fishery example is the virtually incidental exploitation to extinction of the skate in the North Sea while trawling continued for more abundant and biologically less vulnerable species.

Perhaps the most general property of these deterministic models is that they predict reversible changes from one steady-state condition to another, as their parameters are altered by externally imposed events such as exploitation by humans, and natural or human-caused environmental changes.

With this background we are in a position to look more deeply into whether models with such properties can be plausible representations of the real, natural world for the purposes of conservative and precautionary management of extractive uses of these kinds of resources. But first, a diversion to the origins of the power of the otherwise disreputable MSY.

In the early discussions of the phenomenon of overfishing to which I have referred, the existence of maxima in catch curves was noted but not usually taken to suggest desirable objectives. The focus in Britain and other European countries was generally improvement of an undesirable situation. Improvement, for the longer term, not optimization, was the dominant theme. In the post–World War II period Graham and his European colleagues emphasized the benefits to fishermen and fish consumers of restraining the growth of fishing power (more, bigger, and more powerful ships and gears) and hence ensuring higher and more valuable catches over time at lower cost. In North America the emphasis was entirely different. In consideration of the freedom of anyone who wished to go fishing, discussion of limiting fishing power was generally ruled out of order, for political/ideological reasons, and management options were largely reduced to limiting the deployment of fishing power (by, e.g., restricting the legal fishing season) and setting upper limits to catches by fleets.

In the run-up to the post–World War II International Fisheries Conferences of 1958, convened by the FAO and the United Nations itself, the United States pressed very strongly for attainment of MSY to be recognized as the internationally acceptable reference point for management. It was argued by many—including the FAO Fisheries Division at the time—that attainment of MSY was, apart from its other disadvantages, economically suboptimal. Beverton and Holt's curves of net economic gain against fishing intensity crudely showed this, and later analyses by the Canadian mathematician Colin Clark (1971), taking discount rates into account, revealed just how undesirable MSY was as a management objective for the improvement of fisheries.[7] Nevertheless, the United States and other delegations argued that economic criteria could not in practice guide management decisions in the international arena. In considering the implications of that position it should be remembered that at that time most of the main fishing nations strongly opposed any extensions of national jurisdictions beyond the traditional three-mile territorial sea, while only a few others would be prepared to discuss six- or twelve-mile limits.

There was, however, an overwhelmingly strong political reason for U.S. insistence on the MSY formula. By far the most valuable fisheries along the west coasts of North America were for the several species of Pacific salmon; only much later was their economic, and hence political, dominance to be replaced by the growth of the tuna industry. The salmon were taken in rivers and estuaries and in waters close to the American coasts. However, there being few salmon runs on the Pacific coasts of Asia, the Asiatic fishing countries—primarily Japan and the U.S.S.R.—were anxious to catch Pacific salmon on the high seas. The United States and Canada sought to oppose this on grounds of efficiency: surely it made sense to exploit salmon at low cost in the rivers rather than expensively in the open ocean. The North Americans also pointed out that they went to the trouble and expense of keeping the rivers in which the salmon breed clean, open, and productive. But these "reasonable" arguments carried little political weight internationally.

At this point the U.S. authorities produced the idea that where a resource was already being "fully utilized," and its utilization properly managed by coastal nations, others should refrain from exploiting it. This abstention principle was built into the treaty setting up the International North Pacific Fisheries Commission (INPFC) then being negotiated by Canada, the United States, and Japan. But the question arose: How is full utilization to be defined? The (apparently) simple and quantifiable answer was: when MSY is being taken. Japan very reluctantly accepted this position, though the U.S.S.R., not being party to the INPFC, could hardly be expected to comply with it.[8]

The spin-off from this decision has been fateful. Eventually the concept of national jurisdictions extending 200 nautical miles from the coast was universally accepted. Nevertheless, the hold of MSY has been extraordinarily tenacious. It was accepted as a "marker" when the IWC tried to bring some order into commercial whaling in the 1970s, despite the absence of any reference to it in the 1946 International Convention for the Regulation of Whaling, under which the IWC operates. It has been argued by Japan that failure to reduce a resource down to its MSY level constitutes unacceptable waste of that resource. Ultimately it was introduced in a modified but essentially inoperable form in the new UN Convention on the Law of the Sea. It is an objective for high seas fisheries; it is also a marker for fisheries in exclusive economic zones (EEZs) in the sense that if coastal states are not fully utilizing the resources in their zones, they are expected to be ready to license others to take the "surplus." In international law we are for the moment stuck with it, although no one seems to be taking it seriously in ongoing international fisheries negotiations and disputes.

I now return to the implications of the form and properties of the simple population models I have briefly described. It has long been known that matters change dramatically when two species, each described by such models, interact, as in predator-prey relationships. Vito Volterra (1926, 1931),[9] an Italian mathematician, and Alfred J. Lotka (1925), an American mathematical biologist, revealed that when there are such interactions, cyclic and pseudocyclic variations in the abundance of both species could be expected. Later it became clear that similar cycles could be expected in single-species models, depending on the parameter values found or assumed and such features as delays in reproduction and other time lags in the system. In what is now regarded as a classic seminal paper, Robert May (1976) demonstrated that simple models can have very complicated behavior, including closed cycles and chaotic, unpredictable behaviors, depending on their parameter values.

More recently Lars Witting has radically challenged the basic assumption in all previous models that the intrinsic population growth rate when the population is small is exponential (1997, 2000). Witting's new formulation, derived from genetic considerations, leads to population behaviors including intrinsic long cycles of abundance. He has successfully applied his radical theory to explain the observation that the gray whales of the northeastern Pacific have recovered under protection, after intensive exploitation resulting in their near extermination, to an abundance apparently very much higher than that which prevailed before commercial whaling began in the nineteenth century, despite the great reduction in their breeding areas, especially lagoons in southern California and along Baja California (Witting 2003).[10] Witting's convincing analysis has quite dramatic implications for the conservation of this species. An assessment made by the IWC Scientific Committee, using traditional methods, led to the conclusion that the current replacement yield of this population (the number that could be killed during the year and leave the population next year unchanged, a sort of subsustainable catch) is several hundred animals. Witting's result is that the replacement yield now is in fact negative; that is, the population will begin, or is beginning, to decline naturally, and that any catch from it will simply increase the rate of decline.

Economists, businessmen, and accountants also have problems with cycles—the so-called business cycles. Conventional economic models assume that a perfect market can track changes that represent transient states between presumptive steady states of the economic system that, in practice, are never reached. But they cannot successfully predict the business cycles that will eventually modulate the behavior of investors, each of whom is assumed to act as an independent agent. To explain the business cycles, systematic variations in external ("environmental") factors are necessarily invoked. But even then the cycles are irregular and essentially unpredictable. Unconventional models, in which it is assumed that each agent's behavior is affected by the behavior of some others, do much better; they generate fluctuations that appear as irregular cycles without the need to invoke externalities. This does not mean that the expected cycles can be successfully predicted, but at least they can be understood post facto, and wise investors can avoid the erroneous hope that they might be predictable.[11] By the same token we should expect flaws in population models that implicitly assume that all animals are independent agents, and that they do not modulate their feeding behavior in accordance with their knowledge of what others in the population are doing.

Fisheries biologists in the first half of the twentieth century, especially those studying small pelagic fish such as herring, sardines, anchovies, and the like, were practically obsessed with the problem of understanding large year-to-year fluctuations in abundance that were mainly due to even wider fluctuations in the numbers of recruits. Great efforts were made to predict these fluctuations; correlations were found with various environmental measures, which almost invariably broke down when applied for the coming year. Little attention was given to the sustainability of such fisheries—until they crashed, sometimes as a consequence of large-scale oceanographic changes, probably more often as a result of the opening of new industrial fisheries that massively removed the younger animals. The struggle to introduce conservation measures for these species posed the critical question: How long is sustainable? In perpetuity? For a fish or human generation? Or two or three? Depending on discount rates? I shall return to these questions later, but I note here that in the literature on conservation and sustainable use the matter of the time frame is rarely mentioned explicitly and practically never discussed in depth. Yet how to deal with variations through time (including cycles and what I have more generally referred to as "vibrations" internally generated within the system) is a core problem in efforts to manage for sustainability.

I turn now to a general problem of another kind. Considerable attention has been given to the shapes of curves of sustainable yield against population size or exploitation rate around the middle of the range of population size and toward its upper limit, sometimes referred to—misleadingly—as "carrying capacity" or, in the more colorful lingo of whaling, "virgin stock." Depletion of the resource is most commonly expressed in terms of relative or absolute distance from—that is, below—the "natural" asymptote. But if that asymptote is naturally varying periodically (vibrating), what is the appropriate measure of depletion or the index of conservation status?

In the context of self-regenerating population models, attention has equally been given to the shapes of the upper parts of stock-recruitment curves. A popular alternative to the Beverton-Holt asymptotic curve of 1957 has long been a domed curve proposed by W. E. Ricker (1954). In 1982 J. G. Shepherd published a model with one more parameter, of which the Beverton-Holt and Ricker models, as well as another proposed by D. H. Cushing, are special cases. Several other three-parameter models, all giving domed curves at least for some sets of parameter values, have been offered, and occasionally used, by a number of workers (reviewed by Quinn and Deriso 1999). With few exceptions these models predict a continuous and accelerating decline in recruit numbers as the population is pushed down toward the origin (zero population, zero recruits), that is, below the population level of the dome. One exception is the Gamma model proposed by Reish et al. (1985). This has the Ricker and Cushing models as special cases; it cannot, however, capture the asymptotic property of Beverton-Holt. It does, however, show for certain parameter values an inflection toward its lower end. This signifies the existence of depensation, meaning that the rate of recruitment increase is not at its maximum very close to the origin but, rather, at a distinctly higher level. This phenomenon is sometimes referred to in the literature as the Allee effect (Allee 1931). Another exception is due to G. G. Thompson (1993), who generalized the Beverton-Holt model in such manner that with certain parameter values depensation arises.

Turning now to the properties of surplus production models of the logistic and similar families, I note that they generally have similar properties to those of most of the stock-recruitment models I have mentioned—domed curves with no inflections or downturns to the left of the peak, and thus lacking depensation. Occasionally depensation has been introduced in such models by offsetting the population size axis by a constant number, so that the domed curve passes not through the origin but through a finite population size (Quinn and Deriso 1999). The introduction of a specific absolute threshold seems to me to be a clumsy and unsatisfactory way to introduce depensation. However, the Beverton-Holt self-regenerating model (akin to a surplus production model but with age- and sex-structure incorporated) can—and for most parameter sets investigated usually does—exhibit depensation even though the simple stock-recruitment function in it does not itself have an inflection. In this model the sustainable yield curve always passes through the origin.

There have been some attempts to detect depensation in fisheries data sets, notably by Ransom A. Myers et al. (1995) and by M. Lierman and R. Hilborn (1997). Detection in surplus production models is practically impossible for lack of reliable data, so attention has focused on the stock-recruitment relationships. Myers et al. and Lierman and Hilborn looked at data for 128 fish stocks and found it difficult to detect depensation in any but a few cases, one of which was the cod (*Gadus morhua*) in the Gulf of Maine, published by the U.S. National Research Council in 1998.

A problem with this approach is that the absence of depensation is not necessarily, in management situations, an appropriate null hypothesis. The existence of depensation can have four interconnected consequences. First, if a relatively low level of fishing is maintained when the population has been depleted, the population can be driven to extinction if the population is reduced to or below a minimum viable population level. Second, if fishing ceases, or is sharply reduced in intensity, then the population, if it begins to recover, will do so at a slower rate than non-depensatory models will predict. Third, the process of population increase and decrease may not be reversible. Fourth, if the condition for critical depensation exists (that is, where the relative population growth rate goes to zero and even becomes negative at some positive population size), the population may proceed to extinction even if fishing ceases (Clark 1985). As Clark points out, "More realistic models of extinction would of course be based on probabilistic considerations," and he cites Ludwig (1974) as one of the earliest researchers studying that matter (1985, 109). Evidently, the practical consequences of either of the hypotheses (depensation vs. no depensation) being correct or incorrect are strongly asymmetric. Under the usual conditions of considerable uncertainty as to population dynamic characteristics and parameter values, an assumption that depensation is absent, when it is actually present, can lead to excessive depletion and other undesirable changes. On the other hand, an assumption that depensation exists when it does not leads only to allowable catches being less than they could be from relatively depleted stocks.

When thresholds for designating endangerment of species or wild populations have been considered by scientific groups seeking to provide advice on national or international regulations concerning trade, or moratoria on exploitation (e.g., in the context of the Convention on International Trade in Endangered Species of Wild Fauna and Flora, IUCN–World Conservation Union "red books," the U.S. Endangered Species Act, and the like), a simplistic approach has usually been taken. Such approaches include proposing a minimum absolute population size at some assumed threshold of viability, or a depletion level measured as the population size relative to some previous unexploited population, or some combination of these, and occasionally with a time element inserted. An example with which I am familiar is the New Management Procedure (NMP) of the IWC, which was in effect from 1976 until implementation of an indefinite moratorium on all commercial whaling in 1986. This mandated that exploitation of a whale stock should cease if and when it had been reduced to less than 54% of its original number, that is, its size before exploitation began. This figure was entirely arbitrary, derived from an assumption that MSY would be obtained from a stock reduced to 60% of the original number, which was an assumption with very weak scientific basis. The fundamental aim of the NMP was to optimize exploitation, not merely to avoid risk of extinction or other irreversible phenomena, but it was assumed that the population level at which such risks would be manifest was below the 54% threshold, and probably far below it. This was a presumption that was never tested, primarily because the parameters of the basic population model—a modified logistic with age- and sex-structure inserted—could not be estimated reliably, anyway.

There have been several approaches to modeling depensation, usually by adding another parameter to a simple compensation model. Failure to use these in fisheries assessments perhaps derives from a misunderstanding of Occam's razor, since the additional parameter appears to add another element of complexity. But introduction of depensation does not necessarily involve inserting more parameters. An important example has been elaborated by the Russian physicist Sergei Kapitza in papers that appear to be little known outside the field of human demography.[12] Kapitza's model for the growth of human populations over many past millennia, and projected some decades and even centuries into the future, gives an apparently "ordinary" sigmoid curve eventually reaching an upper asymptote. It has only two variable parameters and one constant. Depending on the parameter value the maximum absolute growth rate may occur at population levels well above or well below half the asymptotic levels, so in that sense it can mimic the Pella-Tomlinson models in the fisheries field.

In this model depensation can occur at a population level not far below what would be the MSY level in fisheries terms.

The Kapitza model has important special characteristics. In very small populations the intrinsic growth rate is hyperbolic, in contrast with the exponential, the logistic, and the "hyperexponential" of Witting's models. This, consistent with the depensatory process, generates very slow growth in small populations, such as is seen in the slow recovery of depleted wild animal populations that are protected or subjected to a reduced exploitation rate intended to permit restoration.

I now look at some of the practical implications of using population models to provide management advice, of which neither the structures nor their parameter values can be justified empirically. After many decades of observations and experiments the actual processes of density dependences in reproduction, mortality, and individual growth rates remain largely unknown. Particular processes that seem to provide hard evidence are uncommon in the literature. For example, in the study of whale populations several researchers have apparently shown that as a population is depleted, or as the food supply per capita is otherwise caused to increase, the age at sexual maturity declines, providing a force that may tend to enhance reproduction.[13] Such studies are confounded by biases in time series of age composition data coming from examination of commercial catches that are selective. Even if such biases are somehow removed we are left not knowing, for example, what other density dependences or counterprocesses might be operating to change fecundity or natural mortality rates of the young animals—pre-recruits, in fisheries terminology. Therefore such isolated partial observations provide little insight to the integrative operation of any or all density-dependent processes.

In an analysis of the properties of the IWC's NMP, William de la Mare (1986) showed by computer simulation (the first time such a method was used in fisheries studies) that this procedure would not be conservative of whale populations even if the model was structurally correct and its parameters perfectly estimated. This revelation was instrumental in encouraging the IWC scientists to look at completely different possible ways of managing future whaling for sustainability and with due precautions and safeguards. The method eventually adopted was to invent an algorithm for computing annual catch limits that could be shown by simulation (using population models) to provide numbers that would meet prespecified management goals. In the case of the IWC these goals were, in order of priority: (1) avoid accidental reductions of the population at any time during the simulation period that might bring it below a specified threshold level; (2) obtain as high a cumulative catch as possible over a specified time interval consistent with the first goal, but also allow maximum possible recovery of depleted populations toward levels close to the presumed pristine levels; and (3) reduce the need to vary catch limits greatly from one year to the next unless circumstances showed it to be absolutely necessary (this goal is, of course, purely to meet the requirements of industry and commerce). Appropriate algorithms were produced by de la Mare himself, Douglas Butterworth, and Justin Cooke (1994), but, in a competitive process, Cooke's was favored by the IWC's Scientific Committee and by the IWC (described in a series of publications defining a catch limit algorithm [CLA] within a Revised Management Procedure [RMP]).[14]

An essential part of the CLA development process was the testing of candidate algorithms using data generated from a number of different population models,[15] and from selected models with various sets of parameter values (which, for example, might specify an MSY level of 70% of the unexploited level or, alternatively, 30%). However, all the models and parameter sets tested had some common characteristics. None manifested depensation, and none had intrinsically driven population cycles, as in Witting's models. So the level chosen, below which the stock should not be unintentionally driven, was entirely arbitrary, more or less as in the old NMP, and recovery from low population levels would theoretically be relatively fast. These features could have very undesirable management consequences if in fact the true population behavior was more like that shown in Kapitza's study.[16]

Another feature of the CLA development process was a decision to carry out simulations over a period of one hundred years. This decision was made primarily because of the limitations of available computing resources, but also with an element of consideration for the life spans of whales and of the human predators hunting them. However, the subsequent application by Witting of his model to the gray whale showed this species (and presumably at least other baleen whales) to have a natural oscillation of roughly the same frequency as the chosen simulation period. Thus limitation of the duration of the simulations could mislead management considerations. And here is a dilemma. The natural, intrinsic population cycles can be long, with important consequences whatever definition of sustainability might be adopted, yet the time period over which it could be reasonable to expect the application of a management procedure to be continued, essentially unchanged, is now surely, in terms of social stability and continuity of legal and administrative norms and arrangements, much less than a century.[17]

This dilemma forces us to confront a feature of discussion about conservation that is almost never acknowledged explicitly: the notion of sustainability is essentially meaningless except within a specified and appropriate time frame. It cannot, for all sorts of reasons, be forever, whatever that means.[18] And, further, a precautionary approach to the attainment of sustainable use of a wild living resource necessarily involves an assumption of possible depensation and intrinsic oscillation in the population. If the management regime is to be based on the method first examined by the IWC, the simulations of possible population models must provide the possibility of these features.

I close with a couple of questions, for further contemplative thought and scholarly exploration. Who, other than conservationists and administrators, and maybe also lawyers with tidy minds, really desires sustainability and sustainable use, and what exactly do they think they mean by that? Likewise, who truly seeks "optima" and "maxima"? Science does not prescribe them, and the progress of ecological science tends to confuse and complicate rather than clarify these issues. Economists can provide insights involving the concepts of discount rates and time lags, business cycles and externalities, but these generally move us further from an ability to define good management.

The most common confusion in the minds of the general public and, it seems, politicians, is between biological sustainability and economic sustainability. Long ago Colin Clark's work showed that, provided a finite and limited time frame is acknowledged, the latter might be obtained by orderly extermination of the resource. On the other hand, while a limited continuing catch of minke whales in the Antarctic, under the precautionary RMP, might be biologically sustainable, it would most likely not be as economically sustainable—that is, profitable—as a high-seas, distant-water operation with one or more factory ships and their squads of catcher boats operating thousands of miles from their home bases.

This, I think, explains the need for what is left of the Japanese commercial whaling industry to be subsidized, indirectly, by subventions from the government for "scientific research." It also provides a strong motivation for engaging in more intensive unsustainable exploitation, under the guise of culling the species supposedly for population control purposes. In these circumstances, as I believe is happening now in both Japanese and Norwegian minke whaling operations, it becomes important for operators to adjust the scale of their catches, not in accordance with biologically based calculations but rather to balance and optimize commodity production, taking into account the sensitivity of price to supply levels (i.e., the elasticities of supply and demand). I suspect that that is precisely the sort of calculation recently made by both the Norwegian and Japanese industries and authorities, which have, while the commodity price is very high—too high for many younger potential consumers— and unsold stocks of meat accumulate in freezer warehouses, decided to dramatically increase their annual catches of minke whales and, in the case of Japan, to add a few catches of other endangered species such as fin and humpback whales.

In this chapter I have so far remarked exclusively on single species management, and not touched some important aspects even of that. One such aspect is the contraction of the geographical range of a population as it is reduced by exploitation. This has been explored rather thoroughly in a marine fisheries context by Alex MacCall (1990) and I shall not pursue the subject further here. His basin theory takes into consideration the fact that as the range contracts, and animals are reduced relatively more at the edges (i.e., the less productive or otherwise less suitable habitats), indexes of average density will decline relatively less than the population number or biomass. This has, of course, practical consequences where only average density is estimated rather than population size (as in most studies of catch-per-unit fishing effort) but also introduces complications into the consideration of how density dependence is to be modeled.

Proper consideration of multispecies management, the currently fashionable ecosystem management (more appropriately referred to as the ecosystem approach to management), and what I would like to call the biosphere approach to management would call for another essay. However, some connecting features are worth noting here.[19] In the report of the First Airlie House Workshop on Sustainability written by Lee Talbot and myself (published as Holt and Talbot 1978), we referred to the optimization of sustained yield while not wasting other resources. That conditional statement has been misunderstood and it was surely too crisp, even opaque, as presented. One such waste that we had in mind at the time was the use of excess energy and other resources needed to support continued fishing effort in taking a yield that, while it might be biologically sustainable, perhaps even at maximum level, would be only marginally greater than a yield that could be taken with significantly less effort and probably with greater profit, as would be the case if the curve of sustainable yield against effort was domed but rather flat-topped. Such input resources do not usually originate within the ecosystem being exploited. This was a way of expressing Michael Graham's dictum that fishing effort should be conserved as well as fish, that overfishing meant using too much effort to get too little yield.

Modeling multispecies and ecosystem management involves, at least, hypothesizing and quantifying the multitude of relationships—in current practice mostly predator-prey relationships—between virtually innumerable pairs of elements in the multispecies/ecosystem network. Most such hypotheses have to date been extraordinary simplistic, leading, I suspect, to the same kinds of errors that have arisen from uncritical application of the simple logistic in single species assessments. Thus most models contain the assumption that an animal will eat a variety of foods in its diet in proportion to the relative abundances of those items in the immediate environment, an assumption that is clearly at variance with the results of modern studies of animal behavior. In some current ecosystem models it is assumed that all that really matters is the movement of energy and biomass through the network, yet we know from research on human nutrition and on domesticated animals that quality matters as much as quantity, ranging from such matters as ingestion of the right proportions and types of proteins, carbohydrates, and fats, through access to essential trace elements, vitamins, and even to the deliberate ingestion of medicinal organisms, in which several animal species have been shown to engage. Thus I would argue that we have a long way to go before we are ready to engage seriously in such forms of management.

A worrying feature of current debates about sustainability, conservation, and wise use of living resources is the ease with which the ecosystem approach can be perverted to justify unsustainable exploitation of some of them, usually the most valuable ones, in terms of market price per unit commodity weight. This is already evident in relation to the sealing and whaling industries, and there are signs of it in discussion about deliberate overfishing, in the traditional sense, of other large predators such as cod, tuna, and sharks in order to save the prey and maintain balance in the ecosystem. Some whaling interests are even arguing that after all the largest whales have been reduced to virtually negligible numbers, the remaining smaller species should likewise be depleted in order to restore a mythical pristine balance; this, of course, rather than allowing the depleted species to recover, if biologically possible. Let us note that the notions of sustainability and sustainable use, now being thoroughly respectable, are embedded in national laws and international agreements. The economic and social forces favoring unsustainability are alive and kicking and looking for loopholes in the scientifically based conservation matrix.

Notes

1. A historical review of this subject, possibly little known among fisheries scientists, can be found in an essay by Sharon Kingsland (1982).

2. Johan Hjort was the grand old man of fisheries science in Norway, and a major luminary in northern and western Europe. Putting his name on a paper gave it great weight at the time, but the application discussed here was probably devised by P. Ottestad (1933). Hjort and his colleagues referred to the difference between reproductive and

natural mortality rates as regeneration, a difference that IWC scientists have commonly referred to as net reproductive rate.

Additionally, note that Graham's 1935 and 1939 works, along with several other relevant papers published between 1935 and 1983, are conveniently reproduced in a volume compiled by D. H. Cushing (1983). That volume has useful linking commentaries by the editor, whose own important 1983 contribution on stock and recruitment closes the anthology.

3. The original formulation in the fisheries context was by Jerry Pella and P. K. Tomlinson. This particular modification of the simple logistic was originally devised by F. J. Richards in a different, botanical context.

4. The foreword to the fourth imprint (2004) of this work contains my review of events in what Graham termed the *theory of fishing* since the original publication. The 1993 imprint of the book contains forewords by T. I. Pitcher and Daniel Pauly (unfortunately not contained in the 2004 imprint) that are worthy of consultation, as well as corrections to the original compiled by J. Hoenig.

5. In an otherwise perceptive essay about progress in and practice of research on fisheries management, Michael Holden (1995) wrote critically that aiming for maximum yields was the basis of the Beverton and Holt book. (He was reviewing the third imprint of the book, containing a list of corrections and a foreword by Pauly). This was an extraordinary and profound misunderstanding of our aim; we noted the existence of maxima but certainly did not advocate seeking any of them as management objectives.

6. The author of this review was J. F. Caddy. It includes reference points suggested as early as 1973, and adopted since by J. A. Gulland and L. K. Boerema (1973).

7. Clark has subsequently published many papers and several books expounding on this theme.

8. Although what was going on was realized by several "outsiders" at the time (including senior staff members of the FAO and the United Nations), the details are now in the public domain by virtue of release of documents by the U.S. State Department and some other authorities under the fifty-year rule. I am indebted to Mary Carmel Finley for drawing my attention to this and providing me with copies of some of the relevant papers.

9. See also Volterra (1938) for further discussion of the properties of the logistic.

10. Witting's 2003 work is a published, revised version of a 2001 IWC document. See also Witting (2001).

11. A comprehensible explanation of such "interacting agent" models is given by Paul Ormerod (1998). Ormerod's appendix 3 outlines the mathematical model, and he gives references to the mathematically difficult original publications. Among these a survey of the field by Alan Kirman of the Santa Fe Institute is recommended; this is Kirman's chapter in an edited volume (1997). Kirman's own seminal study was published in 1993.

12. For more information, see various versions of S. P. Kapitza's paper on world population growth (1992a, 1992b, 1993, 1994). It is interesting that this new approach to population modeling, like Verhulst's original work on the logistic, has originated in the field of human demography, and in a strongly political context.

13. Ray Gambell, for example, sought to draw such conclusions about change in age at maturity in fin and sei whales (1975). He noted that the *apparent* age at maturity of both fin and sei whales had declined (although he also detected no significant changes in pregnancy rate), and that this had begun to happen before the intense exploitation of the latter species had begun (in the mid-1960s, after the much larger blue and fin whales had been depleted). This led to speculation by Gambell and others that the apparent changes were connected with an overall decline in total baleen whale biomass in the region. But it seems to have escaped their notice that while fin whales in the Southern Ocean subsist

almost entirely on krill, the sei whales—feeding mostly at lower latitudes—feed mainly on copepods, so serious direct competition between them and the fins (and blues and minkes) is not entirely plausible.

Later, Y. Masaki concluded that the age at maturity of male minke whales in the Southern Ocean declined from about twelve years in 1945 to about six years in 1970; for females the apparent decline was from about thirteen years to about five years in the same period (from his figure 8 in his 1979 work). In that period the total biomass of baleen whales was supposed to have halved. However, the biomass also halved, apparently, from the onset of Antarctic whaling to the beginning of World War II, when Antarctic whaling practically paused. Yet the transition layer data, from waxy earplugs of the older whales in 1970s Japanese catches, did not show any trend in age at maturity of either sex in the prewar period, and therefore do not support his hypothesis of strong interspecific interactions between these species populations of baleen whales. Nevertheless, Masaki's observations have been used by the Institute for Cetacean Research in Tokyo and Japanese official authorities to calculate that the minke whales became many times more abundant in the early years of pelagic whaling in the Antarctic than they are now, thereby seeking to justify the reopening of commercial whaling and the unsustainable reduction of the species that is supposedly impeding the recovery of the nearly exterminated blue whale.

14. For more on this topic, see also Cooke (1999) and "The 'C' Procedure for Whale Stock Management" in a 2005 special issue of the *Journal of Cetacean Research and Management*.

15. The RMP is composed of the chosen CLA together with a procedure for applying it safely in the usual situation where there may be two or more stocks, or supposed stocks, with unclear geographic boundaries and possible mixing.

16. The management development process initiated by the IWC Scientific Committee has since begun to be applied to fisheries. For an example, see L. T. Kell et al. (1999).

17. In this connection it should be mentioned that the simulation trials of candidate CLAs to test their "robustness" and efficiency included circumstances in which environmental factors possibly affecting recruitment and natural mortality were changed, assuming either a sudden, persistent change or a gradual change over a prolonged period. Naturally the results of such trials depend to some extent on the length of the simulation period as well as the assumed pattern and degree of environmental change.

18. In his delightful *The Infinite Book* (2005), John D. Barrow reminds us of a remark attributed to the dramatist Tom Stoppard: "Eternity's a terrible thought. I mean, where's it all going to end?" (5). Barrow continues, "'Immortality,' it has been said (by Caius Glenn Atkins, 1989, *General Thoughts on Immortality*), 'is the bravest gesture of our humanity towards the unknown.' This is not an obvious response to the nature of everyday reality" (5). And so on to a deeper examination—by a mathematical physicist—of the notion of immortality, and that "time goes on for us when others die" (6).

19. The various characteristics of and relations between single species, multispecies, and ecosystem models and procedures were closely examined in 1984 at an international workshop in Berlin held under the auspices of the Dahlem Conferenzen Foundation. See R. M. May (1985).

References

Allee, W. C. 1931. *Animal aggregations*. Chicago: University of Chicago Press.

Barrow, J. D. 2005. *The infinite book: A short guide to the boundless, timeless, and endless.* New York: Pantheon.

Beverton, R. J. H., and E. D. Anderson. 2002. Reflections on 100 years of fisheries research. In *One hundred years of science under ICES: Papers from a symposium held in Helsinki, 1–4 August 2000,* ed. E. D. Anderson, 453–463. ICES Marine Science Symposia, no. 215. Copenhagen: International Council for the Exploration of the Sea.

Beverton, R. J. H., and S. J. Holt. [1957] 2004. *On the dynamics of exploited fish populations.* Ministry of Agriculture, Fisheries and Food, Fishery Investigations Series II, vol. XIX. Fourth reprinting, with new foreword by S. J. Holt. Caldwell, NJ: Blackburn Press.

Clark, C. W. 1971. Economically optimal policies for the utilization of biologically renewable resources. *Mathematical Biosciences* 12:245–260.

Clark, C. W. 1985. *Bioeconomic modeling and fisheries management.* New York: Wiley.

Cooke, J. G. 1994. The International Whaling Commission's revised management procedure as an example of a new approach to fishery management. In *Whales, seals, fish, and man: Proceedings of the International Symposium on the Biology of Marine Mammals in the North East Atlantic, Tromsø, Norway, 29 November–1 December 1994,* ed. A. S. Blix, L. Walløe, and Ø. Ulltang, 647–657. Developments in Marine Biology, no. 4. Amsterdam: Elsevier.

Cooke, J. G. 1999. Improvement of fishery-management advice through simulation testing of harvest algorithms. *ICES Journal of Marine Science* 56(6): 797–810.

Cushing, D. H., ed. 1983. *Key papers on fish populations.* Washington, DC: IRL.

de la Mare, W. K. 1986. Simulation studies on management procedures. *Reports of the International Whaling Commission* 36:429–450.

Food and Agriculture Organization of the United Nations Fisheries Department. 1997. Reference points for fishery management: Their potential application to straddling and highly migratory resources. *FAO Fisheries Circulars,* no. 864. Rome: FAO Fisheries Department.

Gambell, R. 1975. Variation in reproduction parameters associated with whale stock sizes. *Reports of the International Whaling Commission* 25:182–189.

Graham, M. 1935. Modern theory of exploiting a fishery, and application to North Sea trawling. *Journal du Conseil International pour l'Exploration de la Mer* 10(3): 264–274.

Graham, M. 1939. The sigmoid curve and the overfishing problem. *Rapports et Procés-Verbaux des Réunions du Conseil International pour l'Exploration de la Mer* 110:15–20.

Gulland, J. A., and L. K. Boerema. 1973. Scientific advice on catch levels. *Fisheries Bulletin of the United States* 71(2): 325–35.

Hjort, J., G. Jahn, and P. Ottestad. 1933. The optimum catch: Essays on population. *Hvålradets Skrifter* 7:92–127.

Holden, M. 1995. Beverton and Holt revisited. *Fisheries Research* 24(1): 3–8.

Holt, S. J. 2006. The notion of sustainability. In *Gaining ground: In pursuit of ecological sustainability,* ed. D. M. Lavigne, 43–81. Guelph, ON: International Fund for Animal Welfare.

Holt, S. J., and L. M. Talbot. 1978. *New principles for the conservation of wild living resources.* Wildlife Monographs, no. 59. Bethesda, MD: Wildlife Society.

Kapitza, S. P. 1992a. A mathematical model for global population growth. *Mathematical Modeling* 4(6): 65–79.

Kapitza, S. P. 1992b. World population growth as a scaling phenomenon and the population explosion. In *Climate Change and Energy Policy,* ed. L. Rosen and R. Glasser. New York: AIP.

Kapitza, S. P. 1993. World population growth. Paper presented at the 43rd Pugwash Conference on Science and World Affairs, Sweden.

Kapitza, S. P. 1994. The population imperative and population explosion. In *Proceedings of the 42nd Pugwash Conference on Science and World Affairs, Berlin, 1992*, 822. Singapore: World Science.

Kell, L. T., C. M. O'Brien, M. T. Smith, K. T. Stokes, and B. D. Rackham. 1999. An evaluation of management procedures for implementing a precautionary approach in the ICES context for North Sea plaice (*Pleuronectes platessa* L). *ICES Journal of Marine Science* 56(6): 834–845.

Kingsland, S. 1982. The refractory model: The logistic curve and the history of population ecology. *Quarterly Review of Biology* 57(1): 29–52.

Kirman, A. 1993. Ants, rationality, and recruitment. *Quarterly Journal of Economics* 108(1): 137–156.

Kirman, A. 1997. The economy as an interactive system. In *The economy as an evolving complex system II*, ed. W. B. Arthur, S. Durlauf, and D. Lane, 491–532. Reading, MA: Addison-Wesley.

Larkin, P. A. 1977. An epitaph for the concept of maximum sustainable yield. *Transactions of the American Fisheries Society* 106(1): 1–11.

Lavigne, D., R. K. Cox, V. Menan, and M. Wamithi. 2006. Reinventing wildlife conservation for the 21st century. In *Gaining ground: In pursuit of ecological sustainability*, ed. D. M. Lavigne, 379–406. Guelph, ON: International Fund for Animal Welfare.

Lierman, M., and R. Hilborn. 1997. Depensation in fish stocks: A hierarchic Bayesian meta-analysis. *Canadian Journal of Fish and Aquatic Science* 54:1976–1984.

Lotka, A. J. 1925. *Elements of physical biology*. Baltimore, MD: Williams and Wilkins.

Ludwig, D. A. 1974. *Stochastic population theories: Lecture notes in biomathematics*. Vol. 3. Berlin: Springer-Verlag.

MacCall, A. D. 1990. *Dynamic geography of marine fish populations*. Books in Recruitment Fisheries Oceanography. Seattle: University of Washington Press.

Masaki, Y. 1979. Yearly change of the biological parameters for the Antarctic minke whale. *Reports of the International Whaling Commission* 29:375–395.

May, R. M. 1976. Simple population models with very complicated dynamics. *Nature* 261:459–467.

May, R. M., ed. 1985. *Exploitation of marine communities*. Life Sciences Research Report, no. 32. New York: Springer.

Myers, R. A., N. J. Barrowman, J. A. Hutchings, and A. A. Rosenberg. 1995. Population dynamics of exploited fish stocks at low population levels. *Science* 269(5227): 1106–1108.

Ormerod, P. 1998. *Butterfly economics: A general theory of social and economic behaviour*. London: Faber and Faber.

Ottestad, P. 1933. A mathematical model for the study of growth. *Hvålradets Skrifter* 7:30–54.

Pella, J. J., and P. K. Tomlinson. 1969. A generalized stock production model. *Bulletin of the Inter-American Tropical Tuna Commission* 13(3): 419–496.

Pitcher, T. J. 1998. A cover story: Fisheries may drive stocks to extinction. *Reviews in Fish Biology and Fisheries* 8(3): 367–370.

Quinn, T. J., and R. B. Deriso. 1999. *Quantitative fish dynamics*. Oxford: Oxford University Press.

Reish, R. L., R. B. Deriso, D. Ruppert, and R. J. Carroll. 1985. An investigation of the population dynamics of Atlantic menhaden (*Brevoortia tyrannus*). *Canadian Journal of Fish and Aquatic Science* 42:147–157.

Richards, F. J. 1959. A flexible growth function for empirical use. *Journal of Experimental Botany* 10(2): 290–300.

Ricker, W. E. 1954. Stock and recruitment. *Journal of the Fisheries Research Board of Canada* 11:559–623.

Schaefer, M. B. 1954. Some aspects of the dynamics of populations important to the management of commercial fisheries. *Bulletin of the Inter-American Tropical Tuna Commission* 1(2): 26–56.

Schaefer, M. B., O. Sette, and J. Marr. 1951. *Growth of the Pacific coast pilchard fishery to 1942*. U.S. Fish and Wildlife Service Research Report, no. 29. Washington, DC: U.S. Fish and Wildlife Service.

Shepherd, J. G. 1982. A versatile new stock-recruitment relationship for fisheries, and the construction of sustainable yield curves. *Journal du Conseil International pour l'Exploration de la Mer* 40: 67–75.

Thompson, G. G. 1993. A proposal for a threshold stock size and maximum fishing mortality rate. In *Risk evaluation and biological reference points for fisheries management*, ed. S. J. Smith, J. J. Hunt, and D. Rivard, 303–320. Canadian Special Publication of Fisheries and Aquatic Science, no. 120. Ottawa: National Research Council of Canada.

U.S. National Research Council. 1998. *Review of northeast fishery stock assessments*. Washington, DC: National Academy Press.

Verhulst, P. F. 1838. Notice sur la loi que la population suit dans son accroissement. *Correspondance Mathématique et Physique* 10:113–121.

Verhulst, P. F. 1845. Recherches mathématiques sur la loi d'accroissement de la population. *Nouveaux Memoires de l'Academie Royale des Sciences et Belles-Lettres de Bruxelles* 18:1–38.

Verhulst, P. F. 1847. Deuxieme mémoire sur la loi d'accroissement de la population. *Memoires de l'Academie Royale des Sciences, des Lettres et des Beaux-Arts de Belgique* 20:1–32.

Volterra, V. 1926. Variazioni e fluttuazioni del numero d'individui in specie animali conviventi [Variations and fluctuations in the number of individuals in animal species living together]. *Memorie della Regia Accademia Nazionale dei Lincei* 2:31–113.

Volterra, V. 1931. *Leçons sur la théorie mathématique de la lutte pour la vie*. Paris: Gauthier–Villars.

Volterra, V. 1938. Population growth, equilibria and extinction under specified breeding conditions: A development and extension of the logistic curve. *Human Biology* 10(1): 1–11.

Witting, L. 1997. *A general theory of evolution: By means of selection by density dependent competitive interactions*. Aarhus, Denmark: Peregrine.

Witting, L. 2000. Population cycles caused by selection by density dependent competitive interactions. *Bulletin of Mathematical Biology* 62(6): 1109–1136.

Witting, L. 2001. *On inertial dynamics of exploited and unexploited populations selected by density dependent competitive interactions*. IWC Document SC/D2K/AWMP6 (rev.). Cambridge, UK: IWC.

Witting, L. 2003. Reconstructing the population dynamics of eastern Pacific whales over the past 150–400 years. *Journal of Cetacean Research Management* 5(1): 45–54.

13

Marine Mammal Conservation

John R. Twiss Jr., Robert J. Hofman,
& John E. Reynolds III

Much of the last thirty years' history of marine mammal conservation is reflected in the background, content, and implementation of, and changes to, the U.S. Marine Mammal Protection Act (MMPA) of 1972. The act was one of a series of federal environmental laws enacted in the United States in the late 1960s and early 1970s in response to the then-growing awareness that human activities were threatening the natural resources and ecosystems upon which human welfare depends. In addition to the MMPA, those laws included the Wild and Scenic Rivers Act of 1968, the National Environmental Policy Act of 1969, the Clean Air Act Extension and the Coastal Zone Management Act of 1972, the Endangered Species Act of 1973, and the Fishery Conservation and Management Act of 1976. In the years leading to passage of the MMPA, only one issue—the Vietnam War—generated more mail from the public to the members of the U.S. Congress.

Three issues were of particular concern to Congress, the scientific community, and the public at the time the MMPA was being formulated. They were:

1. The killing of hundreds of thousands of dolphins each year in the eastern tropical Pacific Ocean as a consequence of setting purse seines around dolphin schools to catch the yellowfin tuna that associate with the dolphins
2. The failure of the International Whaling Commission (IWC) to prevent the overexploitation and near extinction of virtually all stocks of large whales throughout the world
3. The clubbing and skinning of tens of thousands of newborn (baby) harp seals each year in the ice fields of the North Atlantic for the international fur market

Since passage of the act, a broad spectrum of additional issues has surfaced. These issues include declines of additional species and stocks in both U.S. and international waters—for example, West Indian and African manatees (*Trichechus manatus* and *T. inunguis*), California and Alaska sea otters (*Enhydra lutris nereis* and *E.l. lutris*), Steller sea lions (*Eumetopias jubatus*), Hawaiian and Mediterranean monk seals (*Monachus schaunslandi* and *M. monachus*), killer whales or orcas (*Orcinus orca*),

Gulf of California harbor porpoises or vaquita (*Phocoena phocoena*), and Chinese and Amazon river dolphins (*Platanista* sp.). Other issues include unintentional taking incidental to offshore oil and gas development and commercial fishery activity; the taking of bowhead whales (*Balaena mysticetus*) and other marine mammals by Alaskan Native Americans for subsistence and handicraft purposes; increases in some populations of harbor seals (*Phoca vitulina*) and California sea lions (*Zalophus californianus*) and corresponding calls by fishermen and fisheries groups to cull the populations to limit their predation on commercially valuable fish stocks; unusual mortality events such as the massive die-off of bottlenose dolphins (*Tursiops truncatus*) that occurred along the U.S. mid-Atlantic coast in 1987 and 1988; and increasing threats associated with point and nonpoint sources of ocean pollution, lost and discarded fishing gear and other types of persistent marine debris, ship strikes, human sources of ocean noise, and ecosystem changes due to climate change and global warming.

The MMPA was unique in several respects:

- It was the first legislation anywhere in the world to mandate an ecosystem approach to the conservation of marine living resources.
- It established the concept of optimum sustainable populations (OSP).
- It was the first U.S. legislation to shift the burden from resource managers to resource users to show that proposed taking of marine living resources would not adversely affect the resources or the ecosystems of which they are a part—that is, it prohibited the hunting, killing, capture, or harassment of marine mammals for other than scientific research, public display, or subsistence uses by Alaskan Native Americans unless the advocate of the activity could provide reasonable evidence that the activity would not cause the affected species or stock to be reduced below its optimum sustainable level.
- It directed the relevant federal agencies to seek corresponding changes in international agreements such as the Whaling Convention and the North Pacific Fur Seal Convention.
- It established an independent overview body and scientific advisory group—the Marine Mammal Commission and its Committee of Scientific Advisors—to overview implementation of the act and to advise Congress and the responsible regulatory agencies of needed actions.
- It has been amended periodically to respond to problems that were unforeseen when it was enacted.

Lee M. Talbot, who at the time was senior scientist and director of international affairs at the Council on Environmental Quality, played a lead role in advising Congress and formulating administration policy regarding the content and implementation of the MMPA. He was responsible, in no small measure, for many of the new and innovative concepts in the act, including the OSP concept, the ecosystem approach to marine mammal conservation and management, and the establishment of an independent scientific advisory body (cf. National Oceanic and Atmospheric Administration [NOAA] National Marine Fisheries Service 2002).

KEY PROVISIONS OF THE MMPA

In formulating the MMPA, the lawmakers determined that:

- Certain species and population stocks of marine mammals were in danger of extinction and depletion as a result of human activities.
- Such species and stocks should not be permitted to diminish below the level at which they cease to be significant functioning elements in the ecosystems of which they are a part and, consistent with this principal objective, should not be permitted to diminish below their OSP level.
- Marine mammal species and population stocks should be encouraged to develop to the greatest extent feasible consistent with sound policies of resource management, and the primary objective of their management should be to maintain the health and stability of the marine ecosystem.

Before enactment of the MMPA, states were responsible for conserving and regulating the take of marine mammals in their adjacent coastal waters. The Department of State was responsible for conserving and regulating the take of marine mammals on the high seas through international agreements (e.g., the International Whaling Convention and the North Pacific Fur Seal Convention). Many marine mammals, such as the great whales, were viewed as commodities, like fish and shellfish, and were managed to obtain maximum sustainable yields, an outdated single-species management concept (cf. Holt and Talbot 1978). Others, like harbor seals and California sea lions, were viewed as vermin, competing with fishermen for fish and shellfish resources, and were the subject of bounty programs and unrestricted hunting.

The MMPA established a moratorium on the taking of marine mammals in U.S. waters and the importation of marine mammals and derived products into the United States. It assigned responsibility for whales, dolphins, porpoises, seals, and sea lions to the Department of Commerce, which in turn assigned most of those responsibilities to the National Marine Fisheries Service (NMFS, now known as NOAA Fisheries). Responsibility for walruses, polar bears, manatees, dugongs, and sea otters was assigned to the Department of the Interior, which in turn assigned most of its responsibilities to the Fish and Wildlife Service (FWS). The secretary of state was directed to seek new international agreements and amendments of existing agreements to further the purposes and polices of the act.

Congress recognized that there were legitimate uses of marine mammals and marine mammal products and that states like Alaska had vested interests in controlling the taking of marine mammals in their coastal waters and land areas. Consequently the MMPA included provisions for both waving the moratorium on taking and returning management authority to states. Likewise, it recognized the importance of marine mammal research and public education, and it provided that permits could be issued by the responsible regulatory agencies—the NMFS and the FWS—authorizing the taking of marine mammals and importation of marine mammals and marine mammal products for scientific research and public display. It also recognized that marine mammals often were caught unintentionally in commercial fisheries and provided that permits could be issued to authorize such taking if it would not "disadvantage" the affected species or stocks. Furthermore, the act recognized that many Native Americans residing along the Alaska coast were dependent upon marine mammals for food and other subsistence needs. It therefore exempted from the moratorium on taking the hunting of marine mammals by Alaskan Native Americans for subsistence and handicraft purposes, provided that taking did not threaten the continued existence of the affected species and stocks.

SPECIFIED FUNCTIONS AND COMPOSITION OF THE MARINE MAMMAL COMMISSION AND ITS COMMITTEE OF SCIENTIFIC ADVISORS

As indicated previously, Congress, the scientific community, and the general public were of the opinion that the responsible regulatory agencies had failed to deal effectively with the tuna-dolphin problem, regulation of commercial whaling, and other management problems responsible for the declines in the late 1960s of many marine mammal stocks worldwide. Consequently, also as indicated previously, Congress, through the MMPA, established independent overview and scientific advisory bodies—the Marine Mammal Commission (MMC) and the Committee of Scientific Advisors on Marine Mammals (CSA), respectively—to conduct a continuing review of all federal activities affecting marine mammals and to advise both Congress and the responsible regulatory agencies of actions needed to further the purposes and provisions of the act.

The MMPA specified that the MMC be made up of three members and that each member (1) be knowledgeable in the fields of marine ecology and resource management; (2) not be a political appointee, an employee of the federal government, or in a position to profit from the taking of marine mammals; and (3) be appointed by the president from a list of qualified individuals submitted to him by the chairman of the Council on Environmental Quality, the secretary of the Smithsonian Institution, the director of the National Science Foundation, and the chairman of the National Academy of Sciences. To help ensure compliance with these specifications, the act was amended in 1981 to require Senate confirmation and the unanimous agreement of all four agencies on the names of individuals nominated by the president for appointment to the commission.

The act directed the MMC to establish the nine-member CSA, composed of scientists knowledgeable in marine ecology and marine mammal affairs. It specified that the members be appointed by the chairman of the MMC after consultation with the other commission members and with the heads of the Council on Environmental Quality, the Smithsonian Institution, the National Science Foundation, and the National Academy of Sciences to confirm the qualifications of the nominees. The past and current members of the MMC and CSA are identified in the commission's annual reports to Congress.[1]

The MMC was given no regulatory authority. However, the act specified that any recommendations made by the MMC to a federal official must be responded to within 120 days, and, if a recommendation is not followed, that the commission be advised in writing of the reason or reasons why the recommendation was not followed. It also specified that the commission must consult with the CSA on

all science-related issues—for example, applications for research permits and recommendations to other agencies regarding needed scientific research—and that
any CSA recommendations not accepted by the commission must be forwarded
to the relevant federal agencies and congressional oversight committees, along
with a detailed explanation as to why the recommendations were not followed.
Additionally, the act directed the commission to report its activities annually to
Congress, including recommendations made to other agencies and their responses
to those recommendations.

IMPLEMENTATION AND EVOLUTION OF THE OPTIMUM CONCEPT

Among other things, the original MMPA stated that "the primary objective of their
[marine mammal] management should be to maintain the health and stability of the
marine ecosystem, [and] whenever consistent with this primary objective, it should
be the goal to obtain an optimum sustainable population, keeping in mind the optimum carrying capacity of the habitat" (§ 2, para. 6).

In the original act, the term *optimum sustainable population* was defined with
respect to any population stock as "the number of animals which will result in the
maximum productivity of the population or the species, keeping in mind the optimum carrying capacity of the habitat and the health of the ecosystem of which they
form a constituent element" (§ 3).

The term *optimum carrying capacity* was defined as "the ability of a given habitat to support the optimum sustainable population of a species or population stock
in a healthy state without diminishing the ability of the habitat to continue that
function."

Subsequently it became clear that different interest groups had different interpretations of optimum carrying capacity, as well as the terms *maximum productivity*
and *health of the ecosystem*. As an example, state and federal fish and marine mammal biologists generally viewed maximum productivity as analogous to the then
generally accepted management goal of maximum sustainable yield. They therefore
interpreted the optimum carrying capacity and ecosystem health to mean the habitat conditions that would maintain marine mammal populations at their maximum
sustainable yield levels. Environmental groups, however, generally viewed the terms
to mean the greatest numbers of animals that could be supported by the habitat in its
pristine state. The uncertainty was resolved as a result of a 1974 lawsuit regarding a
permit issued by the NMFS to the American Tuna Boat Association authorizing the
taking of unspecified numbers of dolphins in the eastern tropical Pacific (ETP) tuna
purse seine fishery.

As mentioned earlier, the deaths of hundreds of thousands of dolphins each year
in the ETP as a consequence of setting purse seines around dolphin schools to catch
yellowfin tuna was one of the issues that led to the MMPA. Also, as mentioned
earlier, the act provided that permits could be issued by the responsible regulatory
authority—the NMFS in this case—to authorize the taking of marine mammals
incidental to commercial fisheries, provided the taking would not "disadvantage" the
affected marine mammal species or stocks.

THE *COMMITTEE FOR HUMAN LEGISLATION* DECISION
AND THE LA JOLLA WORKSHOP

In September 1974 the NMFS issued regulations to govern the taking of dolphins by U.S. vessels engaged in the tuna purse seine fishery in the ETP. Subsequently the agency issued a permit to the American Tuna Boat Association authorizing the encirclement and associated mortality of unspecified numbers of dolphins in the 1975 tuna fishing season. Following these actions, several environmental groups filed a lawsuit in the federal district court of Washington, DC, claiming that the regulations and the permit issued to the tuna boat association violated the MMPA because the NMFS had not established a limit on the number of dolphins that could be encircled and killed, and had not determined the size or status of the affected dolphin stocks relative to their OSP levels.

On May 11, 1976, Judge Charles R. Richey issued his findings in the lawsuit (*Committee for Humane Legislation, Inc. v. Elliot L. Richardson*). Among other things, Judge Richey found that the NMFS had violated the intent and provisions of the MMPA by not establishing a limit on the species and numbers of dolphins that could be killed in the fishery, and by not providing estimates of the sizes and optimum sustainable levels of the affected dolphin stocks. He issued an order voiding the regulations and the permit issued to the American Tuna Boat Association. In partial response to that order, the NMFS convened a group of experts, including the chairman of the MMC, to review available information and provide assessments of the sizes and OSP status of the affected dolphin stocks.

The workshop was held at the NMFS's Southwest Fisheries Science Center in La Jolla, California. The participants identified eleven species and twenty-one stocks of dolphins subject to taking in the fishery. They estimated the then-current sizes of the species and stocks most affected using the results of a pilot aerial survey done by the NMFS in 1974 (Smith 1974). They also estimated the stock sizes prior to the beginning of the purse seine fishery in the late 1950s by back-calculating from the current estimates, using estimates of the annual fishery-related mortality and estimates of the likely maximum annual replacement rates. They concluded that the three stocks most affected by the fishery—the offshore stock of spotted dolphins (*Stenella attenuata*) and ETP stocks of eastern spinner and white-belly spinner dolphins (*S. longirostris*)—were approximately 64%, 54%, and 76%, respectively, of their pre-fishery or pre-exploitation sizes.

As indicated earlier, the MMPA's definition of OSP was ambiguous in that its references to maximum productivity and optimum carrying capacity could be interpreted in different ways. The workshop participants therefore developed and used the following interpretive definition of the term to avoid the ambiguity: "Optimum sustainable population is a population size which falls within a range from the population level of a given species or stock which is the largest supportable within the ecosystem to the population level that results in maximum net productivity. Maximum net productivity is the greatest net annual increment in population numbers or biomass resulting from additions to the population due to reproduction and/or growth less losses due to natural mortality."

Finally, the participants concluded that the maximum net productivity levels of the dolphin stocks likely were between 50% and 70% of their carrying capacity levels and that 60% would be a prudent approximation when available information was insufficient, as in these cases, to determine the actual maximum net productivity level. Thus the eastern spinner stock was below its OSP range, and the offshore spotted stock was approaching the lower limit of its OSP range.

These workshop findings had three long-lasting effects on implementation of the MMPA: (1) the interpretive definition of OSP was adopted by both the NMFS and the FWS for regulatory purposes; (2) 60% of the estimated carrying capacity level was adopted, in the absence of information to the contrary, as the lower limit of the OSP range; and (3) back-calculation using estimates of current population size and annual mortality rates was accepted as a reasonable means for estimating pre-exploitation sizes or "optimum" carrying capacity.

It subsequently was recognized that there were redundancies in the MMPA's original definitions of OSP and optimum carrying capacity. Therefore, in the 1981 MMPA amendments, the definition of optimum carrying capacity was eliminated and the reference to "optimum carrying capacity" in the definition of OSP was changed to "carrying capacity."

Details of these and other actions regarding the tuna-dolphin problem can be found in Gosliner (1999) and in the annual reports of the MMC, the NMFS, and the Inter-American Tropical Tuna Commission. As indicated therein, the numbers of dolphins killed annually in the ETP tuna purse seine fishery have declined from more than 400,000 in 1972, when the MMPA was enacted, to fewer than 5,000 since the early 1990s. However, it appears that the depleted stocks of spotted and spinner dolphins are not recovering, due possibly to unobserved or unreported mortality and/or stress caused by chase and capture.

THE *KOKECHIK* DECISION AND THE 1988 MMPA AMENDMENTS

In May 1987 the NMFS issued a permit to the Japanese Salmon Fisheries Cooperative Association authorizing the take of up to 2,942 Dall's porpoises (*Phocoenoides dalli*) annually, for a period of three years, incidental to salmon drift-net fishing in the 200-mile U.S. fishery conservation zone off Alaska. Shortly after the permit was issued, several Alaskan Native American fishing groups and environmental organizations filed lawsuits claiming, among other things, that the permit violated the MMPA because it applied only to Dall's porpoise when it was virtually certain that other marine mammals would also be taken, including northern fur seals (*Callorhinus ursinus*) from the depleted populations on St. Paul and St. George islands. The court ruled in favor of the plaintiffs and issued a preliminary injunction voiding the permit. The NMFS appealed the decision. However, the appellate court upheld the decision.

The decision in this case (*Kokechik Fishermen's Association v. Secretary of Commerce*) cast doubt on the ability of the NMFS to issue incidental take permits for other fisheries, including many domestic fisheries, for which there was insufficient information to reasonably conclude that all species and populations likely to be affected were at or above their maximum net productivity levels. Also, as noted earlier, the MMPA, with three exceptions, prohibited taking from endangered, threatened, and depleted species and stocks, even in cases where the taking would have little or no effect on the recovery of those species or stocks.

Both the environmental community and the fishing industry, as well as the state and federal regulatory agencies, recognized that a total prohibition on the incidental taking of marine mammals would have severe economic impacts on a number of U.S. fisheries. In addition, it was clear that available information was insufficient in most cases to reliably assess and determine how to avoid or mitigate the adverse effects of interactions on the affected marine mammals and fisheries. Consequently, Congress amended the MMPA in 1988 to provide a five-year exemption to the act's permit and small-take requirements for U.S. and certain foreign fisheries, other than the ETP tuna purse seine fishery covered by other provisions of the act. The basic purposes of the five-year exemption were to provide time to (1) compile and analyze data on the types, levels, and biological and socioeconomic implications of marine mammal–fishery interactions in U.S. waters, and (2) develop a new regime to govern interactions that could both avoid adverse effects on marine mammals and minimize impacts on fisheries. Among other things, the amendments required that:

- Owners of vessels engaged in fisheries that take marine mammals more than rarely in U.S. waters register with the NMFS and report all incidents of inter-actions with marine mammals
- By March 23, 1989, the NMFS, in consultation with the MMC and after opportunity for public comment, develop and then annually update lists iden-tifying fisheries that take marine mammals frequently, occasionally, and rarely
- Twenty-nine to thirty-five percent of fishing vessels engaged in Category 1 fisheries—fisheries identified as taking marine mammals frequently—be monitored by onboard NMFS observers
- A volunteer observer or alternative observation program be developed by the NMFS to obtain statistically reliable information on the species and numbers of marine mammals being taken incidentally in fisheries for which observers are not required or are not available
- The NMFS design and implement an information management system capable of processing and analyzing incidental take and related data provided by fishermen, observers, and others
- The MMC, in consultation with the CSA, develop and provide to the NMFS recommended guidelines to govern the taking of marine mammals incidental to commercial fisheries in U.S. waters after October 1, 1993, when the interim exemption was scheduled to expire
- The NMFS provide to Congress by January 1, 1992, its recommendations for a new regime to govern marine mammal–fishery interactions and a proposed schedule for implementing the regime

The amendments themselves and the subsequent efforts by the MMC and the NMFS to implement them resulted in several practical and philosophical changes to the regulation of marine mammal–fishery interactions. For example, the amend-ments directed that the guidelines for a new regulatory regime to be developed by the MMC take into account, among other things, the status and trends of the affected marine mammal species and stocks. In its recommended guidelines, provided to the

NMFS in July 1990, the commission indicated that there was no compelling bio-logical reason to have a categorical prohibition on the taking of endangered, threat-ened, and depleted species and stocks. It recommended, among other things, that the incidental take of marine mammals listed as endangered or threatened under the Endangered Species Act or depleted under the MMPA be authorized if the tak-ing would not cause a further decline or impede recovery of the affected species or stocks.

The amendments also recognized, and the subsequent mandatory and voluntary observer programs implemented by the NMFS confirmed, that placement of suf-ficient numbers of trained observers aboard fishing vessels is necessary to obtain reli-able information on fishing practices and on catches of both target and nontarget species. Also, while most of the NMFS's funding for and decisions regarding marine mammal research and management had previously been delegated by the agency to its regional management and science centers, following the 1988 amendments much of the marine mammal decision-making and funding authority was vested in the Office of Protected Species at NMFS headquarters in Silver Spring, Maryland. Among other things, this minimized regional differences in perceptions of, and efforts to deal with, marine mammal research and management problems. Finally, the new regulatory regime proposed to Congress by the NMFS in December 1992 suggested a new and simpler conceptual means for assuring that incidental take in commercial fisheries does not cause any marine mammal species or stock to be reduced or to be maintained below the lower limit of its OSP range, as described previously. That concept, calculation of potential biological removal (PBR) levels, was incorporated in the 1994 MMPA amendments as described below.

THE 1994 MMPA AMENDMENTS

Several significant changes to the MMPA's provisions regarding marine mammal–fishery interactions were enacted in 1994. Those changes reflected input from the fishing industry and the environmental community as well as the MMC, the NMFS, and the FWS.

A new section added to the act (§ 117) required the preparation and periodic update of status reports for all marine mammal stocks in U.S. waters. It directed that each stock assessment (1) describe the geographic range of the stock; (2) provide a minimum abundance estimate, assessments of the stock's current and maximum net productivity rates and current trend, and a description of the information used to make those determinations; (3) estimate by source the level of annual human-caused mortality and serious injury, including for strategic stock (see below) factors in addi-tion to fishery-related mortality and injury that may be causing a decline or impeding recovery; (4) describe the commercial fisheries that interact with the stock, including the number of vessels in each fishery, fishery-specific estimates of mortality and seri-ous injury levels and rates, any seasonal or area differences in incidental mortality or serious injuries, and whether the level of mortality and serious injury has achieved or is approaching the zero rate goal; (5) assess whether the level of mortality and serious injury is or is not likely to cause the stock to be reduced below the lower limit of its

OSP range, and whether the stock should be classified as a strategic stock; and (6) indicate the PBR level for the stock and the information used to do the calculation (see below). This new section also directed the NMFS to establish regional scientific review groups for Alaska, the Pacific coast including Hawaii, and the Atlantic coast including the Gulf of Mexico in order to assist in preparing and updating the stock assessments.[2]

Another new section (§ 118) established the replacement regime to govern the taking of marine mammals incidental to commercial fishing operations. Among other things, it mandated, with minor changes, the continuation of the vessel registration and observation programs established in accordance with the 1988 amendments. Furthermore, it directed that take reduction plans be developed for each strategic stock that interacts with a Category 10 or Category 2 fishery—fisheries that frequently or occasionally kill or seriously injure marine mammals—and that take reduction teams, composed of scientists and representatives of the various fishery and environmental interest groups, be constituted to draft the plans. The immediate goal of these plans is to identify measures that will reduce, within six months, fishery-related mortality and serious injury to less than the PBR levels calculated in the stock assessments. The long-term goal is to reduce incidental mortality and serious injury to insignificant levels, approaching a zero rate, taking into account the economics of the fishery, existing technology, and applicable state or regional fishery management plans. To date, the NMFS has established take reduction teams and take reduction plans for a variety of species and species groups, including right whales (*Eubalena glacialias*) and other large whales in the northwestern Atlantic, harbor porpoises along parts of both the Atlantic and Pacific coasts, bottlenose dolphins along the mid-Atlantic coast, and sperm whales (*Physeter macrocephalus*) and other cetaceans off the Pacific coast.[3]

The term *strategic stock* was defined to mean a marine mammal stock (1) for which the level of direct human-caused mortality exceeds the PBR levels; (2) which, based on the best available scientific information, is declining and is likely to be listed as a threatened species under the Endangered Species Act within the foreseeable future; or (3) which is listed as a threatened species or endangered species under the Endangered Species Act or is designated as depleted under this act (the MMPA). The term *potential biological removal level* was defined to mean the maximum number of animals, not including natural mortalities, that may be removed from a marine mammal stock while allowing that stock to reach or maintain its optimum sustainable population. The PBR level is the product of the following factors: (1) the minimum population estimate of the stock, (2) one-half the maximum theoretical or estimated net productivity rate of the stock at a small population size, and (3) a recovery factor between 0.1 and 1.0.

Details of the amendments are described in the MMC's report to Congress for calendar year 1994. Ongoing efforts to implement the amendments are described in subsequent commission reports.

The intent of the PBR concept clearly was to provide conservative estimates of the numbers of marine mammals that could be removed annually from U.S. waters without causing the affected species and stocks to be reduced or maintained below the lower limit of the previously defined OSP range. However, some aspects of the definition were ambiguous. For example, nowhere in the amendments or the associated legislative history was there any indication of what was envisioned by the term *minimum population estimate*, or how the specified recovery factors were to be applied. Therefore, on June 27–29, 1994, the NMFS convened a workshop of knowledgeable scientists and representatives of the NMFS, the MMC, and the FWS to consider and provide advice on the most appropriate interpretations of the variables in the formula for calculating PBR levels.

Among other things, the workshop participants recommended that either an actual minimum count or the twentieth percentile of a lognormal distribution based on the best available population estimate be used as the estimate of minimum population size. They noted the importance of having both reliable and up-to-date population estimates and recommended that calculated PBR levels be reduced by 20% per year when the minimum population estimates are more than five years old.

The workshop participants recommended that default values of 0.12 be used for pinnipeds and sea otters and 0.04 be used for cetaceans and manatees when available information is insufficient to estimate their actual maximum net productivity rates (R_{max}). With respect to recovery factors, the participants recommended using different values depending on the status of the stock—for example, 0.1 for endangered species and 1.0 for species and stocks well within their OSP range. In cases where stock discreteness is unknown or uncertain, the participants recommended that the stocks be defined initially based on the smallest unit approaching that of the area of take, unless evidence of possible smaller subdivisions exists. With regard to the last point, the participants pointed out that a risk-averse strategy requires that small stock groupings be "lumped" only when there is a compelling biological reason to do so. Follow-up workshops were held in 1996 and 2003 (see Barlow et al. 1995; Wade 1994).

Although not without some controversy, the system, established by the 1994 MMPA amendments to govern marine mammal–fishery interactions, has worked effectively to regulate the taking of marine mammals incidental to commercial fisheries in U.S. waters, as well as to minimize the impacts of the regulations on the affected fisheries.

DEVELOPMENT OF THE ECOSYSTEM APPROACH
TO MARINE CONSERVATION

In the 1960s, trawlers from the (then) Soviet Union and Japan began exploratory fishing for krill, *Euphasia superba*, in the seas around Antarctica (Sahrhage 1985). This species is a keystone in the Antarctic marine food web. It is the primary food of fin whales (*Balaenoptera physalus*), blue whales (*B. musculus*), humpback whales (*Megaptera novaeangliae*), minke whales (*B. acutorostrata*), crabeater seals (*Lobodon carcinophagus*), Antarctic fur seals (*Arctocephalus sp.*), chinstrap penguins (*Pygoscelis antarctica*), macaroni penguins (*Eudyptes chrysolophus*), rockhopper penguins (*E. chrysocome*), several other species of seabirds, and several species of fish and squid. Some of these species are eaten in turn by sperm whales, killer whales, leopard seals (*Hydrurga leptonyx*), and other higher-order predators (Beddington and May 1982; Hofman 1985).

Knowledgeable scientists expressed concern that, if the fishery grew and was not regulated effectively, it could prevent or impede recovery of depleted stocks of krill-eating whales, as well as impact the broad range of other species dependent directly and indirectly on krill. In response, the representatives of the parties to the Antarctic Treaty recommended at the Ninth Consultative Meeting in London in 1977 that "a definitive regime for the Conservation of Antarctic Marine Living Resources should be concluded before the end of 1978 and that a Special Consultative Meeting should be convened for that purpose." Australia offered to host the special meeting, the first session of which was held in Canberra from February 27 to March 16, 1978.

Prior to the 1978 negotiating session in Canberra, several of the Antarctic Treaty consultative parties circulated draft conservation regimes for consideration. Each of the drafts had as its central tenet the goal of maximum sustainable yield. During interagency preparations for the negotiations, the MMC pointed out that consultations and workshops sponsored by the Council on Environmental Quality and others in 1974 and 1975 had concluded that maximum sustainable yield was an outdated management concept because it failed to consider the possible effects of harvesting on dependent and associated species. The commission advocated an ecosystem approach as recommended in the report of the consultations and workshops, titled *New Principles for the Conservation of Wild Living Resources* (Holt and Talbot 1978). Among other things, that report states:

> The consequences of resource utilization and the implementation of principles of resource conservation are the responsibilities of the parties having jurisdiction over the resource or, in the absence of clear jurisdiction, with those having jurisdiction over the users of the resource. The privilege of utilizing a resource carries with it the obligation to adhere to the following four general principles:
>
> 1. The ecosystem should be maintained in a desirable state such that
>   a. Consumptive and nonconsumptive values could [can] be maximized [optimized] on a continuing basis,
>   b. Present and future options are ensured, and
>   c. Risk of irreversible change or long-term adverse effects as a result of use is minimized.
> 2. Management decisions should include a safety factor to allow for the fact that knowledge is limited and institutions are imperfect.
> 3. Measures to conserve a wild living resource should be formulated and applied so as to avoid wasteful use of other resources.
> 4. Survey or monitoring, analysis, and assessment should precede planned use and accompany actual use of wild living resources. The results should be made available promptly for critical public review. (Holt and Talbot 1978, 5)

The commission's views regarding the necessity of an ecosystem approach were endorsed by the Department of State and included in the U.S. negotiating position for the Canberra meeting. Those views ultimately were incorporated with minor modifications in the Convention for the Conservation of Antarctic Marine Living Resources (CCAMLR), which entered into force in 1981. As examples, articles I and II of the CCAMLR read as follows:

Article I [Scope and Definitions]

1. This Convention applies to the Antarctic marine living resources of the area south of 60° south latitude [the area to which the Antarctic Treaty applies] and to the Antarctic marine living resources of the area between that latitude and the Antarctic Convergence which forms part of the Antarctic marine ecosystem.
2. Antarctic marine living resources means the populations of fin fish, mollusks, crustaceans [e.g., krill], and all other species of living organisms, including birds found south of the Antarctic Convergence;
3. The Antarctic marine ecosystem means the complex of relationships of Antarctic marine living resources with each other and with their physical environment.

Article II [Objectives]

1. The objective of this Convention is the conservation of Antarctic marine living resources.
2. For the purpose of this Convention, the term "conservation" includes rational use [e.g., commercial fisheries].
3. Any harvesting and associated activities in the area to which this Convention applies shall be conducted in accordance with the provisions of this Convention and with the following principles of conservation:
   a. Prevention of decrease in the size of any harvested population to levels below those which ensure its stable recruitment. For this purpose its size should not be allowed to fall below a level close to that which ensures the greatest net annual increment;
   b. Maintenance of the ecological relationships between harvested, dependent and related populations of Antarctic marine living resources and the restoration of depleted populations to the levels defined in sub-paragraph (a) above; and
   c. Prevention of changes or minimization of the risk of changes in the marine eco-system which are not potentially reversible over two or three decades [a human generation], taking into account the state of available knowledge of the direct and indirect impact of harvesting, the effect of the introduction of alien species, the effects of associated activities on the marine ecosystem and the effects of environmental changes, with the aim of making possible the sustained conser-vation of Antarctic marine living resources.

Many of the details concerning the negotiation of the CCAMLR and actions taken subsequently to implement it are summarized in the MMC's reports to Congress for calendar years 1978 through 1999.[4] The CCAMLR has served as a model for a number of more recent international agreements, including the 1995 Food and Agriculture Organization Code of Conduct for Responsible Fisheries and the 1995 UN Agreement on Conservation and Management of Straddling Fish Stocks and Highly Migratory Fish Stocks. Many of the CCAMLR principles are also reflected in the 1991 Protocol to the Antarctic Treaty on Environmental Protection.

REGULATION OF COMMERCIAL WHALING

As noted earlier, the declines and near extinction of most stocks of large whales due to the failure of the IWC to effectively regulate commercial whaling were among the concerns that led to the MMPA. That concern was shared by many of the other IWC member countries and led to a 1982 IWC decision to suspend commercial whaling, pending review of the status of the affected whale stocks and assessment of the procedures for setting and ensuring compliance with catch quotas. The suspension was implemented by setting commercial catch limits at zero effective with the 1986 coastal and the 1985–1986 pelagic whaling seasons (IWC 1983).

The IWC's scientific committee then initiated a "comprehensive assessment" of the status and trends of all previously exploited whale stocks (Donovan 1989). In addition, the scientific committee evaluated a number of alternative procedures for establishing sustainable catch levels for baleen whales, and in 1991 recommended to the commission adoption of a Revised Management Procedure (RMP) to replace the procedure that had been used since 1975 to set catch limits (IWC 1991, 1992). The goal of the RMP is to establish a transparent system for establishing catch limits, with minimum data requirements, that will enable rebuilding of depleted stocks to their maximum net productivity levels, estimated to be 72% of their pre-exploitation sizes, and ultimately to obtain maximum, long-term sustainable yields, assuming that commercial whaling will be resumed. The recommended RMP was adopted by the commission with minor modifications in 1995 (IWC 1995).

Although the RMP recommended by the scientific committee has been adopted, there remain substantial differing views within the IWC as to whether, and under what conditions, commercial whaling should be resumed. Some members, such as Australia and New Zealand, categorically oppose resumption of commercial whaling largely on ethical grounds. Others, such as Japan and Norway, believe that application of the RMP would effectively eliminate the risk of overexploitation as occurred in the past and that the commission should lift the suspension on commercial whaling. A major point of contention between those advocating and those opposing resumption of commercial whaling is the system of observation and inspection needed to ensure compliance with authorized catch levels if the suspension is lifted. Other points of contention include minimum data standards, progress on development of humane killing methods, the direct and indirect effects of ocean pollution, and the relative economic value and effects on "nonconsumptive" whale watching. To date, the pro-whaling nations have been unable to achieve the three-quarters majority needed to lift the suspension.

The International Whaling Convention provides that commission members may object to, and consequently not be bound by, measures adopted by the three-quarters majority vote of the commission. Norway objected to the suspension of whaling agreed to in 1982 and is not bound by that measure. Believing that the RMP provides a fully adequate means for preventing overexploitation, Norway has authorized its nationals to take increasing numbers of minke whales in the North Atlantic, using the RMP as a guide for establishing catch limits.

The International Whaling Convention also provides that member countries may authorize their nationals to take unspecified numbers of whales for scientific purposes without the endorsement of the IWC's scientific committee or the approval of the commission. Japan, which initially filed but subsequently withdrew an objection to the 1982 agreement suspending commercial whaling, has since 1987 authorized its nationals to take whales, principally in the Antarctic, for purported scientific purposes. Japan also has repeatedly sought IWC authorization to waive the suspension to allow resumption of "subsistence" whaling by its coastal communities, which it asserts depended historically on the take of small numbers of whales for subsistence purposes. Japan's efforts to end the suspension of commercial whaling and to get a waiver for the purported subsistence whaling have thus far been unsuccessful. Its authorized taking of whales for purported scientific purposes has generated much controversy (see, e.g., Gales et al. 2005).

Ray Gambell (1999) provides an overview of the history of commercial whaling and the ongoing controversy as to whether and under what conditions commercial whaling should be resumed. At present, it is not clear whether the controversy can be resolved, or whether the IWC will survive.

SPECIES OF SPECIAL CONCERN

A number of marine mammal species and populations in U.S. waters are threatened or in danger of extinction and have been listed accordingly under the Endangered Species Act. They include right whales in both the North Atlantic and the North Pacific, the Florida manatee, the Hawaiian monk seal, sea otter populations in both California and Alaska, and Steller sea lion populations in the Aleutian Islands and southeastern Alaska. The MMC has worked with the NMFS and the FWS, as appropriate, to develop and periodically update recovery plans for theses species and stocks as required by the Endangered Species Act. The commission also has advocated development, adoption, and regular review and updating of implementation or action plans by the state and federal agencies, industry groups, and environmental organizations with related interests and responsibilities. Moreover, the commission has advocated and taken the lead in developing conservation plans for a number of unlisted species in Alaska (see, e.g., Lentfer 1988).

The status and ongoing efforts to protect and promote recovery of these species and populations of special concern are described in the MMC's annual reports to Congress. The principal threats and ongoing efforts to address them are summarized below.

Right Whales

Right whales were so named because they floated when killed and thus were the "right" whale to hunt. All stocks were hunted to near extinction and, although hunting was banned in 1935 by the predecessor to the IWC, only the Southern Hemisphere stocks have shown signs of recovery (IWC 2001). The northwestern Atlantic stock, which inhabits the waters along the eastern seaboard of the United States and Canada, numbers about 300 individuals and has changed little since monitoring and recovery efforts were initiated in the 1980s (Katona and Kraus 1999). The eastern Pacific stock was thought to be extinct until recently and may number no more than a few tens of individuals. Ship strikes and entanglement in fishing gear are known causes of right whale mortality and injury, and may be one of the reasons, if not the only or principal reason, that the northwestern Atlantic stock is not recovering (Laist et al. 2001). Recovery plans have been developed for both stocks (NOAA

NMFS 2005). Substantial effort and funding have been invested by the NMFS, the Coast Guard, state agencies, and nongovernmental organizations to document and monitor the size, movements, productivity, and critical habitats of the northwestern Atlantic stock, and to identify, assess, and attempt to minimize and mitigate human sources of mortality and injury. To date, however, there are no indications that the mortality reduction efforts are having any effect on population size or productivity. Efforts to date in the northeastern Pacific have been limited largely to opportunistic sighting surveys aimed at determining where and how many whales remain.

Florida Manatees

The Florida manatee is a subspecies of the West Indian manatee. Its geographic distribution and seasonal movements are determined largely by water temperature. In winter, most of the population inhabits areas in the southern part of the state near warm-water sources—both naturally occurring warm-water springs and warm water effluents from power plants. As temperatures increase in the spring, animals disperse throughout the state's coastal waters, with small numbers migrating north to Georgia and the Carolinas, and west along the Gulf Coast as far as Texas and northern Mexico. The principal threats to this subspecies are boat strikes and habitat degradation and loss, both of which have increased as the state's human popula- tion and corresponding urban development and recreational boat use have increased (Reynolds 1999). In the five years from 1999 through 2003, over 400 animals, of a total population numbering somewhat more than 3,000, were killed by boat strikes (MMC 2004). Much of the population bears propeller scars from past encounters with watercraft. In the last decade, there also have been two recorded die-offs caused by natural biotoxins associated with red tides: one in 1996 that killed at least 149 indi- viduals, and one in the spring and fall of 2003 that killed at least 168 individuals.

The federal FWS and the Florida Fish and Wildlife Conservation Commission share responsibility for protecting and conserving manatees in state waters. The FWS, in consultation with the commission and others, has developed and periodi- cally updated both recovery and implementation or action plans (U.S. FWS 2001). One of the principal conservation measures has been the establishment of a statewide education program alerting schoolchildren, boat owners, and the general public of the natural history of manatees and the threats posed by boats, habitat degradation, and the like. The principal protective measures have been restricting marina con- struction and establishing and enforcing boat speed zones in areas commonly inhab- ited by manatees. These efforts have had some success, as populations in some areas have been growing slowly. There is a growing controversy, however, as to whether the slow population growth means that some of the protective measures can be relaxed. The future of this subspecies clearly depends on the continuing success of efforts to protect critical habitats and to limit human sources of mortality and serious injury.

Hawaiian Monk Seals

The Hawaiian monk seal is one of three distinct monk seal species, the other two being the Caribbean (*M. tropicalis*) and Mediterranean species. The Caribbean monk seal likely is extinct, and the Mediterranean species, like the Hawaiian species, is in danger of extinction. The Hawaiian monk seal is the most endangered pinniped in U.S. waters and currently numbers only about 1,300 individuals. It occurs principally in the northwestern Hawaiian Islands, although in recent years there have been increasing sightings in the main Hawaiian Islands. Current abundance is reasonably stable, but is less than half what it was in the late 1950s when the first comprehensive surveys were done (MMC 2004; Ragen and Lavigne 1999). The cause of the decline from the 1950s to the early 1990s is uncertain. There has been at least one major die-off, due apparently to an unusual outbreak of a ciguatoxin-producing dinoflagellate in the vicinity of Laysan Island. Other known sources of mortality include entanglement in marine debris (principally lost and discarded fishing gear ostensibly carried by currents from fishing grounds in the North Pacific); shark predation, principally on naive pups; and "mobbing" of females by multiple males attempting to copulate with them (possibly a consequence of an imbalanced sex ratio). Although not demonstrated, reduction of some key prey species by commercial fisheries in the vicinity of some of the principal pupping atolls may have contributed to the decline and be impeding recovery.

The NMFS has lead authority under both the MMPA and the Endangered Species Act for conservation and protection of monk seals. However, responsibility for the land areas where monk seals haul out and pup are divided between the state and the FWS, which manages the National Wildlife Refuge System. Thus, the future of this species will depend largely on the ability of these three entities to cooperatively control potentially harmful fishery, recreational, and urban developments.

Sea Otters

Historically, an estimated 150,000 to 300,000 sea otters occurred in coastal waters around the rim of the North Pacific Ocean from northern Japan to Baja California, Mexico. Hunting by fur hunters in the 1700s and 1800s exterminated all but a few small groups in isolated areas of Alaska, British Columbia, and California (see Kenyon 1969). Hunting was prohibited by the 1911 International Fur Seal Convention, and a number of the surviving groups or remnant populations began to recover. In central California, a remnant group, which may have numbered as few as fifty individuals, grew to about 1,000 individuals and had reoccupied about 200 miles of its former California range by 1972 when the MMPA was enacted. In Alaska, remnant groups had grown to near historic levels and reoccupied much of their former range in the Aleutian Islands and Alaska Peninsula by the 1980s. In the late 1960s and early 1970s, several hundred otters were captured and moved from Amchitka Island and Prince William Sound in Alaska in an attempt to reestablish populations in southeastern Alaska and along the coasts of Washington and Oregon. Populations were successfully established in Washington and southeastern Alaska, but not Oregon.

Because of its small size and limited range, and because of the increased risk of oil spills from burgeoning tanker traffic along the California coast, the recovering California population was designated in 1977 as threatened under the Endangered Species Act. Although it was assumed that the population size and range would continue to grow, the population growth stopped in the late 1970s and early 1980s, due apparently to incidental mortality in coastal gill and trammel net fisheries (Wendell, Hardy, and Ames, 1986). Beginning in 1982, the state enacted a series of regulations prohibiting gill and trammel net fisheries in areas where seabirds, sea otters, and other marine mammals were likely to be caught and killed. Thereafter, the population growth resumed, until the mid-1990s when growth stopped again due to uncertain causes. The possibilities include incidental mortality in new pot and trap fisheries, previously unknown diseases, and exposure to naturally occurring biotoxins (see Estes et al. 2003; see also figure 11 and accompanying discussions in MMC [2004]).

Recognizing that expansion of the California sea otter range would result in competition between sea otters and commercial and recreational fisheries for clams, abalones, and sea urchins, the MMC recommended in December 1980 that the FWS adopt a zonal management strategy—a strategy whereby one or more sea otter colonies would be established outside the then-existing California range, and at the same time prevent sea otters from recolonizing areas where substantial shellfish fisheries had developed in the more than one-hundred-year absence of otters. From 1987 to 1990, 140 sea otters were moved from the mainland California range to San Nicolas Island as part of a zonal management program developed by the FWS in consultation with the MMC and the California Department of Fish and Game. Many of the animals returned to the mainland, and those that remained did not survive or reproduce as expected. As a consequence, the FWS has considered possible alternative management strategies, and in February 2003 approved a revised recovery plan (U.S. FWS 2003).

As indicated earlier, by the mid-1980s the remnant sea otter populations in Alaska were thought to have recolonized much of their former range and to be approaching pre-exploitation levels (see Rotterman and Simon-Jackson 1988). However, populations in the Aleutian Islands have declined precipitously in recent years (Doroff et al. 2003). In some areas, recent counts are less than 10% of counts done in the mid-1960s (MMC 2004). The cause or causes of the decline have not been documented. One possibility is increased predation by killer whales as a consequence of decreases in Steller sea lions (described below) that also have occurred in recent years (Estes et al. 1998). In 2004, the FWS proposed listing the Aleutian Islands sea otter as an endangered species in accordance with the Endangered Species Act. The future of sea otters in Alaska, particularly the Aleutian Islands, Bristol Bay, and Kodiak Island populations, is uncertain at best.

Steller Sea Lions

Historically, Steller sea lions occurred in the Bering Sea, the Sea of Okhotsk, and along the rim of the North Pacific Ocean from northern Japan to the California Channel Islands. In the 1950s, total abundance was estimated to be from 240,000 to 300,000. In the last twenty-five to thirty years, there has been a gradual decrease in abundance, particularly in the Aleutian Islands and Gulf of Alaska, averaging about 4% per year. In many areas current abundance is less than 10–15% of historic abundance. The cause or causes of the decline, like the cause or causes of the Alaska sea otter decline described earlier, have not been determined and are the subject of much speculation. Possibilities that have been hypothesized include environmental pollution, as well as changes in key prey species due either to commercial fisheries or alteration of the marine food web in the North Pacific due to natural variation or climate change.

In 1990, the NMFS listed the Steller sea lion as threatened under the Endangered Species Act. Subsequently the agency appointed a recovery team, and in December 1992 published a species recovery plan (NOAA NMFS 1992). Because of the precipitous decline in the Aleutians and Gulf of Alaska, the agency designated the population west of 144° west longitude as endangered under the Endangered Species Act. Currently, the agency is preparing a revision of the recovery plan to reflect the change in listing status.

STRANDINGS AND UNUSUAL MORTALITY EVENTS

Marine mammals that strand alive and that wash up dead on coastal beaches provide important sources of information on the distributions, regional abundance, anatomy, physiology, general condition, and diseases of marine mammals. Following an MMC workshop in 1977, the NMFS fostered the development of regional networks of volunteers to respond to and collect data on both live and dead marine mammal strandings (Wilkinson and Worthy 1999). These volunteer networks have produced a large database on the species, numbers, general condition, and causes of marine mammal strandings in the United States (Geraci, Harwood, and Lounsbury, 1999; Reynolds and Odell 1987). They also have served as a model for establishing stranding-response programs in other parts of the world (Geraci and Lounsbury 1993).

The volunteer networks have been instrumental in detecting and investigating the increasing numbers of unusual mortality events or die-offs worldwide (Geraci, Harwood, and Lounsbury, 1999). These events include the deaths of more than 400 harbor seals in New England between December 1979 and October 1980 due to an avian influenza virus (Geraci et al. 1982); the previously noted deaths of hundreds of manatees in Florida due to red tides (Bossart et al. 1998); the deaths of fourteen humpback whales in Cape Cod Bay in November 1987 as a consequence of the whales eating mackerel containing saxitoxin, the neurotoxin responsible for paralytic poisoning (Geraci et al. 1989); the deaths of 700 bottlenose dolphins along the U.S. mid-Atlantic coast between June 1987 and January 1988 due to a previously unknown morbillivirus similar to the ones that cause distemper in dogs, measles in humans, and rinderpest in hoofed animals (Geraci 1989; Lipscomb et al. 1994); the deaths of more than 17,000 harbor seals in the North Sea in 1988 and more than 1,000 striped dolphins in the Mediterranean Sea in 1990–1991 due to morbilliviruses similar to the ones that killed bottlenose dolphins in U.S. waters in 1987–1988 (Aguilar and Raga 1993; Duignan et al. 1992; Osterhaus et al. 1990); the deaths of hundreds of bottlenose dolphins along the Florida panhandle and hundreds of sea lions along the central California coast in 1998 and 1999 associated with toxic algal blooms (Gulland 2000); the deaths of more than 600 gray whales (*Eschrichtius robustus*) along the west coast of North America in 1999 and 2000 due to causes that could not be determined (Gulland et al. 2005); and strandings of dozens of beaked whales in different parts of the world due possibly to exposure to military sonars (MMC 2005).

Because of the difficulties encountered in responding to and uncertainties concerning the cause of the dolphin die-off along the mid-Atlantic coast in 1987–1989, Congress in 1992 enacted the Marine Mammal Health and Stranding Response Act (Title IV of the MMPA). Among other things, this legislation directed the NMFS to (1) establish an expert working group to provide advice on measures necessary to better detect and respond to unusual mortality events; (2) develop a contingency plan to help ensure prompt and effective response to unusual mortality events; (3) establish a fund to compensate individuals and organizations for certain costs incurred in responding to unusual events; (4) develop objective criteria for determining when rehabilitated, live stranded marine mammals can be returned to the wild; (5) continue development of the National Marine Mammal Tissue Bank initiated at the National Institute of Standards and Technology following the 1987–1988 dolphin die-off; and (6) establish and maintain a central database for tracking and accessing data concerning marine mammal strandings. In response to this directive, the NMFS, among other things, has constituted and staffed the expert advisory group and has developed a National Contingency Plan for Response to Unusual Marine Mammal Mortality Events (Wilkinson 1996).

Funding of the regional stranding networks has been a problem, and in December 2000 Congress enacted the Marine Mammal Rescue Assistance Act, directing the NMFS and the FWS to initiate grant programs to improve the effectiveness of the stranding networks. The grants are intended to provide financial assistance for recovery and treatment of live-stranded animals, collection and archiving of data from both live- and dead-stranded animals, and the operational costs directly related to those activities. Grants may be awarded for up to three years with a cumulative total of $100,000 per eligible participant per year.

CONTAMINANTS, NOISE, AND OTHER ENVIRONMENTAL THREATS

As noted earlier, a number of threats to marine mammals and other marine biota were not apparent or widely recognized when the MMPA was enacted in 1972. These include entanglement in lost and discarded fishing gear and other types of marine debris; disturbance and possible injury and mortality associated with loud sounds from human sources; and introduction of increasing amounts and varieties of fertilizers, pesticides, herbicides, pharmaceuticals, and other chemical contaminants into the world's oceans.

Marine Debris Pollution

The marine debris problem is largely a product of the development and use of persistent, nonbiologically degradable plastics and other synthetic materials for the manufacture of fishing nets and lines, as well as packaging materials such as garbage bags and soda and beer six-pack holders (Laist, Coe, and O'Hara, 1999). It was first recognized as a potentially significant marine conservation problem in the late 1970s and early 1980s when northern fur seal and Hawaiian monk seal populations were found to be declining coincident with observations of increasing numbers of animals in pupping colonies entangled in bits of fishing net and line and other types of marine debris. Subsequently it was learned that unknown but potentially significant numbers of sea turtles, seabirds, and fish, as well as marine mammals, were mistaking floating plastic bags, deflated balloons, bits of styrofoam, and other synthetic materials for food items, and were dying because the items were indigestible and either clogged their digestive tracts or poisoned them.

Because of the apparent role of entanglement in the decline of fur seals in the Pribilof Islands, the MMC recommended in 1982 that the NMFS convene an international workshop to assess the magnitude and sources of the problem and to determine what could be done to address it. The workshop was held in Honolulu, Hawaii, in November 1984 and led to a worldwide effort to document and eliminate the causes of the problem. Details of these and subsequent follow-up actions can be found in the MMC's annual reports to Congress, beginning in 1981.[5]

Ocean Noise Pollution

The first indications that human sources of ocean sound might be a problem surfaced in the late 1970s and early 1980s, when studies done in Alaska and Canada found that the distributions, movements, and behavior patterns of ringed seals, beluga whales (*Delphinapterus leucas*), and bowhead whales were affected by sounds associated with offshore oil and gas exploration, sometimes at distances in excess of ten kilometers. W. J. Richardson et al. (1995) provide a comprehensive review of these and subsequent studies to assess the effects of anthropogenic sound on marine mammals.

The effects on marine mammal behavior found in the early Alaska studies were thought to be biologically insignificant. Consequently, a new section was added to the MMPA in 1981 (§ 101(a)(5)(A)). This addition directed the secretaries of commerce and the interior to authorize, for periods up to five years, the unintentional taking, including the accidental killing, of small numbers of nondepleted marine mammals incidental to activities other than commercial fisheries (covered by other provisions of the act), when the taking involved only small numbers of marine mammals, would have negligible impacts on the affected species and stocks, and the responsible regulatory agency (the NMFS or the FWS) issued regulations specifying when, where, what, how, and how many marine mammals were authorized to be taken. This provision was amended in 1986 to authorize the unintentional taking of small numbers of depleted as well as nondepleted marine mammal species—for example, endangered bowhead whales—when the population level effects would be negligible, and there would be no unmitigable effects on the availability of the affected species or stocks for taking by Alaskan Native Americans for subsistence purposes.

Although the effects of offshore oil and gas development have continued to be subjects of controversy and study, much of the concern and controversy in the last fifteen years has been focused on activities conducted or supported by the U.S. Navy. Those activities include (1) the Heard Island Feasibility Study and follow-up Acoustic Thermometry of Ocean Climate Program supported by the Defense Department's Advanced Research Projects Agency; (2) legislatively required shock testing of new classes of Navy surface vessels and submarines under simulated combat conditions; (3) the development and planned use of low-frequency active sonar to detect and track new classes of quiet submarines at distances of 200 miles or more in deep, offshore waters; (4) the development and testing of additional active sound sources as part of the Littoral Warfare Advanced Development program to detect and track submarines in shallow coastal waters where neither standard tactical sonars, low-frequency active sonar, nor passive listening systems can function effectively; and (5) the stranding and deaths of at least seventeen cetaceans, including fourteen beaked whales, in the northern Bahama islands in March 2000 in apparent response to a Navy antisubmarine exercise involving several ships using standard, mid-frequency tactical sonars. Robert J. Hofman (2004) reviews these and related actions.

Because of the concern and controversy concerning the possible effects of anthropogenic sound on marine mammals, the National Research Council was asked for, and has conducted, four separate studies to assess and identify uncertainties and suggest means for addressing the problem (1994, 2000, 2003, 2005). Also, because of the concern and controversy, Congress in the Omnibus Appropriations Act of 2003 directed the MMC to "fund an international conference or series of conferences to share findings, survey acoustic 'threats' to marine mammals, and develop means of reducing those threats while maintaining the oceans as a global highway of international commerce." In response, the MMC constituted an advisory committee, made up of knowledgeable scientists and representatives of the agencies, industry groups, and environmental organizations with related interests, to consider and, as possible, come up with consensus views on the critical uncertainties, what would be required to resolve them, and what measures should be taken to minimize possible adverse effects pending resolution of the uncertainties. The results of that ongoing effort are expected to be made known to Congress and the public in late summer or early fall 2005.[6]

Chemical Pollution

The greatest long-term threats to marine mammals, sea turtles, fish, seabirds, and other marine organisms may well be nonpoint source ocean contamination—that is, herbicides, pesticides, fertilizers, road tars, pharmaceuticals, and so on that are carried by rain runoff and sewage into rivers and ultimately the world's oceans. Such contaminants have been found in the tissues of marine mammals and other marine organisms from the Arctic to the Antarctic, and virtually everywhere in between. Participants in a 1998 workshop convened jointly by the MMC, the Biological Resources Division of the U.S. Geological Survey, the Environmental Protection Agency, and the National Fish and Wildlife Foundation pointed out that, while there is a growing database concerning the types and levels of contaminants present in the tissues of marine mammals and other marine organisms in many areas, very little is known about the effects of the contaminants, either singularly or collectively, on the growth, longevity, or reproduction of the affected biota (O'Shea, Reeves, and Long, 1999). To date, however, there has been only limited progress in documenting the sources and effects of various contaminants, and how introduction of harmful contaminants into the world's oceans can be minimized (cf. O'Shea 1999). These clearly are topics meriting more attention.

SUMMARY AND CONCLUSIONS

This chapter has provided a broad overview of past and current marine mammal conservation issues, along with several examples of steps that have been and are being taken to address those issues. It has pointed out that conservation must be a dynamic process, taking into account both socioeconomic and biological-ecological factors. It is intended, in part, to call attention to the continuing evolution of conservation laws and policies, both domestically and internationally. It has illustrated the interactions and the important roles that the general public, the scientific community, environmental and industry groups, Congress, and the courts play in formulation and implementation of conservation policies and laws in the United States. It has also illustrated the important roles that marine mammals, the Marine Mammal Protection Act, and the Marine Mammal Commission and its Committee of Scientific Advisors have played in instituting the "optimum" concept and the ecosystem approach to marine resource conservation embodied in the MMPA.

Since the MMPA was enacted in 1972, there has been substantial progress in addressing a number of marine mammal and marine ecosystem conservation issues. For example, the mortality and serious injury of dolphins in the ETP tuna purse seine fishery have been reduced dramatically; commercial whaling currently does not pose a threat to large whales and, if resumed, is likely to be better regulated than in the past; research and regulatory programs undertaken cooperatively by the NMFS and the Alaska Eskimo Whaling Commission have promoted the recovery of the bowhead whale population in the western Arctic, while ensuring the continuing availability of the whales to meet Eskimo subsistence needs; more manatees exist today in Florida than when the MMPA was enacted; the eastern Pacific stock of gray whales has recovered to the point that it was removed from the U.S. endangered species list; the optimum concept and the ecosystem approach to marine resource conservation have been incorporated in a number of international agreements, including the Convention for the Conservation of Antarctic Marine Living Resources, the Food and Agriculture Organization's Code of Conduct for Responsible Fisheries, and the UN Agreement on Conservation and Management of Straddling Fish Stocks and Highly Migratory Fish Stocks; and the PBR and related concepts incorporated in the 1994 MMPA amendments have provided practical solutions to the difficult problems of assessing and monitoring the status of marine mammal stocks in U.S. waters.

Conversely, several issues have resisted solution, and a number of new or previously unrecognized and sometimes controversial issues have arisen. Examples include the continuing failure to identify and take steps necessary to facilitate recovery of the highly endangered right whale population in the northwestern Atlantic; the recent and ongoing declines of sea otters and Steller sea lions in Alaska and the failure to date to determine and eliminate the cause or causes of the declines; uncertainty concerning the cause or causes of the apparent failure of depleted stocks of ETP dolphins to recover now that mortality and serious injury associated with the tuna purse seine fishery have been reduced to what should be biologically insignificant levels; the increases in harbor seal and sea lion populations in certain areas and the resulting proposals by some fishermen and fishery managers to cull the populations to limit perceived predation-related impacts on fisheries and fishery resources; the escalating controversy concerning the effects of anthropogenic sound on marine mammals and other marine organisms, and the regulatory measures necessary to avoid or mitigate adverse effects; and uncertainty concerning the direct and indirect effects on marine mammals of ocean pollution, without a doubt the greatest long-term threat to marine mammals worldwide.

Students, lawmakers, and others with related interests should ask: What ingredients are necessary for successful resource conservation, and how can conservation efforts be carried out effectively, with minimum cost and controversy? The work of the MMC provides some useful clues.

First, it is important, if not vital, to have a framework and an infrastructure in which to conduct activities, and a legal mandate to do so. Those factors exist in the United States and in many other countries around the globe. However, even the best infrastructure and legal mandates are unlikely to be effective unless appropriately qualified, informed, dedicated, and ethical individuals are involved. Simply stated, people make a difference.

A survey by R. L. Wallace and K. A. Semmens (2005) indicates that the MMC is widely viewed as an agency peopled with individuals characterized by hard work, fairness, knowledge, creativity, thoroughness, tenacity, ethics, and substance over process. Successful resolution of difficult multi-stakeholder conservation issues requires such attributes, and must be encouraged.

Another perhaps rarer and more innate attribute exemplified by the commission has been its ability to look forward and attempt to proactively address issues before they reach the crisis stage, rather than responding reactively only after the crisis stage is reached. The ability to be farsighted and anticipatory, and then to work with all affected interest groups to seek balanced and effective solutions to problems, has possibly been the key to the MMC's success. An anticipatory approach, taking into account relevant socioeconomic as well as biological-ecological variables, is almost always more cost-effective than the reactive one.

In the opening years of the twenty-first century, it has become clear that litigation often dictates agency priorities and responses to living resource conservation issues here in the United States and in some other countries. This is a consequence of competing interests and values regarding resource use and promotes the inevitable cycle of ineffective and costly crisis management. That is, failure to anticipate resource conservation issues well in advance leads almost inevitably to (1) overutilization and depletion of the resource or uncertainty concerning the effects of ongoing activities; (2) the need to restrict or limit the activity in question to enable recovery of the resource or to resolve the uncertainty concerning its possible adverse effects; (3) actual or perceived socioeconomic impacts if the activity is prohibited or restricted; (4) lobbying of Congress and the responsible regulatory agency, and the threat of litigation due to competing and polarized views as to the appropriate course of action; (5) no or delayed action by the responsible regulatory agency leading to further depletion of the resource and/or escalation of the controversy concerning the necessary conservation measures; (6) utilization of limited agency financial and personnel resources to avoid or respond to lobbying and lawsuits, reducing the funding and personnel available for dealing with the actual problem; and (7) often ineffective solutions as a consequence of attempting to satisfy or seek a balance between conflicting interests.

To break this cycle, problems must be anticipated and databases must be developed to identify and evaluate the pros and cons of alternative management approaches before crises develop. Toward this end, Congress in 2003 directed the MMC to undertake consultations with knowledgeable scientists, expressly to identify the most critical long-term research needs regarding marine mammal conservation, and the means for proactively meeting those needs. The results of this process (Reeves and Ragen 2004; Reynolds et al. 2005) should pay huge dividends in the future. The issues judged to be most pressing in the foreseeable future were the direct and indirect effects of fisheries, environmental contaminants, harmful algal blooms, disease, underwater noise, habitat degradation and destruction, climate change, difficulty identifying optimal management units, and ineffective management strategies.

The future of marine mammal conservation depends on the willingness and ability of governments, government agencies, international organizations, affected industries, and public interest groups to work together to anticipate and find solutions to conservation problems that are both biologically and economically sound.

ACKNOWLEDGMENT: Much of the information in this chapter has been derived from a report being done for the Marine Mammal Commission on the history and accomplishments of the MMPA and the commission itself.

Notes

1. The MMC's annual reports to Congress are matters of public record and can be requested from the commission (current mailing address 4340 East-West Highway, Room 905, Bethesda, Maryland 20814). Descriptions of ongoing activities, publication lists, etc., can be found at the commission's Web site, http://www.mmc.gov.

2. The latest versions of the stock assessment reports can be obtained on the NOAA Fisheries Office of Protected Resources Web site, http://www.nmfs.noaa.gov.

3. The take reduction plans, like the stock assessment reports, can be obtained through the NOAA Fisheries Office of Protected Resources Web site, http://www.nmfs.noaa.gov.

4. The reports of the meetings of the commission and scientific committee established by the CCAMLR can be obtained from the headquarters of the commission in Hobart, Tasmania. Additionally, Hofman (1985, 1993) reviews the key features of the CCAMLR.

5. Further information on entanglement can be found on the Web site of the Ocean Conservancy (formerly the Center for Marine Conservation). See http://www.oceanconservancy.org. The conservancy has played a lead role in calling attention to and enlisting the public in efforts to resolve the problem.

6. Details of these and other noise-related issues can be found in the commission's annual reports and on the commission's Web site. See http://www.mmc.gov.

References

Aguilar, A., and J. A. Raga. 1993. The striped dolphin epizootic in the Mediterranean Sea. *Ambio* 22(8): 524–528.

Barlow, J., S. L. Swartz, T. C. Eagle, and P. R. Wade. 1995. *U.S. marine mammal stock assessments: Guidelines for preparation, background, and a summary of the 1995 assessments.*

NOAA Technical Memorandum, no. NMFS-OPR-95-6. Washington, DC: U.S. Department of Commerce, NOAA National Marine Fisheries Service.

Beddington, J., and R. May. 1982. The harvesting of interacting species in a natural ecosystem. *Scientific American* 247(5): 62–69.

Bossart, G. D., D. G. Baden, R. Y. Ewing, B. Roberts, and S. C. Wright. 1998. Brevetoxicosis in manatees (*Trichechus manatus latirostris*) from the 1996 epizootic: Gross, histologic, and immunohistochemical features. *Environmental Toxicologic Pathology* 26(2): 276–282.

Committee for Humane Legislation, Inc. v. Elliot L. Richardson. 414 F. Supp. 297 (1976).

Donovan, G. P. 1989. Preface: The comprehensive assessment of whale stocks; The early years. Special issue, *Reports of the International Whaling Commission* 11:iii–v.

Doroff, A. M., J. A. Estes, M. T. Tinker, D. M. Burn, and T. J. Evans. 2003. Sea otter population declines in the Aleutian archipelago. *Journal of Mammalogy* 84(1): 55–64.

Duignan, P. J., J. R. Geraci, J. A. Raga, and N. Calzada. 1992. Pathology of morbillivirus infection in striped dolphins (*Stenella coeruleoalba*) from Valencia and Murcia, Spain. *Canadian Journal of Veterinary Research* 56(3): 242–248.

Estes, J. A., B. B. Hatfield, K. Ralls, and J. Ames. 2003. Causes of mortality in California sea otters during periods of population growth and decline. *Marine Mammal Science* 19(1): 198–216.

Estes, J. A., M. T. Tinker, T. M. Williams, and D. F. Doak. 1998. Killer whale predation on sea otters linking oceanic and nearshore ecosystems. *Science* 282(5388): 473–476.

Gales, N. J., T. Kasuya, P. J. Clapham, and R. L. Brownell Jr. 2005. Japan's whaling plan under scrutiny. *Nature* 435(7044): 883–884.

Gambell, R. 1999. The International Whaling Commission and the contemporary whaling debate. In *Conservation and management of marine mammals,* ed. J. R. Twiss Jr. and R. R. Reeves, 179–198. Washington, DC: Smithsonian Institution Press.

Geraci, J. R. 1989. *Clinical investigation of the 1987–88 mass mortality of bottlenose dolphins along the U.S. central and south Atlantic coast.* Final report to the National Marine Fisheries Service, Office of Naval Research, and Marine Mammal Commission, Washington, DC.

Geraci, J. R., D. M. Anderson, R. J. Timperi, D. J. St. Aubin, G. A. Early, J. H. Prescott, and C. A. Mayo. 1989. Humpback whales (*Megaptera novaeangliae*) fatally poisoned by dinoflagellate toxin. *Canadian Journal of Fisheries and Aquatic Sciences* 46(11): 1895–1898.

Geraci, J. R., J. Harwood, and V. J. Lounsbury. 1999. Marine mammal die-offs: Causes, investigations, and issues. In *Conservation and management of marine mammals,* ed. J. R. Twiss Jr. and R. R. Reeves, 367–395. Washington, DC: Smithsonian Institution Press.

Geraci, J. R., and V. J. Lounsbury. 1993. *Marine mammals ashore: A field guide for strandings.* Galveston: Sea Grant College Program, Texas A&M University.

Geraci, J. R., D. J. St. Aubin, I. K. Barker, R. G. Webster, V. S. Hinshaw, W. J. Bean, H. L. Ruhnke, J. H. Prescott, G. Early, A. S. Baker, et al. 1982. Mass mortality of harbor seals: Pneumonia associated with influenza A virus. *Science* 215(4536): 1129–1131.

Gosliner, M. L. 1999. The tuna-dolphin controversy. In *Conservation and management of marine mammals,* ed. J. R. Twiss Jr. and R. R. Reeves, 120–155. Washington, DC: Smithsonian Institution Press.

Gulland, F. 2000. *Domoic acid toxicity in California sea lions (*Zalophus californianus*) stranded along the central California coast, May–October 1998: Report to the National Marine Fisheries Service Working Group on Unusual Marine Mammal Mortality Events.* NOAA Technical Memorandum, no. NMFS-OPR-17. Washington, DC: NOAA National Marine Fisheries Service.

Gulland, F. M. D., M. H. Pérez-Cortés, J. Urbán, L. Rojas-Bracho, G. Ylatalo, J. Weir, S. A. Norman, M. M. Muto, D. J. Rugh, C. Kreuder, et al. 2005. *Eastern North Pacific gray whale (*Eschrichtius robustus*) unusual mortality event, 1999–2000.* NOAA Technical Memorandum, no. NMFS-AFSC-150. Seattle, WA: U.S. Department of Commerce.

Hofman, R. J. 1985. The Convention on the Conservation of Antarctic Marine Living Resources. In *Antarctic politics and marine resources: Critical choices for the 1980s; Proceedings from the eighth annual conference held June 17–20, 1984, Center for Ocean Management Studies, University of Rhode Island,* ed. L. M. Alexander and L. Carter Hanson, chapter 8. Kingston: Center for Ocean Management Studies, University of Rhode Island.

Hofman, R. J. 1993. The Convention for the Conservation of Antarctic Marine Living Resources. *Marine Policy* 17(6): 534–536.

Hofman, R. J. 2004. Marine sound pollution: Does it merit concern? *Marine Technology Society Journal* 37(4): 66–77.

Holt, S. J., and L. M. Talbot. 1978. *New principles for the conservation of wild living resources.* Wildlife Monographs, no. 59. Bethesda, MD: Wildlife Society.

International Whaling Commission. 1983. Chairman's report of the 34th annual meeting. *Reports of the International Whaling Commission* 33:20–42.

International Whaling Commission. 1991. Comprehensive assessment of whale stocks: Progress report on development of revised management procedures. *Reports of the International Whaling Commission* 41:213–218.

International Whaling Commission. 1992. Report of the Scientific Committee. *Reports of the International Whaling Commission* 42:51–86.

International Whaling Commission. 1995. Chairman's report of the 46th annual meeting. *Reports of the International Whaling Commission* 45:15–52.

International Whaling Commission. 2001. Report of the workshop on status and trends of western North Atlantic right whales. Special issue, *Journal of Cetacean Research and Management* 2:61–87.

Katona, S. K., and S. D. Kraus. 1999. Efforts to conserve the North Atlantic right whale. In *Conservation and management of marine mammals,* ed. J. R. Twiss Jr. and R. R. Reeves, 311–331. Washington, DC: Smithsonian Institution Press.

Kenyon, K. W. 1969. *The sea otter in the eastern Pacific Ocean.* North American Fauna, no. 68. Washington, DC: U.S. Department of the Interior, Bureau of Sport Fisheries and Wildlife.

Kokechik Fishermen's Association v. Secretary of Commerce. 839 F.2d 795 (D.C. Cir. 1988).

Laist, D. W., J. M. Coe, and K. J. O'Hara. 1999. Marine debris pollution. In *Conservation and management of marine mammals,* ed. J. R. Twiss Jr. and R. R. Reeves, 342–366. Washington, DC: Smithsonian Institution Press.

Laist, D. W., A. R. Knowlton, J. G. Mead, A. S. Collet, and M. Podesta. 2001. Collisions between ships and whales. *Marine Mammal Science* 17(1): 35–75.

Lentfer, J. W., ed. 1988. *Selected marine mammals of Alaska: Species accounts with research and management recommendations.* Washington, DC: Marine Mammal Commission.

Lipscomb, T. P., F. Y. Schulman, D. Moffett, and S. Kennedy. 1994. Morbilliviral disease in Atlantic bottlenose dolphins (*Tursiops truncatus*) from the 1987–1988 epizootic. *Journal of Wildlife Diseases* 30(4): 567–571.

Marine Mammal Commission. 2004. *Annual report to Congress 2003.* Bethesda, MD: MMC.

Marine Mammal Commission. 2005. *Annual report to Congress 2004.* Bethesda, MD: MMC.

Marine Mammal Protection Act. Public Law 92–522. *U.S. Statutes at Large* 86:1027, codified at *U.S. Code* 16, § 1361–1407. Amended 1981: Public Law 97–58. *U.S. Statutes at Large* 95:979. Amended 1988: Public Law 100–711. *U.S. Statutes at Large* 102:4755. Amended 1994: Public Law 103–238. *U.S. Statutes at Large* 108:532.

Marine Mammal Rescue Assistance Act. Public Law 106–555. *U.S. Statutes at Large* 114:2765, codified at *U.S. Code* 16, § 1421f-1.

National Oceanic and Atmospheric Administration National Marine Fisheries Service. 1992. *Recovery plan for the Steller sea lion (*Eumetopias jubatus*).* Report prepared by the Steller Sea Lion Recovery Team. Silver Spring, MD: NMFS, Office of Protected Resources.

National Oceanic and Atmospheric Administration National Marine Fisheries Service. 2002. Special issue commemorating the 30th anniversary of the Marine Mammal Protection Act (1972–2002). *MMPA Bulletin* 22.

National Oceanic and Atmospheric Administration National Marine Fisheries Service. 2005. *Recovery plan for the North Atlantic right whale (*Eubalaena glacialis*).* Silver Spring, MD: NMFS, Office of Protected Resources.

National Research Council (Committee on Low-Frequency Sound and Marine Mammals). 1994. *Low-frequency sound and marine mammals: Current knowledge and research needs.* Washington, DC: National Academy Press.

National Research Council (Committee to Review Results of ATOC's Marine Mammal Research Program). 2000. *Marine mammals and low-frequency sound: Progress since 1994.* Washington, DC: National Academy Press.

National Research Council (Committee on Potential Impacts of Ambient Noise in the Ocean on Marine Mammals). 2003. *Ocean noise and marine mammals.* Washington, DC: National Academies Press.

National Research Council (Committee on Characterizing Biologically Significant Marine Mammal Behavior). 2005. *Marine mammal populations and ocean noise: Determining when noise causes biologically significant effects.* Washington, DC: National Academies Press.

O'Shea, T. J. 1999. Environmental contaminants and marine mammals. In *Conservation and management of marine mammals,* ed. J. R. Twiss Jr. and R. R. Reeves, 485–564. Washington, DC: Smithsonian Institution Press.

O'Shea, T. J., R. R. Reeves, and A. K. Long. 1999. *Marine mammals and persistent ocean contaminants: Proceedings of the Marine Mammal Commission Workshop, Keystone, Colorado, 12–15 October 1999.* Bethesda, MD: Marine Mammal Commission.

Osterhaus, A. D. M. E., J. Groen, H. E. M. Spijkers, H. W. J. Broeders, F. G. C. M. UytdeHaag, P. de Vries, J. S. Teppema, I. K. G. Visser, M. W. G. van de Bildt, and E. J. Vedder. 1990. Mass mortality in seals caused by a newly discovered morbillivirus. *Veterinary Microbiology* 23(1–4): 343–350.

Ragen, T. J., and D. M. Lavigne. 1999. The Hawaiian monk seal: Biology of an endangered species. In *Conservation and management of marine mammals,* ed. J. R. Twiss Jr. and R. R. Reeves, 224–245. Washington, DC: Smithsonian Institution Press.

Reeves, R. R., and T. J. Ragen. 2004. *Future directions in marine mammal research. Report of the Marine Mammal Commission Consultation, Portland, Oregon, August 4–7, 2003.* Bethesda, MD: Marine Mammal Commission.

Reynolds, J. E., III. 1999. Efforts to conserve the manatees. In *Conservation and management of marine mammals,* ed. J. R. Twiss Jr. and R. R. Reeves, 267–295. Washington, DC: Smithsonian Institution Press.

Reynolds, J. E., III, and D. K. Odell, eds. 1987. *Marine mammal strandings in the United States: Proceedings of the Second Marine Mammal Stranding Workshop, Miami, Florida,*

December 3–5, 1987. NOAA Technical Report, National Marine Fisheries Service, no. 98. Seattle, WA: U.S. Department of Commerce, NOAA NMFS.

Reynolds, J. E., III, W. F. Perrin, R. R. Reeves, T. J. Ragen, and S. Montgomery, eds. 2005. *Marine mammal research: Conservation beyond crisis.* Baltimore, MD: Johns Hopkins University Press.

Richardson, W. J., C. R. Greene Jr., C. I. Malme, and D. H. Thompson. 1995. *Marine mammals and noise.* San Diego, CA: Academic Press.

Rotterman, L. M., and T. Simon-Jackson. 1988. Sea otter, *Enhydra lutris.* In *Selected marine mammals of Alaska: Species accounts with research and management recommendations,* ed. J. W. Lentfer, 237–275. Washington, DC: Marine Mammal Commission.

Sahrhage, D. 1985. Fisheries overview. In *Antarctic politics and marine resources: Critical choices for the 1980s; Proceedings from the eighth annual conference held June 17–20, 1984, Center for Ocean Management Studies, University of Rhode Island,* ed. L. M. Alexander and L. Carter Hanson, chapter 7. Kingston: Center for Ocean Management Studies, University of Rhode Island.

Smith, T. D. 1974. *Estimates of porpoise population sizes in the eastern tropical Pacific, based on an aerial survey done in early 1974.* Southwest Fisheries Center Administrative Report, no. LJ-79–41. Washington, DC: U.S. Department of Commerce, NOAA NMFS.

U.S. Fish and Wildlife Service. 2001. *Florida manatee recovery plan (*Trichechus manatus latirostris*), third revision.* Atlanta, GA: U.S. Fish and Wildlife Service.

U.S. Fish and Wildlife Service. 2003. *Final revised recovery plan for the Southern sea otter (*Enhydra lutris nereis*).* Portland, OR: U.S. Fish and Wildlife Service.

Wade, P. R. 1994. *Managing populations under the Marine Mammal Protection Act of 1994: A strategy for selecting values for* N_{min}*, the minimum abundance estimate, and* F_r*, the recovery factor.* Southwest Fisheries Center Administration Report, no. LJ-94–9. Washington, DC: U.S. Department of Commerce, NOAA NMFS.

Wallace, R. L., and K. A. Semmens. 2005. *Oversight and effectiveness: An appraisal of the Marine Mammal Commission.* A report to the Marine Mammal Commission. Collegeville, PA: Ursinus College.

Wendell, F. E., R. A. Hardy, and J. A. Ames. 1986. *An assessment of the accidental take of sea otters,* Enhydra lutris, *in gill and trammel nets.* California Department of Fish and Game Marine Resources Technical Report, no. 54. Sacramento: California Department of Fish and Game.

Wilkinson, D. M. 1996. *National contingency plan for response to unusual marine mammal mortality events.* NOAA Technical Memorandum, no. NMFS-OPR-9. Washington, DC: U.S. Department of Commerce, NOAA NMFS.

Wilkinson, D. M., and G. W. J. Worthy. 1999. Marine mammal stranding networks. In *Conservation and management of marine mammals,* ed. J. R. Twiss Jr. and R. R. Reeves, 396–411. Washington, DC: Smithsonian Institution Press.

14

Marine Wildlife Policy: Underlying Ideologies

Michael L. Weber

The period of the late 1960s and early 1970s was a golden age for marine conservation, as it was for environmental protection generally. During this time, Congress passed the Endangered Species Act, the Marine Mammal Protection Act, the Clean Water Act, and the Marine Protection, Research, and Sanctuaries Act. Internationally, the campaign to place a moratorium on commercial whaling began. And the Convention on International Trade in Endangered Species of Wild Fauna and Flora was finalized and ratified by the United States and dozens of other countries.

Behind all these advances were Lee M. Talbot and a small number of colleagues. The kinds of dramatic change that Talbot and his colleagues generated are a rare event; more commonly, problems and solutions are identified and re-identified without any measurable change. There are many possible reasons for such lack of change. One is the subtle but pervasive influence of ideology. Ideology may be defined as the body of doctrines, myths, and beliefs that guides an individual, social movement, institution, class, or large group. I use the word *ideology* because it includes that mixture of myth and fact that guides how we view the world and how we make decisions. The more an ideology is repeated, the more it takes on a life of its own. Indeed, when we embrace an ideology, we often shape reality to fit it.

Until the 1970s, marine wildlife policy was dominated by an ideology of abundance. Passage of the Marine Mammal Protection Act in 1972 introduced a different ideology that I call the ideology of scarcity. Initially, this new ideology benefited just marine mammals and endangered and threatened species. As I describe later in this chapter, the ideology of scarcity began influencing the management of marine fisheries only after the collapse of America's hallmark fisheries.

THE IDEOLOGY OF ABUNDANCE

The belief in the limitless abundance of the seas goes back to at least the seventeenth century, when the Dutch lawyer Hugo Grotius first articulated the theory of the freedom of the seas. The belief persisted well after the rise of science. In 1969, for instance, the Stratton Commission (created by Congress) concluded that the oceans could produce 400–500 million metric tons of catches annually—a tenfold increase over catches at the time (Weber 2002). These predictions, which proved wildly optimistic, gave scientific credibility to the ideology of abundance.

The ideology of abundance also reflected a conviction that scientists could, through rational management, manipulate nature so that it produced the greatest good for the greatest number of people. This ideology emphasized maximum use. An unused resource, whether water, forest, or fish, was a wasted resource.

Practically speaking, the belief in abundance removed any need for caution. Until the late 1980s, policy makers believed that U.S. landings of fish were limited only by the size and sophistication of U.S. fleets. A principal goal of government policy then was to encourage the expansion of U.S. fleets. Likewise, the apparent abundance of some fish populations and the discovery of new populations suggested that there was a greater danger of underexploiting fish populations than of overexploiting them. The burden of proof rested with those who urged restraint.

When I first became involved in marine wildlife conservation in the early 1970s, a rival ideology was just surfacing. This ideology was captured by 1974's *Mind in the Waters*, which summarized the much broader view of marine wildlife that had fueled the earlier passage of the Marine Mammal Protection Act. No longer were marine mammals viewed simply as commodities or as passive players in mathematical models. Instead, they were intelligent, social creatures that offered humans a source of spiritual inspiration. Joan McIntyre, the compiler of this book, recognized that conservation of these creatures would not be achieved simply by appeals to their special nature. At the end of the book, she included critiques of the policy and science that had allowed the International Whaling Commission to oversee the overexploitation of several species of great whales. One of these essays, "The Great Whales and the International Whaling Commission," served as my introduction to the world of Lee Talbot.

The opening lines of the essay captured the shift in values that was under way: "More than any other form of life, whales have come to epitomize the problem of managing living resources" (Talbot 1974, 232). This opening statement reflects Talbot's choice to raise general concerns about marine wildlife by focusing on creatures that people already viewed as more than sources of food or materials. As I read on, I found other insightful statements, some of which saddened me greatly. One in particular described a pattern that I later found in many other fisheries— serial depletion: "The total history of whaling to date is one of an industry which has overexploited and driven species after species into commercial extinction. As one species became so reduced that it was no longer economically exploitable, the industry shifted its attention to the next most economically desirable one" (Talbot 1974, 234).

NEW PRINCIPLES AND THE MARINE MAMMAL PROTECTION ACT

Several years later, John Twiss Jr., the executive director of the Marine Mammal Commission, gave me a copy of a policy statement that Talbot had had a major hand in preparing. I have since given as gifts many copies of *New Principles for the Conservation of Wild Living Resources* (1978), which Talbot authored with Sidney J. Holt. This statement of principles was a response to discussions at the United Nations regarding a new Law of the Sea. At the time, it appeared that the new treaty would embrace the goal of maximum sustainable yield, a fisheries management model whose shortcomings had contributed to the decline of the great whales.

Talbot brought more than technical ability to organizing the group of scientists who developed the statement. He showed his political acuity when he recruited Sidney Holt, a principal author of the maximum sustainable yield model that dominated fisheries management for decades, to the group. Holt's addition as lead author gained the statement instant credibility among a broader group of managers and scientists.

The new principles were a comprehensive alternative to prevailing policy on wildlife. Some of these principles have been incorporated into U.S. law or into international treaties. All of the principles remain relevant today.

1. The ecosystem should be maintained in a desirable state such that
    a. Consumptive and nonconsumptive values could [can] be maximized on a continuing basis,
    b. Present and future options are ensured, and
    c. Risk of irreversible change or long-term adverse effects as a result of use is minimized.
2. Management decisions should include a safety factor to allow for the fact that knowledge is limited and institutions are imperfect.
3. Measures to conserve a wild living resource should be formulated so as to avoid wasteful use of other resources.
4. Survey or monitoring, analysis, and assessment should precede planned use and accompany actual use of wild living resources. The results should be made available promptly for critical public review. (Holt and Talbot 1978, 5)

This statement of principles reflected much of the thinking behind the Marine Mammal Protection Act (MMPA) when it was passed in 1972. The MMPA represented a dramatic shift away from the ideology of abundance (Weber 2002). Among other things, the act advanced other values of marine mammals than the market value of these animals' parts, such as oil, fur, or baleen. It recognized such untraditional wildlife management values as aesthetics and scientific investigation. The MMPA also insisted upon taking an ecosystem perspective in managing marine mammal populations. This view stood in stark contrast to the prevailing view of wildlife management, which focused on single species and ignored ecological interactions and factors such as pollution and habitat loss that influenced the capacity of an ecosystem to support a species. Additionally, the MMPA shifted the burden of proof. In the past, the burden of proof was upon managers to demonstrate a need to restrain human activities. Under the MMPA, the burden of proof was reversed and required a demonstration of no significant harm to a marine mammal population.

How did so dramatic a change come about? Research and interviews that I conducted for a book (Weber 2002) have led me to conclude that a number of conditions underlay the shift away from the prevailing ideology in this instance. First, promoters of the shift to the new ideology underlying the MMPA benefited from having a set of principles based on a credible, scientific alternative to the prevailing ideology. Second, grassroots support provided the political momentum for change. The clubbing of baby seals, the drowning of dolphins in tuna nets, and the bloody business of commercial whaling had enraged many members of the public. Congressman Edward Garmatz of Maryland testified to the public's outrage when he said at a congressional hearing in September 1971: "During my 24 years as a member of Congress, I have never before experienced the volume of mail I have been receiving on the subject of ocean mammals" (Regenstein 1974, 74). Third, there was a small group of very knowledgeable people in influential positions. Lee Talbot was on the staff of the new Council on Environmental Quality and had received approval to pursue reform of marine mammal management from the council's executive director, Russell Train. At the same time, Frank Potter was serving as senior staffer to Congressman John Dingell, then chairman of the House Committee on Merchant Marine and Fisheries, in which much of the MMPA was drafted. Talbot and Potter knew each other well, and they collaborated in fashioning the MMPA with the blessing of their superiors. From his position at the Council on Environmental Quality, Talbot was able to avoid the involvement and opposition of the U.S. Fish and Wildlife Service and the National Marine Fisheries Service.

Finally, political opportunity of two kinds provided favorable conditions for change. The immediate opportunity was created by the defeat of a bill on the floor of the House of Representatives that represented traditional views of wildlife managers both inside and outside the government. The defeated bill did not impose a moratorium on "taking" marine mammals as reformers had wanted. The surprising rejection of the bill led Congressman Dingell to rethink his position and to support some form of moratorium on taking marine mammals. The larger political opportunity arose from the desire of presidential advisors John Ehrlichman and John Whitaker to find issues that would give President Nixon a political boost domestically.

A quite opposite kind of movement was transforming the federal management of marine fisheries; however, a wide variety of factors led to a deepened commitment to the ideology of abundance in the management of marine fisheries.

MARINE FISHERIES: THE PERSISTENCE OF THE OLD IDEOLOGY

Despite financial, technical, and marketing assistance to the fishing industry, U.S. landings of fish remained more or less at World War II levels into the 1970s (Weber 2002). The United States lost its leadership as a fishing power and became a net importer of seafood. Then, in the late 1960s, Soviet and European trawlers began fishing heavily off New England and the Pacific coast, on fishing grounds that U.S. fishermen regarded as their own. The resulting frustration combined with the long-held desire of federal officials to impose rational management on fisheries, and with remarkable swiftness, Congress took action and passed the Fishery Conservation and Management Act in 1976.

The decade after passage of the Fishery Conservation and Management Act saw an unprecedented burst in modernization by the industry and in domestic landings.

The act contributed to the boom in two ways. First, it laid the basis for excluding foreign fleets from within 200 miles of U.S. shores. In doing so, it created an enormous opportunity for domestic fishermen to increase fishing effort. Second, and as important, the act imposed weak restraints on the expansion of fisheries.

At the same time, financing for fishing vessels was relatively easy to come by, directly through subsidized loans or indirectly through the tax law. The investment tax credit, for instance, attracted investment by doctors, lawyers, and others who used fishing vessels as tax write-offs. Other programs more directly sheltered fishermen's profits from taxes or fostered more favorable loan conditions. These programs enabled American fishermen to take advantage of the opportunity created by the Fishery Conservation and Management Act's enclosure of waters within 200 miles, and to expand and modernize fishing fleets that could catch what had once been caught by foreign vessels. As a result, the decade after passage of the act saw the construction of more than 40% of all fishing vessels constructed in the last half of the twentieth century (figure 14.1).

Although growth in fleets led to increased catches in other regions, the most spectacular growth occurred in the Alaska pollock fishery—one of the world's largest fisheries (figure 14.2). Even in this enormous fishery, the fishing fleet soon grew beyond what the fishery could support, and the economics of the fishery deteriorated.

In New England, the exclusion of foreign fishermen and the burst in vessel construction fueled heavy fishing on haddock. After years of heavy fishing that had driven haddock stocks down, the International Commission for North Atlantic Fisheries (ICNAF) adopted restrictions, largely at the urging of the U.S. Bureau of Commercial Fisheries. These restrictions reduced catches dramatically in the late 1960s and early 1970s (figure 14.3). However, soon after passage of the Fishery Conservation and Management Act, U.S. fisheries increased their effort on haddock, leading to increased catches. These increases were short-lived, as catches soon

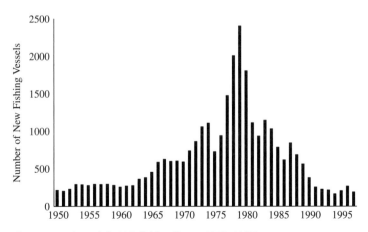

Figure 14.1 Growth in U.S. fishing fleets: 1950–1997.

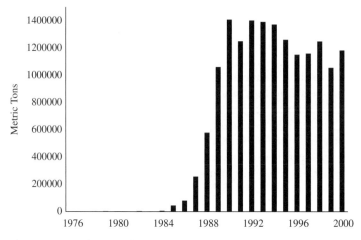

Figure 14.2 Alaska pollock landings: 1976–2000.

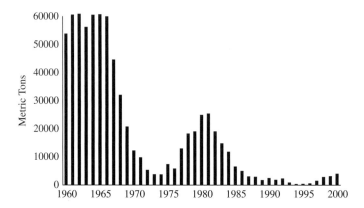

Figure 14.3 Haddock landings: 1960–2000.

resumed their decline. New England groundfish fleets also increased fishing for cod. Landings remained quite high for a dozen years, and warnings from scientists were ignored yet again. Then, in the late 1980s, cod populations went into a free fall (figure 14.4). The federal government declared the fishery a disaster.

MARINE FISHERIES AND THE IDEOLOGY OF SCARCITY

By the 1990s, the decline of many major fisheries exposed the shortcomings of the ideology of abundance. Conservationists, scientists, and progressive fishermen began pressing for change (Weber 2002). Drawing upon principles first fashioned in the

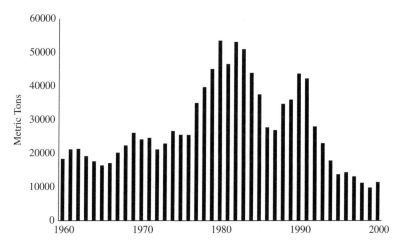

Figure 14.4 Atlantic cod landings: 1960–2000.

1970s, reformers pursued an alternative ideology—an ideology of scarcity. Rather than assuming endless abundance, this ideology assumed limits to the ocean's productivity. The oceans did not annually produce 400–500 million metric tons of surplus life, at least of the type that we humans are interested in consuming. Adherents of the ideology of scarcity urged an ecosystem perspective rather than the focus on single species of traditional fisheries management. They also abandoned the conviction that rational, scientific management could reliably and continuously wring maximum production out of wild populations. In the late 1980s, it also became clear that the buildup of fishing fleets had shifted the balance between the productivity of fish populations and the catching power of fishing fleets. Rather than being limited by the size and sophistication of fishing fleets, catches were limited by the size of fish populations; rather than building up fleets, it appeared that the preeminent need was to build down fleets. Moreover, while the ideology of abundance placed the burden of proof upon those arguing for restraint, the ideology of scarcity reversed the burden.

Reformers burst onto the fisheries scene in the early 1990s and reached a peak of influence with passage of the Sustainable Fisheries Act (SFA) in 1996. One significant substantive change in the SFA was a slight, but critical, shift toward precaution in the definition of optimum yield in a fishery, which fisheries managers used as a benchmark for setting quotas. Instead of allowing optimum yield to be shifted upward to account for economic and ecological considerations, the SFA allows downward shifts only.

Another significant change was an open recognition of overfishing. Under the SFA, fishery managers are required to develop criteria for identifying overfishing and overfished fisheries and to adopt rebuilding plans for overfished fisheries. This is a very different agenda from the development agenda of the mid-1970s and 1980s. It is too early to tell whether the ideology of scarcity has taken hold. The political challenges will be enormous, as we have learned recently on the Pacific Coast.

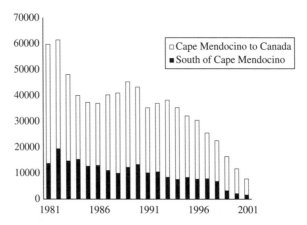

Figure 14.5 Commercial landings of rockfish along the Pacific coast.

On the Pacific Coast, we thought ourselves as being above the sad melodrama that has marked the collapse of the New England groundfish fishery. In the late 1990s, however, overfishing became a common topic in newspapers and discussions by fisheries managers as the decline of Pacific Coast groundfish has become inescapable (figure 14.5). In 2003, nearly all of the continental shelf off California was closed to any type of fishing that might catch any of several overfished species of rockfish (genus *Sebastes*). For some species, recovery will require decades of no catches.

Besides undergoing the same kind of hyperdevelopment of fleets as other fisheries, the Pacific groundfish fishery suffered from less common problems. First, the group of fish that fishermen pursued with trawls, longlines, pots, and other types of fishing gear included sixty-five different species of rockfish with very different life histories; the longevity among rockfish species, for example, differs dramatically (figure 14.6). For this and other reasons, some species are more vulnerable to overfishing than others. Second, rockfish species tend to share habitats, so that species that grow to great age, and often are rarer, mix with species that are relatively short-lived and common (figure 14.7). On land, a steward could reasonably require that a hunter pick the target species from among other species in the same area in order to avoid overexploitation of the more vulnerable species. However, in the Pacific groundfish fishery, the dominant fishing gears—trawls and longlines—are not selective in the species they catch. Nor are fishermen able to observe what their gear is encountering and to guide their gear toward one species rather than another. As a result, it is something of a guessing game as to what will be found in the net or on the longline when it is hauled aboard. Because rockfish generally do not survive being pulled from the depths, any landed fish is lost to the population, regardless whether or not it is the intended catch.

Fisheries managers responded to the diversity of rockfish with a simplifying assumption, lumping all sixty-five species together and managing them as one unit until recently. Although the assumption was convenient for managers and fishermen, it was devastating for some species. For instance, bocaccio (*Sebastes paucispinis*)

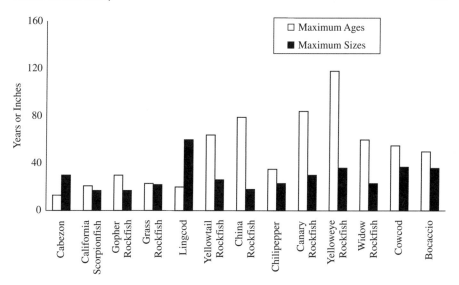

Figure 14.6 Maximum ages and sizes of some Pacific Coast groundfish.

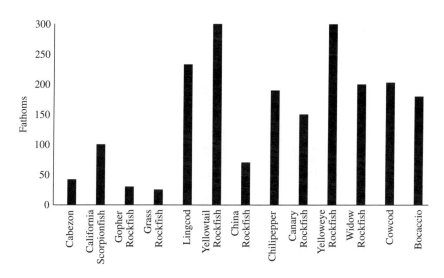

Figure 14.7 General depth range of some Pacific Coast groundfish.

grow to an age of fifty years and an overall length of thirty-six inches. Once one of the most abundant of all rockfish, bocaccio were at no more than 7% of their virgin biomass in 2001 (figure 14.8). Cowcod (*S. levis*) may reach fifty-five years of age and a length of thirty-seven inches. Never very abundant, cowcod were hard-hit by the heavy fishing on other species with which they share deepwater habitat (figure 14.9).

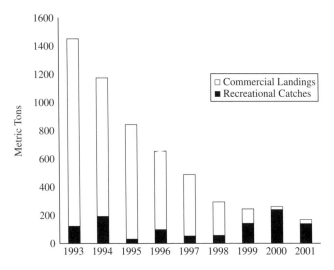

Figure 14.8 Bocaccio: Commercial landings and recreational catches.

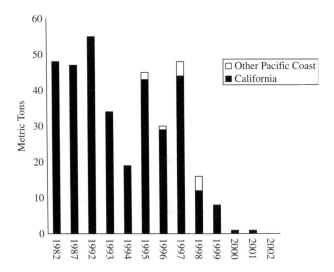

Figure 14.9 Commercial landings of cowcod.

In 2002, a 5,700-square-mile area off southern California was closed to most types of fishing in order to avoid the incidental catch of even small numbers of cowcod. Finally, yelloweye rockfish (*S. ruberrimus*) may reach a maximum age of 118 years and a length of thirty-six inches. Although more abundant at one time than cowcod, heavy fishing over the last two decades drove Pacific populations to abysmally low levels (figure 14.10).

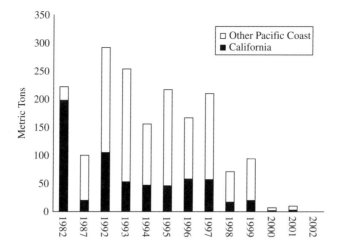

Figure 14.10 Commercial landings of yelloweye rockfish.

Overfishing of these species came as a shocking surprise to everyone but a few scientists whose warnings had been ignored for years. Now managers face the daunting task of rebuilding the populations. Unlike the rebuilding of Atlantic cod and haddock, which is expected to take a decade or less, rebuilding of these Pacific Coast groundfish populations will take several decades. Figure 14.11 gives some idea of the challenges we face in rebuilding these populations. The figure shows the estimated unfished biomass, the current biomass, and the target biomass, which generally is about 40% of the unfished biomass. The figure also indicates the year by which

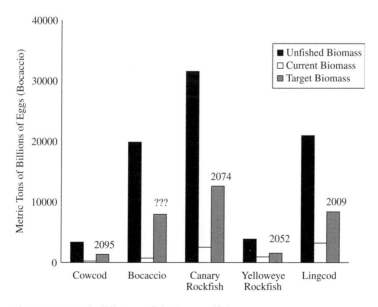

Figure 14.11 Rebuilding overfished groundfish.

biologists expected each population could be rebuilt to the target. These dates range from 2009 to 2095. It is unclear whether bocaccio can ever be rebuilt.

A single figure cannot convey the many ramifications of trying to rebuild these populations. Management restrictions aimed at protecting remaining fish have closed large areas to most types of fishing, sending shock waves through fishing fleets and coastal fishing communities. Between 1995 and 2001, the amount of money paid to fishermen for their groundfish catches fell from nearly $45 million to less than $14 million (figure 14.12). Bankruptcies sent the number of licensed vessels in the fleet tumbling from more than 4,000 vessels in 1989 to little more than 1,000 in 2001 (figure 14.13). We have only a dim idea of the biological havoc we have wrought.

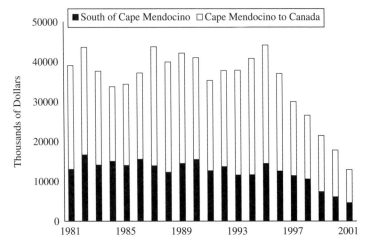

Figure 14.12 Value of commercial landings of rockfish along the Pacific coast.

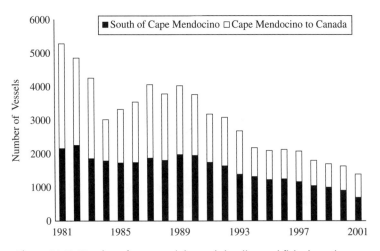

Figure 14.13 Number of commercial vessels landing rockfish along the Pacific coast.

We can gain some conceptual sense of the damage by recalling that Pacific Coast groundfish and fisheries have far more in common with tropical reef fish and fisheries than they do with New England groundfish and fisheries.

CONCLUSION

There is still much to be done in making the New Principles for the Conservation of Wild Living Resources a practical reality in marine fisheries. While we contemplate how far we have yet to go, we must not forget the accomplishments of the past, such as the Convention on International Trade in Endangered Species, the Marine Mammal Protection Act, the National Environmental Policy Act, and the moratorium on commercial whaling. These and other accomplishments by small groups of people, including Lee Talbot, have made an enormous difference.

References

Commission on Marine Science, Engineering, and Resources. 1969. *Our nation and the sea*. Washington, DC: Government Printing Office.

Fishery Conservation and Management Act. Public Law 94–265. *U.S. Code* 16, § 1801–1882.

Holt, S., and L. M. Talbot. 1978. *New principles for the conservation of wild living resources*. Wildlife Monographs, no. 59. Bethesda, MD: Wildlife Society.

Love, M. S., M. Yoklavich, and L. Thorsteinson. 2002. *The rockfishes of the northeast Pacific*. Berkeley: University of California Press.

Marine Mammal Protection Act. Public Law 92–522. *U.S. Statutes at Large* 86:1027, codified at *U.S. Code* 16, § 1361–1421h. Amended 1981: Public Law 97–58. *U.S. Statutes at Large* 95:979. Amended 1988: Public Law 100–711. *U.S. Statutes at Large* 102:4755. Amended 1994: Public Law 103–238. *U.S. Statutes at Large* 108:532.

McIntyre, J., ed. 1974. *Mind in the waters: A book to celebrate the consciousness of whales and dolphins*. New York: Scribner.

National Oceanic and Atmospheric Administration, National Marine Fisheries Service. 2003. Annual commercial landing statistics. Fisheries Statistics Division automated data summary program. http://www.st.nmfs.gov/st1/commercial/landings/annual_landings.html.

Regenstein, L. G. 1974. A history of the Endangered Species Act of 1973, and an analysis of its history; strengths and weaknesses; administration; and probable future effectiveness. Master's thesis, Emory University.

Sustainable Fisheries Act. Public Law 104–297. *U.S. Statutes at Large* 110:3559.

Talbot, L. M. 1974. The great whales and the International Whaling Commission. In *Mind in the waters: A book to celebrate the consciousness of whales and dolphins*, ed. J. McIntyre, 232–236. New York: Scribner; San Francisco: Sierra Club Books.

Weber, M. L. 2002. *From abundance to scarcity: A history of U.S. marine fisheries policy*. Washington, DC: Island Press.

PART V

ENVIRONMENT AND DEVELOPMENT: CONSERVATION AND MANAGEMENT

Megan M. Draheim

The conservation community has by and large embraced the term *sustainable development* for over two decades. However, the authors in part V argue that in practice few development projects have truly been sustainable. The following chapters explore the concept of, obstacles to, and strategies for achieving sustainable development.

Common themes of this part include the following:

1. Scientists must remain involved in all stages of a development project. Too often they are involved in the planning stages, but their involvement recedes as the implementation stage starts. Monitoring and assessing projects is vital to their success; without scientists performing these tasks, there is no way to know if the projects have met their goals.
2. There are no simple, one-size-fits-all, silver bullet solutions to the problems of sustainable development, as the issues involved are very complex.
3. Funding for conservation projects (especially in the developing world) should be considered payments for valuable ecosystem services, not charity or aid.
4. Consumption patterns of the developed world fuel natural resource destruction, although much of this destruction currently occurs in the developing world. Therefore, the developed world holds some responsibility for this environmental degradation.

In chapter 15, Leif E. Christoffersen stresses the importance of integrating the environmental sciences with the social sciences, as well as the fact that this rarely happens. By definition, sustainable development implies an interdisciplinary approach. Yet, Christoffersen argues, we have made little progress in developing the framework for actual interdisciplinary work. Since sustainable development became a goal, there has been more awareness of the breadth of research and analysis that comes from the various disciplines involved in both the conservation and development fields; Christoffersen would deem this awareness only multidisciplinary rather than interdisciplinary. There are several reasons for this failure, including the fact that professionals are rarely trained in a truly integrated fashion. Thus a common analytical framework for practitioners of the various disciplines is not well developed, while the very idea of sustainable development assumes that such a framework exists.

While Christoffersen believes that most policy professionals have embraced an integrated approach to development and conservation, conservation scientists have lagged behind and need to become involved in all stages of sustainable development projects. In most conservation projects, scientists play an important role in the planning stages. However, their involvement lessens or even ends completely during implementation. There are a few exceptions, such as projects that Lee Talbot has been involved with, that show the value of continued involvement of scientists. Scientists must monitor and assess program activities and results; otherwise, it will not be clear whether or not goals are being met in the most efficient manner possible.

Christoffersen closes by describing the parallels between the lack of integration of social and conservation sciences and between economic and social development. An integrated framework for economic and social issues, similar to that which is needed for sustainable development, has not been developed. Consequently, the interdisciplinary approach became unpopular in the development community. In fact, the international development system was founded on this "sharp divide" between economic and social development, as is demonstrated by a reading of the World Bank's founding Articles of Agreement, which restrict the group's activities to those focused on economic development with no regard to social development. This can be seen as a cautionary tale for practitioners of sustainable development and indicates an obstacle to sustainable development implementation and financing.

In chapter 16, Mohamed T. El-Ashry points out that those who are harmed the most by environmental degradation are the world's poor. This holds true because the poor are more directly dependent on local natural resources and are unable to afford a buffer from direct environmental harm. As the gap between the world's rich and poor increases, this fact will become even more magnified.

El-Ashry discusses several of the reasons why sustainable development has been such an elusive goal, noting that even as our knowledge level has increased, the political will to pursue necessary goals has decreased. The reasons for this are numerous, including consumption patterns in the developed world, the debt burden of the developing world, political instability, and perverse subsidies, especially for extractive industries such as agriculture and timber.

At the 1992 UN Conference on Environment and Development, held in Rio de Janeiro, Brazil, signing nations agreed to devote 0.7% of their gross domestic products to international aid. However, most countries have not followed through on their promise. El-Ashry stresses that the developed world should not consider these payments to be charity; instead, they should be considered payments for the ecosystem services provided by developing countries, from which the rest of the world also benefits. As this paradigm shift occurs (and presumably as international aid payments increase accordingly), he proposes policies that can help lead the way to sustainable development—for example, an international carbon tax that could provide funds for sustainable development initiatives.

El-Ashry closes by stating that the world needs to develop new ways of thinking to meet the challenges facing us. The basic principles that currently define economic practices need to be fundamentally changed—most important, the accounting that does not take into account the loss of natural resources and ecosystem services.

In chapter 17, Agnes Kiss asks why, when most agree that a healthy environment is essential to human health and well-being, we act in ways that degrade the environment on a daily basis. Kiss describes four human traits at the root of this disconnect between how we know we should act and how we actually act, including:

1. Humans tend to be attracted to investments that give short-term, concrete results, rather than making small steps toward the resolution of more complicated problems. We are also better at responding to emergencies and crises than to problems whose impacts accrue very slowly over time.

2. Given the option, most people would rather receive a known value from an object than an uncertain future value. For example, the value of chopping down a tree and selling the timber is generally more attractive than the value one would receive by leaving the tree standing and benefiting from such related ecosystem functions as carbon sequestering and flood protection. In order to convince people to accept the more uncertain values, there must be incentives equal to (or more likely greater than) the known value of the resource. Special attention needs to be paid to opportunity costs, which are usually harder to value than direct costs.

3. The well-known "tragedy of the commons" is another human tendency that poses an obstacle to sustainable development. If one person leaves a tree standing, who is to say that someone else would not come along, cut it down, and benefit from selling its wood? Most would choose to benefit directly by cutting down the tree before anyone else does.

4. People tend to ignore costs that are incurred at a distance as a result of their actions. For example, pollution from a factory might pollute a stream downstream but not directly impact the factory's functions. Because of this, there is little incentive for the factory to clean up its pollution.

All of these human tendencies result in cost-benefit analyses that usually favor environmental degradation. If protecting the environment does not negatively impact short-term gains, most are happy to spend money and support the efforts. But once protections require people to act counter to one of these disconnects, they become much more resistant.

The key, Kiss argues, is to make sure that incentives are in place to reward the desired behavior. This needs to be done with care, as most people are skilled at working around incentives with catches by receiving the benefits without changing the behavior in question. In addition, echoing earlier authors, practitioners must guard against trying to simplify complex problems so that they fit simple solutions.

There are two major approaches to preventing environmental degradation in the developing world: (1) separating people from the biodiversity in question and (2) giving people incentives to protect the biodiversity. Two common approaches to the separation strategy are protected areas and community-based resource management; two approaches to the incentive strategy are support for alternative livelihoods or sustainable use, and direct payments for environmental services (PES).

Kiss favors the practice of direct payments for conservation more than the other practices she outlines. The entire world benefits from the ecosystem services provided by parts of the developing world; however, the individuals who own and reside in these ecosystems would individually benefit more by directly harvesting the resources. Therefore, PES should be granted as payment for a valuable public service. Although this path will not be self-sustaining, many (if not most) projects whose goal is to become self-sustaining have never achieved this. The conservation of biodiversity will always need to be compensated in some manner. Otherwise, people will find different ways of benefiting from it.

In chapter 18, Jason Clay discusses how indigenous nations have affected conservation efforts, and how conservation efforts have affected indigenous nations. Nations are distinct from states, he writes; they are composed of culturally distinct groups, usually different from the majority culture in the country where they reside. States tend to be recent creations, while nations have existed for hundreds or thousands of years. Often, there is a language difference between the majority in the state and its nations.

Unfortunately, the world's states by and large have not supported the nations that reside within their boundaries. This is a loss to the entire world, as by refusing to acknowledge and empower their nations, states have eliminated vast stores of knowledge about the sustainable use of the earth's resources. Nations can be the sources of great knowledge about specific sites, as most are tied to a particular region. Ironically, the greatest threat to nations is the destruction of natural resources in their traditional territories. As Clay points out, with only 10% of the world's population but some 25–30% of the earth's resources, nations will almost inevitably come into conflict with those who want to exploit natural resources.

As the technology to extract natural resources has become more efficient, the number of nation groups at risk of extinction has greatly increased; more nations have been destroyed during the twentieth century than during any other period of human history. Although many of these conflicts took and are taking place in the developing world, the consumption patterns of the developed world fuel much of the destruction.

Clay calls the recent trend of bio-prospecting the "last great resource grab of the twentieth century." Bio-prospecting, the search for commercially valuable resources in biodiversity, takes place through many different institutions: universities, corporations, nongovernmental organizations, and governments. The Convention on Biological Diversity gave the rights to genetic and biochemical resources to states, not nations. In fact, nations frequently get shut out of the picture entirely, even when the findings are directly related to the traditional use of resources by indigenous groups. Because this is seen as cultural knowledge, not individual innovation, these uses are not protected as intellectual property. Clay does not argue that nations should have the exclusive rights to this information; however, he proposes that they should be allowed to have at least the same rights as scientists and corporations.

While acknowledging that nations' relationships to nature tend to be romanticized by outsiders, Clay does feel that they are for the most part conservationists (not to be confused with preservationists—they modify their environment, but in ways that ensure there will be resources in the future as well). Cultural destruction often precedes or coincides with biodiversity destruction, tying the two issues together. Therefore, their survival might be an integral part of conserving biodiversity throughout the world.

In chapter 19, Thayer Scudder describes his work (along with Lee Talbot) on the Panel of Environmental and Social Experts (POE) for Laos' Nam Theun 2 (NT2) dam. This chapter can be thought of as a case study that illustrates several of the concepts discussed in the earlier chapters.

Although generally regarded as unsustainable, large dams are thought to be a necessity in development projects. Therefore, the goal should be to make them as sustainable as possible, for both the environment and people.

The Mekong River basin, the watershed in question, is one of the most biodiversity-rich areas left in Southeast Asia. There are also ten ethnic minorities living there, for a total of about 6,000 people. During the 1970s, the government of Laos forcibly moved these villages into existing communities; the ethnic minorities are now in danger of extinction. In addition, more recent immigrant groups are further degrading the environment, as their traditional farming methods are not in sync with it.

The Nakai Plateau, part of the affected region, was greatly hurt by the Vietnam War. Since then, it has been further degraded through land conversion and logging. Residents of the plateau make only about 60% of Laos' national average income. Most of these people will be relocated close to the reservoir. The resettlement action plan specifies that every household must benefit from the project, not simply be left with the same standard of living as before the project. (The plan also ensured that the intended residents were involved in planning their new villages, so that their physical and cultural needs would be satisfied.)

Innovative institutions have been developed as a result of the NT2 project. For example, the Watershed Management and Protection Authority was established to protect the NT2 watershed—not an easy task given rampant poaching—as well as to promote development for indigenous populations. The Nakai Plateau Village Forestry Association was founded to assist new villages in sustainably profiting from their forests.

Earlier in this part, several authors stressed that scientists must remain involved in projects beyond the planning stages. Scudder affirms that the POE was involved not only in the planning stages of NT2 but also in the building stage, and he notes that the POE is responsible for monitoring and assessing the project to ensure that goals will be met.

There are two possible threats to the NT2 project: (1) lack of political will to implement promised social and environmental efforts and (2) the inability of the government and agencies to implement the promised efforts.

Scudder believes that the government does have the political will to make this a model project, but that the second threat holds more danger. For example, the Watershed Management and Protection Authority must find ways to control illegal takes of biodiversity. Other challenges include coordination of the relocation schedule to coincide well with the dam construction schedule and the lack of relevant nongovernmental organizations, as the government does not permit their establishment.

The planning process has been very strong, Scudder concludes, and this project has the potential to become a model of sustainability. However, the implementation phase has just started, so it remains to be seen whether or not the project is a success.

15

Overcoming Barriers to Interdisciplinary Approaches to Sustainable Development

Leif E. Christoffersen

It has been said that rational resource management requires knowledge about two different kinds of systems—ecological systems as well as systems of human activities—and the interface between the two (Underdal 1989).

With this as a starting point for discussion, we are immediately faced with a difficult question: How do we arrive at ecological systems approaches that have broad scientific and technical credibility and acceptance (Talbot 1996)?

A similar question relates to systems within which human activities can be described and analyzed. For example, how can the economic growth objectives of developing countries be met while social progress and sound environmental objectives are being ensured?

Even if we manage to sort out the answers to these two questions, we are still faced with the awesome challenge of how to interface the two systems. The term *sustainable development* describes what we hope to eventually achieve. The term holds hope that economic, social, and environmental issues can be integrated into a common and mutually reinforcing set of policy principles and development results.

Sustainable development is a unifying concept for much of today's international development assistance. Early milestones in the development of the term are the pioneering work of the World Conservation Strategy, produced in 1980 by IUCN (the International Union for the Conservation of Nature and Natural Resources, now the World Conservation Union); the 1987 report of the World Commission on Environment and Development (the Brundtland Commission); and the 1992 UN Conference on Environment and Development, also known as the Earth Summit, held in Rio de Janeiro, Brazil (Holdgate 1999). More recently, political acceptance of this concept was reconfirmed at the World Summit on Sustainable Development, held in Johannesburg, South Africa, in 2002.

Can we realistically achieve this goal? It is no doubt a highly ambitious and laudable objective, but we need to be careful about raising expectations that might not be met in the immediate future.

The head of the Wildlife Conservation Society, Steven Sanderson, is not so encouraging. In a recent article, he argues that over the last decade the economic and nature conservation issues have been pursued in separate forums without good inter-connection: "If development has ignored conservation, conservation has paid little attention to development. Economic policy makers have concentrated on growth, developers on the distribution of the benefits of growth, and conservationists on the costs and consequences of growth for nature and the environment" (Sanderson 2002, 163). The results have been to proceed along different paths. "In practice," he says, "the concept of sustainable development has proven less a viable middle ground than an empty rhetorical shell" (163).

Sanderson believes the fault is widely shared, but he notes that poverty alleviation is a powerful theme on this international agenda. "If poverty is a basic root cause for environmental degradation, then developmental goals must be pursued force-fully to get them out of the poverty trap" (Sanderson 2002, 167). He is severe in his warnings: "If conservationists simply criticize development and poverty allevia-tion without offering realistic alternatives, they will consign themselves to perpetual irrelevance. If they cannot connect short-term human betterment with conservation for long-term sustainability, they will lose the opportunity to influence the future of global public policy" (167). Yet Sanderson holds out some hope. He believes that conservationists are well positioned to make this case, "although it will require a focus on the rural poor and more realistic expectations about dramatic changes in aggregate economic performance on the ecological frontier" (167).

I do not entirely share his pessimism that sustainable development is an empty rhetorical shell, because I know of many valiant efforts around the world to bring it meaningful content. However, I do agree that today, more than two decades after the launch of the World Conservation Strategy, there is still too much fragmentation of discussion and of analytical assessments among the different disciplines on this topic. Sanderson's article may therefore be a useful wake-up call for many of us.

What are the barriers that need to be overcome? Let me focus on three points that I consider to have some immediate importance: (1) the still-weak knowledge base for interdisciplinary approaches to sustainable development, (2) the links between theory and practice, and (3) the institutional and policy consequences we can now observe around the world that may have been caused, at least in part, by our inability to find clarity of purpose as to how sustainable development objectives can be effec-tively carried out.

First, while at the policy level the international community has embraced the integrated approach that the term sustainable development implies, the scientific community seems to have been slow in providing its scientific and technical under-pinnings. For as long as I have been involved in development work, there have been many occasions when the need for an integrated approach, or a more holistic approach, has been promoted. The idea of scientific and technical knowledge being brought into a more inclusive and comprehensive context is appealing and seems sensible. Why is it so slow in coming?

Much has been written about the fragmented organization of research and the specialization-focused academic traditions that stimulate and foster research. Sadly, there are few institutional or financial incentives to conduct interdisciplinary research. Recent discussions on this topic have emphasized the difference between multidisciplinary and interdisciplinary research. Multidisciplinary may perhaps best describe the kind of progress we have seen over the last three decades—different disciplines taking note of each other and being willing to work together, but without having a common analytical framework for integration of the knowledge that each discipline can bring to bear. Interdisciplinarity is a much more ambitious and demanding concept since it assumes the existence of such a commonly accepted theoretical framework. We are still far from reaching this goal. As a recent article by Elisabeth Molteberg, Cassandra Bergstrom, and Ruth Haug notes: "The amount of research done on interdisciplinarity is relatively modest—there is a body of interdisciplinary theory but without ample empirical underpinnings. This is one of the challenges faced by interdisciplinary science—to document and analyze experience more systematically" (2000, 328–329). This point is quite important. There is need to deepen the empirical underpinnings of a rather weak base of available scientific theories, particularly by analyzing and documenting experiences in this regard. This is a serious and urgent challenge to the academic and research communities, but also one that holds some hope.

At the operational level I have witnessed many critical development issues that have required interdisciplinary approaches as part of their solutions. These are best handled when they are problem-focused or part of a wider participatory process involving both scientists and practitioners. From my own international experiences, I am convinced that there are vast opportunities for scientists to take advantage of this type of engagement and thereby help strengthen interdisciplinary theories. Valuable knowledge can be gained from learning from these practical experiences and processes.

A second barrier to sustainable development relates to the link between research and practice. Oran Young has detailed the agenda-setting function of science. Noting that the formation of some international resource regimes is essentially science-driven, he has suggested that "science and scientists (including social scientists) have much to offer to those concerned with the formation and operation of international resource regimes" (Young 1989, 20).

Scientists have certainly been deeply involved in the formation of many of the major environmental conventions, such as those relating to ozone depletion and climate change. I agree that this is an important role for science. However, from my experience it seems that the role of science has received more attention and emphasis in the formation stages of such regimes as compared with the operational stages. In my view scientists need to be much more actively involved in the follow-up to what happens once an agenda has been set and discussed and action begins to be implemented.

Each of the major global conventions has set up scientific bodies that advise the main bodies of the conventions—the conferences of the parties. Likewise, the main financial mechanism for these conventions, the Global Environment Facility (GEF), also has established such a body in the form of the Scientific and Technical Advisory Panel. However, it has become abundantly clear to me from many recent observations that while science has been involved in the start-up phases of determining the objectives sought by each of the conventions, or by GEF programs and projects, scientists have had very little involvement in what happens thereafter.

As a manager of an operating unit in the World Bank, I often encountered scientists happily delivering scientific findings they considered important. However, on many occasions, they seemed either unwilling or unable to help us set priorities for specific action and implementation. On other occasions, scientists did in fact become effectively involved in the technical teamwork, guiding the implementation of development activities.

It is my firm view that scientists should play a role in monitoring and measuring developmental achievements. For one reason, scientific expertise can contribute to measuring, assessing, and verifying the extent to which the intended benefits will indeed be achieved at the completion of a development project. It would be healthy for scientists to be held more accountable for what they advise. Furthermore, this would open up increased opportunities for scientists to help conduct the kind of empirical documentation I mentioned earlier (Christoffersen 2002).

A third barrier to sustainable development relates to policy and institutional issues in the international development field.

Thirty years ago, the key point in many discussions about international development assistance concerned the need to interrelate and integrate economic and social development objectives. Some of these discussions focused on the need for more holistic and integrated types of rural development projects. At the same time, on the national level, arguments were forcefully made that social issues should be brought into a larger analytical context than just the field of economics, which had traditionally been the main determinant of advice and guidance for development policy.

The new types of integrated projects fostered by these ideas soon became very popular with donors, nongovernmental organizations, and the vocal general public. However, when they did not achieve the anticipated positive results, they fell into disfavor. For example, at the policy level, it was hoped that social issues would get more attention under economic policies that explicitly incorporated distribution issues; this lost some sense of urgency when it began to appear that the hoped-for economic growth in many low-income countries (particularly in Africa) was failing to materialize. In any case, it took considerably more time than had been expected in the 1970s to make good progress toward an integrated analytical framework of economic and social issues. We have still not effectively reached that goal.

The international system set up almost sixty years ago constructed sharp institutional dividing lines between economic and social development—the Bretton Woods institutions were given the mandate for economic development, and other UN institutions pursued social development. This division is illustrated by the fact that for many years the United Nations produced global overview reports on social development that had no links to the key issues presented in the economic reporting of the World Bank. It is further illustrated by the Articles of Agreement that established the World Bank. They restricted the focus of the organization's activities to those that promote economic development. In fact, when programs for poverty alleviation were first introduced in the mid-1970s, considerable concern was expressed by many member countries that the World Bank should not weaken its key mission by getting involved in financing social development.

It is interesting to note that when the environment was finally represented institutionally in the UN system—with the creation of the UN Environmental Programme (UNEP) in 1972—it was decided that the new institution should not be given an operational mandate similar to that of a UN specialized agency such as the Food and

Agriculture Organization, the World Health Organization, or the UN Educational, Scientific and Cultural Organization. Rather, UNEP would focus its primary attention on efforts to work with, influence, and provide guidance to the other institutions within the UN system. Thirty years later it is clear that this has been an elusive goal. The agency has not been in a position to impact significantly the substantive content of the products and services of these institutions. For one thing, its governance system, dominated by the country memberships represented by ministers of environment, has continuously been pushing UNEP to become more directly engaged in country-related activities. While UNEP has made concrete contributions to several important environmental issues, such as the establishment of several global and regional conventions, it has not become the hoped-for integrating factor in the policies and operational programs of other UN agencies.

On a more general level, there has been a great amount of cooperation between the various UN agencies, including the World Bank. The agencies have shared many instances of good division of labor, as well as many less flattering cases of destructive competition. Their analytical approaches, however, are not at this point substantively integrated. Both the policy and the programmatic contents of these important institutions still lack solid evidence of successful scientific and technical interdisciplinarity. Case in point: the numerous social and environmental safety net procedures that prevail in many international development assistance agencies. Each procedure requires a great deal of attention, and most of them are conducted as separate processes. This to me illustrates one of the very basic dilemmas we face today—we have made some progress in "taking note" of each other and being willing to collaborate with each other in our separate and parallel multidisciplinary ways. We have not yet had much success in devising truly interdisciplinary approaches.

It has been said that knowledge may be considered a tool for diagnosing problems as well as for prescribing effective therapy. Many of the development problems we face today could benefit from more proactive scientific leadership, from leaders willing to test, deepen, and expand the horizons for more holistic and interdisciplinary analytical frameworks that can offer, and verify, effective solutions.

Let me conclude on a hopeful note. There are indeed many opportunities, some of which I have referred to above, for scientists to become very actively engaged in sustainable development theory and practice. One scientist who has taken notable advantage of such opportunities is Lee M. Talbot, who has demonstrated the importance of getting deeply involved in both the diagnosis and treatment of development problems. He has provided leadership to important national efforts involving both scientists and practitioners, such as the national conservation strategies initiated by IUCN and the national environmental action plans set in motion by the World Bank. He has never shied away from engaging in problem-solving tasks; he understands the value of empirical underpinnings for strengthening interdisciplinary theories; and he exemplifies the field-tested scientist who is able to communicate effectively and persuasively with those outside his own discipline. Talbot is, in my view, an admirable role model for all who are interested in helping overcome barriers to integrated and solid interdisciplinary approaches to sustainable development. He has been an inspiration to me and many of my associates.

ACKNOWLEDGMENT: Lee Talbot and I met in 1987 when the World Bank was in the midst of a major reorganization, setting up new environmental divisions in each of its four regional structures. It was my good luck to be able to bring into one of the organization's senior positions this very seasoned scientist, who had immense field experience, the ability to work effectively in teams of experts from other disciplines, and a remarkable talent for relating science to decision-making processes and policy work. He wanted to be able to make "a difference" in the World Bank. Through his contributions, he certainly did make a difference—a very big one.

References

Christoffersen, L. E. 2002. Globalizing benefits. *Our Planet* 13(3). http://www.ourplanet. com/, under "Past Issues: 2002: Global Environment Facility."

Holdgate, M. 1999. *The green web.* London: Earthscan.

Molteberg, E., C. Bergstrom, and R. Haug. 2000. Interdisciplinarity in development studies: Myths and realities. *Forum for Development Studies* 6(2): 318–330.

Sanderson, S. 2002. The future of conservation. *Foreign Affairs* 81(5): 162–171.

Talbot, L. M. 1996. *Living resource conservation: An international overview.* Washington, DC: Marine Mammal Commission.

Underdal, A. 1989. The politics of science in international resource management. In *International resource management: The role of science and politics,* ed. S. Andresen and W. Østreng, 253–267. London: Belhaven Press.

Young, O. 1989. Science and social institutions: Lessons for international resource regimes. In *International resource management: The role of science and politics,* ed. S. Andresen and W. Østreng, 7–24. London: Belhaven Press.

16

The Global Challenge of Sustainable Development

Mohamed T. El-Ashry

Pursuing a more sustainable global future in this millennium is less a matter of cost than of conscience, commitment, and cooperation. Humanity, greater in number and more economically active with each passing day, is increasingly playing havoc with the earth's natural systems. Our actions are giving rise to a multitude of critical threats: degradation of soils, water, and the marine resources essential to food production; health-endangering air and water pollution; global climate change that is likely to disrupt weather patterns and raise sea levels everywhere; the loss of habitats, species, and genetic resources, which is damaging both ecosystems and the services they provide; and depletion of the ozone layer.

Despite some progress over the last two decades, the state of human development remains at an unacceptable level. More than fifty countries are poorer today than they were in 1990. According to the UN Educational, Scientific and Cultural Organization (2003), almost three billion people live on less than $2 a day, one billion lack means of access to clean water, more than two billion lack access to basic sanitation, and 6,000 people, mostly children, die every day from water-related diseases. The poor are also the most vulnerable to environmental degradation. They depend on natural resources—soil, water, fisheries, forests—for their sustenance and they suffer disproportionately from poor environmental conditions.

In the twenty-first century, hardly a week goes by without further confirmation that nature and its ecosystems can be our saving grace. We can only guess at the existence of plants and fungi capable of joining penicillin, tamoxifen, quinine, and codeine, among twenty-four others in common use, as nature's gifts to our medicine chest. Similarly, to find even one or two new useful crops among 23,000 unexploited species we know to be edible could save millions from malnutrition. Worldwide, 840 million people do not have enough to eat, and children are hit especially hard.

At a time when nearly one billion people depend on fish as their primary source of protein, some 75% of the world's marine fisheries are judged to be at risk. Forests contain two of every three land species and the highest species diversity of any system; yet the world already has lost half of its forests and, according to the Food and Agriculture Organization of the United Nations (2005), about thirteen million hectares of forests disappear every year. The Intergovernmental Panel on Climate Change ([IPCC] 2007) has reported that temperatures could likely rise by 1.8–4.0°C over the next one hundred years. In the second half of the twentieth century, water withdrawals from rivers, lakes, reservoirs, underground aquifers, and other sources increased by a factor of more than four. By 2025, if current trends are not reversed, nearly two-thirds of the world's population will live in water-scarce areas.

This sorry state of the planet and of sustainable development is not because of lack of international attention. In fact, since 1992, the response of the international community to the challenges of environment and sustainable development has included four international summits, four ministerial conferences, three international conventions, two protocols, and a new financial entity—the Global Environment Facility (GEF).

At the UN Conference on Environment and Development (also known as the Earth Summit) held in Rio de Janeiro, Brazil, in June 1992, world leaders made a number of significant commitments for the pursuit of sustainable development. The resulting Rio Declaration on Environment and Development contained specific reference to the need for integrating environment into economic development thinking and planning. The Conventions on Biodiversity and Climate Change were opened for signature, negotiations for the Desertification Convention were launched, and the GEF was recognized as the funding mechanism for the global environment.

At the UN Millennium Summit, held in September 2000, the Millennium Development Goals (MDGs) were adopted in order to help eradicate extreme poverty and hunger, achieve universal primary education, and ensure environmental sustainability, in addition to other aims. In March 2002, finance and development cooperation ministers met in Monterrey, Mexico, at the International Conference on Financing for Development. A number of commitments on additional finance from donor countries were made in order to achieve the MDGs. Finally, in late August 2002, world leaders gathered again, this time in Johannesburg, South Africa, for the World Summit on Sustainable Development, at which they reviewed and renewed the commitment to sustainable development and adopted a plan of implementation.

On the face of it, these are remarkable achievements. But despite all the high-powered gatherings, agreements, and commitments, little progress has been achieved in improving the environment and pursuing sustainable development. Global environmental trends continue to be negative, and the promise of significant financial resources to address the challenges of environment and development has not materialized. The gap between rich and poor within and between countries continues to widen; ninety million people are added to our global village every year, mostly in developing countries; one person in three still lacks adequate fresh water; greenhouse gases are steadily increasing; ecosystems that are critical for human survival continue to be undermined; and land degradation threatens food security and livelihoods, especially in Africa. It is ironic that as the evidence for environmental degradation becomes more convincing, the political will for action becomes weaker or is lacking entirely.

THE CHALLENGES

There are many reasons for the lack of progress in the pursuit of sustainable development. Key among them: (1) rapid population growth and poverty in developing countries; (2) production and consumption patterns in the Organisation for Economic Co-operation and Development (OECD) countries; (3) economic decisions being made with little input about their environmental impacts; (4) failure to integrate environmental sustainability into development planning; (5) current accounting frameworks for economic analysis that fail to treat natural resources as productive capital; (6) overexploitation of natural resources, which is proving to be unsustainable; (7) debt burden of developing countries ($2 trillion in 2000); (8) perverse subsidies for agriculture in OECD countries, and for water and energy services in developing countries; (9) inadequate finance for environment and sustainable development; (10) weak implementation capacity in developing countries; (11) the widening political divide between developing and developed countries; and (12) fragmentation and ineffectiveness of international environmental governance. We can also add lack of peace, lack of security, and civil strife as factors hindering progress toward global sustainability.

These are the root causes of the deteriorating state of the global environment and the slow pace in the pursuit of sustainable development. They need to be tackled promptly if current trends are to be reversed. The lack of commitment to the environment and to sustainable development cannot be resolved by the harmonization of international environmental agreements, or by simply strengthening international environmental institutions. It is symptomatic of governments' and international institutions' failure to integrate the economic, social, and environmental pillars of sustainable development.

OUR COMMON FUTURE

The report of the 1987 World Commission on Environment and Development was a pioneering attempt to illustrate economic and environmental interdependence. It demonstrated that economic development must be not only economically and financially viable, but also socially acceptable and environmentally sound.

Sustainable development means that people can make their living without destroying the natural resources and ecosystems necessary for meeting their needs and for supporting the diversity of other life on the planet. These are the principles of sustainability that were accepted by all nations at the Earth Summit in 1992 and reaffirmed in Johannesburg in August–September 2002. At stake is far more than nature for nature's sake. As species, as well as the entire ecosystems that support them, disappear, so do the many ecological services and the wealth they provide for people. The greatest impact is upon the poor in developing countries—the people most dependent on natural resources for food, shelter, medicine, income, and employment. Globally, 1.2 billion people live on fragile lands—arid zones, slopes, wetlands, and forests—that cannot sustain them.

The international community needs to move quickly and together to create realistic alternatives for poor people if the growing pressures on the environment are to be minimized. It makes little sense to lift people out of poverty today to have their children and grandchildren sink back into it tomorrow. A successful strategy for poverty reduction must take into consideration the serious degradation of the environment and natural resources that underpin the lives and livelihoods of the world's poor and future generations. Former South African president Nelson Mandela, in his eloquent and wise way, gave the world a message in November 2000. In his words: "We know that political freedom is priceless ... But freedom alone is still not enough if you lack clean water. Freedom alone is not enough without light to read at night, without time or access to water to irrigate your farm, without the ability to catch fish to feed your family. For this reason, the struggle for sustainable development nearly equals the struggle for political freedom. They can grow together, or they can unravel each other" (paras. 20–21).

The challenges seem daunting: In Africa, poverty levels have increased by 43% in the last ten years. At the same time, nearly 40% of Africans live below the poverty line and about 70% of them are in rural areas and depend on agriculture. Yet the basic resources for their existence are threatened by land degradation, which affects 65% of agricultural land, and deforestation, which in the relatively short period of fifteen years has stripped sixty-six million hectares.

Then there is the widening gap between rich and poor countries. The average income in the richest twenty countries is already thirty-seven times that in the poorest twenty countries, and the gap has doubled in the past forty years. Poverty remains a formidable obstacle on the road to global sustainability. The UN MDGs aim at cutting hunger and poverty in half by 2015. Yet poverty and hunger reduction goals will not be met if progress is not made on other aspects of development, including education, health, and the environment.

Environmental and economic inequities are closely interrelated. Poverty pushes the poor into degraded lands and polluted slums, where economic opportunities are minimal. Development that subsidizes sewers, piped water, and electricity for the middle classes and provides next to nothing for the poor is more than unfair. The burdens this places on the poor are reflected in higher health costs, lowered productivity, and the desperate hopelessness that can lead to political instability. A deeper appreciation of this economic and environmental interdependence is essential to tackling the global environmental problems that cloud our future and endanger our health, our security, our natural endowment, and the beauty and splendor of life on Earth.

In the pursuit of global sustainability, there is a need to recognize that it is in developing countries that population growth is high, it is developing countries that are richest in biological resources, and it is developing countries' greenhouse gas emissions that are growing the fastest. At the same time, developing countries simply do not have the financial and technical resources required to achieve sustainable development and environmental protection. Thus the realization of global sustainability depends on the support and cooperation of rich countries in the transfer of environmentally friendly technology (especially for sustainable energy development), in providing additional financing, and in supporting infrastructure development in the developing countries.

SUSTAINABLE ENERGY

Examining a satellite photo of the earth at night reveals a world of light and darkness, of haves and have-nots—an illustration of the fact that almost two billion people still do not have access to modern energy services. This energy poverty, combined with the negative environmental and health impacts of conventional energy sources, has attracted international attention and debate for almost two decades.

Since the Earth Summit, there has been an international consensus that energy is essential to achieving the economic, social, and environmental pillars of sustainable development. Sustainable development requires sustainable energy. Experience has shown that the kind of energy and the way it is used are important determinants in the degree of sustainability achieved.

From 1970 to 1998, global use of primary energy expanded by about 2% annually. According to the UN Development Programme's (UNDP) 2000 World Energy Assessment, if this trend continues, it will mean a doubling of energy consumption by 2035 relative to 1998, and a tripling by 2055. This led to the conclusion in the UNDP's 2004 update to the assessment that "continuing along the current path of energy system development is not compatible with sustainable development objectives" (12).

The challenge is to change course toward a sustainable energy future—one that simultaneously meets the energy needs of a growing global population, enhances people's equality of life, and addresses environmental concerns, especially climate change. In such a carbon-limited energy future, the only hope for satisfying basic human needs in the developing world and for sustaining the standard of living in the developed world is to accelerate the pace of development and dissemination of renewable energy sources. Large increases in renewable energy use, combined with higher levels of energy efficiency and the development of carbon sequestration technologies, can go a long way toward a global sustainable energy path.

Recognizing that decisions taken now will be decisive for a transition toward a sustainable energy future, world leaders gathering in Johannesburg for the World Summit on Sustainable Development agreed to substantially increase, with a sense of urgency, the global share of renewable energy sources in the total energy supply.

Clearly, the substantial expansion of renewable energy:

- Is a win-win proposition for developed and developing countries alike
- Provides opportunities for poverty eradication and for satisfying the energy needs of the poor, particularly in rural and remote regions
- Generates employment and local economic development opportunities through the creation of decentralized markets and the introduction of new capital and innovation
- Helps curb global warming induced by human, energy-related activity
- Contributes to the protection of human health caused by air pollutants
- Enhances energy security and the reduction of energy imports and foreign exchange burden through reliance on domestic energy sources (biomass, hydro, wind, solar, geothermal)

- Has the potential, together with increased energy efficiency, to become a most important and widely available future source of energy
- Offers a new paradigm for cooperation among all countries

Realizing such a vision, however, requires an enabling environment based on the establishment of viable markets for renewable energies, the expansion of financing options, and the development of the human and institutional capacity required to transform the energy markets.

In this regard, coherent regulatory and policy frameworks that remove barriers, leveling the playing field for renewable energy sources that internalize external costs, and provide incentives for market development are essential to realizing the potentials for renewable energy technologies and to creating favorable conditions for public and private investments in renewable energy. No single policy instrument, however, can ensure solutions for every application, sector or subsector, sociopolitical situation, or country.

As renewable energy sources are increasingly becoming economically viable, commercial sources of finance can be tapped. By enabling regulatory frameworks and appropriate incentive structures, we can improve competitiveness, reduce finance-related risks, and increase commercial finance opportunities.

Awareness of the benefits and applicability of renewables is one of the key requirements for their accelerated deployment. This includes: (1) strengthening educational efforts and awareness; (2) building capacity for policy analysis and technology assessment in developing countries; (3) raising awareness among government decision makers and financiers of the benefits of renewables; (4) supporting development of marketing, maintenance, and other service capacities; and (5) strengthening regional and international collaboration and stakeholder participation, including that of women's groups, in order to facilitate access to, and sharing of, relevant information and good practice.

FINANCE FOR SUSTAINABLE DEVELOPMENT

Historically, official development assistance (ODA) has supported economic growth and financial stability in developing countries. In the last decade, it has also supported social and human development. Addressing current and emerging environmental threats will require a greater commitment from the donor community in the coming decades. Donors did not live up to commitments they made at the 1992 Earth Summit to devote 0.7% of their gross domestic products (GDPs) to ODA. During the 1990s, total ODA flows stagnated in nominal terms—and actually fell in relation to the donor countries' GDPs, from 0.33% in 1992 to 0.22% in 2000.

Significant additional resources must be found to meet today's expanded sustainability agenda. The transfer of financial resources from north to south should not be viewed as "charity." If, as the evidence shows, tropical forests and other ecosystems provide a global service in terms of carbon fixation and biological diversity, then there must be willingness to pay a fee for such services. The Kyoto Protocol's Clean Development Mechanism is an important step in that direction. In the meantime, a number of options to enhance public sector resource mobilization have been proposed. They include taxing energy and CO_2 emissions, taxing international travel and tourism, and phasing out subsidies that encourage the inefficient use of natural resources.

An international carbon tax based on the consumption and carbon content of fossil fuels could provide a new source of funding for global environmental initiatives. A recent UN technical note concluded that a carbon tax on gasoline applied universally—equal to 4.8 cents per U.S. gallon—has the potential to raise $125 billion a year (Annan 2001).

International aviation is not included in international agreements on quantitative reductions of greenhouse gas emissions. Levies of taxes and fees on international travel could provide incentives to reduce emissions and also help finance investments in environmental protection. The revenue-generating potential of such taxes could be substantial; it is estimated that a 1% tax on passenger tickets and air freight would yield approximately $2.2 billion per year.

In many developed and developing countries, government subsidies lower the prices of resources such as energy and water, thereby encouraging excessive use. For example, subsidies associated with the production and sale of electricity were estimated at $112 billion per year in 1996 in developing and transitional economies. These huge subsidies encourage waste and contribute not only to local air pollution but also to global climate change. Phasing out electricity subsidies could produce a "win-win" effect. A reduction of 10% would generate annual revenues of $11 billion and reduce greenhouse gas emissions.

There are hidden energy subsidies in OECD countries as well. But the most significant are subsidies to farmers, to the tune of $365 billion a year or $1 billion a day. Reducing these agricultural subsidies by 15% would double current levels of ODA and bring the world closer to meeting the MDGs.

With regard to the private sector, foreign direct investment (FDI) in developing countries in 2002 totaled $112 billion—the lowest in the last ten years but still more than twice the size of ODA. The average in the 1990s was $185 billion and the high was $267 billion. However, the majority of FDI continues to be invested in about fourteen countries, and FDI has not supported social or environmental activities. Given this reality, two key concerns are how to increase capital flows to low-income countries, particularly those that have been largely ignored by investors, and how to maximize the use of foreign investment to help solve developing countries' social and environmental problems.

The GEF

The GEF was established in 1991 as a three-year pilot program in anticipation of the Earth Summit and the negotiations on the conventions of biodiversity and climate change. To secure the participation of developing countries in the implementation of the conventions, financial resources had to be made available by OECD countries.

The GEF emerged from the Earth Summit as a key component of the financial package that sealed agreement by developing countries on Agenda 21 and the conventions. Over twelve years later, the GEF remains the only financial accomplishment arising from the summit. Following the summit, the GEF was restructured and became independent from its three implementing agencies (the World Bank, the UNDP, and the UN Environment Programme), governed by a council and with a voting mechanism that balances the voice of developing and developed countries. It was also replenished by $2 billion. Four years later, it was replenished by another $2 billion.

Since its restructuring, the GEF has grown from fewer than thirty members during its pilot phase to 176 member countries, and it has become the largest single source of funding for the global environment. In 2002, donor countries replenished the GEF trust fund by $3 billion, the largest such increase for the fund ever. While the mandate of the GEF is the global environment, it was recognized that this objective can best be realized in the context of national sustainable development. After all, sustainable energy, sustainable forestry, and sustainable fisheries are the means for capturing global benefits related to climate change and biodiversity.

Without additional or incremental finance, the GEF would not have survived. Its resources, however, are small compared to global needs. That is why those resources have been used to leverage additional investments from private and public sources. The GEF's current portfolio of about $17 billion consists of close to $5 billion in GEF grants and more than $12.5 billion in cofinancing.

POLICIES FOR SUSTAINABLE DEVELOPMENT

As part of its contribution to the preparations for the World Summit on Sustainable Development, the GEF convened the Ministerial Roundtable on Financing the Environment and Sustainable Development, which brought together ministers of finance, development cooperation, and environment. The ministers concluded that the international community should pursue policies to:

- Phase out subsidies that encourage economic inefficiency and excess use of natural resources
- Create an enabling environment and public policies that shape markets so as to attract greater FDI
- Create conditions for socially responsible investment in developing countries
- Promote public-private partnerships to fully integrate long-term sustainability into poverty eradication and economic development

Aside from finance issues, the questions that need to be asked are as follows: In the face of ninety million additional people being added to our global village each year, how can we ensure sustainable development by bringing growth and environmental protection into harmony? What will it take to protect our biological heritage, avoid devastation from climate change, and sustain the soil and water that give support to life, protect human health, and reduce the scourge of poverty and hunger? How do we summon and perpetuate the necessary political will?

First, we must forsake single-sector approaches that rely too heavily on short-term technical fixes and end up causing long-term environmental degradation. We need to address policy and institutional issues such as the pricing of natural resources and counterproductive subsidies. We need to spread knowledge about sustainability in agriculture, water-saving irrigation, energy use, and other commonsense practices that work. We need to employ science and modern information technologies and tap traditional and indigenous wisdom. We need systems of land tenure that do justice to land and the people who work it, especially women.

Looking ahead, the next decade presents a unique opportunity to ensure that environmental sustainability is fully and effectively integrated into actions designed to achieve sustainable development and fulfill the aspirations of poor and rich people in the world. This will require new ways of acting and thinking—we need to reform the current accounting framework for economic analyses, which fails to treat natural resources as productive capital and does not record the loss of this natural capital as disinvestment or capital consumption. We need to remove the barriers faced by the least developed countries seeking trade and investment, and we need to provide resources for legal, policy, and institutional reforms that would help attract private investments. And we need a stronger involvement of the science and technology community in the societal and political processes that are shaping the global sustainability agenda.

CAPACITY BUILDING FOR SUSTAINABLE DEVELOPMENT

Knowledge sharing, whether on a specific technology or on policy reform, and building developing countries' institutional capacity and their abilities to integrate the environment in economic planning and thinking are essential to the realization of sustainable development. Common capacity-building needs include:

- Low levels of awareness and knowledge, which limit the ability for decision making and action
- Lack of information management, monitoring, and observations, which hampers policy and decision making
- Inadequately developed incentive systems and market instruments, which are common in the least developed countries
- Institutional mandates that either overlap or have gaps, and key institutions that are not involved in planning and decision making
- Institutional effectiveness that is hampered by weak management and resource constraints
- Science and technology that are ineffectively mobilized in support of policy and decision making

The scientific complexity of underlying problems and of potential solutions requires close links at all levels between the institutions driving the sustainable development agenda and the scientific community. Unfortunately, disparities between countries of the north and south in the generation of scientific information and its use make it difficult for the south to participate fully in actions for global sustainability. Developed countries representing 20% of humanity have more than 90% of the world's share of scientific publications and more than 90% of research and development expenditures.

Whether developing countries will, in the long run, become deeply engaged with sustainability issues such as global climate change depends on a clear understanding of the risks of "business as usual" to both their own citizens and future generations. Current scientific knowledge and global modeling exercises have yet to pinpoint the regional impacts, let alone national impacts, of most global environmental issues. Yet these problems cannot be resolved without local scientific capacity to analyze and define specific impacts on individual nations. Local scientists must be able to provide their policy makers with the advice they need to formulate strategic directions and to press for their adoption; otherwise, there can be little progress toward sustainability.

In this regard, the GEF has contributed in various ways to improving capacity in sciences related to the global environment in developing countries. For example, it has facilitated the participation of scientists from developing countries in the activities of the IPCC, and it is enabling more than 120 nations to draft strategies and action plans for further actions on biological diversity and climate change. The GEF's Scientific and Technical Advisory Panel (STAP), which includes distinguished representatives from developing countries, provides critical scientific input on the GEF's programs and priorities. STAP has also been involved in an outreach effort targeting networks of scientists at local, regional, and international levels. Much more, however, needs to be done.

CONCLUSION

In many ways, we have entered one of the most creative phases in human history. Science, technology, and communications are advancing at breathtaking speed and offering unmatched opportunities for responsible action. We have new tools and a vastly increased understanding that our strength lies in working together across the globe to overcome the threats facing the planet. Interdependence means that all of us, whatever the stage of our development, are traveling in the same boat, floating and sinking together. "It is time to break away from the obsolete images of the world of the 1960s, the 1970s, and the 1980s. That world no longer exists. Now is a rare moment, a clearing horizon of historic opportunity, for all nations to promote peace, liberty, and global prosperity through partnerships." These words are from the 1992 report of the Carnegie Commission on Science, Technology, and Government (§ 2.4). They are as valid today as they were a decade ago.

References

Annan, K. 2001. *Existing proposals for innovative sources of finance.* Global Policy Forum Technical Note, no. 3. New York: Global Policy Forum.

Carnegie Commission on Science, Technology, and Government. 1992. *Partnerships for global development: The clearing horizon.* New York: Carnegie Commission.

Food and Agriculture Organization of the United Nations. 2005. *Global forest resources assessment 2005: Progress towards sustainable forest management.* FAO Forestry Paper 147. Rome: FAO.

Intergovernmental Panel on Climate Change. 2007. *Climate change 2007: The physical science basis*. Cambridge: Cambridge University Press.

Mandela, N. 2000. Beyond freedom: Transforming "ngalamadami" into "sithi sonke." Remarks given at the launch of the final report of the World Commission on Dams, London, November 16.

UN Development Programme. 2000. *World energy assessment: Energy and the challenge of sustainability*. New York: UNDP.

UN Development Programme. 2004. *World energy assessment: Overview; 2004 update*. New York: UNDP.

UN Educational, Scientific and Cultural Organization. 2003. *Water for people, water for life*. The UN World Water Development Report. Paris: UNESCO/World Water Assessment Programme.

17

Biodiversity Conservation in the Real World: Incentives, Disincentives, and Disconnects

Agnes Kiss

A healthy and stable environment is essential to our long-term survival, and destroying the natural assets on which our lives and economies depend is, ultimately, suicidal. This basic truth has been articulated in countless international conferences, workshops, and classrooms and in technical journals and the popular press for many years. Yet while we can point to specific examples of action taken here and there to protect or restore selected areas or species, ecosystems worldwide continue to deteriorate as a direct result of our collective misuse and mismanagement (Millennium Ecosystem Assessment 2005). If we understand the importance of maintaining the environment, and if in many cases we have the necessary technology to do this, why do we do such a poor job of it?

THE DISCONNECTS THAT DRIVE ENVIRONMENTAL ABUSE

A large part of the answer probably lies in some fundamental "disconnects" between the costs and benefits of good environmental management and the realities of human nature and behavior.

Tangible and Certain Costs versus Hard-to-Measure and Uncertain Costs and Benefits

The benefits of environmentally destructive economic activities are usually easy to grasp, both conceptually and physically. Wood can be used to build shelters or burned for fuel, wildlife and fish can be eaten, cleared land can be planted to produce crops, and all of the above can be sold to obtain other necessities and luxuries. By contrast, it is often very difficult to define and measure the benefits from good environmental management with any certainty or precision. Even generally accepted linkages such as the connection between water pollution and illness, or deforestation and flooding, can be difficult to quantify or sometimes even to demonstrate, and can vary greatly from one place to another (e.g., Calder 2004; Center for International Forestry Research/Food and Agriculture Organization of the United Nations 2005; Smith et al. 1995). The person who cuts a tree and uses or sells the wood is also more certain of benefiting from it than is the one who leaves it standing. Who knows whether it will be cut down tomorrow by someone else, or destroyed by fire or pests?

Even economists recognize that not all benefits we care about are material or can be translated directly into cash. Economic analyses routinely refer to intangible aesthetic, cultural, legacy, and existence values, though usually without attempting to quantify them. These values certainly apply to biodiversity and natural ecosystems, and many areas are protected for such purposes. But when nature is destroyed or displaced it is almost always for economic reasons, ranging from basic survival to improving quality of life to accumulating wealth. Therefore, to prevent or reverse the destruction, people must be provided with viable (and socially acceptable) alternatives that can be directly translated into roughly equal economic value. In fact, because adopting such alternatives often involves risk, extra initiative and effort, or changes in well-entrenched customs and behavior, a simple break-even result is usually not sufficient. Additional incentives are often required to trigger a change in customary practices.

A wrinkle in this disconnect is the difference between direct costs and opportunity costs (the potential value one gives up by choosing not to carry out a particular action). While direct costs may seem to be more tangible and therefore more compelling, they are in fact often both lower and easier to deal with than opportunity costs. While there is almost never enough money to satisfy all the needs, it is at least usually easy to determine how much money should be paid, and to whom, in order to carry out activities such as building sewage treatment plants, monitoring air quality, or guarding national parks. Some recent analyses (e.g., Balmford et al. 2003) have proposed that it is more efficient and cost-effective to protect and maintain biodiversity and protected areas in developing countries than in industrialized countries because costs such as guards' salaries and supplies are so much lower. But it is acknowledged that these analyses do not really take opportunity costs into account. If they did, their conclusions would likely be quite different, since the real opportunity costs of forgoing consumptive and extractive activities—such as farming, logging, mining, hunting, and the like—are typically greater in developing countries, where the demand for arable land can be very high and the exporting of natural resources such as minerals or timber provides a large part of the national budget.

There are also critical but complex issues regarding who pays the costs of biodiversity conservation, and who decides whether it should be paid. The direct costs are typically met by governments or private contributors on a voluntary basis, while indirect costs (e.g., damage to crops and livestock) and opportunity costs are often imposed on local communities, who may have little real choice in the matter. Assigning and calculating opportunity costs can also be difficult. For example, if a new protected area is created, who should be regarded as losing the opportunity to farm or log or hunt in that area? Is it only the people who are actually doing these things now, or also everyone who did so in the past or might conceivably have intended to do so in the future? And does the loss encompass what they might have produced and earned during the next year, or the next ten years, or the next ten generations? Because these questions are so intractable, the tendency is to try to compensate for opportunity costs in kind rather than in cash—for example, by finding people new land to farm or new ways to earn a livelihood. This is inevitably more difficult, expensive, and time-consuming than cash payment. It frequently does not work, and people return to using the protected area's natural resources, either legally or illegally.

Individual versus Collective Costs and Benefits

The benefits of environmental degradation or overexploitation of natural resources are often captured by individuals or small groups of people, giving them strong motivation to continue their activities. By contrast, the costs are often spread over large numbers of people; individually, these people may not be sufficiently strongly motivated to take action to stop the resource exploitation, at least initially. By the time the negative impacts have become severe enough to stimulate collective action, the damage to the environment is likely to be very difficult and expensive, if not impossible, to repair.

On-Site versus Off-Site Costs and Benefits

Because of the fluid and mobile nature of environmental goods and services, the negative consequences of destructive economic activities often have their greatest impact on people and places a considerable distance away. One classic example is that of downstream pollution and siltation caused by upstream factories and deforestation; another involves damage to remote forests and crops from acid rain originating from urban and industrial centers. Conversely, possession of the benefits of clean water and air often depends on economic sacrifices upstream and upwind, either through limiting damaging activities or investing in cleanup. Altruism is a laudable but limited commodity in human nature. And individual, family, and national interests usually take precedence over the needs and welfare of others, particularly others who are far away.

Results of Disconnects

Because of these disconnects, the cost-benefit analyses that we are constantly making as individuals and societies generally come out in favor of degrading the natural environment rather than preserving it. Nevertheless, on some level we do recognize the importance of protecting and managing the environment, and as a global society we invest considerable time and resources in trying to do so.[1] The success of these efforts depends on the extent to which they come into conflict with the short-term, concrete, individual priorities of powerful stakeholders. It is also influenced by two important characteristics of the human psyche: the law of unintended consequences and the "keep it simple (and technical)" axiom.

First, we tend to be good at identifying our own interests, and quite persistent and creative in pursuing them. People generally need incentives to change environmentally destructive behavior, though, particularly if that behavior is deeply rooted or if the change is seen as potentially risky. So it can be tricky to design incentives that actually result in the desired outcomes. People frequently find ways to capture the incentive without doing their part ("free riders"). The system also often ends up rewarding unhelpful or even perverse behavior, either inadvertently or through active manipulation by the recipients. For example, one commonly heard suggestion is that the price paid to developing countries for biodiversity products such as hard-wood timber should be increased, because (1) it would make the trees more valuable and thus create an incentive to manage them sustainably and (2) fewer logs would need be cut to generate the same level of income. But in the absence of strict quotas or regulations (as is usually the case), an equally or more likely outcome is that more people will go into the business of cutting more logs because it has become an even more lucrative business. The end result is more logging rather than less.

Second, we tend to distill multidimensional issues down to narrowly defined, tangible problems that lend themselves to technological solutions. Historically this process has opened the way for remarkable creativity and invention; skill at solving technical problems has been a hallmark of the human species since its beginning. Unfortunately, we ignore real-world complexities such as human behavior, market forces, and power relationships at our peril. One frequently cited example is the green revolution. Norman Borlaug won a Nobel Peace Prize in 1970 for leading research to develop high-yielding varieties of staple crops aimed at dramatically increasing food production in developing countries, thus putting dire Malthusian predictions of global starvation to rest. However, the unforeseen and unwanted consequences of this scientific tour de force have included increased pollution and health impacts from pesticides, a dangerous decline in the genetic diversity of major crop species, and the impoverishment of many small farmers who neither have been able to afford the inputs these varieties need to flourish, nor compete in the marketplace with farmers who can (Hazell and Ramasamy 1991; International Food Policy Research Institute 2002; Shiva 1991). It is now generally recognized that the issue is not the total volume of global food production, which has increased dramatically, but rather inadequate local production, food distribution, and poverty (inability to buy food).[2]

STRATEGIES FOR CONSERVING BIODIVERSITY: SEPARATION VERSUS INCLUSION

The conflicts between short-term versus long-term, local versus global, and group versus individual interests are played out in many settings and variations in relation to environmental management and biodiversity conservation. There are countless examples from the United States and other industrialized countries, such as western ranchers' resistance to reintroduction of wolves and to allowing wild bison to graze outside national parks, government-sanctioned oil drilling and mining in wilderness areas, subsidized logging of old-growth forests, overfishing of commercial fish stocks, draining of wetlands for housing or industrial development, and so on. But the dilemma of poor rural communities in developing countries attracts the greatest amount of attention in the conservation community. At the 1972 UN Conference on the Human Environment, held in Stockholm, Sweden, Indira Ghandi memorably remarked that poverty is the greatest polluter of all. In fact much, if not most, of the environmental degradation around the world can be attributed to the greed of the wealthy and powerful rather than the struggles of the poor and marginalized. Nevertheless, the sheer numbers of the rural poor across the world make them a major factor in the ecological equation. The areas of the greatest and most intractable rural poverty also overlap strongly with the greatest concentrations of the world's remaining biodiversity—in the forests, plains, and coastal zones of tropical developing countries. As a result, significant financial and human resources, and countless pages of professional books and articles, are devoted to trying to find ways to save as much as possible of the world's remaining biodiversity while at the same time responding to the needs and aspirations of rural communities in developing countries.

Conservationists have two main strategies for dealing with this dilemma: either try to separate the people from the biodiversity (physically and economically), or try to give them a stake in preserving it. The "separation" strategy includes the creation of exclusive protected areas and bans on hunting, fishing, and tree cutting. The alternative "incentive" approach usually involves trying to use the biodiversity in an economically rewarding yet ecologically sustainable way. However, it can also mean paying people directly for providing the service of conserving biodiversity (Ferraro and Kiss 2002). Numerous approaches and variations have evolved within each of these two broad strategies as conservation supporters have adapted projects to local situations. Each strategy has strengths and weaknesses, and there is no silver bullet that will be successful and effective in all circumstances. Most conservation projects these days seek to combine elements of both strategies.

Lasting success will ultimately depend on whether or not we can bring about significant changes in the choices and decisions that people make—particularly the people who control use of the land where biodiversity is found and those who make their living or their wealth from harvesting it. These choices and decisions in turn are strongly influenced by the characteristics of human nature described above. Below I examine the theory and practical experience associated with four common approaches to conservation in developing countries in this context: (1) protected areas, (2) support for alternative livelihoods, (3) sustainable use or community-based natural resource management, and (4) direct payment for environmental services. In principle, the protected area and alternative livelihoods approaches mainly represent the separation strategy, while sustainable use and payment for environmental services represent the incentive strategy. In practice, these distinctions have become somewhat blurred.

Protected Areas (PAs)

The effectiveness of PAs for conserving biodiversity is well demonstrated, despite limitations such as size, isolation, and inadequate budgets (Bruner et al. 2001; Kramer, van Schaik, and Johnson, 1997; Struhsaker, Struhsaker, and Siex, 2005). The main issue is their sustainability, particularly in view of the threat posed by growing populations with growing aspirations. This analysis therefore focuses on community and political support for PAs.

By definition, a PA protects biodiversity assets by excluding, or at least limiting the access of, people who would otherwise be exploiting them. In the traditional "fortress conservation" model (Hulme and Murphree 2001), establishing PAs was a pure separation strategy: alienate the land, fence it off (literally or figuratively), and impose and enforce a total ban on encroachment or poaching. The Soviet system was perhaps the most extreme, with strict nature reserves (*zapovedniks*) that were closed to all use, including tourism and recreation. Even management interventions such as firefighting or control of alien invasive species were prohibited.

In recent years, however, the term PA has come to cover a wide range of management regimes, as codified in the widely adopted categorization system produced by IUCN—The World Conservation Union (table 17.1).[3]

Only categories Ia, Ib, and II correspond to the traditional concept of a PA, with entry restrictions and prohibition of residence, conversion of land use, and consumptive use of biodiversity. Expanding the term to include the others (particularly category VI) obscures the distinction between the separationist PA approach and the sustainable use approach. But even for category Ia, Ib, and II PAs, the concept of separation has been eroding, as it has come to be accepted that PAs must address the needs and interests of local communities if they are to be successful in the long run. It is now expected that communities should benefit economically from the presence of neighboring PAs so that they have a stake in them and an incentive to support them. Ideally this entails mechanisms that involve a direct linkage with the existence and success of the park (e.g., distributing a share of entry fees). Sometimes it means allowing continued but regulated (in principle, sustainable) use of some of the biodiversity resources within the PA. Often the benefits come instead in the form of general economic development assistance, or support for development of alternative sources of goods and income, provided by PA supporters in the hope of winning community support and reducing pressure on biodiversity resources. Thus PAs are

Table 17.1. IUCN Categorization System.

Category	Management Objectives
Ia Strict Nature Reserve	mainly for science
Ib Wilderness Area	mainly for wilderness protection
II National Park	mainly for ecosystem protection and recreation
III Natural Monument	mainly for conservation of specific natural features
IV Habitat/Species Management Area	mainly for conservation through maintaining habitats or other requirements of specific species
VI Managed Resource Protected Area	mainly for the sustainable use of natural areas; use of biological resources is to be balanced between the objectives of maintaining biodiversity and of meeting local community needs

now usually just one component of a broader program that includes one or more of the other approaches, which will be discussed below. Here the focus is on the PA itself as a conservation tool.

Creating a PA generally involves imposing costs that are on site (primarily affecting the local communities), short term (with immediate impact),[4] and tangible (restrictions on access to economically valuable land and resources). The benefits are on site and short term from an ecological perspective but to a large extent off site (even global) and long term from a human perspective. There may be local human benefits as well, such as preserving streams or providing a refuge and source for wildlife to repopulate depleted surrounding areas (particularly important in the case of marine reserves). But these benefits are likely to be collective, compared with the individual benefits that previously came with direct access to the area. The PA approach triggers all the disconnects I have mentioned, then, so it is not surprising that it tends to be resisted by local communities. Where communities voluntarily support the establishment of PAs or even create their own, these are likely to be of the category VI variety. Alternatively, local people may support more restrictive PAs if they are convinced that they will receive significant, tangible, short-term benefits as a result. These benefits are usually expected to come from sharing in park revenues, from tourism, and from complementary community development and "alternative income" projects that people recognize are only coming their way because an external donor wants their support for the PA.

The problem is that economic benefits linked to PAs are often slow in coming and fall short of expectations. PA proponents often unintentionally exacerbate this by raising local expectations to unrealistic levels in an effort to capture interest and cooperation. The reality is that most PAs in developing countries generate relatively little revenue, particularly in the beginning, and income from tourism is usually considerably smaller in amount, longer in coming, less certain, and harder to earn than local people anticipate (Alexander 2000; Kiss 2004a; Wunder 2000). Their patience wears thin, and they come to regard the compensation they do receive as inadequate. PA supporters react by front-loading the benefits to communities, thus delinking them from the financial success of the PA. This triggers a shift from the desired partner relationship to a donor-beneficiary relationship. From that point, the community's interest is not in whether or not the PA is successful, but in how they can extract more benefits from the project sponsors (Kiss 2004b). There is also the risk that the PA will attract an influx of immigrants drawn by the economic opportunities (real or perceived) and social services that the project is providing (i.e., the "magnet effect").

As agriculture and other development spreads, PAs often come to be isolated islands of natural habitat and the only remaining sources of products such as building materials, medicinal plants, and wildlife, which are very important to local livelihoods. This can be a significant and perennial source of conflict between PAs and local communities. PA managers often try to reduce the conflict by allowing limited and regulated harvesting of selected unendangered species. While the motivation is understandable, and while this type of arrangement should theoretically be possible, in practice it is difficult to implement in a way that satisfies both parties. For example, a well-known initiative, often cited as a model, is the grass-cutting program in the Royal Chitwan National Park in Nepal. Its purpose is to allow local communities controlled access to a grass that is highly valued for building material; this grass grows abundantly inside the park but is now rare on the outside. When it began, the program was critical to gaining local people's acceptance of the park, but it has not stood the test of time. Steffen Straede and Finn Helles (2000) reported that in 1999 the total gross economic value of grass removed during the ten-day open season was over $1 million, but that illegal firewood was the single most important product extracted from the park and accounted for half of the total quantity and economic value of all resources collected. They argued that the core conservation area is being exploited unsustainably to support local development and that park authorities have not solved the conflict with local communities, merely postponing it at the expense of conservation objectives.

The PA approach, as it is implemented today, is sensitive to local community needs and no longer isolationist. It encompasses measures aimed at compensating for the strong cost-benefit disconnects from the community perspective. Unfortunately, the gap is difficult to bridge, and nearly impossible if the PA management has to rely solely on its own resources and internally generated revenues. The needs and demands of local communities can be large—in fact, they are effectively endless. At the same time, many PAs in developing countries are located in relatively remote, economically depressed areas where people have few options. In such cases, even small benefits, such as employment of a few people, a small annual contribution of park revenues to a community trust, or a modest investment in social infrastructure, can go a long way. Such expenditures are increasingly becoming accepted as a regular part of the operating costs of a PA because they can reduce PA-community conflicts and therefore the financial and political costs of enforcing PA regulations. However, social expenditures can only complement, not replace, basic enforcement (Kramer, van Schaik, and Johnson, 1997).

Alternative Livelihoods

This approach aims to convince people to abandon traditional activities based on destruction of biodiversity by offering them better alternatives. The alternatives can take many forms, including:

- Raising or cultivating wild species—for example, guinea fowl, cane rats, medicinal plants, woodlots, and medicinal plant gardens—instead of harvesting them from the wild
- Substituting domestic species for the wild ones—for example, using poultry as a source of protein instead of bush meat
- Introducing different ways of meeting basic needs—for example, substituting electricity, liquid gas, or biogas for firewood; introducing pressed concrete, adobe, and other building materials to replace fire-baked bricks; supporting construction of fiberglass boats instead of hollow log canoes; or providing vaccines and other modern medicines

The alternative livelihoods approach can also mean helping local people develop entirely new sources of income with which to buy some of the necessities and luxuries of life instead of producing them. Intensified agriculture, handicrafts, and tourism are among the favorite income prospects. Together with support for social infrastructure, these types of activities represent the development side of what are frequently referred to as integrated conservation and development projects (ICDPs). And an extreme form of the alternative livelihoods approach involves providing training and support for people to develop specialized work skills (e.g., auto repair skills) in an effort to encourage them to leave remote, biodiversity-rich areas and emigrate to urban centers. Often, this may in fact be the best solution for them as well as for the biodiversity, but it is rarely explicitly proposed because of the risk of appearing insensitive to people's social and cultural heritages, and of appearing to favor wildlife over humans.

The alternative livelihoods approach is a very popular element of biodiversity conservation projects around the world—so popular that one might presume that it has been a proven success, therefore worth replicating wherever possible. In reality, however, its prevalence is more a reflection of wishful thinking than success. There is little concrete evidence that ICDPs generate significant and lasting conservation, or even development, benefits (McShane and Wells 2004; Wells et al. 1999). The model actually seems to be based on some fundamentally flawed premises.

The first such premise is that there are viable but as yet untried alternative livelihoods and enterprises available, that all people need in order to pursue them are kick-starts in the form of introducing the idea, providing some technical assistance and training, and maybe offering a bit of start-up capital. In reality, there are often substantial obstacles to such enterprises, such as unavailability or high cost of inputs, incompatible social conventions and preferences, and lack of market linkages and competitive advantage. Thus there may be good reasons why people are not already engaging in these activities. (Although some of the new alternative economic activities have the potential to become viable eventually, they will need to be subsidized for long periods of time. This conflicts with the relatively short time frames of most conservation projects.)

A second false premise is that finding substitutes for a few key biodiversity products, or providing modest cash income, can alleviate the driving forces behind biodiversity loss. The main source of biodiversity loss around the world is not harvesting of biodiversity products, but rather the conversion of natural habitats to other land uses; in developing countries, aside from coastal areas, this is primarily due to agricultural expansion. It is difficult to imagine an economic activity that can substitute for subsistence agriculture as a means of feeding large numbers of people with few assets, limited education, and few modern work skills. Most developing countries also still look to agriculture as a growth sector and expect it to absorb a large portion of the growing numbers of unemployed people. And new products and income sources are risky, and people are risk-averse. They will not easily abandon tried-and-true activities in favor of new ones even if they are given start-up grants and other incentives. In any case, revenues from the alternative livelihoods will usually only be sufficient to serve as an incremental source of income, not to replace traditional economic activities. Furthermore, the hunting and harvesting of wild plants is often not only an economic necessity but also a highly valued cultural activity. Many projects have tried to introduce home-rearing of poultry and other livestock, only to find that the women wind up with this additional task while the men continue to hunt because it is more prestigious and lucrative.

The third and most important false premise is that people are satisfied with the status quo when it comes to their lifestyles, assets, and incomes. The alternative livelihoods approach assumes that people will substitute new products or income sources for old ones. Rather than substitute, though, they are much more likely to add the new activities to their existing ones. The extra income earned from new sources can in fact provide the means to further expand traditional, ecologically destructive activities. For example, the Communal Areas Management Program for Indigenous Resources (CAMPFIRE) in Zimbabwe was created to help rural communities earn income from safari hunters as a means of encouraging them to value and preserve wildlife on their lands. James C. Murombedzi (1999) found that villagers were instead using the new income to purchase inputs and labor in order to expand their agricultural fields further into the wildlife areas. Similarly, supporting intensification of agriculture by subsidizing inputs and helping producers link to commercial markets can result in people increasing their encroachment into PAs or wild habitats in order to take full advantage of this increasingly profitable enterprise (Angelsen and Kaimowitz 1999; Helmuth 1999).

In principle the alternative livelihoods approach measures up well against our potential disconnects, as its aim is to maintain the existing levels of local, concrete, short-term, individual benefits, simply deriving these benefits from different sources. The approach falls short in practice because it does not require, or create an incentive for, people to abandon their current (biodiversity-unfriendly) activities. Their interest usually lies in doing both the old and the new, if possible. This approach also reflects the tendency to apply relatively simple solutions to very complex situations: a wide range of social, cultural, and economic factors go into determining people's choices about something as fundamental as how they make a living. Support for alternative livelihoods is, however, an important complement to the PA approach, as it is no longer considered acceptable to establish or enforce PAs without compensating people for the costs this imposes on them. Assisting people in local communities to raise poultry, plant woodlots, or make and sell handicrafts will not in itself stop them from hunting or cutting firewood or clearing new agricultural plots in the forest. But it can help to make it practically, socially, and politically possible to establish and enforce restrictions on land use and the harvesting of wild biodiversity.

Sustainable Use

As a tool for biodiversity conservation, sustainable use is both extremely popular and extremely controversial. The concept is to get people to stop "mining" biological resources for short-term benefit in favor of harvesting the resources at a rate that is consistent with their natural regeneration capacity—in other words, to allow them to achieve their potential as renewable resources. Most community-based natural resource management (CBNRM) involves efforts to develop sustainable consumptive use of communally owned or controlled biological resources, although the term also encompasses aspects such as nonconsumptive uses of biodiversity (e.g., tourism) and managing nonliving resources (particularly water).

Supporters say that, given the heavy dependence of people—particularly rural communities in developing countries—on biological resources for survival and economic advancement, sustainable use is the only realistic option for maintaining the great majority of the world's biodiversity, which is found outside PAs. Critics counter that sustainable use, like the notion of alternative livelihoods, is built on a shaky foundation of wishful thinking. The debate centers on two basic questions: Is

economically viable, ecologically sustainable consumptive use of biodiversity techni-
cally possible?[5] If it is technically possible, is there a way to get people to practice it?
An in-depth discussion of the first question is beyond the scope of this chapter (the
reader is referred to Alvard et al. 1997; Barrett and Arcese 1995; Bennett et al. 2002;
Kramer, van Schaik, and Johnson 1997; Redford 1992; Redford and Richter 1999;
Robinson and Redford 1994a, 1994b).

To examine the merits of the sustainable use approach with respect to our cost-
benefit disconnects, we must make the assumption that it is at least technically possible. That is, we will assume that there are species that overproduce offspring or
biomass to such an extent that an economically significant part of the annual production can be removed without endangering the viability of either that species or others
in the ecosystem that normally consume it. With this assumption, the key question
becomes, what circumstances or incentives will cause people to limit their off-take to
that ecological excess?

One answer is to give them no other option. This may be the case when controlled harvesting of certain species is permitted within a PA, or when quotas are
set and enforced by an outside party, usually a government agency. In some cases,
local communities may be granted rights to land or resources only on the condition
that they adopt sustainable management practices (e.g., the well-known social forestry programs in India, Tanzania, and other countries). In all cases, success depends
on the controlling party's ability to calculate the sustainable off-take level, which
may vary from season to season or from year to year, and to enforce it, both logisti-
cally and politically. In reality, these conditions have usually been difficult to meet.
Sustainable off-take levels are estimated based on (usually limited) data and research
regarding population size and dynamics, and on monitoring of the impacts of past
harvesting (e.g., decreasing catch per unit effort in the fishing industry indicates
that fish have been harvested at unsustainable levels). In keeping with the "wake me
when it's an emergency" philosophy, gradual declines in populations are not always
recognized, as over time people forget what the baseline was. Jeremy B. C. Jackson
et al. (2001) demonstrated this fact in relation to marine resources, pointing out that
most ecological research has been carried out after the 1950s and has involved local
field studies lasting only a few years. They note that our concern over declining fish
stocks and populations of sea mammals is based on this narrow window and does
not take into account the fantastically large historical abundances of large marine
species, like sea turtles, whales, manatees, monk seals, and large fish species—many
of which are now nearly extinct. Resources devoted to research and monitoring are
rarely adequate to ensure a high degree of accuracy and the ability to anticipate problems. The ability of responsible authorities to enforce quotas is similarly constrained
by a lack of money and personnel.

Even when monitoring data are available and mechanisms for enforcement are
in place, willingness to enforce quotas is often lacking. Lee Talbot (1972) gave the
example of rangeland management projects in Kenya that failed when Masai families
refused to abide by the quotas for numbers of cattle. On a very different scale, under
international treaties (e.g., in the North Atlantic), quotas for harvesting ocean fisheries are established based on monitoring data, but then they are constantly overridden
or ignored by both fishermen and, as a result of political pressure, the authorities who

are supposed to enforce them. On rare occasions, nature itself might provide a constraint that makes it very difficult or impossible to overharvest a species, facilitating the establishment of a sustainable use regime (see, e.g., Norris and Chao 2002).

The second answer is to give people a direct incentive to manage resource sustainability. The classic proposed solution is to create the incentive through ownership (individual or communal). The argument behind this solution is based on eliminating the "tragedy of the commons," which may be simplistically summarized to mean that individuals have an incentive to seize open access resources as quickly as possible, before someone else does. Conversely (it is assumed), when one is able to prevent others from taking the resources, he or she has an incentive to protect them. This is particularly true if a resource represents a person's main economic asset, which he or she expects to pass on to children and grandchildren so that they may secure their livelihoods. In fact, however, even if one owns the forest or the herd of gazelles, it may still be in his or her interest to mine it rather than manage it sustainably. The short-term versus long-term disconnect still arises, even if the resource user is relatively certain of having exclusive access to the resource in the long term. After all, the income from harvesting and selling the whole lot at once might be urgently needed, or could potentially be invested in an enterprise that offers prospects of better mid- or long-term returns. Sven Wunder (2000) found that in the highlands of Ecuador, greater security of land tenure actually increased deforestation rates because people with more secure land tenure were more likely to be able to capture the long-term economic benefits associated with logging, followed by cultivation and then conversion to pasture. Overall, ownership or security of tenure over the resource is probably necessary, but far from sufficient, to provide an incentive for sustainable management.

A relatively new approach, which has captured the imagination of the conservation community, is that of offering price premiums for sustainably harvested products. Timber certification (e.g., by the Forest Stewardship Council) is the best-known example and has provided the model for others (e.g., the Marine Stewardship Council). In order for this approach to work, the price premium must be sufficient to overcome the strong intrinsic preference for short-term benefits over long-term benefits. The choices made by resource owners and users may not be solely driven by financial discount rates. For instance, some logging companies have a "triple bottom line" policy that places value on environmental stewardship or a long-term strategy based on a sustainable harvesting cycle. However, these policies face real-world constraints; Deborah B. Jensen, Margaret Torn, and John Harte (1990) noted that California timber companies practicing sustainable management were ripe for takeover by companies whose business model is to extract resources quickly. Price premiums for sustainability also depend on establishing and maintaining a credible (independent) verification/certification system and effective marketing to influence consumer choices. These costs must be factored into the equation.

Direct Payment for Conservation

Projects involving setting aside PAs (in the narrow sense) and assisting people to find alternative ways of making a living have the basic objective of separating biodiversity from people's economic interests. Any connection between the biodiversity and economic benefit for local stakeholders is indirect, sometimes so indirect as to be virtually nonexistent. Projects involving sustainable use aim to encourage and support people to obtain economic benefits directly from harvesting biodiversity, while hoping to control the extent of the exploitation. The payment for environmental services (PES) approach also tries to link biodiversity to economic benefit, but with a key difference. It is based on the premise that the existence value of biodiversity is higher than its use value. Thus, a living tree playing its role in the ecosystem is more valuable than the timber or firewood the tree would yield if cut down.

The question, of course, is more valuable to whom? The answer generally is, to society as a whole, but not necessarily to the individual who owns the land or uses the land. If an individual with the opportunity to harvest a tree or bag an antelope chooses to leave it standing instead, this is a service he or she provides to society at his or her own expense. Some people are willing and able to accept this expense for aesthetic, cultural, or other reasons. But most people demand to be compensated for performing significant public services, and those who control the fate of forests and other biodiversity around the world are no different. The challenge is to capture the value that people around the world place on maintaining biodiversity, translating it into an efficient and practical form of compensation to those who can render this particular service.

Again, a detailed discussion of the strengths and weaknesses of indirect compensation, such as ICDPs, versus direct payment is beyond the scope of this chapter (the reader is referred to Ferraro and Kiss 2002; Ferraro and Simpson 2000; Heal 2000; McShane and Wells 2004; and Simpson and Sedjo 1996). To summarize, in some cases direct payment can be the most efficient and effective way to provide an incentive for landowners or resource users to preserve rather than destroy biodiversity, because (1) it eliminates indirect steps and questionable assumptions about what drives people's decisions and (2) it can be designed to reward precisely the behavior or outcome that is sought. Designing the incentive system is not always as easy as it sounds, however. A classic example is that of a project in Tanzania, which sought to engage local communities in helping to reduce wildlife poaching by paying people a bounty for every snare they turned in. The project sponsors eventually became suspicious as to why the wire on the snares they received often looked quite new. It turned out, of course, that people were buying inexpensive wire, twisting it into snares, and turning them in for the reward. A similar example is the bounty that U.S. cities used to offer for every rat tail that was turned in. It was meant to be a rat control program, but as rat tails were such a valuable commodity, enterprising boys soon began breeding rats.

The main criticisms usually leveled against the PES approach are that it can be socially disruptive and inequitable and that it is not sustainable, in the sense that it can only work as long as some external party is willing to continue paying for the conservation service. As Paul J. Ferraro and I argue (2002), these observations are both true, but they apply equally to other conservation approaches. Any effort by outsiders to influence the lifestyles and behavior of a social group can cause disruption, and where economic or other incentives are involved there is always the possibility that powerful individuals will try to capture as much of the benefit as possible for themselves. Nevertheless, this is the purpose of the whole business of international development.

As to sustainability, there are very few examples of alternative livelihood enterprises initiated through conservation projects that have become financially self-sufficient. A case in point is the Biodiversity Conservation Network program, which selected enterprises to support specifically on the basis of their good prospects for financial success. A detailed analysis of the program showed that nearly all the initiatives required substantial subsidization for long periods and that few if any ultimately seemed likely to become financially self-sustaining (Salafsky et al. 2001). The same is true of most community-based ecotourism projects, even those frequently cited as success stories (Kiss 2004a). The simple truth is that the destruction of biodiversity for economic gain is not a problem that can be solved or expected to come to an end. It is an ongoing, perennial threat that can only be countered through ongoing financial support for conservation from a society that values biodiversity. The question therefore becomes, if biodiversity conservation must be paid for on a continuing basis, is it more efficient to do so through indirect mechanisms, such as subsidizing an ecotourism lodge, or through direct means, such as by paying people for providing the stewardship service?

Like the alternative livelihoods approach, PES scores well on addressing the main disconnects: it takes the long-term, intangible, global value of biodiversity and translates it into immediate, concrete, local benefits. PES supporters would say that, unlike alternative livelihoods, it does so based on a clear recognition of the factors that really drive people's economic choices. Making the payments conditional on actually achieving concrete, measurable biodiversity benefits can go a long way toward avoiding the problem of free riders and the inadvertent creation of perverse incentives. The extent to which PES addresses the issue of individual versus collective benefits depends on the nature of the landowner or end user, and on the mechanisms used to distribute the payment. It is simplest is negotiate a PES contract with a single individual or corporation who owns and controls a biodiversity-rich property. For the Wildlife Conservation Leasing project operating in the Kitengela area adjacent to the Nairobi National Park in Kenya, the Wildlife Foundation has signed one-on-one contracts with about eighty individual Masai heads of households, each of whom owns about 100–200 acres of land (Dawson 2004). The greater challenge lies in communally owned lands and resources, where some type of community institutional structure is needed to negotiate contracts, accept and distribute payments on behalf of the group, and ensure that the group members uphold their part of the contract. The need to help communities establish an organization that is acceptable both to them and to the prospective donor is often the main stumbling block and source of delay in PES initiatives.

CONCLUSION

The existence of disconnects between costs and benefits is an inescapable reality of environmental management in general, and biodiversity conservation in particular. There are intrinsic direct and opportunity costs involved, which largely affect a limited group of people. Sometimes these are wealthy, powerful, connected individuals and corporations focused on mining resources for short-term profits, but often they are poor rural communities dependent on clearing new agricultural land and harvesting biodiversity for their daily sustenance. Either way, the actions they take to meet their individual needs and interests are the cause of the continuing and escalating destruction of biodiversity around the world. As Jackson et al. (2001) have demonstrated, this is not a modern phenomenon. Rather, it goes back to the very early days of human society. And as Talbot (1972), Jared Diamond (2005), and many others have pointed out, it is equally true of "primitive" peoples who are often said to live (or to have lived in earlier times) in equilibrium with nature. To the extent that their lifestyles were sustainable, it was for the most part not because of a conscious and prudent choice to sacrifice their own immediate needs in the interests of posterity and society at large, but because they were few in number compared with the scale of available resources, and because their technical destructive capacity was more limited (spears and snares instead of automatic rifles and dynamite, hoes instead of tractors) than ours.

All conservation projects and measures have the basic objective of altering the destructive behavior of resource users. This means finding ways to address these disconnects and bridge the gap between short-term, concrete, individual costs and long-term, often intangible, society-wide benefits. Traditional, exclusive PAs circumvented the issue by simply eliminating or restricting people's access to the biodiversity. In the past this was done without consultation and with little or no consideration of the impacts on local communities. In recent decades this has become unacceptable, and it is now the general practice (particularly if funding from donor organizations or international conservation organizations is involved) to consult and negotiate with local communities and to obtain their agreement before a PA is created. Thus, even the PA approach can no longer escape the challenge of tackling cost-benefit disconnects.

The alternative livelihoods approach aims, by providing better alternatives, to get people to stop depending on natural ecosystems and biodiversity resources for their essential products and income. In principle this should be an effective way to address the main cost-benefit disconnects, but in practice it frequently fails because it does not provide a real incentive for conservation. Community-based ecotourism is a very popular choice for such projects precisely because of the belief that it provides such an incentive. However, tourism is a difficult and competitive business with a high failure rate, and it rarely generates sufficient revenues to replace or make a significant dent in existing economic activities (especially if practiced in a truly biodiversity-friendly manner). The main limitation of the alternative livelihoods approach is that, given the option, people are much more likely to add new economic activities to their existing activities rather than substitute one for the other. While support for alternative livelihoods cannot replace access restriction (e.g., PAs) as a strategy, it can help to make restriction more feasible and acceptable by compensating for people's costs.

The sustainable use approach aims to get people to stop mining biodiversity resources for short-term benefits, and instead to adopt long-term strategies. Aside from the debate over whether ecologically sustainable use is even technically possible, this approach epitomizes the disconnect between short-term costs and long-term benefits. As a result, it is only likely to work in cases where the resource users' short-term needs are not very pressing, or where people are only given access to the resource as long as they agree to follow sustainable practices. Even then, the tendency is for people to "cheat" and continue overexploitation, and it is difficult to take away rights once they have been given.

The PES approach is much less common than the other three but has begun to gain momentum in recent years, perhaps because it addresses the major cost-benefit disconnects so effectively. It is similar to the alternative livelihoods approach, but it avoids the key shortcoming of a lack of direct linkage with biodiversity outcomes, and thus the lack of an actual incentive to conserve biodiversity. In this case the new economic activity is stewardship of biodiversity, and the revenues flow only if concrete conservation results are achieved. Like the alternative livelihoods approach, the PES approach is most likely to be effective where the opportunity cost of maintaining biodiversity (giving up biodiversity-destructive activities) is relatively low. And like all approaches to biodiversity conservation, it will only be sustainable as long as society at large is willing to pay to maintain biodiversity in the world.

Notes

1. For example, Conservation International estimates that the international community (governments, multilateral development banks, and conservation groups) spends at least half a billion dollars each year on conserving biodiversity in the tropics (Hardner and Rice 2002). Michael Satchell (2000) notes that about $4 billion has been spent on conservation over the past decade; a recent report from the Organisation for Economic Co-operation and Development (2000) estimated that in 1998, $778 million of bilateral official development assistance in sectors such as agriculture, forestry, water, and general environmental protection sectors had a biodiversity focus.

2. For example, in 1997 world grain production was estimated at 4.3 pounds of food per person per day, and 78% of all malnourished children under the age of five were living in countries with food surpluses, including some countries that had benefited most from the green revolution, such as Mexico, India, and the Philippines (Lappe, Collins, and Rossett, 1997; World Resources Institute 1998).

3. For more information on the categorization system, see the Web site of the UN Environment Programme's World Conservation Monitoring Centre at http://www.unep-wcmc.org/protected_areas/categories/index.html.

4. The term *short term* implies an impact that begins to be felt right away; it does not preclude the possibility that negative impacts may then continue for a long period of time.

5. The distinction is somewhat arbitrary, but for the purposes of this chapter, nature-based tourism is treated as an alternative income rather than a form of sustainable use of biodiversity.

References

Alexander, S. E. 2000. Resident attitudes towards conservation and black howler monkeys in Belize: The community baboon sanctuary. *Environmental Conservation* 27(4): 341–350.

Alvard, M. S., J. G. Robinson, K. H. Redford, and H. Kaplan. 1997. The sustainability of subsistence hunting in the neotropics. *Conservation Biology* 11(4): 977–982.

Angelsen, A., and D. Kaimowitz. 1999. Rethinking the causes of deforestation: Lessons from economic models. *World Bank Research Observer* 14(1): 73–98.

Balmford, A., K. J. Gaston, S. Blyth, A. James, and V. Kapos. 2003. Global variation in terrestrial conservation costs, conservation benefits, and unmet conservation needs. *Proceedings of the National Academy of Sciences* 100(3): 1046–1050.

Barrett, C. B., and P. Arcese. 1995. Are integrated conservation-development projects (ICDPs) sustainable? On the conservation of large mammals in sub-Saharan Africa. *World Development* 23(7): 1073–1084.

Bennett, E., H. Eves, J. Robinson, and D. Wilkie. 2002. Why is eating bushmeat a biodiversity crisis? *Conservation in Practice* 3(2): 28–29.

Bruner, A. G., R. E. Gullison, R. E. Rice, and G. A. B. da Fonseca. 2001. Effectiveness of parks in protecting tropical biodiversity. *Science* 291(5501): 125–127.

Calder, I. R. 2004. Hydrology and land use: Forest and water interactions. Presentation at the IIED Workshop on Developing Markets for Watershed Protection Services and Improved Livelihoods, Charity Centre, London.

Center for International Forestry Research/Food and Agriculture Organization of the United Nations. 2005. *Forests and floods: Drowning in fiction or thriving on facts?* Forest Perspectives, no. 2. Bogor Barat, Indonesia: CIFOR/FAO.

Dawson, J. 2004. Can the lion lie down with the lamb? *Earth Island Journal* 19(2). http://www.earthisland.org/eijournal/new_articles.cfm?articleID=878&journalID=79.

Diamond, J. 2005. *Collapse: How societies choose to fail or succeed.* New York: Viking.

Ferraro, P. J., and A. Kiss. 2002. Direct payments to conserve biodiversity. *Science* 298(5599): 1718–1719.

Ferraro, P. J., and R. D. Simpson. 2000. *The cost-effectiveness of conservation payments.* Resources for the Future Discussion Paper, no. 00-31. Washington, DC: Resources for the Future.

Hardner, J., and R. Rice. 2002. Rethinking green consumerism. *Scientific American* 286(5): 88–95.

Hazell, P. B. R., and C. Ramasamy. 1991. *The green revolution reconsidered: The impact of high-yielding rice varieties in south India.* International Food Policy Research Institute Food Policy Statement, no. 14. Baltimore, MD: Johns Hopkins University Press.

Heal, G. M. 2000. *Nature and the marketplace: Capturing the value of ecosystem services.* Washington, DC: Island Press.

Helmuth, L. 1999. Economic development: A shifting equation links modern farming and forests. *Science* 286(5443): 1283.

Hulme, D., and M. W. Murphree, eds. 2001. *African wildlife and livelihoods: The promise and performance of community conservation.* Oxford, UK: James Currey.

International Food Policy Research Institute, 2002. *Green revolution: Curse or blessing?* Washington, DC: International Food Policy Research Institute.

Jackson, J. B. C., M. X. Kirby, W. H. Berger, K. A. Bjorndal, L. W. Botsford, B. J. Bourque, R. H. Bradbury, R. Cooke, J. Erlandson, J. A. Estes, et al. 2001. Historical overfishing and the recent collapse of coastal ecosystems. *Science* 293(5530): 629–637.

Jensen, D. B., M. Torn, and J. Harte. 1990. In our own hands: A strategy for conserving biological diversity in California. California Policy Seminar, University of California, Berkeley.

Kiss, A. 2004a. Is community-based ecotourism a good use of biodiversity conservation funds? *Trends in Ecology and Evolution* 19(5): 232–237.

Kiss, A. 2004b. Making biodiversity conservation a land use priority. In *Getting biodiversity conservation projects to work: Towards more effective conservation and development,* ed. T. O. McShane and M. P. Wells, 98–123. New York: Columbia University Press.

Kramer, R., C. van Schaik, and J. Johnson. 1997. *Last stand: Protected areas and the defense of tropical biodiversity.* Oxford: Oxford University Press.

Lappe, F. M., J. Collins, and P. Rossett. 1997. *World hunger: Twelve myths.* London: Earthscan.

McShane, T. O., and M. P. Wells, eds. 2004. *Getting biodiversity conservation projects to work: Towards more effective conservation and development.* New York: Columbia University Press.

Millennium Ecosystem Assessment. 2005. *Ecosystems and human well-being: Synthesis.* Washington, DC: Island Press.

Murombedzi, J. C. 1999. Devolution and stewardship in Zimbabwe's CAMPFIRE programme. *Journal of International Development* 11(2): 287–293.

Norris, S. N., and N. L. Chao. 2002. Buy a fish, save a tree: Safeguarding sustainability in an Amazonian ornamental fishery. *Conservation in Practice* 3(3): 30–35.

Organisation for Economic Co-operation and Development. 2000. *Aid targeting the Rio conventions: First results of a pilot study.* Paris: OECD.

Redford, K. H. 1992. The empty forest. *BioScience* 42(6): 412–422.

Redford, K. H., and B. D. Richter. 1999. Conservation of biodiversity in a world of use. *Conservation Biology* 13(6): 1246–1256.

Robinson, J. G., and K. H. Redford. 1994a. Community-based approaches to wildlife conservation in neotropical forests. In *Natural connections: Perspectives in community-based conservation,* ed. D. Western, R. M. Wright, and S. C. Strum, 300–319. Washington, DC: Island Press.

Robinson, J. G., and K. H. Redford. 1994b. Measuring the sustainability of hunting in tropical forests. *Oryx* 28(4): 249–256.

Salafsky, N., H. Cauley, G. Balachander, B. Cordes, J. Parks, C. Margoluis, S. Bhatt, C. Encarnacion, D. Russell, and R. Margolis. 2001. A systematic test of an enterprise strategy for community-based biodiversity conservation. *Conservation Biology* 15(6): 1585–1595.

Satchell, M. 2000. Hunting to extinction: A wildlife crisis is forcing conservationists to rethink their tactics. *U.S. News and World Report,* October 9, 2000.

Shiva, V. 1991. The green revolution in the Punjab. *The Ecologist* 21(2): 57–60.

Simpson, R. D., and R. A. Sedjo. 1996. Paying for the conservation of endangered ecosystems: A comparison of direct and indirect approaches. *Environment and Development Economics* 1(2): 241–257.

Smith, N. J. H., E. Adilson, S. Serrão, P. T. Alvim, and I. C. Falesi. 1995. *Amazonia: Resiliency and dynamism of the land and its people.* UNU Studies on Critical Environmental Regions. Tokyo: United Nations University Press.

Straede, S., and F. Helles. 2000. Park-people conflict resolution in Royal Chitwan National Park, Nepal: Buying time at high cost? *Environmental Conservation* 27(4): 368–381.

Struhsaker, T. T., P. J. Struhsaker, and K. S. Siex. 2005. Conserving Africa's rain forests: Problems in protected areas and possible solutions. *Biological Conservation* 123(1): 45–54.

Talbot, L. 1972. Ecological consequences of rangeland development in Masailand, East Africa. In *The careless technology: Ecology and international development,* ed. M. T. Farvar and J. P. Milton, 694–711. New York: Natural History Press.

Wells, M., S. Guggenheim, A. Khan, W. Wardojo, and P. Jepson. 1999. *Investing in biodiversity: A review of Indonesia's integrated conservation and development projects.* Washington, DC: World Bank.

World Resources Institute. 1998. *World resources 1998–99: A guide to the global environment; Environmental change and human health.* Joint project of the World Resources Institute, the United Nations Environment Programme, the United Nations Development Programme, and the World Bank. New York: Oxford University Press.

Wunder, S. 2000. Ecotourism and economic incentives: An empirical approach. *Ecological Economics* 32(3): 465–479.

18

Resource Wars

Jason Clay

This chapter was originally written in August 1994, a time when many were optimistic about the potential of the new world order to reduce the global tensions and conflicts that had been fed by the cold war. Saddam Hussein had been forced to retreat from Kuwait; the United Nations and NATO had made a dramatic statement about the sanctity of states. But the internal cleansing of ethnic and religious minorities and majorities was already beginning in eastern Europe and would only intensify there and elsewhere. And in 2006, the Darfur crisis and other internal conflicts, while tragic, failed to ignite the same condemnation and action as previous conflicts in Angola, Chad, Ethiopia, Guatemala, Myanmar, Mozambique, Nicaragua, the Philippines, Rwanda, Somalia, and Sudan. Those conflicts could be more easily attributed to specific, understandable, and possibly correctable problems—colonial legacies, superpower proxies, famine, and ideological rebellions. There is a growing realization that resolving such conflicts takes considerably more time and money than many countries are prepared to invest, particularly given the fact that lasting solutions are not always possible.

What has also happened since the mid-1990s is that scholars from many different quarters have started to painstakingly piece together, a bit at a time, information about what was destroyed or has largely been lost through colonization and the global consolidation of states. This scholarship is providing a new window on the history of human beings on this planet. As the adage goes, history is written by the winners, but fortunately that seems not to be entirely true. Perhaps the best evidence of the wealth of nations around the world is the beginning of an understanding of what has been lost since Europeans arrived in the Americas. Charles Mann has attempted to consolidate much of that disparate information in his book *1491: New Revelations of the Americas before Columbus* (2005). What he suggests is quite interesting and worth noting in the context of the piece that follows.

- In 1491, there were probably more people living in the Americas than in Europe.
- Cities such as Tenochtitlan had greater populations than any European city of the day, and unlike any European counterpart had botanical gardens, clean streets, and some running water.

- The earliest cities in the Western Hemisphere were thriving before the Egyptians built the Great Pyramids.
- Indians in Mexico had developed corn through a complex breeding process described by science as man's first, and perhaps greatest, feat of genetic engineering.
- Indians knew how to farm the Amazonian rainforest without destroying it—a feat scientists are still trying to understand.
- By the time Europeans arrived in the New World, Native Americans had massively landscaped their environment.

In 1985, I was interviewing Oromo refugees who had fled to Sudan about how and why they had been forced to leave Ethiopia by the Amhara-dominated Marxist government. The Oromo, who with twenty million people are one of the largest nations in Africa and represent more than half of the population of the country, have been largely excluded from political power by dominant Abyssinian groups. I asked one man why the world didn't understand their plight. More to the point, I asked him why Westerners were giving assistance to the government, assistance that was being used to force the Oromo to relocate or to move into villages where they could be more closely watched by the army (for more detail see Clay and Holcomb 1986, as well as well as Clay, Steingraber, and Niggli, 1988). He replied with an old Oromo proverb: "You can't wake a person who's pretending to sleep."

With this backdrop, I hope the chapter will help those who read it wake up to the reality of what has happened—and, as important, is still happening—to nation peoples around the world. It addresses what has happened to indigenous peoples, or distinct nations, since colonization and more specifically since decolonization and the rise of states in the twentieth century.

INTRODUCTION

The world has changed: good-bye Berlin Wall, hello Berlin Mall. Basic assumptions about the role of the state are crumbling as well. The rise of nationalism and recent moves toward democratic pluralism and autonomy throughout the world demonstrate that highly centralized, top-down authority can vanish overnight. But we have also seen that capitalism cannot put all the pieces back together again. Cultural identity, more than ideology and as much as economies, may be the building block for truly democratic, bottom-up political systems.

Nations, the different culturally and linguistically distinct peoples on the planet, are distinct from states or countries. They are challenging notions that states are the building blocks for global peace and environmental security. At stake is not the existence or even legitimacy of states but rather the survival of nations, an issue that involves cultural and religious freedom, language rights, locally significant education, and political autonomy. However, no single issue affects the survival of nations as much as the state appropriation of resources, in particular land, which nations require if they are to survive as societies. It is greed and the global appetite for natural resources that fuel the threats to nations.

NATIONS AND STATES

There are, today, some 200 states in the world, encompassing approximately 6,000 distinct nations. While it often feels like states have existed forever, the majority have been created since World War II, when only fifty or so existed. By contrast, most nations—generally characterized by distinct language, culture and history, and territorial bases and self-government that predate the creation of modern states— have been around for hundreds of years, in some cases even millennia. Nations have been distinguished in the following ways (adapted from Goodland 1982; see also Clay 1991):

- Geographically isolated or semi-isolated
- Unacculturated or only partly acculturated into the societal norms of the dominant society
- Having nonmonetized or only partially monetized economies
- Ethnically distinct from the dominant society
- Nonliterate and without a written language
- Linguistically distinct from the wider society
- Closely identified with a particular territory
- Having an economic lifestyle largely dependent on the specific natural environment
- Having limited participation in the political life of the wider society
- Having few if any political rights in the wider society
- Having insecure tenure of traditional lands and resources and weak powers of enforcement against encroachers

Nation peoples believe that states have only as much legitimacy as is bestowed voluntarily by those incorporated into them. The Oromo in Ethiopia do not think of themselves as Ethiopians; the Kayapo in Brazil do not think of themselves first or even foremost as Brazilians; and until recently the Penan of Sarawak, Malaysia, barely even knew what Malaysia was, much less thought of themselves as part of that country. States mean very little, at least in a positive sense, to many (if not most) distinct nations. For example, there are more than 180 nations in Brazil, 450 in Nigeria, 350 in India, 450 in Indonesia, 300 in Cameroon, and 80 in Ethiopia.

Simply put, there are no nation-states. All states contain more than one nation, try as many of them might to eliminate or assimilate them. Every state is multinational.

Put another way, every state is an empire. Furthermore, modern states, particularly third world states, are ruled like empires—that is, from the top down. This leads many nations to ask if the costs of being part of states outweigh the benefits they get from being incorporated into them. It is the result of analyses like this, community by community, nation by nation, that are at the heart of the rise of nationalism. The result should surprise no one.

The refusal of states to acknowledge their cultural diversity, much less embrace it as a strength, is a major cause of the rise of nationalism and conflict throughout the world. More important, the elimination of nations by states has caused the loss of more information about the earth's resources and how they can be managed sustainably than any other single factor. The knowledge of a single nation about resource management, gathered and honed over centuries, cannot be duplicated even by hundreds of person-years of research by scores of scientists. And the past century of "progress" and "civilization" provides good evidence that more nations have disappeared during this century than during any other century in history. Brazil, for example, has "lost" one Indian nation per year throughout the twentieth century (one-third of the groups existing in 1900) while government officials and planners have done their best to "develop" the Amazon into a wasteland.

At the heart of this matter is the state-building, nation-destroying process, a process founded on the beliefs that indigenous peoples are historical anomalies and that indigenous peoples and states cannot coexist. One way in which states eliminate distinct peoples or nations is to ban their languages or at the very least denigrate them or prohibit their teaching in public schools. It is impossible to know how many self-identified peoples have disappeared—many were unknown to outsiders—but language provides perhaps the best window on this process, as there is more data available on this issue than on others related to nations. While language is admittedly not a perfect reflection of a nation's identity, it is an important window on the world's cultural diversity. More than half the world's 15,000 known languages have disappeared already; linguists predict that only 5–10% of the remaining 6,000 to 7,000 are likely to survive another fifty years.

In their attempts to create cultural homogeneity, states also attack cultural practices, local religions, and community-centered governments. States cannot exploit resources—land, trees, minerals, water—without denying the rights of the local inhabitants, often indigenous nations, who have lived on and maintained the natural resource base and who often hold parts or all of it sacred. To look at this process another way, states are destroying nations even faster than they are destroying the often fragile natural resources these groups have used and maintained for centuries.

These states have rarely, if ever, been created by those who are governed by them. For the past several centuries, men—often white and always culturally different—thousands of miles away have created the boundaries of most new states. After centuries of experimentation, it has become clear that self-governed entities are cheaper than colonial occupation. Make no mistake: Europeans and North Americans are essentially seeking trading partners who can ensure politically stable economic systems and the free flow of goods.

NATIONS AND DEVELOPMENT

Despite efforts to destroy them, there remain about 600 million nation peoples (about 10% of the population on the planet) who retain a strong social and cultural identity as well as an attachment to a specific territory. In this context, nations can be distinguished from ethnic groups who, though much larger in absolute numbers, are usually immigrants in new geographical areas or local groups who have made an accommodation with states by trading away political autonomy for the ability to retain and practice some other cultural beliefs. Recent political changes in eastern Europe and the former Soviet Union have shown that even ethnic groups, however, can push for autonomy, reasserting their national identities as well as their claims for independence. In short, ethnic groups can become nations. Israel is a prime example.

After being decimated through contact, disease, colonization, and wars from 1500 to 1900, the world's nations have increased tremendously in population since 1900 and especially since World War II—precisely the time during which most states were created. Thus, states were established precisely when many nations felt that it was possible to push for more autonomy and regain the political independence that was denied them during colonial rule. Unfortunately, most nations found that the newly independent states were not truly democratic and that many were little more than local colonial empires.

Nations have had many different experiences in their incorporation into postcolonial states. Some nations, like the Kikuyu in Kenya, decided to take their chances in the new states, becoming, if you will, ethnic groups by choice. Other groups, like the Mbundu in Angola, knew that they had no chance of sharing power in the new state, so they took up arms immediately. Still other nations, such as isolated Indian nations in the Amazon, were so untouched that they were unaware of the political significance of decolonization. Finally, some nations, including about a dozen groups in Burma, negotiated local autonomy as a condition of independence, only to see it taken away by military coups sponsored by dominant ethnic or nation groups. Each of these situations can prompt, and often has prompted, numerous groups around the world to take up arms.

Since World War II, many factors have affected the willingness of nations to accept a priori the legitimacy of states. In most instances nation-states have been created in the image of, and are dominated by, only one or a few of the nations that exist in each state. When cooperation between nations breaks down, dictatorships and single-party states become the norm. Sub-Saharan Africa remains a good example of this problem.

The elites who dominate new states, particularly in the third world, can be characterized by a winner-take-all mentality. Those who control states make laws in their own image and in their own interest. They control foreign investment and assistance (both development and military), both of which are used to reinforce their power to rule. More often than not, they fix local commodity prices and control exports. These sources of income, on average, account for about two-thirds of state revenues; the final third of state revenues is derived from taxes, often disproportionately levied on nation peoples. Through all these mechanisms, government in many states has become the biggest game in town. Those who rule decide the national laws, including who owns which natural resources and which traditional resource tenure systems will be honored by the state. Those who rule often dictate religion, language, cultural traditions, and even national cultural holidays. Because such issues are integral to the cultural survival and integrity of nations, these issues often trigger violent confrontations.

Many assumed that the ending of the cold war and the opening up of the political system in third world countries would make them more democratic. The idea was that without superpower rivalries to provide the money and arms to establish puppet states, the groups in developing countries would have to find ways to live together, becoming more socially equal as a result. In places like eastern Europe, the former

Soviet Union, the Middle East, parts of Asia, and most of Africa, this does not seem to have been the case. The new political systems are based on patronage through budgets and civil service. Economic reforms and political liberalization may actually exacerbate conflicts between nations or nations and states. A few recent examples illustrate the point:

- From 1991 to today in the former Yugoslavia, tens of thousands of people have died and hundreds of thousands have been displaced.
- In 1994 in Rwanda, more than one million people, mostly Tutsi, died, with even more displaced.
- From March through June 1994, 350 villages in Ghana were burned, several thousand people were killed, and 150,000 were displaced.
- Through the early 1990s in Togo, several hundred thousand people were displaced by nation versus state conflicts.
- Beginning in 1993 in Kenya, more than 1,000 people were killed and 250,000 displaced from their rich farmland in the Rift Valley by members of President Daniel Moi's tribe, the Kalenjin.

NATIONS AND NATURAL RESOURCES

Nations represent only around 10% of the world's population, but they have traditional claims to 25–30% of the earth's surface area and resources. This is one of the major reasons for conflict between nations and states. In particular, since World War II there has been a growing awareness of the finite nature of the earth's resources. This has led many states to lay legal claim to resources and to invade remote nation areas and appropriate the resources (e.g., Indian land throughout the Amazon, pastoralists' land throughout Africa, oil from the Kurds in Iraq, timber from the Penan of Malaysia). Profits from the sale of resources accrue to those who control the state. Those who can control and sell such resources stand to gain considerably. Thus, traditional resource rights of nations that may have been constitutionally guaranteed or at least assumed as a condition of independence are often subsequently denied. This, too, leads to conflict.

Since World War II more than 3,000 constitutions have been written. Globally, this comes to about one constitution for every three years for every country on the planet. (Of course, some have changed their constitutions more often then others.) These constitutions have not focused on who can or cannot vote, but rather on what the rights of citizens are, particularly with regard to natural resources. Resource rights can be seen as bundles of assets, and for the past fifty years states in developing countries have been breaking apart the bundles. They have done this by separating land rights from subsoil rights, land rights from water rights, trees from land rights, animals from land rights, genetic material from land rights, etc. In this way, states have attempted to appropriate the most valuable resources.

Most of this chapter focuses on the relationship of third world states to nations, but so-called developed countries are not unwitting observers in this process. Rather, they are active participants in the creation of the international state/trading system.

The West created most third world states in order to maintain global stability, facilitate free trade, and reduce the costs of governing far-flung empires. Investments, political intervention, and foreign and military assistance have helped to maintain the dictatorships and single-party states that dominated the world up to the last few years. It is likely that any political or economic support that is given to fledgling pluralistic governments will be used to assist, in turn, the free flow of resources needed by multinational corporations to feed the voracious appetite for consumer goods in developed countries.

Still, the third world political reality is that with or without Western assistance (but usually with it) nations that resist the authority of the state are destroyed or marginalized. It is no accident that more nation groups have been destroyed in the twentieth century than during any century in history because it is only recently that we have had the technology to inventory and exploit the world's natural resources. As global population increases, the consumption of natural resources, too, is higher than ever before in human history. This is not merely a function of population growth, however; per capita consumption in the developed world is neither sustainable nor an appropriate model for the rest of the world.

Some larger nations can defend themselves and their resources in the face of state expansion or at least hamper the invasion of their areas. Smaller nations, however, can rarely defend their homelands. If the rights of these groups are to be protected, they must be defended at the level of state or multistate organizations. Herein lies the contradiction. Are states, or multilateral or bilateral state organizations, genuinely interested in the survival of nations? Or would they rather see them quietly disappear?

Of the world's roughly 6,000 nations, perhaps 500 could physically defend themselves in armed conflict. Many have taken up arms in the past two decades as their political and cultural autonomy, as well as their resource base, has been curtailed. Of the approximately 120 shooting wars in the world today, 75% are between nations and the states that claim them as citizens.

What do nation/state conflicts produce? Victims, most of whom are women, children, and the elderly. Since World War II, at least five million people have been killed as a result of such conflicts; even larger numbers have died as a result of malnutrition and famine. While some fifteen million have officially fled across international borders as refugees (with possibly just as many going unnoticed), more than 150 million others have been forced to flee their homelands and become internally displaced. Most of this displacement has occurred in the name of "national" integration, development, or the appropriation of resources for the greater benefit of all. Displaced nation peoples, whether they cross international boundaries or not, cause environmental degradation and conflict with local groups. For example, the 1984–1986 government-sponsored resettlement program in Ethiopia during the famine led to the clearing of 8% of Ethiopia's remaining forests in 1985 alone. Displaced nation peoples also suffer from malnutrition, disease, and poverty. Ironically, much displacement results from bilaterally and multilaterally funded development programs that displace those groups who are powerless to oppose them.

No single state ideology seems to protect nations or to promote pluralism better than the others. States on both the left and right as well as religious and sectarian states deny the rights of nations—Ethiopia and Guatemala, Indonesia and Burma, Malaysia and Israel, Nicaragua and South Africa, the United States and the former Soviet Union. Those who control states see nations as a threat to national security. They justify even the forced assimilation of nation peoples in the name of progress or for the betterment of the majority.

So what are other products of nation/state conflicts? One is debt. Nearly half of all third world debt is for the purchase of weapons that are used for armed conflict with nation peoples who are supposedly already citizens of the state. Military expenditures in developing countries are greater than expenditures for all social and development programs combined. For example, it was estimated that in 1988 states spent an average of $25,000 per soldier and less than $350 per student. In the 1980s in Africa, expenditure per student declined by one-third. By 1993, one in three Africans went without primary school education and one-third of all college graduates left the continent (Darnton 1994c).

Foreign debts (caused by military expenditures and the capital flight of elites) lead to austerity measures and provided the rationale for the further appropriation of the land and resources of nations. It should be noted in passing that in the 1980s and early 1990s third world elites had foreign assets equal to the entire third world debt. But while everyone is responsible for repaying such debts, only some benefited from the original loans.

There has been much speculation in the press about the positive impact of the end of the cold war and the consequent reductions in global military expenditures. The positive statistics (here, for the period 1987–1991) include a 62% decline in global arms sales, a 10% reduction in total troop numbers, and U.S. defense cuts of about $59 billion. U.S. cuts were greater than the entire 1991 defense budgets of such countries as Great Britain, France, or China. However, while overall spending is declining, most declines are in developed countries. (And in 1991 the U.S. defense budget was still almost as great as those of all other countries combined.) Despite these global trends, developing countries' expenditures actually rose 9% (Darnton 1994a).

The point is that the appropriation of nations' resources leads to conflict, which leads to weapons purchases, which leads to debt, which leads to the need to appropriate more resources. As a result, there is a spiraling escalation of such conflicts with no end in sight. In some parts of the world, particularly Latin America, debt levels have been successfully renegotiated. In Africa, however, debt has tripled since 1980, and debt burden amounted to 110% of gross national product in 1991. Interest on African debt is $10 billion per year, four times more than all public expenditures on health and education (Darnton 1994b).

State control of nation peoples is seen as essential to state survival, yet the measures taken in the name of national security fuel nation hostility, making control imperative. State-sponsored relocation, colonization, resettlement, and villagization programs ensure the control of nations as well as their lands and resources. Food and famine have become weapons in the resulting nation/state conflicts. In the 1980s, for example, food production in Africa dropped to 20% below the 1970 level. The World Bank estimates that it will take forty years for Africa to reach the food production levels of the colonial period (Darnton 1994c). Displacement, malnutrition, environmental degradation, refugees, and genocide have become commonplace.

The bottom line is that state control of nations and the dismantling of nations' sociopolitical organizations create dependent populations from formerly self-sufficient ones. Thus already heavily indebted third world states are faced with increasingly dependent populations that can only look to the state for basic necessities. However, since such groups are systematically discriminated against, they receive little or no help. The international community is asked to assist instead. But the assistance given tends to reinforce the power of the central state. The recent Ethiopian famine is a case in point—all groups involved used the famine to reinforce their power.

Nations' Knowledge and Resource Rights

In the recent rush to discover, inventory, and save the world's biodiversity, considerable bio-prospecting is being done by corporations, scientists, nongovernmental organizations, and governments. This is probably the last great resource grab of the twentieth century. The recent international Convention on Biological Diversity, written and signed by states, recognizes states' sovereignty over resources—including genetic and biochemical resources—as well as the right of states to control the benefits derived from commercializing such resources.

In the name of saving the world, or at least salvaging information before it disappears, shamans, tribes, and groups of people are being interviewed and their information used in attempts to understand the chemical properties of plants and animals used by nation peoples. Basic agreements (e.g., contracts, licensing, or royalty agreements) that would be signed with any Western researcher for information are routinely denied to groups that provide culturally specific discoveries that have taken generations, even millennia, to test and develop.

The knowledge of such groups is not protected as intellectual property precisely because it is cultural knowledge. When a product cannot be shown to be the result of an individual invention, Western legal systems do not recognize the information as being protected as intellectual property. What this means is that designs of indigenous peoples or nations can be stylized by designers and patented for use in place mats, sheets, rugs, and other items even though the nations cannot patent the designs themselves. Similarly, when anthropologists, ethnobotanists, or other researchers interview shamans and traditional curers and publish the information they have been given about specific medicinal and chemical properties made from different substances, this information is in the public domain and, as a consequence, not protected as intellectual property.

This is not to say that nation peoples or individual shamans should have all the rights to genetic or biodiversity materials, or even necessarily to medicines and cures that their ancestors discovered and developed. There are many descendants who would have claims to such knowledge. The point is rather that they, too, should have rights, just as scientists, countries, and corporations already do. Without the cooperation of all these players, few raw materials with interesting biochemical and medicinal properties would become new medicines or other products. However, nation peoples are unique among the players, because they are without exception excluded legally from profiting from their information and the knowledge of their ancestors.

Nations and Resource Management

Recently, another line of argument has been put forth by research scientists and government officials to justify the state control of nations' resources as well as their resource management practices. Many now argue that nation peoples degrade their own resource bases, often fragile ecosystems, and that the world should not allow this to happen. (A quick look in our own backyards might raise questions about what right we should have to tell others how to manage their resources.) Instead, it is said, such resources should be seen as the common heritage of mankind. The next cure for a disease, it is argued, could be going up in smoke in the rainforest. This issue requires discussion, because it is usually couched in the language of the greatest good for the greatest number of people. In fact, the situation is almost always the reverse. It is "good" for only the few who have political or consumer power.

What, then, is the record of nation peoples as conservationists? Do they merely use resources or do they manage them? Anthropologists, who have done most research on the economic activities of nation peoples, frequently err on the side of romanticism in their views of such peoples as "the once and future resource managers" (see, e.g., AMARU IV 1980). Many practices of indigenous and nation peoples are indeed conservationist even though their scientific basis may not yet be understood. It should be noted, too, that nation peoples have domesticated our basic foods. In fact, field trials of new crops and management systems continue throughout the world, and many of these new management systems and the range of traditional seed varieties are the result of the efforts of nation peoples. Globally, it is doubtful that researchers or scientists conduct even 5% of such field trials. The rest are done by nation peoples and other peasant farmers trying to find a better way to make a living or solve a problem.

Nation peoples, because of the romantic views concerning their perceived pristine lifestyles, are often forced to adhere to a different standard than everyone else. Yet it is clear that their resource management systems are in a more sustainable stasis with the environment than our own. Through the centuries, throughout the world, nation peoples have developed sustained-yield subsistence systems that often combine root crops, vegetable crops, and select tree crops that, in turn, provide food, which also improves hunting, fishing, and gathering. Domesticated animals and cash or marketable crops have been added to the mix.

The various worldviews and beliefs about the environment that distinguish nation peoples from us as well as from each other have led to culturally specific systems of resource management. These systems are rarely random or even primarily opportunistic. Nation peoples are not preservationists; they are actively involved in manipulating their environment. But they are conservationists, and they know that they must use their resources but leave enough to guarantee the survival of future generations. Some of their systems have been sustainable over time, others have not. Some are sustainable under certain conditions but become destructive under others. Some individuals are more cautious and conserving of resources than others in the same groups.

Yet nation peoples are becoming extinct at an even faster rate than the regions they have traditionally inhabited. The example of the loss of indigenous groups in Brazil is not isolated—the loss of cultural diversity is much faster than that of biological diversity. Over the past century about 10% of the Amazon has been converted to pasture or other uses. The point, however, is not just that biological diversity fared better in the last century than cultural diversity. Rather, the point is that human rights violations often precede environmental degradation.

Nor is cultural extinction necessarily overtly violent. In many nation or indigenous societies undergoing rapid change, young people no longer want to learn the methods by which their ancestors maintained fragile regions. They would prefer to move to cities or sell natural resources to create and maintain new lifestyles. Little time remains if this information is to be maintained for future generations.

Resource management systems of nation peoples stress sophisticated and extensive knowledge of local environments. They are based on the view that the environment is the source of life for future generations and therefore should not be pillaged for short-term gain and long-term loss. Unlike farmers in mid-latitude areas who depend on machinery, specialized seeds, fertilizers, and pesticides and increasingly view the land as their adversary, nation peoples traditionally see the land and other resources as their lifeblood.

What conditions, then, encourage nation peoples living in fragile environments to conserve resources? The most important factors, it appears, are resource rights (e.g., to land, timber, water, and, at the very least, the ability to oppose the destructive extraction of minerals), the ability to organize themselves to protect their lands and resource bases, and the ability to transform traditional resource management systems to meet modern needs. It is in fact the adaptation of traditional resource management systems, rather than their abandonment in favor of more "advanced" agricultural technologies, that will allow nation peoples to develop more rational and long-term land-use patterns.

By discovering the extent to which traditional management practices can be adapted without degrading the environment, more cash crops can be generated to meet the increasing material needs of nation peoples. Given the often isolated and harsh environments in which they live, the continued existence of nation peoples demonstrates their ability to maintain the earth's resources for centuries without destroying them. While respect for resources is not universal among native cultures, it is common. Respect for resources, with some groups at least, reveals itself in such beliefs as the sacredness of the earth, the spiritual characteristics of aspects of the environment, or taboos on using certain resources at all or at least during crucial times for the species' reproduction.

However, it must be remembered that much of the pressure on nations' resource bases comes not from within but from the insatiable consumption patterns and unsustainable resource utilization practices in industrialized or rapidly industrializing countries. Whether nation peoples will be able to survive, often in fragile habitats, will depend in large part on halting or reducing the practices in the industrialized regions of the world that threaten the world's cultural and biological diversity.

THE SHAPE OF THINGS TO COME

As developed countries begin to reconsider seriously the roots of their debts, unemployment, and social malaise and adopt new policies for getting their own houses in order, they will also reduce their overall assistance to other states. Cutting the umbilical cords to elites who dominate third world states will unleash changes similar to those that have carved up the former Soviet Union and many parts of eastern Europe.

In fact, as developed countries shift their focus to their own internal problems, in all likelihood regional conflicts will become even more violent. Conflicts long thought to be dormant are now being rekindled because of the lack of foreign military assistance to keep ruling elites in power. And conflicts do not require sophisticated weaponry; in Rwanda and Guatemala, machetes, knives, and sticks were used to kill hundreds of thousands of people. Further complicating this situation, the number of shooting wars within states is increasing, precisely at a time when arms manufacturers and NATO and Warsaw Pact countries are trying to dump obsolete weapons, and when third world arms producers are seeking to expand sales to subsidize their own weapons needs and to earn foreign exchange.

Such conflicts will continue to spawn large numbers of refugees and displaced people, not to mention untold environmental degradation. Such conflicts will inevitably disrupt food production, making "development" impossible. More important, children will become even more malnourished and the quality—even the very existence—of the education they so sorely need to help them face the next millennium will suffer. This is not a worst-case, doomsday scenario. Considerable evidence already points to an increase in internal, regional conflicts in Africa, Asia, and the Middle East during the 1990s.

The next millennium is upon us. Within a generation, the struggles between nations and states will be decided. If we cannot invent new forms of states, perhaps in the form of confederations that both embrace and reflect diversity, then the earth will lose most of its biological and cultural diversity. In short, we will lose most of the tools needed to solve the problems we will certainly face in the future.

POSTSCRIPT

Since the early 1990s, many developed countries and their development arms (e.g., the U.S. Agency for International Development [USAID], the Dutch Directorate General for International Cooperation [DGIS], the United Kingdom's Department for International Development [DFID]) and multilateral development organizations (e.g., the World Bank, the International Finance Corporation) have written or strengthened policies that guide loans or grants to developing countries that might adversely affect indigenous peoples. These policies have generally improved the performance of internationally funded development projects and investments vis-à-vis indigenous people.

Unfortunately, developed countries no longer dominate trade and investments with most developing countries. China's growth has made it the largest single investor in developing countries. Its investments now represent more than 50% of the combined total of all multilateral agencies, bilateral agencies, and nongovernmental organizations in the world combined. This is a problem for several reasons, but two stand out: (1) China has no equivalent indigenous peoples' policies and in fact has not respected indigenous peoples' rights within its own borders. (2) More important, China's investment strategy is not just to help countries develop their natural resources, but rather to do so in a way that gives it access and also ownership as a joint investor. In effect, China is laying claim to resources and propping up states and those who control them at the same time. As the economies in Brazil, India, Indonesia, and South Africa grow, they, too, are replacing developed countries as the main investors in natural resources around the world. Just when it appeared that the main economic forces on the planet were becoming more enlightened about indigenous peoples, a whole new wave of investors from countries with poor to nonexistent policies regarding indigenous peoples is making their chances of survival slim indeed.

References

AMARU IV. 1980. *The once and future resource managers: A report on the native peoples of Latin America and their roles in modern resource management.* Washington, DC: AMARU IV.

Clay, J. 1991. *World Bank policy on tribal peoples: Application to Africa.* AFTEN Technical Notes, no. 16. Washington, DC: World Bank.

Clay, J. W., and B. K. Holcomb. 1986. *Politics and the Ethiopian famine, 1984–85.* Cultural Survival Report, no. 20. Cambridge, MA: Cultural Survival.

Clay, J. W., S. Steingraber, and P. Niggli. 1988. *The spoils of famine: Ethiopian famine policy and peasant agriculture.* Cultural Survival Report, no. 25. Cambridge, MA: Cultural Survival.

Darnton, J. 1994a. Africa tries democracy, finding hope and peril. *New York Times,* June 21.

Darnton, J. 1994b. In poor, decolonized Africa, bankers are new overlords. *New York Times,* June 20.

Darnton, J. 1994c. "Lost decade" drains Africa's vitality. *New York Times,* June 19.

Goodland, R. J. 1982. *Tribal peoples and economic development.* Washington, DC: World Bank.

Mann, C. C. 2005. *1491: New revelations of the Americas before Columbus.* New York: Knopf.

19

Conservation and Development: The Nam Theun 2 Dam Project in Laos

Thayer Scudder

No form of development is likely to succeed unless it is part of a comprehensive resource use program which takes into account the ecology of the people and the area.
                    —Lee M. Talbot, quoted in M. T. Farvar and J. P. Milton's
                    The Careless Technology: Ecology and International Development

Since 1996 Lee Talbot and I have been members of the World Bank–required, but
independent, Panel of Environmental and Social Experts (POE) for Laos' Nam Theun
2 (NT2) dam project. Tropical forestry expert Timothy Whitmore previously served
on this panel as well, but he tragically died of cancer in 2002. The panel consisted only
of Talbot and myself until conservationist David McDowell joined us in 2005.

NT2 is a large dam (defined as over 15 m in height), the construction of which will
cost over a billion dollars. Such dams are involved in one of the most contentious issues
in development discourse today. While their construction offers major short- and
medium-term benefits in terms of hydropower generation, irrigated food production,
and urban water supplies, it also entails significant long-term, as well as shorter-term,
costs. Indeed, most of the world's 50,000 large dams are unsustainable. They inevitably
fill up with sediment, but the financing, science, and engineering of their decommissioning continue to be ignored during the planning process. Such dams also have caused
the involuntary resettlement of forty to eighty million people, the majority of whom are
not only poor but also further impoverished by their removal (World Commission on
Dams 2000); still larger numbers of people are impoverished below dams, especially
in late-industrializing countries where economies are more dependent on natural flows
(Scudder 2005). Furthermore, dams on main streams and tributaries have had adverse
and irreversible impacts on the world's river basins, with especially severe effects on
deltas and wetlands. In sum, dams represent an outdated development paradigm based
on the "big project" syndrome (Scott 1998), whereby planners manipulate nature on a
grand scale with little understanding of, or concern for, impacts.

Yet, tragically, because of the disjuncture between the health of the world's major
river basins (60% of which have already been dammed) and the basic needs of an

expanding human population in the world's late-industrializing countries, large dams remain a necessary development option for the immediate future. In India, for example, where rainfall is restricted to a few months a year and shallow reservoirs dry up during droughts, dam storage remains necessary for irrigated food production. In China, aquifer depletion in the North China plain, with its hundreds of millions of residents, will require interbasin transfers from the Yangtze River basin. In Laos and Nepal, exportation of hydropower will provide the major source of the foreign exchange that is needed for poverty alleviation (in the NT2 case, the government has set up a separate institution to receive royalties, which are to be used for that purpose).

Talbot and I were aware of these issues when we agreed to join the NT2 panel. Since the 1960s Talbot has been associated with policy-oriented studies and discussions dealing with the future of the Mekong River basin, of which the Nam Theun is a major tributary. During the second half of the 1980s he reviewed the World Bank's handling of the environmental issues associated with the dams it had financed since 1970. As a result of that review, and from his work as coauthor of *Dams and the Environment: Considerations in World Bank Projects* (Dixon, Talbot, and Le Moigne, 1989), Talbot understood the unacceptable costs associated with large dams. He also understood that large dams would continue to be built, especially in late-industrializing countries. More important, he knew that implementation of state-of-the-art guidelines such as those proposed by the World Commission on Dams could render some previous costs unnecessary, for there were ways to mitigate and enhance river basin habitats and to help affected people become beneficiaries rather than losers. That conclusion also led to Whitmore's and my involvement in the NT2 project, which we believed could be a model of global significance if it was executed as planned. All three of us were convinced, however, that if the NT2 project was not carried out as planned, it would increase rather than decrease rural poverty, and it would seriously degrade the globally recognized biodiversity values of the Nam Theun watershed (NT2 POE 2004).

Why did we think that NT2 had the potential for development that was environmentally, economically, institutionally, and culturally sustainable? We had three primary reasons. The first involved the setting of NT2 in Laos and within the Mekong basin, as well as the potential of the project itself to contribute to national development and, more specifically, to poverty alleviation. The second involved the evolving institutional structure, including the POE, for project planning, implementation, monitoring, and evaluation. The third involved the current plans for dealing with environmental and social issues within the NT2 watershed, the reservoir basin, and downstream areas. There are, of course, major risks that those plans will not be implemented. The potential is there, however, and that warrants examining why NT2, as a case history, is so important in spite of the complexity, controversy, and risks involved.

THE PROJECT'S NATIONAL CONTEXT

The Lao Peoples Democratic Republic (commonly called Laos) is one of the more important countries in the world in terms of biological and cultural diversity.

Approximately 60% of the population belongs to a wide range of ethnic minorities and linguistic communities. This population of roughly 5.9 million in 2005 is one of the poorest in the world. With an estimated annual per capita income of $440, Laos' economy is still predominantly rural; approximately 80% of the labor force is employed in agriculture. Social indicators are closer to those in tropical Africa than in much of Asia, with a life expectancy of about fifty-five years and a 69% literacy rate (World Bank/Asian Development Bank 2007).

As in neighboring Vietnam, the government is run by a Communist regime, which has been experimenting with a market economy since its 1986 implementation of the New Economic Mechanism reform program. During the 1990s the export of unprocessed timber was the country's major source of foreign exchange, at the expense of its forest cover. It is hoped that increasing exportation of hydropower from NT2 and other hydro dams may reduce the pressure on forest cover, as will the government's recent ban on the export of logs.

THE PROJECT'S GEOGRAPHIC AREA

The national, multipurpose NT2 project will affect the future of central Laos between the Annamite Mountains, which form the country's border with Vietnam, and the Mekong River, which serves as the border between Laos and Thailand (figure 19.1). The project will involve interbasin transfers and have important impacts on the development and biocultural diversity of two major Mekong River tributaries: the upper Nam Theun and the middle and lower Xe Bang Fai. The key issue for the POE will be the nature of that development and its relationship to biocultural diversity.

The NT2 dam site is located on the edge of an upland plateau immediately below the river's 4,013 km² watershed, which drains the Annamite range. The site itself meets several of the criteria for "good hydro" (Goodland 1996; Ledec, Quintero, and Mejía, 1999), as it is located on the upper reaches of a tributary that has already been dammed below the NT2 site. On the other hand, the completion—scheduled for 2008—of a dam that is 39 m high will create a large and shallow 450 km² reservoir (with an extensive drawdown area) that will inundate approximately 40% of the Nakai Plateau.

Approximately 40 km up the reservoir, water will be diverted down an escarpment, falling 335 m to a power station with a net generating capacity of 1,070 MW. Over 90% of the energy produced will be exported to Thailand, with the remainder to be delivered to the Lao grid. Outflow from the turbines will enter a regulating reservoir from which water will flow down a 27 km channel and waterway to the Xe Bang Fai approximately 80 km above its junction with the Mekong (figure 19.1).

The NT2 Watershed

The 4,013 km² NT2 watershed supports perhaps the most intact contiguous forest in Southeast Asia. It varies from different types of evergreen broadleaf forest to upper montane forest, in which a valuable cypress grows in scattered pockets above 2,000 m.

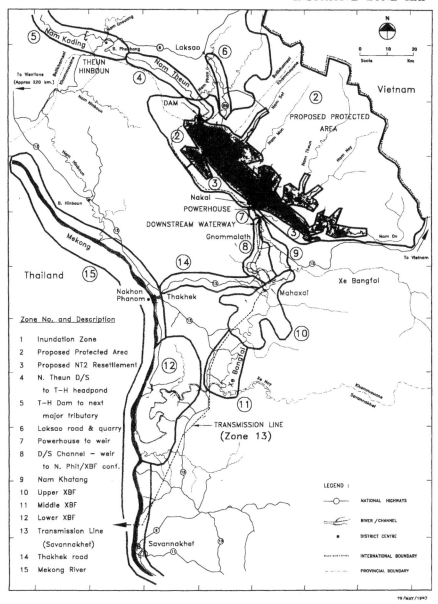

Figure 19.1 The Nam Theun 2 project in Laos. (*Source:* Nam Theun Electricity Consortium 1997.)

Though threatened by commercial hunters from Vietnam, wildlife is still prolific, and the watershed is considered one of the region's most valuable remaining biodiversity areas (IUCN—The World Conservation Union 1997). In addition to at least 430 species of birds, mammals include elephants, tigers, and several recently discovered species; one such species is an endemic bovid, the saola (*Pseudoryx nghetinhensis*), and another is a member of the deer family. In 1993, recognizing the importance of the area, the government incorporated it within its largest national protected area, the Nakai Nam Theun National Protected Area.

At least ten distinct ethnic minorities, totaling up to 6,000 people, live in twenty-six villages within the watershed. The earliest arrivals were Vietic speakers, whose "ethnic diversity represents a critical dimension of biodiversity" (Chamberlain 1997, § 1, 7). Separated by ridges between tributary valleys, they developed different cultures with economies varying from foraging to sedentary cultivation of a variety of crops, including paddy rice. As recently as the early 1970s, three groups of Vietic speakers, each containing fewer than fifty families, were still nomadic foragers with invaluable indigenous knowledge of the upper reaches of the tributaries and associated forests where they lived. In James R. Chamberlain's words, "their intimate relationship between nature and culture" represents "a resource of inestimable value for Laos, a cultural type that is practically extinct in South East Asia, and that is found nowhere else on the planet" (1997, § 1, 8). Tragically, starting in the mid-1970s, the government began rounding these groups up for involuntary settlement in existing communities within and outside the watershed. Mortality rates have been excessive; the groups are threatened with extinction unless they intermarry and return to their former use areas.

The two largest minorities in the watershed area are the Sek and the Brou. The Sek numbered about 600 in the mid-1990s and lived primarily in three villages along the upper reaches of two Nam Theun tributaries. They brought with them a sophisticated system of dry-season rice cultivation, diverting water by small weirs through canals and bamboo piping to their paddies. As with the various communities of Vietic speakers, the Sek economy appears to be in balance with the natural resource base. That is not the case with the Brou, the latest and largest group of immigrants. Numbering over 3,000 individuals, they have established villages along all six watercourses within the watershed, and the basis of their economy is an expansive system of bush fallow (shifting) agriculture. Due to their high rate of population increase within the conservation area, estimated time for a doubling of Brou numbers is less than twenty years. Stabilization of their population and system of agriculture is a major challenge.

The Nakai Plateau

One of the most extensive pine forests in Southeast Asia remains at the upper end of the Nakai Plateau, while mixed deciduous and some riverine forests remain at the lower end. The lower reaches of the catchment hills are also well forested. However, vegetation, wildlife, and the living standards of plateau residents already have been adversely affected by the Vietnam War, postwar land use, and more recent indiscriminate and project-related logging (conducted between 1993 and 1995; thereafter logging outside the reservoir basin and within the watershed was prohibited as a World Bank requirement). As a result, "most of the terrestrial area of the plateau is considered to be substantially modified by human activity and from a biodiversity standpoint substantially degraded from its original status" (NT2 POE 1997, 9). This is a point that critics of NT2 gloss over. Two other commonly ignored issues are the people's isolation and their poverty, estimated per capita income in the Nakai Plateau being only about 60% of the national average.

During the POE's first visit to the plateau in January 1997, we were struck by that poverty, as well as by the scarcity of wildlife and the general degradation of much of the area. We were especially aware of the absence of birds. That reminded me of the civil war–ravaged areas that I had traveled through in Mozambique in 1993.

With the exception of the residents of the town in which Nakai District is head-quartered and one outlying community, the entire population of nearly 6,000 people living in twenty-one villages will require relocation. With few exceptions, they wish to relocate close to the reservoir that will require their removal. This will simplify resettlement, not just because most villages will be able to remain in their customary use areas and spirit territories and benefit from the reservoir, but also because there will be no host population with whom they must compete for land, jobs, and social services.

Aside from a few households, the villagers are indigenous peoples according to World Bank definitions. Though the population can be split "into twenty-eight ethnic sub-groups on the basis of linguistics" (NT2 Electricity Consortium [NTEC] 1997, 4), all have been affected by intermarriage, by the dominant Lao lowland culture, and by war. According to Stephen Sparkes, "One can speak of a Nakai culture, a kind of 'melting-pot' culture consisting of a shared material culture, common socio-religious beliefs and agricultural practices" (1997, 11). That does not mean, however, that there are no distinctive cultural variations within and between communities.

The large majority of the people are impoverished. Agriculture remains the dominant component of each household's diversified economy. Most soils are poor, though, with an estimated cultivation area for the twenty-one villages of only 471 ha—less than 0.10 ha per inhabitant (NTEC 1997, 4–6). Rain-fed rice is the preferred crop; in sixteen villages surveyed by the humanitarian organization CARE in 1996, only 17% of households could produce enough to last a full year, with nearly half having a rice deficiency for over six months. Livestock and nontimber forest products provide the cash or bartering capacity that is needed to acquire rice and other basic needs staples. And aside from a school and a hospital in the district head-quarters, social infrastructure is poor to nonexistent in all villages. Where schools do exist, they are restricted to the lower grades, and morale among the teachers, who are meagerly and infrequently paid, is low.

The Vietnam War caused much of this area's poverty. From 1964 to 1973 the United States dropped over two million tons of ordnance on Laos "during 580,344 bombing missions, or the equivalent of one planeload of bombs every eight minutes around the clock for nine years" (Lao PDR Trust Fund for Clearance of Unexploded Ordnance 1995, 1). Because the Ho Chi Minh Trail traversed the length of the Nakai Plateau, all twenty-one villages would have been under attack from aerial bombardment, with the area around the district headquarters heavily hit. Under such circumstances, reports by villagers that the number of their water buffalo was drastically reduced, with some families losing all their animals, must be taken seriously. Since domestic stock constitutes the villager's bank account—to be drawn upon during times of misfortune such as crop failure, sickness, and death—the war's impoverishing effect was serious.

The Middle and Lower Xe Bang Fai

The middle and lower reaches of the Xe Bang Fai are lowlands that grade into the floodplain of the Mekong River. Aside from Brou living in villages immediately below the escarpment, villagers close to the Xe Bang Fai typically consider themselves to be members of the country's lowland Lao majority. Though the population close to the river probably exceeds 70,000 people, their current lifestyles, including degree of dependence on the river, have been less extensively surveyed than those of the population on the Nakai Plateau and within the NT2 watershed. As elsewhere in the lowlands, the economy is dominated by the cultivation of rain-fed rice, complemented by a slowly increasing amount of dry-season irrigation. Water for this work is gravity fed into one community irrigation project, but lifted for other projects from the Xe Bang Fai by diesel-operated pumps that are being electrified. Where riverbanks are not too steep, flood-recession dry-season gardens are an important source of maize, vegetables, and tobacco. Fish is the major source of protein and is nearly as important a food as rice.

PROJECT POTENTIAL AND EVOLVING INSTITUTIONAL STRUCTURE

Conceived as a hydro project for the export of power to Thailand, NT2 has been slowly evolving not just into Laos' major multipurpose project, but into a magnet for the coordination and integration of other development initiatives. That "magnet" role will be especially important if it results in a major expansion of dry-season irrigation along the Xe Bang Fai. Currently, irrigation in that area suffers from inadequate water supplies and increased pumping costs during the dry season when natural flows are sharply reduced.

The NT2 project is also providing an opportunity to experiment with new institutional mechanisms that can play a role as national pilot projects. For example, the Watershed Management and Protection Authority (WMPA) has been established to take over responsibility for the conservation of the NT2 watershed and the development of its indigenous population. Another example is the creation of the Nakai Plateau Village Forestry Association, which will provide the institutional mechanism for the resettled villages to manage and profit from approximately 5,600 ha of forest resources within the 21,692 ha that have been set aside for resettlement with development purposes. Both associations are supported by prime ministerial decrees. As for project revenues, they are estimated to be US$28–33 million during the first ten years, after which they are expected to rise sharply "to an average of U.S. $73 million from 2020 to 2034" (World Bank 2005, 16). Government arrangements, in the form of a project-specific holding company, have been made to ensure "that NT2 revenues will be used to finance incremental expenditures on poverty reduction and environmental management programs and meet adequate standards of transparency" (16).

The Project Authority and the World Bank

The project authority, Nam Theun 2 Power Company Ltd., is a private sector–government joint venture that will build, own, and operate the project during a twenty-five-year period, after which it will be handed over to the Lao government. The overseeing partner is the parastatal group Electricité de France, which owns a 35% share. A subsidiary of the Electricity Generating Authority of Thailand owns a 25% share, as does Laos' Electricité de Laos, which has the option of increasing its share to 40% after fifteen years. The fourth partner is the Italian-Thai Development Company, which owns 15%. Base project cost is estimated at US$2.5 billion.

Nam Theun 2 Power Company Ltd. has demonstrated its goodwill in regard to implementing social and environment safety net policies in a number of ways. These include a commitment to improving, as opposed to restoring, the livelihoods of all households undergoing resettlement; shifting, at a higher financial cost, the outflow of turbined waters from one tributary system to another to reduce impacts on local villages; halting power generation should wet-season flows in the upper project reaches of the Xe Bang Fai exceed historic flows; financing what are, in my experience, the first downstream transbasin pre-project surveys of an existing fisheries; and providing US$30 million for watershed conservation over a thirty-year period.

The lengthy project planning process was accompanied by considerable uncertainty for all involved, including affected people and the POE, since the private-sector partners had made it clear to the government of Laos in the mid-1990s that their involvement would require provision of a World Bank financial guarantee to cover the risk of government default. The uncertainty did not end until March 31, 2005, when the World Bank's executive board approved US$250 million of guarantees, as well as a US$20 million grant for the Nam Theun 2 Social and Environment Project.

As with other contentious projects, gradually increasing World Bank involvement during the 1990s included a requirement that the Lao government appoint an independent dam safety panel as well as the POE. The World Bank also pioneered activities relevant to the construction of large dams in other countries. It appointed its own international advisory group to provide guidance on how it can improve its handling of environmental and social issues in the hydropower projects it supports worldwide (International Advisory Group, World Bank, 1997,) with NT2 the advisory group's first assignment. The World Bank also increased funding in the NT2 case for standalone projects to help the government deal with social and environmental issues. One would become the Nam Theun 2 Social and Environment Project; another, initiated under a 2004 scoping mission, has become a rural livelihoods project, which includes a portion of the Xe Bang Fai basin. And a pioneering activity initiated by three donors was an agreement that the World Bank, the Asian Development Bank, and the French International Development Agency work together during the later stages of project preparation and during project implementation and monitoring.

The POE and the Government of Laos

The POE was appointed in 1996 by the government of Laos through the Ministry of Industry and Handicrafts (now the Ministry of Energy and Mines). It currently reports to the ministry's Department of Energy Promotion and Development as well as to the standing deputy prime minister. The government selected panel members from a roster of names suggested by the World Bank. Their primary responsibility was to review, and provide guidance on, all planning related to environmental and social issues. During the long planning process the panel's role was advisory; following the commencement of implementation, however, the panel's responsibility was significantly broadened to include deciding when, or if, environmental and social conditionalities in legal agreements have been met and, if not met, to specify as binding requirements what additional activities are necessary before the reservoir can be filled and, thereafter, the resettlement process completed.

Since the end of 1996, panel members have made fourteen trips to Laos, submitted twelve reports and one interim report, and contributed to three other reports. Prior to each visit, the panel and the government discuss terms of reference. The panel, because of its independence, may also look into any other issues that it considers of relevance. After being reviewed by the government, all panel reports become public reports—usually within a three-month period.

Since 1997, the panel has systematically covered the entire project area from the Vietnamese border to the Thai border: walking, rafting, and camping along the Nam Theun and its tributaries above the dam site; walking the gorge below the dam site to the first incoming tributary; and rafting the middle and lower reaches of the Xe Bang Fai. In addition, the panel has flown over the entire project area by helicopter and driven through it where roads exist.

Working closely with all parties, the panel has played an important role in critiquing various revisions of relevant documentation. This documentation includes the three-volume "Environment Assessment and Management Plan," the three-volume "Social Development Plan," the two-volume "Social and Environment Management Framework and 1st Operational Plan for the NT2 Watershed," the "Xe Bang Fai Strategy Paper," and the four-volume "NT2 Downstream Restoration Program, Phase 1," as well as separate reports on environmental, fisheries, health, wildlife, and broader socioeconomic issues.

Specific issues on which the panel has been able to influence policy cover a wide range of topics relevant not just to environmental and social issues, but also to the wider experience of panel members with large-scale development projects. For example, beginning early in the project, the panel successfully emphasized in its reports and meetings the importance of involving the Office of the Prime Minister in NT2, so that there would be an incentive for other government ministries and agencies to work with the Ministry of Energy and Mines. The panel also arranged the first visit with the central government's Department of Irrigation and, early on, emphasized the importance of making the project a multipurpose one.

Another panel recommendation led to a program of annual pre- and post-project completion fishery surveys along the Xe Bang Fai that was started in 2001 and will document what impacts actually occur. In report after report, the members of the panel have applied to the management of the NT2 watershed their international experience as it relates to patrolling, substituting narrow tracks for roads, and involving local communities. The panel has also indicated the need for researching and incorporating the indigenous knowledge of these communities and for implementing policies that allow previously resettled Vietic speakers to return to their customary use areas, if they so wish.

Environmental and Social Planning

Planning for the NT2 project has involved the World Bank since the 1980s; in the project's early stages, the group advised the Lao government to build it on a "build, own, operate, and transfer" basis so as to avoid acquiring a heavy debt load. An initial feasibility study identified NT2 as Laos' most viable hydro project based on technical and economic criteria. However, it did not deal in detail with necessary environmental and social issues.

The subsequent decision of the government to proceed with NT2 before considering environmental and social issues was a mistake. It gave project critics an opportunity to challenge later studies, no matter how competent, as mere justifications for a decision already made. It also put the World Bank Group in an awkward position, because the group had provided earlier funding through the government of Laos for the initial feasibility study. These problems help explain the extreme caution with which the World Bank has proceeded since 1993, as well as its much delayed decision to proceed to project appraisal and project implementation.

Environmental and social issues came to the project's forefront in 1995 when the private-sector firms and the government of Laos asked the World Bank to provide a partial-risk guarantee. A World Bank technical mission visited Laos during November of that year, outlining in its report the terms of reference for various studies dealing with consultation at district, provincial, and national levels and with economic, environmental, and social issues. During the visit, World Bank officials recognized that the project would involve major environmental and social impacts that had yet to be adequately addressed. An ensuing decision led to the POE's creation in 1996.

Also during 1996, the government of Laos contracted three major studies with the help of World Bank funding. One was a study of alternatives for electricity generation; it concluded that NT2 still ranked higher than other hydro and thermal projects, even with environmental and social impacts included. The second study produced a favorable analysis of the economic impacts of the NT2 project. The third was a broad study for formulating an environmental and social management plan for the Nam Theun watershed as well as for a corridor that would connect the Nakai Plateau, via the dam site, with a conservation area to the southwest. That study was contracted to IUCN and the Wildlife Conservation Society. At the same time, Nam Theun 2 Power Company Ltd. funded two major studies of its own. One dealt with reservoir basin resettlement and involved company staff and consultants working with government officials. The other was for an outsourced environmental assessment and management plan.

Due to satisfactory progress with overall NT2 planning, in October 1997 the World Bank decided to proceed from its project identification stage to its project preparation stage. That required new government initiatives, such as creation of the Lao Holding State Enterprise for project revenues. The preparation stage also required further elaboration of Nam Theun 2 Power Company Ltd.'s evolving resettlement action plan and environmental assessment and management plan, as well as government planning for the management of the watershed.

Utilizing a small World Bank grant that included funds for consultation, an operational plan for a watershed authority was submitted through the government's Ministry of Agriculture and Forestry in May 2000 along with a work program and budget. World Bank funding also allowed IUCN to carry out an innovative program in the conservation area from May 1998 through July 1999; it concentrated on conservation and economic development in three pilot villages located on three of the six tributaries. Preliminary results were encouraging, including development of cooperative patrols to conserve biodiversity, with special emphasis on transboundary poaching and nonsustainable collection of nontimber forest products (NT2 POE 2001). Since 1999, unfortunately, funding gaps have disrupted continuity, slowed momentum, and tempered initial village enthusiasm.

The NT2 Watershed

The POE's March 2004 report reflects panel perspective on the planning under way—planning that focuses on the role of the newly established WMPA. The panel applauds a statement in the main planning document for the watershed indicating that conservation is its main priority. Accomplishing that priority will not be easy, though, given ongoing and unsustainable exploitation of wildlife, including uncontrolled poaching from Vietnam. It is absolutely essential that the WMPA implement "a comprehensive patrolling system with a strong central authority responsible for monitoring, coordinating and directing the system" (NT2 POE 2004, 19). Such a recommendation is based on experience in other protected areas "that the density of guards and patrols is by far the most effective determinant of long term conservation success" (19).

The patrolling system must involve both village-based patrols and Lao military patrols. The latter are essential for a number of reasons. Because villagers must also allot time to a variety of livelihood activities, they cannot be relied upon to adequately patrol areas distant from their villages, nor will they have the capacity to deal with possible encounters with heavily armed poaching gangs. However, while military personnel can be effective guards of biodiversity, they can also pose severe threats to it. Consequently there is a need to develop strong conservation leadership and motivation in the various types of military personnel. Effective monitoring of patrolling is essential.

To realize conservation goals, the WMPA must additionally control access by limiting it to trails, tracks, and river and stream crossings designed to allow passage of two-wheeled tractors with trailers but not four-wheeled vehicles. As for activities to improve the living standards of the village population within the watershed, conservation and development activities must be carefully balanced and well integrated.

The Nakai Plateau

The evolving resettlement action plan for the Nakai Plateau is the best such plan that that I have seen for two reasons. Not only does the plan require that all affected households must be project beneficiaries, but it also includes a feasible range of development options for realizing that goal. If implemented as required in the project's concession agreement and other legal documents, the plan has the potential to help the large majority of the population to raise significantly its living standards. The word "significantly" is important. The people's current poverty is so great that implementation of even a mediocre plan could reduce their impoverishment. But they would still be poor. The intent of the resettlement action plan is to help those involved end their poverty.

Villagers were actively involved in planning the resettlement process. Anthropologist Stephen Sparkes (1997) asked separate groups of men and women to map their dream villages. A range of gender-sensitive household livelihood options have been developed; these take into consideration both income generation and household labor resources. Each option is based on a "high degree of diversification and flexibility . . . to allow for changes over time" (NTEC 1997, E-8). Seven major components are involved, none of which is required to generate more than 20% of a household's income.

In addition to cultivation of rice, fruits, and vegetables, the action plan provides for livestock management, wage labor, fishing, collection of nontimber forest products, and participation in the management of community pine and deciduous forests. The reservoir will also be utilized as a resource. Because its average depth is less than 10 m at full storage level, an extensive drawdown area will be available for recession cultivation and grazing. According to the action plan, fishing is to be restricted to members of the resettled communities. As for fish processing and marketing, the POE has recommended that resettlers dominate those activities rather than being restricted to a single concessionaire, as has been the case elsewhere in Laos. Should the various livelihood components be implemented as planned, total family income could be expected to exceed the country's poverty datum line by a factor of almost two.

An all-weather road will be constructed through the resettlement area, which is currently served by tracks (significant portions of which are impassable during the rainy season). Electricity will be available, and a piped water supply, along with wells, will be provided within each village.

The Middle and Lower Xe Bang Fai

The best approach for mitigating project impacts and improving livelihoods in this area is both hydrological and socioeconomic. For mitigation, turbines will be shut down when water levels reach a predetermined height in the river's middle zone so as to avoid increased flooding of irrigated floodplains. In the irrigated areas of the delta the backwater effect of the Mekong will be somewhat lessened due to reduced Mekong flows from the Nam Theun. At an increase in project cost, another mitigation measure has been to select a less densely settled route for channeling the turbined water from the regulating reservoir to the Xe Bang Fai.

If affected communities are to benefit—as the POE believes is essential—it will be necessary to use turbined water to increase significantly the area under dry-season irrigation with increased emphasis on higher-value cash crops. If implemented, existing plans for the use of project water and electricity have that potential in all project zones. Between the power station and the Xe Bang Fai, one-fourth of the 27 km channel will be sufficiently raised above ground level to allow double cropping in the adjacent community irrigation project. Though designed to irrigate 2,400 ha, during the 2000–2001 dry season only 335 ha were cultivated. Outflows from the regulating reservoir adjacent to the power station are also designed to release 15 m³/s into another Xe Bang Fai tributary, which can be used for gravity flow irrigation, fishing, and domestic purposes by twenty riverine villages.

Within the Xe Bang Fai, falling natural regime water levels during the dry season are a major constraint to an estimated dry-season irrigation potential of over 20,000 ha. Even the 3,000 ha that have been developed to date are adversely affected by low water levels, which increase pumping costs and may stop pump operations entirely. Greatly increased project flows during the dry season will not only reduce those pumping costs but also provide the opportunity to increase greatly the area under dry-season irrigation. The Department of Irrigation has already drawn up plans to use turbined waters for that purpose. Both above and below the highway that separates the middle and lower reaches of the Xe Bang Fai, increased electrification of pumps can be expected to follow the ongoing electrification of the area by the World Bank–funded Southern Provinces Rural Electrification Project. The most productive and extensive irrigable land is adjacent to the Mekong River. Within the backwater zone when the Mekong floods, it is enclosed within an inadequate levee system. Should current discussions lead to its upgrading and to installation of sufficient two-way floodgates, significant increases in the productivity of fisheries and the dry-season irrigation of rice and higher-value crops could be expected.

The POE has emphasized that it is very much in the interests of Nam Theun 2 Power Company Ltd. and the government to encourage integration of its plans into an overall regional development plan for the entire Xe Bang Fai basin. Such a plan should also combine aquaculture with irrigation along lines proposed by the Department of Fisheries and address the fishery implications of the new Xe Bang Fai regime. With the potential to raise living standards, such an approach could address what otherwise could be some of the most serious costs of the NT2 project, as well as provide a model for the integrated development of Laos' five other major river basins.

PROJECT RISKS

Unless the plans previously described are adequately implemented, the NT2 project has the potential to be a disaster for the environment of the Nakai Plateau and the Nam Theun watershed, as well as for affected communities from the Vietnam border to the Mekong River. Of various risks, two continue to be emphasized by critics. One is lack of political will on the part of the government; the other is lack of implementation capacity. Another, emphasized by the POE in previous reports but no longer a threat, was World Bank withdrawal from the project.

The Previous Risk of World Bank Withdrawal

Should the World Bank Group have withdrawn from involvement in the NT2 project, the POE doubted that the government of Laos would have canceled the project, for that would have required defaulting on its commitment to provide 3,000 MW to Thailand as well as losing what is likely to be the government's largest single source of foreign exchange. More likely, the government would have requested help from a friendly government—China, Korea, and Vietnam being likely choices—to proceed with a less expensive project, without the environmental and social safeguards and plans that currently exist.

While strongly supportive of World Bank Group involvement, the POE has been critical of the length of time that it took the group to move from project identification to project approval. As stated in its January 2001 report, "The POE is aware of no other World Bank-assisted hydropower project where a member country has been asked to meet such a large, indeed escalating, number of requirements over such an extended period of time" (34).

Political Will

With large dam projects, one never knows whether governments and project authorities have the necessary political will until environmental and social plans are actually implemented. Though some critics believe that the Communist government of Laos has no intention of implementing the NT2 project as planned, history shows that how affected communities are dealt with does not appear to be influenced by form of government. The records of Canada and the United States, for example, are poor in regard to planning and implementation of large dams requiring the resettlement of ethnic minorities. On the other hand, two of the few well-planned and implemented projects are associated with a socialist government (Egypt's Aswan High Dam resettlement, in which Nubian resettlers were provided with their own large-scale irrigation project) and a Communist one (China's Shuikou Dam resettlement).

In the NT2 case, of more concern are a number of past and present government policies which, if applied in the Nam Theun project area, would run counter to current plans. One such policy relates to past attempts to resettle ethnic minorities from the uplands to the lowlands, in some cases incorporating them within larger Lao host communities. Few such relocations, if any, can be considered successful from the point of view of the resettling communities (see especially Baird and Shoemaker 2005; Goudineau 1997). Another policy that could impoverish a majority of those living in the NT2 watershed is a still-standing government commitment to eliminate shifting cultivation, rather than to work with communities to capitalize on the strengths of rotational shifting cultivation (which can actually increase biodiversity) and stop the predatory form.

Top-down government control of rural land, water, and other natural resources is another worrisome policy, since it has interfered with the ability of the government to implement World Bank–financed community forestry projects. That inability raises questions as to the government's political will not just to implement such projects within the context of the resettlement action plan for the Nakai Plateau, but also to pursue a strong conservation policy in the watershed and allow local villagers to play a major role in the WMPA.

Capacity

Since we believe that the government and the project consortium are committed to implement existing plans, the most serious risk of NT2 is apt to be inadequate implementation capacity. The most serious capacity problem relates to the WMPA, which must quickly become able to control the currently unsustainable transboundary poaching of wildlife, illegal cutting of rosewood, and overharvesting of nontimber forest products.

Because of the active involvement of Nam Theun 2 Power Company Ltd. and the nature of the resettlement program, staffing capacity on the Nakai Plateau, while less serious than with the WMPA, will still stretch government and company resources. The greatest initial risk involves failure to synchronize the resettlement schedule with the construction schedule. To reduce that risk, the resettlement of all affected villages must be satisfactorily completed before the dam is closed in May–June 2008. Satisfactory resettlement at the time of dam closure involves more than the physical shifting of all households from one village site to another. Minimal requirements include habitable houses for all, adequate water supply and sanitation (no exceptions allowed), available schools and medical services, and forward movement on the various livelihood development components of the resettlement action plan.

Development along the Xe Bang Fai will be largely the responsibility of the staff of the various districts working with provincial and central government Department of Fisheries and Department of Irrigation personnel and with international donor agencies whose help is essential. The government's ongoing prohibition against local and independent environmental and community development nongovernmental organizations is unfortunate since they could play an important role, as in other countries, in capacity-building activities.

In addition to staffing, another capacity problem is coordination of the various government and other agencies involved in the sustainable development of the project areas. The NT2 project falls under a single ministry. In the POE's experience, responsibility for such a complex project with major regional and national development implications is "too much for a single ministry" (NT2 POE 1997, 7). The active involvement of the standing deputy prime minister in NT2 planning and his ongoing interest in the project, as well as the interest of the prime minister, may enable the minister of energy and mines to obtain the necessary cooperation from other ministries and agencies through a central working committee that he chairs. Because that remains to be seen, the panel has recommended that the effectiveness of such an institutional setup should be periodically reevaluated. Project components such as the resettlement action plan, the environmental assessment and management plan, the NT2 downstream restoration program, and the WMPA all have their own coordination problems with other government agencies. The WMPA's effectiveness, for example, will depend in good part on its ability to work with Lao security forces in patrolling the international border and the upper portion of the watershed and with Nakai Plateau resettlers and district staff.

CONCLUSION

As previously mentioned, the NT2 project has the potential to meet its regional and national development goals—including raising the living standards of affected people in the Nam Theun catchment, on the Nakai Plateau, and contributing to the integrated development of the Xe Bang Fai basin—along with the potential to meet its environmental goals. The planning process has been state-of-the-art. Implementation, however, is key, and because it has just started, its final outcome remains to be seen.

The historical record dealing with environmental and social impacts of large dams offers little ground for optimism. Available evidence does show an improvement in environmental and social planning, at least in projects assisted by the World Bank and regional banks. There is scant evidence, however, that improved planning has yet to be associated with improved outcomes. While no large-scale statistical analyses of resettlement outcomes have been carried out, the conclusion of the most detailed (involving forty-six cases for which adequate information for analysis was available) was that there were "no significant changes in implementation outcomes" during "three time periods between 1932 and the present" (Scudder 2005, 62). Environmental impacts, according to the knowledge base of the World Commission on Dams, "are also more negative than positive" (2000, 93).

While adverse environmental impacts to mainstream river systems, and to deltas in particular, are likely to be irreversible, a few replicable cases are known in which the living standards of a majority have improved following their involuntary resettlement. In those cases good planning was satisfactorily implemented. The NT2 project also has that potential, as well as the potential to conserve the biocultural diversity of the project watershed (which would otherwise be at risk) and, at least, to mitigate downstream impacts along the Xe Bang Fai. Realization of that potential is important not just for Laos but globally. For failure to implement what are, to date, the most carefully planned social and environmental safeguard policies for such a project will undercut the rationale behind the 2004 decision of the World Bank to recommence assisting with the construction of what it calls "high-risk/high-reward" large dams.

References

Baird, I. G., and B. Shoemaker. 2005. *Aiding or abetting: Internal resettlement and international aid agencies in the Lao PDR.* Toronto: Probe International.

Chamberlain, J. R. 1997. *Nature and culture in the Nakai–Nam Theun conservation area.* Vientiane, Laos: privately printed.

Dixon, J. A., L. M. Talbot, and G. J.-M. Le Moigne. 1989. *Dams and the environment: Considerations in World Bank projects.* World Bank Technical Paper, no. 110. Washington, DC: World Bank.

Farvar, M. T., and Milton, J. P., eds. 1972. *The careless technology: Ecology and international development.* Garden City, NY: Natural History Press.

Goodland, R. 1996. The environmental sustainability challenge for the hydro industry. *Hydropower and Dams* 3(1): 37–42.

Goudineau, Y., ed. 1997. *Resettlement and social characteristics of new villages: Basic needs for resettled communities in the Lao PDR.* ORSTOM survey, 2 vols. Vientiane, Laos: UN Development Programme.

International Advisory Group, World Bank. 1997. *Terms of reference, appendix 2: World Bank's handling of social and environmental issues in the proposed Nam Theun 2 hydropower project in Lao PDR.* Report of the International Advisory Group. Washington, DC: World Bank.

IUCN—The World Conservation Union. 1997. *Environment and social management plan for Nakai–Nam Theun catchment and corridor areas.* Gland: IUCN.

Lao PDR Trust Fund for Clearance of Unexploded Ordinance. 1995. Vientiane, Laos: Government of the Lao PDR, UN Development Programme, and UNICEF.

Ledec, G., J. D. Quintero, and M. C. Mejía. 1999. *Good dams and bad dams: Environmental and social criteria for choosing hydroelectric project sites.* Washington, DC: World Bank.

Nam Theun 2 Electricity Consortium. 1997. *Nam Theun 2 hydroelectric project: Resettlement action plan.* Vientiane, Laos: NTEC.

Nam Theun 2 Panel of Environmental and Social Experts. 1997. *Report of the international environmental and social panel of experts.* Vientiane, Laos: NT2 POE.

Nam Theun 2 Panel of Environmental and Social Experts. 2001. *Report of the international environmental and social panel of experts.* Vientiane, Laos: NT2 POE.

Nam Theun 2 Panel of Environmental and Social Experts. 2004. *Report of the international environmental and social panel of experts.* Vientiane, Laos: NT2 POE.

Scott, J. C. 1998. *Seeing like a state: How certain schemes to improve the human condition have failed.* New Haven, CT: Yale University Press.

Scudder, T. 2005. *The future of large dams: Dealing with social, environmental, institutional, and political costs.* London: Earthscan.

Sparkes, S. 1997. *Observations relating to the resettlement of people on the Nakai plateau.* 2nd ed. Vientiane, Laos: NTEC.

World Bank. 2005. *Project appraisal document for the Nam Theun 2 hydroelectric project.* Washington, DC: World Bank.

World Bank/Asian Development Bank. 2007. *Update on the Lao PDR Nam Theun 2 (NT2) hydroelectric project.* Washington, DC: World Bank; Manila: Asian Development Bank.

World Commission on Dams. 2000. *Dams and development: A new framework for decision-making.* The Report of the World Commission on Dams. London: Earthscan.

PART VI

ENVIRONMENTAL POLICY
AND MANAGEMENT

H. Paige Tucker

The authors of part VI review the development of environmental policy and management in chapters covering (1) ethics and the natural environment, (2) international environmental law, and (3) forest practices in the United States and around the world. All three authors discuss the rise of environmental awareness, acknowledging that many current policies and management practices are still in the early stages of development and implementation. The authors also stress the importance of continued scientific involvement and coordination with decision-making processes, the role and need for leadership, the need for environmental education and outreach efforts, the need for better understanding and awareness of social and cultural norms within the policy process, and the need for long-term planning and vision.

In chapter 20, Stephen R. Kellert addresses the contemporary environmental crisis as one that requires fundamental societal shifts of values and ethical relations to the natural world. He begins with reference to Lynn White Jr.'s "The Historical Roots of Our Ecologic Crisis" (1967), which portrayed the roots of the contemporary environmental crisis as tied to societal cultures and values. Kellert discusses the important and progressive changes that have occurred in perceptions of nature during the past half century, changes that have resulted in improved stewardship of aspects of the natural world. However, he stresses that these changes are limited and selective and will therefore be insufficient over the long term. Kellert points to the reliance of society on expansion of scientific knowledge, new technologies, and legal and regulatory controls to address environmental issues. He asserts, though, that few major environmental challenges have been resolved and that many others appear to be demonstrably worse, such as the rate of species extinction.

Kellert suggests that a biocultural perspective be taken in response. This response views human values and ethical relations toward the natural world as bounded by species biological requirements, but shaped and influenced by individual and cultural learning and experience. This implies the need for an expanded instrumental environmental ethic that hinges on relating values such as material utility, aesthetics, emotional connectivity, spirituality, and human biology to human free will in constructing society and cultural norms.

Kellert uses the concept of biophilia—a complex of weak biological tendencies to value nature that contributes to human physical and mental well-being—as a biocultural construct that is dependent on learning and experience of individuals and groups, and is expressed in both adaptive and maladaptive manners. He identifies nine biophilic values or inherent tendencies, and examines five of them briefly: utilitarian, aesthetic, scientific, humanistic, and moralistic. A table providing definition and function of all nine values is included for the reader.

Also included is a brief illustration of how significant changes in the nine values may lead to profound shifts in ethics and policy making related to the natural world. Kellert uses the decline of cetacean species to illustrate three points: expanded appreciation for nature can foster equally radical changes in ethical and moral relations; changes in values and ethics can trigger pronounced shifts in legal and regulatory policies; and a transformation in values, ethics, and policies can occur in a shorter time period than often presumed, which suggests the practical significance of an ethical strategy as a basis for advancing significant environmental change.

In chapter 21, Nicholas A. Robinson articulates the evolution of environmental laws and examines current international environmental law. He traces the increase and maturation of the science of ecology and the influence of science on policy processes; he also stresses the current fledgling stage of international environmental policy. For example, many people fail to understand that the field is still in its infancy, with some nations having barely begun to implement the norms of the created environmental policies. As a result, many see the norms of international environmental law as unresponsive to the earth's current environmental crises. Robinson stresses that a lack of leadership must be addressed if we are to improve the situation.

The chapter begins with a review of the evolution of international environmental law, starting in the early 1970s with worldwide recognition of a myriad of threats to the environment. However, the enacting of statutes and treaties alone did not stem the widening patterns of environmental degradation. Many crises remain. Robinson's purpose here is to provide the foundation for new worldwide patterns of environmental stewardship, and he sketches an emerging global framework of international laws and norms that individual nations need to apply.

To this end, Robinson traces the primary sources of international law: (1) general principles of law, (2) customary law and conventional or treaty law, (3) analysis of legal experts who elaborate the principles, and (4) customs or treaty obligation through their commentaries. Robinson takes the reader through the important aspects of the topic by reviewing the following: the emergence of the system of international environmental law and policy, the key institutions involved, the legal principles of global application, the use of a case study reflecting how these principles are made operational for environmental concerns, and the norms for implementing international environmental law nationally and regionally.

The case study considers the ecosystem management approach for land stewardship and is organized around biomes and how to sustain their continuing natural functions. Robinson reviews the complexities of political and socioeconomic inertia working against implementation of reforms. This is a problem that is widespread for many nations across the planet. He discusses grasslands, arid landscapes, and watershed and soil conservation issues through the idea of best management practices. He suggests that there is one principal means by which nations can observe and implement provisions of international environmental law on a decision-by-decision basis—that is, the technique of environmental impact assessments (EIAs).

Robinson explores the development of EIAs in the United States in the late 1960s, the spread of utilization to other nations by the mid-1980s, and EIAs' purpose, use, and benefits.

He stresses the EIA as an essential element of sustainable development for implementation of best management approaches, use of the precautionary principle, evaluation of competing resources uses, identification of alternatives, and preservation of ecosystem services. He discusses two types of environmental assessment methodologies, EIAs and strategic environmental assessments, and their occurrence at differing stages of planning and development procedures.

Robinson concludes with a discussion of means to streamline the complexities of international environmental legal norms by recodification of the norms on a regional basis. By doing so he suggests that guidance can then be focused upon, understood, and adapted to regional geographic conditions. He cites the current efforts in the European Union and those in the Association of Southeast Asian Nations as examples of this process in action.

The chapter review of the international legal norms on land stewardship reveals not only a sound body of international rules and recommendations, but also a need to match the efforts of development of norms with the effort to explain the norms to national constituencies and secure their implementation. Greater focus is needed on the methodologies and tools needed to accomplish this. Institutional and political support is critical to the future expansion of international environmental policy.

In chapter 22, V. Alaric Sample characterizes sustainability as a central and traditional concern within the science and management practices of forestry. He notes that what has changed over time, and led to "sustainable forest management" as defined today, is the range and diversity of resources that forests are seen to represent and the values society places on those resources. Social and economic acceptability of scientific uncertainties and understanding of sustainability in practice are considered ongoing challenges for the future. Sample stresses the need for understanding of the differences between the notions of sustainable yield of timber and the sustainability of all the components and processes necessary to maintain a long-term productive ecosystem. These are not necessarily the same.

This chapter summarizes the historical evolution of sustainability in forest management and the concept of a "regulated forest." Sample then examines the means by which natural resource policy developed in the United States, the influence of the European concept of sustained yield, and the promotion of forest conservation practices, and finally he explores additional possibilities for policy development to achieve sustainable forest management both on local levels and in a global context.

Sample traces the development of polices in the United States from the reservation approach of the Creative Act of 1891 through custodial management practices until the mid-1940s. He describes the depletion of forest reserves during and following World War I and World War II, to the rise of the practices of multiple use and sustained yields. The Multiple-Use Sustained-Yield Act of 1960 is cited as an important turning point in the interpretation of the responsibilities of management for sustainable forest practices. The National Forest Management Act of 1976 is mentioned as another important shift for forestry policy and management within the United States. Inclusion of public comment periods and development of detailed interdisciplinary management plans were an effort toward addressing long-term planning and social concerns in a more detailed and inclusive manner.

One of the most dramatic changes in forest policy noted by Sample is the increasing concern regarding forestry management and scientific understanding on a global scale. In the last decade issues such as the biodiversity crisis, mitigation of global climate change, and deforestation have been considered and debated worldwide. More than at any other time in history, forests are recognized as essential to the long-term well-being of local populations, national economies, and the earth's biosphere. Accordingly, Sample examines the emerging precepts, standards, and certifications that are forming the basis for sustainable forestry management on local, national, and international levels.

Sample concludes with a discussion of a systematic approach to addressing forest management and the need for a shared vision. To accomplish this, he stresses the need for reinforcing the scientific foundation of forest management as well as the need for cooperation between the scientific and management communities on local, national, and international levels. Sample suggests that numerous opportunities exist to create a policy framework, or set of guidelines, that enables and encourages public and private forest land managers to make rational choices that will tend to be consistent with and supportive of identified policy needs. He notes that from a policy-making and operational management perspective, the future sustainability challenge will be to protect the long-term productivity of forest ecosystems without unduly limiting the utilization of forests to meet the current needs of society and to form a biological, social, and economic understanding of the processes.

Reference

White, L., Jr. 1967. The historical roots of our ecologic crisis. *Science* 155(3767): 1203–1212.

20

A Biocultural Basis for an Ethic toward the Natural Environment

Stephen R. Kellert

The historian Lynn White Jr. (1967) in his seminal publication nearly a half century ago suggested that the roots of the contemporary environmental crisis lie in our values and culture, which largely support assumptions such as: (1) a fundamental difference separating humans from nature; (2) humans as inherently superior to the natural world, and (3) people having the right to exercise control over the natural world relatively unrestrained by an ethical concern for the rights of natural objects, although bound by moral obligations to treat nature well to the extent that it affects ethical relations among people. He concluded that the resolution of the current scale of environmental destruction would necessitate a basic change in our values and ethics toward the natural world.

Many critics have taken exception to White's various claims, citing more benign and conservation-oriented traditions in Western culture and in modern society (Farley 2002; Moncrief 1970; Passmore 1974). Still, I believe his thesis is mainly correct and an appropriate diagnosis of the current challenge—that is, we will not effectively resolve the scale of our contemporary environmental crisis until we have fundamentally altered our values and ethical relations to the natural world. Clearly, important and progressive change has occurred in perceptions of nature during the past half century, resulting in improved treatment and stewardship of the natural environment. I believe these changes, however, continue to be limited, selective, and insufficiently effective and comprehensive. Moreover, our society continues to rely mainly on expansion in scientific knowledge, new management technologies, and legal and regulatory controls to address our environmental problems. Despite impressive gains and improvements, the overall inadequacy of this approach is demonstrated by the fact that few major environmental challenges have been resolved and, indeed, many appear to be demonstrably worse. More than ever I would support White's diagnosis that a basic shift in values and ethical relations to the natural world will be required not only to mitigate and avoid various environmental woes but also, just as important, to enhance our physical and mental well-being contingent upon the quality of our relational ties to the natural world. Lest this vision seem hopelessly unrealistic, I would argue that such a change is not only possible but practical, and can occur far more rapidly than generally assumed, as I hope to illustrate later in this chapter.

I want to suggest, however, that the emergence of values that give rise to and sustain a new ethic toward the natural world will need to be based on a greatly expanded understanding of human self-interest and biological dependence on the natural world, although recognizing the particular human capacity to exercise choice and free will in choosing our values and ethical relations. By contrast, I believe the articulation of an environmental ethic that relies on assumptions of a rigid biological and narrow utilitarian dependence on the natural world, or an infinite capacity to construct seemingly right relationships to nature, will ultimately be unsustainable, unconvincing, and, worse, misguided. What is offered here is a biocultural perspective, one that views human values and ethical relations toward the natural world as bounded by the biological requirements of our species but greatly shaped, influenced, and mediated by individual and cultural learning and experience.

To elucidate this position, I need to place the biocultural perspective within the context of two basic ethical arguments or positions—although not being trained as an environmental ethicist you will need to be tolerant of my crude delineations of these positions. The biocultural perspective of environmental values and ethics advanced in this chapter falls roughly within the so-called utilitarian or instrumental ethical viewpoint, although in a somewhat new and greatly expanded sense. The utilitarian perspective generally argues that an environmental action is morally just or right if it contributes to the greatest good for people now and into the future. Thus environmental objects or subjects, rather than being moral ends in and of themselves, are the means to a human end (e.g., just or ethical treatment advances human justice, goodness, fulfillment, happiness, physical and mental well-being, and so on). By contrast, a so-called rights-based or biocentric ethic suggests that environmental objects or subjects are a moral end in themselves, possessing intrinsic or inherent value independent of how they may or may not advance human interest, benefit, or well-being. This position regards nature as morally valuable simply because it exists as the recipient of our love and affection, appreciation of its beauty, or the spiritual qualities it may evoke independent of its utility.

A utilitarian or instrumental environmental ethic has often been associated with harm and injury to the natural world that diminishes human material security and physical well-being. Extinguishing a species or causing pollution is viewed as morally reprehensible because it damages the utility that might be derived from, for example, eventually exploiting the biogeochemistry inherent in any biotic form, or because it potentially inflicts injury to human health often among the most vulnerable, such as children and the poor.

I regard this depiction of utilitarianism as more precisely narrow utilitarianism that may be useful but ultimately is an inadequate basis for advancing a meaningful, accurate, and relevant environmental ethic for several reasons. First, most species and habitats currently do not, and probably never will, generate much material advantage, and environmental pollution (with the exception of global climate change) affects only a relatively small percentage of people and typically can be remedied through technical rather than ethical means. Second, people can, in most circumstances, advance an equally compelling narrow utilitarian ethic that argues for the elimination of a species or the occurrence of some form of pollution to protect the needs of people and society. Third, the seeds of destruction may be sown in any ethical calculus that promotes the value of only a fraction of the natural world, implying the expendability of the rest depending on compelling human circumstances. Fourth, a narrow utilitarian ethic offers only a partial and inadequate understanding of human biological dependence on the natural world for advancing human physical and mental well-being.

Before elucidating this last point, which is the basis for the ethic advocated in this chapter, I want to briefly indicate why I believe an intrinsic rights or biocentric ethic is also equally flawed in generating a convincing and pragmatic environmental ethic. The basic problem of a biocentric position in extending moral worth and standing to the natural world is that it offers little practical guidance and convinces few. For example, how does one from this perspective choose between species, or more important between human welfare and the well-being of species or nature? A biocentric position offers limited assistance in situations where an ethic is most needed—that is, not the choice between good versus bad but between competing and compelling goods. Additionally, a biocentric ethic, by being indifferent to the preferences and needs of most people, convinces few and is politically unrealistic and untenable. I do not want to deny the possibility that in an ideal future an enlightened humanity could be swayed by a biocentric or rights-based environmental ethic. But for the moment I wish to defer its consideration for several reasons, including:

1. A biocentric ethic is difficult to demonstrate and prove.
2. This perspective possesses only a limited ability to convince the environmentally ambivalent and/or noncommitted.
3. This ethical position is marginally practical or politically relevant in rendering difficult policy choices.
4. There is so much more to learn about the human tendency to value nature and how it contributes to human physical and mental well-being.

This last point brings us back to the biocultural position advanced in this chapter, based on a greatly expanded understanding of human biological self-interest influenced and mediated by culture, learning, and experience. This perspective can be viewed as a sort of environmental ethic of the middle way, lodged somewhere between a narrow utilitarianism and a rights-based biocentric position, although I recognize that in these matters one cannot be a little pregnant. Still, I want to suggest that a broad utilitarian-based ethic can encompass many of the arguments traditionally advanced to rationalize and support a biocentric environmental ethic, such as nature defended as a source of affection, love, beauty, and spiritual inspiration. I will thus offer an environmental ethic based on an understanding of human biology and culture that connects human physical and mental well-being not only to material and commodity advantage, but also to a host of equally compelling benefits people derive from their inclination to value nature for its aesthetic, emotional, intellectual, moral, and other qualities. This environmental ethic, in other words, marries a narrow instrumentalism with critical aspects of a rights-based or biocentric perspective. It represents a position occasionally articulated by others, for example Edward O. Wilson:

What humanity is now doing [by the large-scale loss of biological diversity] will impoverish our descendants for all time to come. Yet critics often respond "so what?" The most frequent argument is one of material wealth at risk. This argument is demonstrably true but contains a dangerous flaw—if judged by potential value, species can be priced, traded off against other sources of wealth, and when the price is right, discarded…The species-right argument…, like the materialist argument alone, is a dangerous play of cards…The independent-rights argument, for all its directness and power, remains intuitive, aprioristic, and lacking in objective evidence…A simplistic adjuration for the right of a species to live can be answered by a simplistic call for the right of people to live…In the end, decisions concerning preservation and use of biodiversity will turn on our values and ways of moral reasoning. A sound ethic…will obviously take into account the immediate practical uses of species, but it must reach further and incorporate the very meaning of human existence…*A robust, richly textured, anthropocentric ethic can instead be made based on the hereditary needs of our species, for the diversity of life based on aesthetic, emotional, and spiritual grounds.* (1993, 37; emphasis added)

The case here for a greatly expanded instrumental environmental ethic hinges on relating values such as material utility, aesthetics, emotional connection, spirituality, and more to human biology as well as to the particular capacity of people to exercise choice and free will in constructing personality, culture, and society. In making this ethical argument, I will invoke the concept of biophilia (Kellert 1997; Kellert and Wilson 1993; Wilson 1984), a complex of weak biological tendencies to value nature that includes material, aesthetic, emotional, intellectual, spiritual, and other basic dependencies on the natural world that contribute to human physical and mental well-being. Because biophilia is viewed as an inherent tendency, it is fundamentally rooted in assumptions regarding human biology and evolution and, in effect, an argument for an ethic of care and conservation of nature based on long-term individual and collective self-interest. As the biophilic tendency to value nature is regarded as a weak inherent inclination, it assumes these affinities for the natural world must be learned, although as genetically encoded features they can be taught relatively quickly. In other words, the biophilic values are highly shaped, mediated, and conditioned by experience and culture. Hence biophilia is a biocultural construct in which the inherent tendency to value nature is greatly influenced by human choice, creativity, and free will. Because the biophilic values depend on learning and experience, they are potentially expressed in both adaptive and maladaptive ways.

Nine biophilic values or inherent tendencies to impute worth and importance to the natural world have been identified (Kellert 1996, 1997). Reflecting the influence of learning and culture, each value is highly variable among individuals and groups, but as expressions of human evolution reflects a range of physical and mental benefits when adaptively revealed. All the biophilic values confer advantages but, being reliant on learning and social support, can be potentially distorted and dysfunctionally expressed. The nine values, thus, reflect the richness of the human dependence on the natural world for fitness and security, and when collectively revealed constitute a web of relational dependency so pronounced that an ethic of care and concern for nature may emerge from a profound realization of self-interest.

It would take far more space than available to detail all nine values and the various ways they may potentially contribute to human well-being. For illustrative purposes, five of these values will be briefly described: specifically, a utilitarian value that most closely embraces the narrow instrumental basis for an environmental ethic, as well as four others—the scientific, aesthetic, humanistic, and moralistic values—often associated with a rights-based or biocentric position because of their focus on appreciating nature as a source of knowledge, love, beauty, and spirituality. (One-sentence definitions of all nine values and frequently observed adaptive benefits are noted in table 20.1.)

A utilitarian value reflects the human inclination to affiliate with nature for its material and commodity advantage. The term is somewhat misleading since all the biophilic values are viewed as advancing human welfare. People have always recognized the natural world as an indispensable source of physical sustenance and security. Despite this ancient reliance, modern society often views the domestication of the wild as a measure of progress reflected in industrial agriculture and related large food surpluses, an abundance of technically produced consumer goods, relative physical health achieved through suppressing other organisms, and the massive transformation of natural into human-made buildings and landscapes integral to urbanization. This belief in progress as the measure of our material independence from nature is an illusion. People continue to depend largely on the natural world as an irreplaceable source of food production, medicines, building supplies, and other essential areas of commodity production. Moreover, this utilitarian reliance will expand greatly as a consequence of rapid advances in systematics, molecular biology, and bioengineering, which portend a revolution in new product development. Additionally, people rely even more basically on various ecosystem services such as

Table 20.1. Typology of Biophilic Values of Nature.

Value	Definition	Function
Aesthetic	Physical attraction and appeal of nature	Harmony, security, creativity
Dominionistic	Mastery and control over nature	Physical prowess, self-confidence, mastery skills
Humanistic	Emotional bonding with nature	Bonding, cooperation, companionship
Naturalistic	Exploration and discovery of nature	Order, meaning, connection
Moralistic	Moral and spiritual relation to nature	Curiosity, exploration, discovery
Negativistic	Fear and aversion of nature	Safety, protection, awe
Scientific	Systematic and empirical study of nature	Knowledge, understanding, critical thinking skills
Symbolic	Nature in language and expressive thought	Communication, mental development, analytical skills
Utilitarian	Material and physical exploration of nature	Physical sustenance, material productivity, survival skills

decomposition, pollination, and oxygen and water production to sustain life. Apart from these obvious sources of material advantage, people obtain a host of physical and mental rewards from nurturing their physical dependence on nature in the absence of necessity. They pursue a variety of harvesting activities because these pursuits nourish their ability to extract with skill a portion of their needs from the land. Beyond the material gains, they also reap physical fitness, feelings of independence and self-sufficiency, and self-confidence.

A scientific value reflects the human desire to know the world with understanding and authority. This tendency occurs among all cultures because it has facilitated the development of intellectual and cognitive capacities through systematic study and observation. The natural world provides an extraordinary array of opportunities for sharpening critical thinking skills and problem-solving abilities. Empirically and methodically examining nature builds capacities for acquiring knowledge and understanding, as well as for sharpening analytical and evaluative aptitudes. Other contexts exist in (especially) modern society for advancing these cognitive abilities. But contact with nature provides a stimulating and almost always accessible means for nurturing intellectual competence, particularly for the young and developing person. Moreover, simply by chance, the knowledge gained from intellectual pursuits pursued independent of their immediate utility often yields tangible and practical gains over time. And in exploring the mysteries of nature, people expand their realization of how much they can benefit from comprehending even a fraction of the extraordinary complexity of the biophysical world.

An aesthetic value reflects the human appreciation of nature as a source of physical attraction and beauty. Few experiences in people's lives exert as consistent and powerful an impact as the aesthetic appeal of certain features of the natural world. Even the most insensitive person would likely be unable to resist feelings of attraction to certain aesthetically compelling elements in nature no matter how fitfully expressed. Studies in various cultures have demonstrated a consistent and widespread inclination to respond to the attractiveness of certain landscapes, species, and other features of the natural world (Ulrich 1993). Yet a tendency exists to undervalue the significance of our aesthetics of nature. Even the well-known environmentalist Norman Myers has remarked, "The aesthetic argument for [environmental conservation] is virtually a prerogative of affluent people with leisure to think about such questions" (1979, 46). Still, it appears an aesthetic value of nature occurs universally among humanity and is, thus, genetically encoded, reflecting a tendency that evolutionarily developed because it yielded a variety of functional benefits.

What might be some elements of this adaptive significance? Recognizing beauty in nature can engender an awareness and appreciation of balance, symmetry, harmony, and grace. Unity and order observed in natural features inspire and instruct, offering a kind of quasi-design model and template where through mimetic adaptation analogous qualities of excellence and refinement can be captured in human life. Aesthetic preference for certain natural features can also be linked to the enhanced likelihood of achieving safety, sustenance, and security. People across the globe typically favor landscapes with clean and flowing water, that enhance sight and mobility, that possess bright and flowering colors, and that present other features which over time have proven instrumental in human survival (Heerwagen and Orians 1993;

Hildebrand 2000; Ulrich 1993). And, at a very basic level of experience, the aesthetic appeal of nature reflects being attracted or drawn to the most information-rich environment people will ever encounter (Wilson 1984). Through this attraction, people engage their sense of wonder, curiosity, and imagination and, as a consequence, increase their capacity for exploration, discovery, and creativity, all adaptive capacities in the struggle to survive and thrive.

A humanistic value reflects the ability of the natural world to provoke human affection and emotional attachment. This occurs through the companionship of other animals, but also through special fondness for certain plants and landscapes. These feelings of emotional attachment offer people opportunities for expressing and experiencing intimacy, relationship, connection, and sometimes a feeling of kinship. By contrast, isolation and aloneness represent heavy burdens for most people. With rare exceptions, people crave the companionship and affection of others, and affiliating with other species, even plants and landscapes, can provide an important source of trust and relationship. Bonding with others can be a significant pathway for cultivating capacities for cooperation and sociability, especially functional for a largely social human species. People covet responsibility for others and gratefully receive their affection and allegiance. Caring and being cared for by another creature and, more generally, by nature provides opportunities for expressing affection and building a sense of affiliation and fondness. These feelings accrue under normal circumstances, but become especially pronounced during moments of crisis and disorder. The caring and intimacy of other life is often mentally and physically restorative, whether expressed in the giving and receiving of flowers, contact with companion animals, or the experience of gardens, seashores, and other habitats.

A moralistic value reflects nature's ability to be a source of moral and spiritual inspiration. The philosopher Holmes Rolston remarked, "Nature is a philosophical resource, as well as a scientific, recreational, aesthetic, or economic one. We are programmed to ask why and the natural dialectic is the cradle of our spirituality" (1986, 88). This spiritual insight is often derived from the perception of a seeming similarity that unites life despite its extraordinary diversity, reflected in some 1.4 million classified and an estimated 10 to 100 million extant species (Wilson 1992). Despite this remarkable variability, most people recognize living creatures as often sharing analogous circulatory and reproductive features, parallel bodily parts, and common genetic structures. The perception of this unity and connection suggests an underlying order that often provides a cornerstone for spiritual and moral belief. Discerning universal patterns in creation intimates that at the core of human existence exists a fundamental logic, order, and even harmony and goodness. Faith and confidence are nurtured by recognizing an underlying unity that transcends and mutes our individual separation, isolation, and aloneness.

Five of nine biophilic values have been briefly described. Each value reflects weak genetic tendencies or prepared learning rules to affiliate with the natural world that developed over evolutionary time because of their proven instrumental significance in advancing human physical and mental well-being. The nine values collectively reflect the richness of the human reliance on nature as a basis for adaptive fitness and security. Together, they provide the basis for an environmental ethic rooted in a greatly expanded realization of self-interest that encompasses not only a conventional utilitarian understanding of material and commodity advantage, but also the functional importance of nature as a source of beauty, love, intellect, spiritual inspiration, and more. When these values are functionally expressed, they comprise a web of relational dependency that rationalizes and supports an ethic of care and responsibility for the natural world. Yet this is a difficult achievement. It requires the functional and adaptive expression of most if not all of the nine values, none so weakly evident as to atrophied or so strongly evident as to be inordinately exaggerated. Each value represents a piece of the relationship to nature that needs to occur in a balanced and adaptive fashion. A powerful ethic toward the natural world depends on most if not all of these pieces occurring in what can be called "right relationality." To add to this complexity, legitimate variation occurs among individuals and cultures as a consequence of developing these weak tendencies. This variability is bounded by human biology, underscoring again the biophilic values as biocultural phenomena.

Before concluding, I would like to offer a brief illustration of how significant changes in the nine values can lead to profound shifts in ethical relationships and policies toward the natural world. This historic example supports three important points in this chapter. First, it suggests that an expanded appreciation for nature can foster equally radical changes in ethical and moral relations. Second, it reveals how changes in values and ethics can trigger pronounced shifts in legal and regulatory policies. Third, it argues that such a transformation in values, ethics, and policies can occur in a shorter time period than often presumed, underscoring the practical significance of an ethical strategy as a basis for advancing significant environmental change.

This historic illustration involves profound changes in values, ethics, and policies toward large cetaceans during the latter half of the twentieth century. Like all examples, there is danger in losing sight of the general point in examining the specific case. Keep in mind that it matters little if you are interested in whales or whether or not you agree or disagree with current policies toward these creatures. This case is offered to illustrate the relationship between values, ethics, and policy, and to demonstrate how pronounced shifts in all three can occur in a surprisingly brief and relevant period of time.

This will be a cursory review, unable to examine in detail the historic decline and recent, uneven recovery of many cetacean species. The most obvious cause of the decline of the great whales was their excessive commercial exploitation with, as recently as 1960, whales comprising approximately 15% of the world's so-called fish catch (Kellert 1996; Lavigne, Scheffer, and Kellert, 1999). The endangerment of most species was fueled by assumptions regarding the inexhaustibility of their populations and the relatively easy product substitution of one species for another. Additional important factors in the decline were large capital equipment expenditures, large and reinvested surplus profits, absent property rights in the open ocean, a tendency to manage all species in the same manner, ineffectual regulatory practices, the enormous efficiency of new harvesting and processing technologies and, of course, widespread scientific ignorance. Underlying and motivating all these factors was a narrow set of exaggerated values that rendered the excessive and often cruel exploitation of whales both morally justifiable and ethically acceptable. These creatures were, in effect, viewed and treated from the perspective of three values—an exaggerated utilitarianism, an inordinate desire to master and dominate them, and a tendency to see them as monstrous fish. By the twentieth century most people, of course, recognized that whales were not fish, but they continued to be treated in this way.

Important attitudinal changes in values and perceptions toward large cetaceans mainly occurred following World War II. These changes probably not coincidentally happened in parallel with the rise of the modern conservation movement, prompting Gilbert Grosvenor, then head of the National Geographic Society, to observe, "The whale has become a symbol for a new way of thinking about our planet" (1976, 721). A sense of impending catastrophe loomed as the world contemplated the purposeful elimination of the largest creature the planet had ever known. The marine biologist Kenneth Norris proclaimed, "No other group of large animals has had so many of its members driven to the brink of extinction" (1978, 320). Significant advances in marine science following World War II also resulted in vastly expanded assumptions regarding the advanced intelligence, social behavior, and communication abilities of whales. These creatures suddenly seemed far less like fish and, indeed, more like people; this perceptual shift was fostered by almost mythic depictions in popular music, literature, and film. Highly successful captive aquarium displays, and the development of a major whale-watching industry generating more than one billion dollars annually and involving more than three million participants, further reflected a change from consumptive to nonconsumptive uses and values of whales. These changes in attitudes and behaviors fueled major shifts in policy, most particularly the passage of the revolutionary U.S. Marine Mammal Protection Act in 1972, and major regulatory shifts in the International Whaling Convention.

The motivation and political will behind these profound changes were fundamental shifts in values relating to whales, which eventually rationalized a new ethic toward the welfare of these creatures. The postwar period witnessed the dramatic rise of aesthetic, naturalistic, humanistic, scientific, and moralistic values, and a corresponding decline in utilitarian, dominionistic, and negativistic perspectives of whales. Aesthetically, large cetaceans were viewed as creatures of wonder and beauty; naturalistically, as the focus of outdoor recreational interest for millions to enjoy in the wild or in captivity; humanistically, as the subjects of strong emotional attachments and feelings of kinship and personal identification; scientifically, as highly complex and important biological organisms; and moralistically, as subjects of pronounced concern for their suffering and preservation. In effect, a profound shift occurred in what can be called valuational chemistry, especially in nations such as the United States, Great Britain, and Germany. That other nations and peoples still viewed these creatures with a different set of perceptual and ethical lenses can be seen in the views of Norwegian whaling advocate Arne Kalland: "Rational discussions of whaling, a 'fishery' which remains important for social, economic, cultural, and dietary reasons in some societies, is emotionally clouded by the popular conception of whales as a special class of animals. This presumed 'special' nature of whales derives from a widespread belief that whales are intelligent, endangered, killed by methods that are cruel, and the products they provide are no longer needed" (1993, 129).

This pejorative reference to the emotional and seemingly irrational thinking of those who regard whales as a "special class of animals" fails to note the ubiquity and logic of such attributions given certain value assumptions. As another example, one might take the reaction among most people in our society to the notion that rather than killing and incinerating millions of surplus cats and dogs, we treat them as edible protein, sending the meat to hungry millions in countries like North Korea and Somalia.

Independent of personal interest in and opinion about whales, this case illustrates how radical shifts in values toward a component of the natural world can foster pronounced changes in ethical regard and regulatory treatment. The seminal development by Sidney Holt and Lee Talbot (1978) of fundamental principles for the management of wild living resources provides another example of how shifts in values and ethics toward cetaceans helped drive basic public policy. This case also reveals how radical shifts in values, ethics, and policies can sometimes occur in a surprisingly short period of time. By comparison, consider the pace of policy shifts involving global climate change, the management of fisheries and commercial forests, or ecosystem protection. Indeed, this case may indicate that altering people's values and ethics toward nature, rather than being impractical and idealistic, is a highly relevant strategy for advancing significant change in environmental policy.

CONCLUSION

This illustration and the theoretical framework that preceded it have sought to reveal how a bioculturally based ethic rooted in inherent human tendencies to value the natural world can be both demonstrated and related to major policy change. Limited information has been provided regarding how biophilia and associated values constitute varying strands of relationships between people and the natural world that may confer significant physical and mental benefits. Each value represents a vital thread of connection to an ethic of care and concern for the natural world based on a broad understanding of human self-interest. This environmental ethic relies less on feelings of charity, kindness, and altruism, and more on a biocultural understanding of how individual and collective welfare can be advanced through a multiplicity of inherent ties to the natural world. People can see in their ethical connections to the natural world a moral posture of caring for the health and integrity of environmental systems that originates in a powerful realization of physical and mental well-being. Like Ishmael in *Moby Dick,* they can recognize in their relation to nature "the precise situation of every mortal that breathes; [how] he, one way or other, has this Siamese connection with a plurality of other mortals" (Melville [1851] 1941, 294).

Our inherent values toward nature remain an unrivaled means for nourishing the human body, mind, and spirit. The values of biophilia represent the genetic substrate of an ancient evolutionary dependence, which is molded and shaped by human choice and free will, as individuals and groups through the agency of learning and culture engender the means for expressing their ties to the natural world, in either adaptive or maladaptive ways. This biocultural complexity is discussed by the Nobel Prize–winning biologist René Dubos:

Conservation of nature is based on human value systems that rather than being a luxury are a necessity for the preservation of mental health. Above and beyond the economic reasons for conservation there are aesthetic and moral ones which are even more compelling. We are shaped by the earth. The characteristics of our environment in which we develop condition our biological and mental health and the quality of our life. Were it only for selfish reasons, we must maintain variety and harmony in nature. (1980, 126)

More poetically, the writer Henry Beston arrived at much the same conclusion when he suggested more than half a century ago:

Nature is a part of our humanity, and without some awareness and experience of that divine mystery man ceases to be man. When the Pleiades and the wind in the grass are no longer a part of the human spirit, a part of very flesh and bone, man becomes, as it were, a kind of cosmic outlaw, having neither the completeness nor integrity of the animal nor the birthright of a true humanity. ([1949] 1971, vi)

References

Beston, H. [1949] 1971. *The outermost house: A year of life on the great beach of Cape Cod.* Repr. New York: Ballantine.

Dubos, R. 1980. *Wooing of the earth.* London: Athlone Press.

Farley, M. 2002. Religious meanings for nature and humanity. In *The good in nature and humanity: Connecting science, religion, and spirituality with the natural world,* ed. S. R. Kellert and T. J. Farnham, 103–112. Washington, DC: Island Press.

Grosvenor, G. 1976. Editorial. *National Geographic* 150:721.

Heerwagen, J., and G. Orians. 1993. Humans, habitats, and aesthetics. In *The biophilia hypothesis,* ed. S. R. Kellert and E. O. Wilson, 138–172. Washington, DC: Island Press.

Hildebrand, G. 2000. *The origins of architectural pleasure.* Berkeley: University of California Press.

Holt, S. J., and L. M. Talbot. 1978. *New principles for the conservation of wild living resources.* Wildlife Monographs, no. 59. Bethesda, MD: Wildlife Society.

Kalland, A. 1993. Management by totemization: Whale symbolism and the anti-whaling movement. *Arctic* 46(2): 124–133.

Kellert, S. R. 1996. *The value of life: Biological diversity and human society.* Washington, DC: Island Press.

Kellert, S. R. 1997. *Kinship to mastery: Biophilia in human evolution and development.* Washington, DC: Island Press.

Kellert, S. R., and E. O. Wilson, eds. 1993. *The biophilia hypothesis.* Washington, DC: Island Press.

Lavigne, D., V. Scheffer, and S. Kellert. 1999. The evolution of North American attitudes toward marine mammals. In *Conservation and management of marine mammals,* ed. J. R. Twiss Jr. and R. R. Reeves, 10–47. Washington, DC: Smithsonian Institution Press.

Melville, H. [1851] 1941. *Moby Dick; or the white whale.* Repr. New York: Dodd, Mead.

Moncrief, L. W. 1970. The cultural basis for our environmental crisis: Judeo-Christian tradition is only one of many cultural factors contributing to the environmental crisis. *Science* 170(3957): 508–512.

Myers, N. 1979. *The sinking ark: A new look at the problem of disappearing species.* New York: Pergamon Press.

Norris, K. 1978. Marine mammals and man. In *Wildlife and America: Contributions to an understanding of American wildlife and its conservation,* ed. H. P. Brokaw, 315–325. Washington, DC: Council on Environmental Quality.

Passmore, J. A. 1974. *Man's responsibility for nature: Ecological problems and Western traditions.* New York: Scribner.

Rolston, H. 1986. *Philosophy gone wild: Essays in environmental ethics.* Buffalo, NY: Prometheus Books.

Ulrich, R. 1993. Biophilia, biophobia, and natural landscapes. In *The biophilia hypothesis,* ed. S. R. Kellert and E. O. Wilson, 73–137. Washington, DC: Island Press.

White, L., Jr. 1967. The historical roots of our ecologic crisis. *Science* 155(3767): 1203–1207.

Wilson, E. O. 1984. *Biophilia: The human bond with other species.* Cambridge, MA: Harvard University Press.

Wilson, E. O. 1992. *The diversity of life.* Cambridge, MA: Belknap Press.

Wilson, E. O. 1993. Biophilia and the conservation ethic. In *The biophilia hypothesis,* ed. S. R. Kellert and E. O. Wilson, 31–41. Washington, DC: Island Press.

21

Evolving Earth's Environmental Law: Perspectives on a Work in Progress

Nicholas A. Robinson

Before Darwin, there was no science of evolutionary biology or ecology. The law of nations is still largely pre-Darwinian, grounded in a nineteenth-century system of sovereign nation states that controlled colonies and amassed wealth from natural resources while also pursuing the Industrial Revolution. Neither the science nor the law of the environment featured in this legal system. When major elements of this regime collapsed after World War II, nations established the United Nations, which ushered in an era of decolonization, recognition of human rights, and international cooperation. Environment was not a part of the organization's 1945 charter. All the while, as pollution levels increased, the science of ecology matured. Beyond earlier concerns for fish and game, ecology explained the plight of species and habitats.

So a mere thirty years ago, in response to the increased knowledge produced by the environmental sciences, nations tepidly began to launch legal measures to cope with environmental problems. The United Nations convened its Conference on the Human Environment in 1972, establishing the UN Environment Programme (UNEP). Human society started to perceive that it had become a force of nature in its own right, altering Earth's living conditions. To manage themselves, nations began to fashion international environmental law. This body of law has evolved and expanded along with the refinements and discoveries of the ecological sciences.

International environmental law today refers to the general principles of law, the obligations and rules of treaties (known as "hard law"), and the strong intergovernmental guidance of the policy declarations of nations (known as "soft law"). Like most international law, these environmental legal provisions must be implemented through national environmental laws, and then enforced. The complexity of the environmental treaty obligations, and the soft-law policy guidance, often makes it difficult for nations to understand what the international community expects of them. Analysis of the legal aspects of most environmental issues engulfs even the specialist in a complex set of rules that apply in many socioeconomic sectors and entail detailed administrative rules.

Nations interact with one another and thereby come to understand how each implements the norms of international environmental law; as a result, most nations are gradually enacting similar patterns of administrative environmental law. Despite this emerging commonality of content, the very complexity of the rules often poses a barrier to their implementation. Most diplomats, judges, government officials, and politicians have never studied environmental law and can scarcely comprehend it. Moreover, some doubt that the current body of international environmental law is effective, since scientists continue to report Earth's ever more difficult environmental challenges, such as species extinction, habitat loss, climate alteration, and higher sea levels. For many, the norms of international environmental law seem unresponsive to Earth's gathering environmental crises.

This chapter examines international environmental law at its present stage of evolution. Although national governments have reflected upon the knowledge that ecology and other environmental sciences have adduced about the changes in Earth's natural systems, their establishment of this field of law must be deemed to be at an early stage. Many fail to understand that this field is still in its infancy because it appears extensive in its content and complexity. It is already so vast that its content cannot be restated or encapsulated here. Moreover, perhaps because of inexperience with the field, nations have scarcely begun to implement its norms.

A glimpse of international environmental law today can be found by briefly reviewing (1) the emergence of the system of international environmental law and policy; (2) the key institutions involved; (3) the legal principles of global application; (4) a case study showing how those principles are made operational, as, for instance, in managing grasslands, arid lands, and waters; (5) the norms for implementing international environmental law nationally, as in environmental impact assessment techniques); and (6) the norms that have been adapted and applied regionally across nations, as in a new generation of integrative treaties such as the 2003 African Convention on the Conservation of Nature and Natural Resources. Because nations make international environmental law through agreement on specific agreed texts, it is instructive to quote some of these texts, to illustrate their complexity and dependence on the environmental sciences. Reference to the legal texts reveals challenges that remain to be confronted in integrating ecology and law. When comparing the legal texts to the environmental problems that they address, it is evident that nations are far from having effective systems of stewardship for Earth's natural systems.

THE EVOLUTION OF INTERNATIONAL ENVIRONMENTAL LAW

In 1972, the United Nations recognized explicitly for the first time that the world's national governments faced myriad threats to the environment. At the UN Conference on the Human Environment, held in Stockholm, Sweden, representatives of the organization's founding nations were joined by those of the former colonies that were then sovereign states, plus the People's Republic of China, which replaced Taiwan as a permanent member of the UN Security Council and the General Assembly. Participants at the conference confronted new scientific evidence that pollution of the seas was now a global phenomenon and that air pollutants had spawned acid rain. The growing extinction of species, documented since 1948 by the International Union for the Conservation of Nature and Natural Resources (IUCN, now the World Conservation Union), also troubled the diplomats. Nonetheless, the assembled governments did not yet feel a sense of urgency about the deterioration of nature. The needs of the growing human population for socioeconomic development were the more pressing political priority of the day. But deteriorating trends in environmental quality were troubling enough to justify the establishment of the first UN body to monitor scientific environmental trends and organize international cooperation to cope with environmental problems. Later in 1972, UNEP was established, and the new nations voted to locate its headquarters in Kenya, symbolic of their new voting strength in the UN General Assembly.

Throughout the last quarter of the twentieth century, nations went on to establish their first environmental protection agencies, considering seriously the findings of the environmental sciences. New tools, such as remote sensing from satellites and ever more sophisticated computer modeling of natural conditions and trends, provided knowledge about conditions on Earth that informed decision making in these environmental agencies. At the same time, however, entrenched patterns of energy consumption, land development, habitat loss, and pollution persisted. In response, nations adopted national environmental legislation that depended upon and was guided by environmental sciences, which were also advancing rapidly in these same decades.

Developing nations were encouraged by UNEP to enact such national laws; UNEP also proactively promoted a range of innovative environmental treaties. Under the stewardship of the program's long-serving executive director, Mustafa Tolba, nations ushered in a new period of international law and policy. At the urging of UNEP, IUCN, and the scientists across the earth who documented environmental problems, nations agreed to new treaty obligations that created the world's first framework of principles and norms to govern international environmental cooperation among nations. For instance, nations negotiated and signed the UN Convention on the Law of the Sea, part of which constituted the world's first environmental charter for the oceans. Treaties on subjects such as endangered species, the stratospheric ozone layer, wetlands, or migratory species were prepared. In the brief span of two decades, from 1972–1992, an entirely new field of law emerged. Sovereign states established policies and treaties to shape a body of international environmental law that complemented a growing body of national environmental statues and regulations.

Yet enactment of statutes and adherence to treaties alone could not stem the widening patterns of environmental degradation across the earth. The UN World Commission on Environment and Development, chaired by Norway's prime minister, Gro Harlem Brundtland, clearly and succinctly focused the world on Earth's gathering environmental crisis through its report, *Our Common Future*, in 1987; in response, the UN General Assembly convened its preparatory committee for an international conference on environment and development. The 1992 UN Conference on Environment and Development (UNCED), informally known as the Earth Summit, convened in Rio de Janeiro, Brazil, under the leadership of Ambassador Tommy Koh of Singapore (who had also chaired the negotiations that produced the UN Convention on the Law of the Sea). The UNCED summit meeting was the largest ever held, before or since. The UN General Assembly endorsed

UNCED's Rio Declaration on Environment and Development—accompanied by a 600-page action plan, Agenda 21—as a set of principles to guide nations. Both these texts were accepted by consensus of all states. At UNCED, nations also signed both the Convention on Biological Diversity and the Framework Convention on Climate Change, the culminations of two years of intensive negotiations.

In the years following UNCED, the body of environmental law continued to grow, but compliance and implementation lagged at best. Evidence of environmental degradation accumulated in all regions, and growing populations in developing nations encountered durable patterns of economic depravation and poverty as well. By the time the UN General Assembly convened its 2002 World Summit on Sustainable Development in Johannesburg, South Africa, nations expressed distress that too little had been done to implement Agenda 21. The Johannesburg Declaration on Sustainable Development, and its Plan of Implementation, proposed measures to enhance international cooperation to make measurable gains in the realms of both environment and development.

Coming decades will tell whether nations will successfully embrace, implement, and secure the benefits of this global body of agreements on the norms and rules of global environmental law. It is uncertain whether or not these laws can, in fact, facilitate stabilization of Earth's carbon cycle or hydrologic cycle or secure its biodiversity. Environmental law has yet to address the nitrogen cycle; persistent organic pollutants bioacccumulate, and new genetically modified organisms have been introduced into the dynamic of evolution. It is evident that new environmental management systems remain to be enacted once these phenomena are better understood. Although it appears inevitable that anthropogenic changes to Earth's physical and natural changes will alter life for all living beings, it is environmental law that will determine how, and to what extent, humans can ameliorate such change, stem its negative aspects, or restore once-impaired biological systems.

The elements of environmental law and policy described here aim to provide the foundations for new, worldwide patterns of environmental stewardship. In a system of still sovereign nations, international norms are necessarily implemented through systems of national and regional environmental laws. This chapter sketches the emerging global framework of international law and the norms that nations need to apply.

THE INTERNATIONAL LAW FRAMEWORK

International environmental law is a branch of public international law. Some argue that environmental law, or ecological law, is an entirely new field of law in its own right, but this view is not yet accepted by nations.[1] As an acknowledged part of the law of nations, international law regulates the activities of entities possessing international personality. Prior to World War II, this meant the conduct and relationships of nation states. However, it is now well established that international law also concerns the structure and conduct of international organizations, such as those in the UN family of institutions, IUCN, or regional bodies. To a still emerging degree, through international human rights agreements it also affects individual human beings and nongovernmental organizations (NGOs), and through commercial agreements it governs multinational or national incorporated business companies.

There are four primary sources of international law: general principles of law, customary law, and conventions or treaty law, plus the analysis of legal experts who identify and elaborate these principles, customs, or treaty obligations through their commentaries. General principles of fundamental norms are acknowledged by the civilized nations and have come to be deeply embraced, such as the abhorrence of human slavery. Customary international law is created when states consistently follow certain practices or conduct out of an acknowledged sense of legal obligation to do so. Judgments of international tribunals, as well as scholarly works, have traditionally been looked to as persuasive sources for custom in addition to direct evidence of state behavior. Most traditional international law derives from written international agreements or treaties and may take any form that the contracting parties agree upon. International agreements create law for the parties of the agreement, thus treaty law is like a contract between the parties, which must be obeyed (*pacta sunt servanda*). Treaties may also lead to the creation of customary international law when they are intended for general adherence, are implemented, and are, in fact, widely accepted as being binding whether or not a nation has adhered to the treaty.

General principles common to systems of national law constitute a fundamental part of international law, which cannot be abrogated by custom or treaty. These principles arise out of the human condition and are not dependent on governmental action for their creation. There are situations where neither conventional nor customary international law can apply, and in this case a general principle may be invoked as a rule of international law because it is a general principle common to the major legal systems of the world. Most nations prefer to rely on treaty law and are reluctant to embrace new principles. In *Oposa v. Factoran*, the Supreme Court of the Philippines found that a human's right to a balanced environment was such a principle, predating constitutions or governments (Craig, Robinson, and Lian, 2002).

An expression of principles and customs, in addition to providing a source of international law, is described in texts as soft law. The term "soft law" refers to written declarations and other legal texts that do not have binding force, or whose binding force is somewhat weaker than the binding force of conventional or customary law. The resolutions of the UN General Assembly express such soft law, for example the 1986 World Charter for Nature resolution. Other examples of soft law are "declarations" and "statements of principle," certain types of resolutions produced by other international organizations (e.g., resolutions of the IUCN World Conservation Congress), and various kinds of quasi-legal instruments such as "codes of conduct," "guidelines," and "communications." Soft-law sources can be persuasive in identifying which nations acknowledge that they are bound by customary international law.

International Environmental Law Instruments

International environmental law operates through a framework of principles of law set forth in both treaties and soft-law instruments, as well as specific norms and obligations in multilateral environmental agreements (MEAs) and other treaties. More than 300 international environmental treaties exist today; the major global ones are called MEAs. Many basic principles of law, and core precepts, are set out in declarations, resolutions, and other decisions of nations. These documents are known as soft law, because although government representatives have negotiated them, they have not been formally adhered to by national ratification procedures as is required for treaties.

The principal provisions of international environmental law cover many fields of law, from transboundary movements of hazardous waste to nuclear energy rules. International laws for Antarctica or for use of the oceans and seabeds are voluminous. A sense of the scope of international environmental law in each subject area of concern can be illustrated by a focus on the biological norms relevant to life on Earth. Among the score of treaties and four score of soft-law instruments relevant to international biological conditions are the following international environmental law instruments.

Agenda 21

Agenda 21, a soft-law instrument, is the comprehensive plan of action to be taken globally, nationally, and locally by organizations of the UN system, by governments, and by NGOs and individuals in every area in which humans impact the environment. Agenda 21, the Rio Declaration on Environment and Development, and a "nonbinding" Statement of Principles for the Sustainable Management of Forests were all adopted by more than 178 governments at the UNCED meeting from June 3–14, 1992 and subsequently endorsed by the UN General Assembly.

Rio Declaration on Environment and Development

The Rio Declaration is a short, soft-law statement of principles produced at UNCED in 1992. It consists of twenty-seven principles intended to guide future sustainable development around the world. One key principle, Rio Principle 2, was first stated as Principle 21 of the Declaration of the UN Conference on the Human Environment in 1972, and it provides that no nation may act so as to harm the environment on the territory of another nation, or in the high seas and other common areas such as the atmosphere.

Convention on Biological Diversity

Opened for signature at UNCED in 1992, the first international agreement to provide an umbrella framework for the biological integrity of life on Earth is the CBD. (More specific treaties on conservation of biological resources include the Bonn Convention on the Conservation of Migratory Species of Wild Animals, or the Convention on International Trade in Endangered Species of Wild Fauna and Flora [CITES].) The CBD aims to protect the planet's biodiversity, which it defines as "the variability among living organisms from all sources including, *inter alia*, terrestrial, marine and other aquatic ecosystems and the ecological complexes of which they are part; this includes diversity within species, between species and of ecosystems" (article 2). It has three main goals: (1) conservation of biological diversity (or biodiversity), (2) sustainable use of its components, and (3) fair and equitable sharing of benefits arising from genetic resources. The CBD established a conference of parties, which deals both with thematic biodiversity programs such as agricultural biodiversity and marine biodiversity, and with cross-cutting issues such as bio-safety, access to genetic resources, and intellectual property rights.

UN Framework Convention on Climate Change

Opened for signature at UNCED in 1992, along with the CBD, the UNFCCC aims to reduce emissions of greenhouse gas in order to combat global warming. The convention, as originally framed, set no mandatory limits on greenhouse gas emissions for individual nations and contained no enforcement provisions. Instead, it included provisions for supplemental agreements, called "protocols," that would set mandatory emission limits. To date, the principal such agreement has been the Kyoto Protocol. The UNFCCC's signatory parties agreed in general that they would recognize "common but differentiated responsibilities," with greater responsibility for reducing greenhouse gas emissions in the near term on the part of developed/industrialized countries, which were listed and identified in annex I of the UNFCCC.

UN Convention to Combat Desertification

Signed in Paris in 1994, the UNCCD is an agreement to combat desertification and mitigate the effects of drought. The conference of parties is the governing body of this convention, which required negotiations after UNCED and has received less attention than either the CBD or the UNFCCC. The UNCCD provides for national action programs (NAPs); these are one of the key instruments in the implementation of the convention, strengthened by action programs on subregional and regional levels. The NAPs are developed in the framework of a participative approach involving local communities. They spell out the practical steps and measures to be taken to combat desertification in specific ecosystems.

Convention on International Trade in Endangered Species of Wild Fauna and Flora

Negotiated and signed in Washington, D.C., in 1973, CITES is an international agreement between governments to ensure that international trade in specimens of wild animals and plants does not threaten their survival. The agreement subjects international trade in species or the products or specimens of selected species to specified controls. All import, export, re-export, and introduction from the sea of species covered by CITES must be authorized through a licensing system. Each party to the convention must designate one or more management authorities in charge of administering that licensing system and one or more scientific authorities to advise them on the effects of trade on the status of the species. The species covered by CITES are listed in three appendixes, according to the degree of protection they need.

The Bonn Convention on the Conservation of Migratory
Species of Wild Animals

The Bonn Convention acknowledges the importance of migratory species being conserved. The range states (i.e., nations through whose territory a migratory species can be found) agree to take action to this end whenever possible and appropriate, paying special attention to those migratory species for which the conservation status is unfavorable, and to take, individually or in cooperation, appropriate and necessary steps to conserve such species and their habitat.[2]

Basel Convention on the Transboundary Movements
of Hazardous Wastes and Their Disposal

The Basel Convention, signed in 1989, provides that nations shall use environmentally sound management of hazardous wastes or other wastes by taking all practicable steps to ensure that hazardous wastes or other wastes are managed in a manner that will protect human health and the environment against the adverse effects which may result from such wastes.[3]

The World Heritage Convention

The UN Educational, Scientific and Cultural Organization's (UNESCO) World Heritage Convention establishes a process for identification of the cultural sites and natural sites nominated by nations as being of importance as part of the international heritage of the earth. IUCN provides technical services to UNESCO with respect to World Heritage sites, namely national parks and other protected natural areas.

Part XII of the UN Convention on the Law of the Sea

This article of the Convention on the Law of the Sea (UNCLOS) states that protection and "preservation" of the marine environment are the "obligation" of nations. The article governs land-based sources of pollution, global and regional cooperation, and environmental impact assessments and scientific and technical cooperation, as well as detailed enforcement provisions of any international environmental agreement.[4]

Regional Air Pollution and Acid Rain Agreements

There is no global treaty on air pollution. The Association of Southeast Asian Nations (ASEAN) Haze Agreement (Robinson 2004) and the UN Long-Range Transboundary Agreement and Protocols are currently the most elaborate treaty systems for controlling air pollution across borders; still, these agreements do not closely correlate to impacts on natural systems, biota, or even human health. The treaty law for pollution of the atmosphere, and contamination of the hydrologic cycle, remains rudimentary.

Convention on Wetlands of International Importance

This convention, signed at Ramsar, Iran, in 1971 (and therefore known as the Ramsar Convention on Wetlands), ensures cooperation to designate and manage consistently large wetlands areas, which are important for migratory birds. IUCN provides the convention's national members with a secretariat, and the agreement now guides national management of wetlands for all their ecological functions.

The IUCN Statutes

Nations become members of IUCN based on recognition and implementation of a duty to cooperate to conserve nature and natural resources. As a hybrid international organization, with both state and NGO members, IUCN brings together 82 states, 111 government agencies, and more than 800 NGOs. Its mission is to influence, encourage, and assist societies throughout the world to conserve the integrity and diversity of nature and to ensure that any use of natural resources is equitable and ecologically sustainable. IUCN's statutes are adhered to as a treaty or international agreement by its members, who are admitted under the terms of its statutes and article 60 of the Swiss civil code. Soft law and policy is made through the IUCN World Conservation Congress, which convenes every four years. Established in 1948, IUCN is the oldest and largest international environmental organization to promote conservation of nature. Its Commission on Environmental Law promotes the progressive development of environmental law at national and international levels, and its autonomous IUCN Academy of Environmental Law advances university education and research on environmental law.

Each of these formal conventions is binding only on the national governments that have signed and ratified the agreements, and the membership of nations in these treaties is not yet universal. The CBD, CITES, and the UNFCCC include the majority of UN member states, but the UNCCD and the air pollution agreements do not. Even where agreement is widespread, there can be gaps in the legal coverage; for instance, UNCLOS is nearly universally accepted, although the U.S. Senate has declined to ratify it (despite repeated requests to do so by presidents since 1982, including a request by President George W. Bush). In light of the widespread agreement about how the UNFCCC codified the prior norms about use of the high seas, the U.S. Department of State formally has acknowledged that the UNCLOS norms have become customary law of nations and that the United States acknowledges UNCLOS as binding upon it. In another example, the U.S. Senate has ratified the UNFCCC, which President George H.W. Bush signed at the 1992 UNCED summit, but President George W. Bush and the Senate have refused to ratify the Kyoto Protocol, which implements cutbacks on emission of greenhouse gases.

Most international treaties are usually not self-implementing. Each nation is expected to enact national legislation to implement the agreement in accordance with its national administrative laws and traditions. At the international level, each agreement usually establishes a secretariat to serve the states that have ratified the agreement. These states gather at meetings to make further decisions about the agreement through a conference of parties (COP). The COP assigns duties to the secretariat, and to any subsidiary bodies, and the secretariat provides the services needed by the COP's member states and any other bodies operating under the given agreement. Because many MEAs have been created, many COPs have their head-quarters in separate secretariat.

The proliferation of international environmental agreements and agencies has produced calls for consolidation. France has urged establishment of a new UN Environmental Agency, analogous to the World Trade Organization. Others have proposed that UNEP should provide services to all the separate environmental treaty organizations. To date, however, each agreement is administered separately, and any synthesis among the environmental norms is established in the national implement-ing legislation and not at the level of international environmental law.

International Organizations Responsible for Shaping International Environmental Law

Progressive elaboration of international environmental law norms and implementa-tion of treaties are advanced by the several international organizations that nations have entrusted with such duties. Institutional actors in the field of international envi-ronmental law of general importance to the topic of land degradation include the organizations detailed below.

Central to promoting cooperation among nations for environmental management are the units of the UN system. Their roles can be summarized as follows.

UN General Assembly

The Second Committee of the UN General Assembly annually holds policy debates and advances soft-law decisions on the law of the sea and on the law of forest manage-ment. The Sixth Committee of the UN General Assembly examines issues of inter-national law, including international environmental law such as the UN Commission on International Law's principles on international rivers. The UN General Assembly as a whole receives, debates, and funds the work of its programs, including UNEP. On average, the assembly itself addresses some fifteen agenda items about the envi-ronment annually.

The UN Commission on Sustainable Development

The CSD was established in December 1992 to ensure effective follow-up of the UNCED agreements. Specifically, the CSD is responsible for reviewing progress in the implementation of Agenda 21 and the Rio Declaration, as well as for provid-ing policy guidance to follow up the Johannesburg Plan of Implementation at the local, national, regional, and international levels. This commission reports to the UN Economic and Social Council (ECOSOC).

UN Regional Economic Commissions

The UN Economic Commission for Europe (UNECE) and the UN Economic Commission for Asia and the Pacific (UNECAP) make policy recommendations to various UN bodies concerning economic policy and development. Since environmental factors are deemed one of the three pillars of sustainable development, along with social and economic factors, these regional commissions further integrate policy and practice in the geographic areas of their state memberships.

UN Development Programme

This organization, better known as the UNDP, helps countries in their efforts to achieve sustainable human development by assisting them to build their capacity to design and carry out development programs. It gives top priority to building programs for poverty eradication, but it also assists with employment creation and sustainable livelihoods, the empowerment of women, and the protection and regeneration of the environment. The UNDP is a UN subsidiary body, operating under the UN secretariat and guided by UN member states.

UN Environment Programme

This program provides an integrative and interactive mechanism through which a large number of separate efforts by intergovernmental, nongovernmental, national, and regional bodies are reinforced and interrelated in the service of the environment. A UN subsidiary body, UNEP operates under the UN secretariat, with a governing council of UN member states to guide its work.

In addition to the United Nations itself, there are a number of separate treaty organizations, independent of the UN General Assembly and established by their separate agreements, with independent secretariats. The UN family of specialized agencies includes the following international organizations, each of which is organized under its own separate treaty.

Food and Agriculture Organization

The FAO leads international efforts to defeat hunger. Serving both developed and developing countries, it acts as a specialized forum where all nations can meet as equals to negotiate agreements and debate policy. It is also a source of knowledge and information, helping developing countries and countries in transition modernize and improve agriculture, forestry, and fisheries practices and thereby ensuring good nutrition for all.

World Health Organization

The WHO is the United Nations' specialized agency for health. It addresses issues related to the impact of the environment on human health, including pollution, access to safe drinking water, nutrition, and diseases such as SARS and avian influenza.

UN Educational, Scientific and Cultural Organization

This organization, which provides the secretariat for the World Heritage Convention, is charged with scientific development, in particular the Man and the Biosphere program. Moreover, in accordance with General Assembly Resolution 57/254, UNESCO is charged with developing an action plan for the UN Decade of Education for Sustainable Development.

In addition to the UN organizations, several international financial institutions are working globally or regionally on issues that relate to the environment. These institutions have been established by their own agreements and are independent of the UN system.

World Bank

The World Bank Group is composed of five international organizations responsible for providing financial assistance and advice to countries for economic development and poverty reduction, as well as for encouraging and safeguarding international investment. The group and its affiliates have their headquarters in Washington, D.C., with local offices in 124 member countries.

International Monetary Fund

The IMF is the international organization entrusted with overseeing the global financial system by monitoring exchange rates and balance of payments, as well as by offering technical and financial assistance as requested.

Regional Development Banks

Regional development banks exist around the world. For instance, the Asian Development Bank (ADB) is a multilateral developmental financial institution headquartered in Manila, the Philippines, and owned by sixty-four members, forty-six from the region and eighteen from other parts of the globe. Its mission is to help its developing member countries reduce poverty and improve the quality of life of their citizens. Other regions also have intergovernmental development banks, such as the Inter-American Development Bank, the African Development Bank, or the European Bank for Reconstruction and Development (EBRD, which is organized on a somewhat different model for nations whose economies are in transition from centrally planned socialist economy to regulated market economy).

The Global Environment Facility

The GEF, established by donor governments, UNEP, and the World Bank in 1991, helps developing countries fund projects and programs that protect the global environment. This organization grants support to projects related to biodiversity, climate change, international waters, land degradation, the ozone layer, and persistent organic pollutants. The GEF is run from the World Bank.

World Trade Organization

Economic relations with respect to liberalized international trade are structured through the WTO. It promotes the negotiated elimination of national barriers to free trade among nations. Its member states agree to submit disputes about trade to WTO tribunals; these tribunals have had occasion to rule on the conflict between a national environmental legal provision and rules for free trade, for example, involving efforts to protect turtles and dolphins from injury in fishing by catch incidents. Article XX of the organization's General Agreement on Tariffs and Trade (GATT) provides that a nation's phytosanitary laws cannot be a barrier on trade, but there remain misunderstandings about what provisions of environmental law are encompassed within the purview of the phytosanitary exemption. (Regional free trade organizations also exist, including the North American Free Trade Agreement [NAFTA], which also has dispute resolution tribunals designed to removed national barriers to free trade between Canada, Mexico, and the United States).

In addition to the UN system and the international and regional development banks, there are a number of other regional organizations of nations. Unique among these is ongoing negotiation of a federative association on a pan-European basis: The European Union (EU) is an intergovernmental and supranational union of twenty-five member states that was established through a number of European regional agreements, including the 1992 Maastricht Treaty. The union's activities cover all areas of public policy, from environmental, health, and economic policy to foreign affairs and defense. The other regional alliances of nations are not federative, but have been established for cooperation and harmonization of their policies. For instance, the aforementioned ASEAN is a political, economic, and cultural organization of ten member states located in Southeast Asia. In North America, the Commission on Environmental Cooperation (CEC) provides environmental harmonization as a side agreement to NAFTA.

In addition to intergovernmental international organizations, such as those described above, there are also international and transnational NGOs. These include expert bodies, many of whom have joined the IUCN in its nongovernmental chamber, as well as advocacy organizations such as the Sierra Club or Friends of the Earth. The UN ECOSOC maintains a consultative status with such organizations, including the International Council of Environmental Law (ICEL), a worldwide organization established in Switzerland and with headquarters in Bonn, Germany. Among other such international environmental expert groups are the following:

- The Environmental Law Institute (ELI), based in Washington, D.C., is an internationally recognized independent research and education center, advancing environmental protection by improving law, policy, and management.
- The International Network for Environmental Compliance and Enforcement (INECE) is a network of governmental and nongovernmental enforcement and compliance practitioners from more than 150 countries. Its goals are to raise awareness of compliance and enforcement, to develop networks for enforcement cooperation, and to strengthen capacity to implement and enforce environmental requirements.

- The International Law Association (ILA) has as its objectives, under its constitution, the "study, elucidation and advancement of international law, public and private, the study of comparative law, the making of proposals for the solution of conflicts of law and for the unification of law, and the furthering of international understanding and goodwill."
- The International Federation of Surveyors (FIG) is an international NGO whose purpose is to support international collaboration for the progress of surveying in all fields and applications, such as the global systems for assessing environmental land use patterns and land degradation. There are several hundred such organizations, but only a small subset address environmental law and policy.

These various intergovernmental and international NGOs interact with national governments to shape international environmental law and policy. Legal commentaries appear often, restating the content of such soft law and hard law, but from the perspective of the earth and the health of its natural systems, the focus must be on whether or not these laws are being observed, complied with, and implemented. One must have recourse to the techniques of comparative environmental law to assess whether, or how well, nations do implement their international obligations.

In doing so, the basic principles of international environmental law can be briefly set forth. These legal principles and rules are both the measure of how far the community of nations has advanced in establishing a body of law for the shared natural systems of the earth, and the current foundation for harmonization of national laws and practices.

INTERNATIONAL ENVIRONMENTAL LAW AND SUSTAINABLE DEVELOPMENT: MILLENNIUM DEVELOPMENT GOALS

The United Nations currently subsumes international environmental law under a broad policy priority—the eradication of poverty. As the world's population grows, so too have the number of human beings whose basic needs and rights are not being met. The UN Millennium Development Goals (MDGs) form a blueprint for dealing with this; the goals have been agreed to by all of the world's countries and by all of its leading development institutions. The UN resolutions based on the MDGs are a form of soft international law. Included in these goals are pledges to eradicate extreme poverty and hunger and ensure environmental sustainability. Currently, to be considered relevant and be taken seriously within the UN systems (even within UNEP) or the international economic systems described above, plans or goals must relate environmental law to the socioeconomic aspects of poverty elimination and sustainable development. Environmental topics are embedded in many issues related to these broad statements of commitment; therefore, environmental stewardship is central to the commitment to the MDGs. Environmental law and policy figure particularly in the following critical international issues:

1. *Poverty:* Pollution and land degradation weakens the ability of communities to depend on their environment for their livelihoods. The rural poor are the first to feel the effects of environmental degradation.

2. *Sustainable development:* Environmental degradation affects a significant portion of the earth's urban and arable lands, decreasing the wealth and preventing economic development of nations.

3. *Security:* As potable water becomes scarce, pollution worsens, natural resources become less productive, water and food security is compromised, and competition for dwindling resources increases. The seeds of potential conflict are sown.

4. *Loss of biodiversity:* Species diversity is lessened and often lost as natural lands are cleared and converted to other uses, and as agricultural lands are converted to nonagricultural uses.

5. *Climate change:* As the conversion of forests, wetlands, and other biologically robust regions of photosynthesis (conversion of carbon dioxide to oxygen) are lost, there are negative impacts on carbon sequestration, which contributes to the dynamics of global climate change.

6. *International waters:* The degradation of the land base, vegetation cover, and soil quality affects the quality and quantity of river basins and rivers flowing across national borders, ultimately impacting global water resources. Land-based pollution of waters is a growing, global problem.

7. *Ecosystem services:* Combating land degradation is needed in order to secure and enhance the benefits of ecosystem utilities and renewable resources.

The MDGs cannot be realized unless environmental laws become operational across the entire range of their application. The general principles of international environmental law that can advance the MDG, by sustaining a sound measure of environmental quality across Earth's regions, are restated in brief below.

GENERAL PRINCIPLES OF INTERNATIONAL ENVIRONMENTAL LAW

Nation-states, through and with international intergovernmental organizations, have developed a patchwork of principles and rules to govern Earth's environment. This body of law is incomplete in terms of what science tells us about anthropogenic impacts on natural systems. It is also inconsistent, since different political concessions were made at different times in order to reach a consensus about the principles accepted (and acceptable) to national governments in power at any given moment of decision. Notwithstanding these shortcomings, a restatement of international environmental law would include the following elements:

1. *Right to development:* The human right to development includes the full realization of the right of peoples to self-determination. All human beings have a responsibility for development.

2. *Principle of common concern of humankind:* Global environmental change is not only a matter of concern among discrete states, but also a matter of concern to the entire international community. There is a general recognition that humankind has common interests in protecting and managing the climate system, the ozone layer, the rainforests, and biological diversity for both present and future generations.

3. *Principle of intergenerational equity:* The human species holds the natural environment of our planet in common with all members of our species: past generations, the present generation, and future generations. As members of the present generation, we hold the earth in trust for future generations.
4. *Precautionary principle:* The precautionary principle holds that when an activity raises threats of harm to human health or the environment, precautionary measures should be taken even if some cause-and-effect relationships have not been fully established scientifically.
5. *Duty to assess environmental impacts:* The duty to assess environmental impacts is based on the premise that rational planning constitutes an essential tool for reconciling development and environment. The environmental impact assessment process provides an important modality for the implementation of the precautionary principle.
6. *The "polluter pays" principle:* The "polluter pays" principle holds that the entity responsible for causing pollution is responsible for the economic costs of that pollution, including damages, compensation, and mitigation costs.
7. *Right of access to information on the environment and the right of public participation:* These rights are based on the assertion that environmental issues are best handled with the participation of all concerned citizens. Citizens must have access to information, must be entitled to participate in decision making, and must have access to justice in environmental matters. Citizens must have the right to assemble and work through NGOs as a part of civil society.
8. *The human right to a sound environment:* Individual human beings have a right to be born into and live in environmental conditions appropriate for healthy and balanced natural life.

These general principles are all contained in soft-law instruments and may also be set forth in hard-law instruments, such as treaties. Various international judicial tribunals have recognized some of these principles. Some nations still refuse to acknowledge a number of these principles; for example, the United States continues to resist the formulation of the precautionary principle, and China and the Russian Federation (which has not adhered to the Aarhus Convention, discussed in a later section) have inconsistent laws toward recognizing civil society's right to environmental information and public participation.

These principles have been elaborated, and made into specific rules, in various treaties and other legal instruments. While the scope of this chapter does not permit examination about how this has been done in every environmental sector, the process can be described in the content of land stewardship. International environmental law is still far from embracing a "land ethic," such as articulated by Aldo Leopold (1949), but important components of that value system do exist and provide a basis for further decisions among nations toward such an objective. Human society is rooted in the soil and dry lands, and the rules for preventing land degradation, and for promoting sustainable practices and wise land stewardship, have been an objective of the conservation of nature and natural resources since the late 1800s.

International Environmental Law and Land Stewardship

Illustrative of the way international law takes on environmental themes are the legal instruments that shape the law of land stewardship. It is approached both from the worst-case scenario of land degradation, through the UNCCD, and from the more detached realm of managing biodiversity, through the UN Convention on Biodiversity. Further legal norms are shaped by soft law or protocols and other international agreements in order to build content between these two hard-law bookends. Early diplomatic proposals to shape a methodology around concepts of ecosystem management are examples of such efforts to build overarching norms that can encompass many aspects of land stewardship.

On February 9, 2004, in furtherance of the objectives of the UNCCD, the UN General Assembly passed Resolution 58/211, which declared 2006 the International Year of Deserts and Desertification. In this soft-law instrument, the United Nations "calls upon all relevant international organizations and Member States to support the activities related to desertification, including land degradation, to be organized by affected countries, in particular African countries and the least developed countries."

This commitment thereafter was reaffirmed in General Assembly Resolution 60/200. In addition, General Assembly Resolution 57/254 declares the ten-year period beginning January 1, 2005, to be the UN Decade of Education for Sustainable Development. This chapter has been written at the outset of that decade and during the International Year of Deserts and Desertification, and it is a direct response to the international call for action. Such measures are a call for cooperation among nations to address common objectives. They fall within the broad aim of the MDGs described above.

Traditional attempts to address the impacts of land degradation have been based upon a sector-by-sector approach, which has resulted in fragmentation of policies, institutions, and interventions. Such approaches have not achieved optimum results because legal, economic, and social sectors have largely ignored or compromised the linkages and interactions among natural systems, as well as with people. Consequently, there is an urgent need for the adoption of management systems embracing comprehensive and cross-sectoral approaches.[5]

International environmental law can be applied toward such an integrated implementation process at the national level. This can be illustrated by setting forth the relevant treaty and soft-law principles that would guide such a process. At the national level, and also the regional level where ecosystems cross borders, it would be necessary to embrace an "ecosystem management" approach in order to do so. The treaty and soft-law principles would need to be incorporated into national legislation, state or provincial laws, and the by-laws of local authorities. Because land use patterns are shaped at local and regional levels of societal activity, and not just at national levels, the international legal rules would need to be implemented at each of the levels at which government is organized, as well as within each of the socioeconomic sectors of human activity.[6] If much remains to be done in this regard, one of the reasons is that this is an enormous task. Political and socioeconomic inertia work against implementing such reforms all of Earth's nations.

ECOSYSTEM MANAGEMENT APPROACHES FOR IMPLEMENTING
INTERNATIONAL ENVIRONMENTAL LAW: A CASE STUDY

Implementation of international environmental laws for land stewardship can be organized around biomes, around how their continuing natural functions can be sustained. This focus allows natural science to set the stage and limits for human use; this is quite different from the focus on maximizing human harvests or exploitation, an approach which gradually succumbs to what ecologist Garrett Hardin and others call the "tragedy of the commons." Equally important, a biome-oriented focus fosters education of political leaders and civil society about the natural systems being addressed. No one will protect soil as a living resource if it is considered simply to be "real estate" or a surface to be paved over and built upon, with no collateral consequences. Today, most national property laws, land cadastres, and regulations (town and country planning or zoning laws) foster the view that all land surfaces can be economically converted to developed uses, and they require neither an ecosystem assessment before classifying lands to protect their ecological values nor an environmental impact assessment before allowing conversions to nonnatural uses.[7]

International environmental laws can be applied through an ecosystem management approach with respect to grasslands, arid lands that are easily eroded, watersheds, and shorelines. There are similar, but even less well elaborated, international environmental law provisions regarding development for forests and agricultural lands.[8] Both silviculture and agriculture entail commercial activities, whose proponents are less willing to see international norms displace their domestic opportunities for profitable endeavors.[9] This may well explain why nations have encouraged the development of international environmental law for restoring lands and water resources that are no longer readily available for profitable activities. Still, an examination of how an ecosystem management approach might be used for biologically significant areas such as grasslands, arid lands, or wetlands can provide useful insights into how international environmental legal principles become (or can become) operational.

Grasslands

The grasslands and prairies of the earth have been converted to other uses for as long as humans have had the capacity to exploit them. Studies of such experiences provide the basis for making sound judgments about ongoing land degradation issues surrounding grasslands. Land degradation is accelerated by conversion of grasslands to cropland because it diminishes the vegetative cover (resulting in increased erosion of the soil) and biodiversity of the land. Conversion to urban uses contributes to land degradation because it increases impervious surface area, which reduces the capacity of the land to absorb and retain water and increases water and land pollution due to runoff. Grasslands support biodiversity essential for agriculture, animal husbandry, and crop pollination; loss of grassland biodiversity due to land degradation not only threatens the grassland ecosystem but also drastically diminishes the benefits that can be garnered from these systems. Poor soil and water management, and wind- and water-driven erosion, accelerate such degradation of grasslands. And the overstocking of grasslands with livestock results in overgrazing and trampling of vegetation, which in the short term reduces soil fertility and the size of the herd that can be sustained, and in the long term destroys the capacity of the soil to support vegetation, resulting in desertification.

Climate change will further cause a shift in the boundaries of vegetation zones. Semiarid lands, including grasslands, will be most immediately affected by encroachment of arid desert environments. All these phenomena have led to the articulation of a set of international environmental law principles and best management practices (BMPs). Environmental law would support the following BMPs.

1. *Ecosystem approach:* The ecosystem approach is a strategy for the integrated management of land, water, and living resources that promotes conservation and sustainable use in an equitable way. It is based on the application of appropriate scientific methodologies focused on levels of biological organization, which encompass the essential processes, functions, and interactions among organisms and their environment. It recognizes that humans, with their cultural diversity, are an integral component of ecosystems.

2. *Sustainable livelihoods:* A livelihood is environmentally sustainable when it maintains or enhances the local and global assets on which the livelihood depends, and has net beneficial effects on other livelihoods. A socially sustainable livelihood can cope with and recover from stress and shocks, as well as provide for future generations.

3. *Community-based management practices:* Community-based management practices are systems for the management of a local resource. They are developed through processes that involve the participation of all stakeholders, and they incorporate the knowledge and experience of indigenous and local communities.

The international environmental law provisions that can be organized through such BMPs are derived from the CBD. This treaty states that "ecosystem means a dynamic complex of plant, animal and micro-organism communities and their non-living environment interacting as a functional unit" (article 2) and that "subject to its national legislation, [each Contracting Party shall] respect, preserve and maintain knowledge, innovations and practices of indigenous and local communities embodying traditional lifestyles relevant for the conservation and sustainable use of biological diversity and promote their wider application with the approval and involvement of the holders of such knowledge, innovations and practices and encourage the equitable sharing of the benefits arising from the utilization of such knowledge, innovations and practices" (article 8(j)).

Further support for these BMPs is found in the decisions taken by the conference of the parties to the CBD, specifically decision V/23. Here, the member states undertook to "establish a program of work on the biological diversity of dryland, Mediterranean, arid, semi-arid, grassland, and savannah ecosystems, which may also be known as the program on 'dry and sub-humid lands,' bearing in mind the close linkages between poverty and loss of biological diversity in these areas" (para. 1). They further agreed that "the elaboration and implementation of the program of work should aim at applying the ecosystem approach adopted under the Convention on Biological Diversity. Implementation of the program of work will also build upon the knowledge, innovations and practices of indigenous and local communities consistent with Article 8(j) of the Convention" (annex 1, part I(3)) and that "promotion of specific measures for the conservation and sustainable use of the biological diversity of dry and sub-humid lands, through, (a) The use and the establishment of additional protected areas and the development of further specific measures for the conservation of the biological diversity of dry and sub-humid lands, including the strengthening of measures in existing protected areas; investments in the development and promotion of sustainable livelihoods, including alternative livelihoods; and conservation measures" (annex 1, part II(11)).

There is a substantial additional body of international decision making on grasslands, undertaken through the many international intergovernmental organizations with competence to address grassland issues.[10] Since these decisions are also based on scientific study of the biome, they are generally consistent with each other, as one would expect as the result of analysis proceeding from a scientific premise, rather than a political or utilitarian premise.

Arid Lands

Degradation of arid lands—easily eroded by action of wind, water, or human disturbance—falls under the rubric of desertification. Drylands susceptible to desertification cover 40% of the earth's surface and put at risk more than one billion people who are dependent on these lands for survival. Large-scale desertification can influence climate change by reducing the carbon sink and increasing the earth's reflectivity. Many factors trigger desertification, including the unpredictable effects of drought, fragile soils and geological erosion, livestock pressures, nutrient mining, growing populations, inadequate/ambiguous property and tenure rights, landlessness and an inequitable distribution of assets, poor infrastructure and market access, neglect by policy makers and agricultural and environmental research systems, and the failure of markets to reward the supply of environmental services.

International environmental law provides international principles and BMPs for improving human stewardship of such arid lands; many of these have been elaborated by the UNCCD. Arid-land BMPs can be set forth as follows.

1. *Integrated approach:* Because land resources are used for a variety of purposes, which interact and may compete with one another, it is desirable to plan and manage all uses in an integrated manner. Integration should take place at two levels, considering on one hand all environmental, social, and economic factors, and, on the other, all environmental and resource components combined (i.e., air, water, biota, land, and geological and natural resources).

2. *National action programs (NAPs) under the CCD:* The NAPs provide the blueprint for implementation of the CCD. They must be created through participatory mechanisms and integrated with existing programs, and they must address the dual goals of combating desertification and eradicating poverty. Included in NAPs would be the following elements: (a) management practices that include broad stakeholder participation, including the important role of women; (b) land-use practices that include carbon sinks,

reforestation, and other methods to combat climate change/desertification;[11] (c) desertification monitoring and assessment using benchmarks and indicators; and (d) preventative measures to protect lands not already degraded by desertification.

3. *Sustainable livelihoods:* A livelihood is environmentally sustainable when it maintains or enhances the local and global assets on which the livelihood depends, and has net beneficial effects on other livelihoods. A socially sustainable livelihood can cope with and recover from stress and shocks, as well as provide for future generations.

These implementation elements are derived from several sources within international environmental law. For instance, the UNCCD provides that "combating desertification includes activities which are part of the integrated development of land in arid, semi-arid and dry sub-humid areas for sustainable development which are aimed at: (a) prevention and/or reduction of land degradation; (b) rehabilitation of partly degraded land; and (c) reclamation of desertified land" (article 1(b)). The UNCCD agrees that "land degradation" means "reduction or loss, in arid, semi-arid and dry sub-humid areas, of the biological or economic productivity and complexity of rain-fed cropland, irrigated cropland, or range, pasture, forest and woodlands resulting from land uses or from a process or combination of processes, including processes arising from human activities and habitation patterns, such as: (a) soil erosion caused by wind and/or water; (b) deterioration of the physical, chemical and biological or economic properties of soil; and (c) long-term loss of natural vegetation" (article 1(f)). Furthermore, the convention explains that "achieving this objective will involve long-term integrated strategies that focus simultaneously, in affected areas, on improved productivity of land, and the rehabilitation, conservation and sustainable management of land and water resources, leading to improved living conditions, in particular at the community level" (article 2(2)). And "the Parties should ensure that decisions on the design and implementation of programs to combat desertification and/or mitigate the effects of drought are taken with the participation of populations and local communities and that an enabling environment is created at higher levels to facilitate action at national and local levels" (article 3(a)). The UNCCD specifies a role for NAPs within the implementation:

> National action programs shall specify the respective roles of government, local communities and land users and the resources available and needed. They shall, *inter alia:* (a) incorporate long-term strategies to combat desertification and mitigate the effects of drought, emphasize implementation and be integrated with national policies for sustainable development; (b) allow for modifications to be made in response to changing circumstances and be sufficiently flexible at the local level to cope with different socio-economic, biological and geo-physical conditions; (c) give particular attention to the implementation of preventive measures for lands that are not yet degraded or which are only slightly degraded; (d) enhance national climatological, meteorological and hydrological capabilities and the means to provide for drought early warning; (e) promote policies and strengthen institutional frameworks which develop cooperation and coordination, in a spirit of partnership, between the donor community, governments at all levels, local populations and community groups, and facilitate access by local populations to appropriate information and

technology; (f) provide for effective participation at the local, national and regional levels of non-governmental organizations and local populations, both women and men, particularly resource users, including farmers and pastoralists and their representative organizations, in policy planning, decision-making, and implementation and review of national action programs; and (g) require regular review of, and progress reports on, their implementation. (article 10(2))

The UNCCD also provides that "affected country Parties shall consult and cooperate to prepare, as appropriate, in accordance with relevant regional implementation annexes, sub-regional and/or regional action programs to harmonize, complement and increase the efficiency of national programs. The provisions of article 10 shall apply mutatis mutandis to sub-regional and regional programs. Such cooperation may include agreed joint programs for the sustainable management of trans-boundary natural resources, scientific and technical cooperation, and strengthening of relevant institutions" (article 11).

In addition to the UNCCD, the UNFCCC also contains relevant provisions for arid-land management. Nations are to "cooperate in preparing for adaptation to the impacts of climate change; develop and elaborate appropriate and integrated plans for coastal zone management, water resources and agriculture, and for the protection and rehabilitation of areas, particularly in Africa, affected by drought and desertification, as well as floods" (article 4(1)(e)) and "promote sustainable management, and promote and cooperate in the conservation and enhancement, as appropriate, of sinks and reservoirs of all greenhouse gases not controlled by the Montreal Protocol, including biomass, forests and oceans as well as other terrestrial, coastal and marine ecosystems" (article 4(1)(d)).

Beyond these hard-law treaty provisions, there is substantial soft environmental law guidance produced in Agenda 21, the action plan adopted at UNCED in 1992 and subsequently by the UN General Assembly: "The priority in combating desertification should be the implementation of preventive measures for lands that are not yet degraded, or which are only slightly degraded. However, the severely degraded areas should not be neglected. In combating desertification and drought, the participation of local communities, rural organizations, national Governments, non-governmental organizations and international and regional organizations is essential" (ch. 12, para. 12.3). Toward this objective, programs are designated for strengthening the knowledge base and developing information and monitoring systems for regions prone to desertification and drought, including the economic and social aspects of these ecosystems.

The global assessments of the status and rate of desertification conducted by the United Nations Environment Programme (UNEP) in 1977, 1984 and 1991 have revealed insufficient basic knowledge of desertification processes. Adequate worldwide systematic observation systems are helpful for the development and implementation of effective anti-desertification programs. The capacity of existing international, regional and national institutions, particularly in developing countries, to generate and exchange relevant information is limited. An integrated and coordinated information and systematic observation system based on appropriate technology and embracing global, regional, national and local levels is essential for understanding the dynamics of desertification and drought processes. It is also important for developing adequate measures to deal with desertification and drought and improving socio-economic conditions. (para. 12.5)

Other Agenda 21 program recommendations aim at creating programs for "the eradication of poverty and promotion of alternative livelihood systems in areas prone to desertification . . . [where] current livelihood and resource-use systems are not able to maintain living standards. In most of the arid and semi-arid areas, the traditional livelihood systems based on agropastoral systems are often inadequate and unsustainable, particularly in view of the effects of drought and increasing demographic pressure. Poverty is a major factor in accelerating the rate of degradation and desertification. Action is therefore needed to rehabilitate and improve the agropastoral systems for sustainable management of rangelands, as well as alternative livelihood systems" (para. 12.26). Further programs are suggested for "encouraging and promoting popular participation and environmental education, focusing on desertification control and management of the effects of drought":

> The experience to date on the successes and failures of programs and projects points to the need for popular support to sustain activities related to desertification and drought control. But it is necessary to go beyond the theoretical ideal of popular participation and to focus on obtaining actual active popular involvement, rooted in the concept of partnership. This implies the sharing of responsibilities and the mutual involvement of all parties. In this context, this program area should be considered an essential supporting component of all desertification-control and drought-related activities. The objectives of this program are to support local communities in their own efforts in combating desertification, and to draw on the knowledge and experience of the populations concerned, ensuring the full participation of women and indigenous populations. (paras. 12.55–12.56).

Soft-law policy recommendations such as these require establishment of programs to assist national agencies in understanding and developing the capacity to implement them. But the same nations that agreed upon such detailed policy provisions did not provide for follow-up on how nations (or UNEP) would work to turn the soft-law suggestions into hard-law agreements that can be implemented. Such soft law, while important, is a preliminary phase of international environmental law. It remains to be realized in action.

Watersheds, Water Conservation, and Soil Conservation

From a biological perspective, the management of water resources also needs much rethinking and refinement. Mismanagement of the effects of rainfall, whether in flood or drought conditions, is becoming crucial as human populations and settlements grow exponentially in some parts of the world. International environmental law addresses the ways in which national governments must rethink their water and soil conservation programs. Previously, the laws on such matters were either local or provincial, and sometimes national. Now, though, issues of water management are a common concern of mankind, and they require a harmonized, international approach.

Impacts on human and ecosystems from mismanagement of water systems are evident. For example, the destruction of surface soils adversely impacts filtration of water into the groundwater table and can cause siltation and contamination of adjacent water bodies. Destruction and filling of wetlands impairs filtration of waters into the groundwater table, exacerbates flood conditions by eliminating hydrologic absorption, and eliminates other biological utilities of wetlands. Grass and shrubs hold the soil firmly together; their removal due to overgrazing leaves the soil bare and susceptible to wind and water erosion. The problem is exacerbated when the degraded soil is broken up or compacted by hooves. Moreover, poor irrigation and drainage practices result in waterlogging and salinization due to the accumulation of salts from poorly managed irrigation water. Municipal, industrial, and agricultural pollution contaminates soil and water through direct dumping and discharges and through nonpoint source contamination.

To address these common concerns across nations, international environmental law has advanced a number of international principles from which the following BMPs can be restated.[12] All relate to the use of sustainable agriculture, which integrates three main goals—environmental health, economic profitability, and social and economic equity.

1. *Application of integrated soil biological management:* Agricultural practices induce changes in the soil environment, resulting in significant modifications of the ratio and interactions of soil organisms. Integrated soil biological management considers these interactions, and aims to enhance the beneficial aspects of soil biota.

2. *Land use planning:* Environmentally sound, socially fair, and economically feasible land use planning is necessary.

3. *Use of watershed management:* A watershed is the land that water flows across or through on its way to a common stream, river, or lake. Because many resources interact to form a watershed, land use planning should occur at the watershed level and should integrate management of all these resources including climate, geology, hydrology, soils, vegetation, and biota. This requires also the identification, protection, and restoration of wetlands and the mapping of groundwater hydrology, sources, and flows in order to sustain water tables. The BMP needs to manage irrigation carefully to avoid raising water tables and preventing salinization, in order to prevent waterlogged soils from becoming less productive for agricultural purposes. Management should also include planting of shrubs and trees to reduce wind-driven erosion of soils, as well as the mapping of soil types to identify the most ecologically appropriate use for each soil type.

These BMPs can be derived from provisions of international environmental law, primarily in the soft-law sector as nations have yet to address many of these issues in hard international law. (One notable exception is a set of treaties adopted for management of international rivers and lakes in order to cope with and prevent flood conditions, avert pollution, and maintain navigation.[13]) The provisions of a global legal pattern for water stewardship can be envisioned in the following recommendations of Agenda 21. Here, programs are recommended for combating land degradation through, inter alia, intensified soil conservation, and reforestation activities: "An increasing vegetation cover would promote and stabilize the hydrological balance in the dryland areas and maintain land quality and land productivity. Prevention of not yet degraded land and application of corrective measures and rehabilitation of moderate and severely degraded drylands, including areas affected by sand dune movements, through the introduction of environmentally sound, socially acceptable, fair and economically feasible land-use systems. This will enhance the land carrying capacity and maintenance of biotic resources in fragile ecosystems" (ch. 12, para. 12.16).

Agenda 21 also urges that "governments at the appropriate level, and with the support of the relevant international and regional organizations, should … Promote improved land/water/crop-management systems, making it possible to combat salinization in existing irrigated croplands; and to stabilize rainfed croplands and introduce improved soil/crop-management systems into land-use practice" (para. 12.18(c)).

Further support for these sorts of BMPs is derived from the recommendations of the CBD's Subsidiary Body on Scientific, Technical and Technological Advice (SBSTTA):

> The SBSTTA further recommends to the Conference of the Parties that Parties: (a) Encourage the use of integrated watershed management approach as a basis of the planning and making decisions in relation to the use of land and water resources, including biological resources, within river catchment; (b) Encourage the use of low-cost (appropriate) technology, non-structural and innovative approaches to meet watershed management goals, such as using wetlands to improve water quality, using forests and wetlands to recharge groundwater and maintain the hydrological cycle, to protect water supplies and using natural floodplains to prevent flood damage, and to use indigenous species for aquaculture; (c) Emphasize more effective conservation and efficiency in water use, together with non-engineering solutions. Environmentally appropriate technologies should be identified, such as low-cost sewage treatment and recycling of industrial water to assist in the conservation and sustainable use of inland waters; (d) Encourage research on the application of ecosystem-based approaches. (Recommendation III/1, § A (III))

The Ramsar Convention on Wetlands also addresses these issues with hard-law provisions. Specifically, it provides that "each Contracting Party shall designate suitable wetlands within its territory for inclusion in a List of Wetlands of International Importance, hereinafter referred to as 'the List' which is maintained by the bureau established under Article 8. The boundaries of each wetland shall be precisely described and also delimited on a map and they may incorporate riparian and coastal zones adjacent to the wetlands, and islands or bodies of marine water deeper than six meters at low tide lying within the wetlands, especially where these have importance as waterfowl habitat" (article 2(1)). The convention further holds that "the Contracting Parties shall formulate and implement their planning so as to promote the conservation of the wetlands included in the List, and as far as possible the wise use of wetlands in their territory" (article 3(1)), and that "each Contracting Party shall promote the conservation of wetlands and waterfowl by establishing nature reserves on wetlands, whether they are included in the List or not, and provide adequately for their wardening" (article 4(1)).

The importance of water resources management, of course, extends beyond these dimensions. Land degradation diminishes the quality and quantity of water resources and results in the loss of water storage capacity of soils and reservoirs. The management of water resources at the international, regional, national, and local levels is essential to attaining the MDGs, as well as for combating desertification.[14] For land degradation is accelerated by the erosion of soil due to water runoff. Water pollution destroys the biotic food chain within water resources and adversely affects a broad range of living systems including harvested resources. It is detrimental to human health. Although hydroelectric dams can be a valuable source of energy, the location, planning, and assessment of dams have to take into account competing uses for water. In light of such considerations, international environmental law would also support BMPs that could encompass these elements, such as those listed below.

1. *Integrated water resource management:* Surface waters should be managed in conjunction with the management of underground and other waters by a regime that takes into account the interconnections of surface and subsurface waters within the drainage basin. States must also integrate information and policies relating to other resources in the management of the waters of an international drainage basin.
2. *Ecological integrity:* Use of water resources must preserve and protect the ecological integrity of the biotic community of those waters. Without ecological integrity, sustainability is impossible.
3. *Management of in-stream flow:* The term "in-stream flow" is used to identify a specific stream flow at a specific location for a defined time, typically following seasonal variations. In-stream flow must be managed so as to protect and preserve in-stream resources and values such as fish, wildlife, and vegetation.
4. Watercourse use: Utilization of transboundary watercourses must be equitable and sustainable.

Foundations in international environmental law for these additional aspects would require implementing provisions of the UN Convention on Non-Navigational Use of International Waters. This agreement (signed in New York in 1997) has not yet been widely adhered to by nations, but it is authoritative as a statement prepared by international law experts in the UN's official International Law Commission. It provides that watercourse states shall:

- "Individually and, where appropriate, jointly, protect and preserve the ecosystems of international watercourses" (article 20)
- At the request of any of them, enter into consultations concerning the management of an international watercourse, which may include the establishment of a joint management mechanism" (article 24)
- "Cooperate, where appropriate, to respond to needs or opportunities for regulation of the flow of the waters of an international watercourse" (article 25)
- "Individually and, where appropriate, jointly, take all appropriate measures to prevent or mitigate conditions related to an international watercourse that may be harmful to other watercourse States, whether resulting from natural causes or human conduct, such as flood or ice conditions, water-borne diseases, siltation, erosion, salt-water intrusion, drought or desertification" (article 27)

On a regional level, there is further authority for implementing such BMPs as national or local laws. The Convention on the Protection and Use of Transboundary Watercourses and International Lakes (signed in Helsinki in 1992) provides that "parties undertake to ensure that transboundary waters are used with the aim of ecologically sound and rational water management, conservation of water resources and environmental protection; and undertake to ensure conservation and, where necessary, restoration of ecosystems" (article 2(2)). It also provides strict, specific guidelines for such management:

> To prevent, control and reduce transboundary impact, the Parties shall develop, adopt, implement and, as far as possible, render compatible relevant legal, administrative, economic, financial and technical measures, in order to ensure, *inter alia*, that: The emission of pollutants is prevented, controlled and reduced at source through the application of, *inter alia*, low- and non-waste technology; Transboundary waters are protected against pollution from point sources through the prior licensing of waste-water discharges by the competent national authorities, and that the authorized discharges are monitored and controlled; Limits for waste-water discharges stated in permits are based on the best available technology for discharges of hazardous substances; Stricter requirements, even leading to prohibition in individual cases, are imposed when the quality of the receiving water or the ecosystem so requires; At least biological treatment or equivalent processes are applied to municipal waste water, where necessary in a step-by-step approach; Appropriate measures are taken, such as the application of the best available technology, in order to reduce nutrient inputs from industrial and municipal sources; Appropriate measures and best environmental practices are developed and implemented for the reduction of inputs of nutrients and hazardous substances from diffuse sources, especially where the main sources are from agriculture (guidelines for developing best environmental practices are given in annex II to this Convention); Environmental impact assessment and other means of assessment are applied; Sustainable water-resources management, including the application of the ecosystems approach, is promoted; Contingency planning is developed; Additional specific measures are taken to prevent the pollution of groundwaters; The risk of accidental pollution is minimized. To this end, each Party shall set emission limits for discharges from point sources into surface waters based on the best available technology, which are specifically applicable to individual industrial sectors or industries from which hazardous substances derive. The appropriate measures mentioned in paragraph 1 of this article to prevent, control and reduce the input of hazardous substances from point and diffuse sources into waters, may, *inter alia*, include total or partial prohibition of the production or use of such substances. Existing lists of such industrial sectors or industries and of such hazardous substances in international conventions or regulations, which are applicable in the area covered by this Convention, shall be taken into account. (article 3)

The convention further requires a series of follow-up measures and assessments:

> In the framework of general cooperation mentioned in article 9 of this Convention, or specific arrangements, the Riparian Parties shall establish and implement joint programs for monitoring the conditions of transboundary waters, including floods and ice drifts, as well as transboundary impact. The Riparian Parties shall agree upon pollution parameters and pollutants whose discharges and concentration in

transboundary waters shall be regularly monitored. The Riparian Parties shall, at regular intervals, carry out joint or coordinated assessments of the conditions of transboundary waters and the effectiveness of measures taken for the prevention, control and reduction of transboundary impact. The results of these assessments shall be made available to the public in accordance with the provisions set out in article 16 of this Convention. For these purposes, the Riparian Parties shall harmonize rules for the setting up and operation of monitoring programs, measurement systems, devices, analytical techniques, data processing and evaluation procedures, and methods for the registration of pollutants discharged. (article 11)

Soft international environmental law guidance for a more holistic approach to water resources management is found in Agenda 21:

1. "Integrated water resources management is based on the perception of water as an integral part of the ecosystem, a natural resource and a social and economic good, whose quantity and quality determine the nature of its utilization. To this end, water resources have to be protected, taking into account the functioning of aquatic ecosystems and the perenniality of the resource, in order to satisfy and reconcile needs for water in human activities. In developing and using water resources, priority has to be given to the satisfaction of basic needs and the safeguarding of ecosystems. Beyond these requirements, however, water users should be charged appropriately" (para. 18.8).
2. "Integrated water resources management, including the integration of land- and water-related aspects, should be carried out at the level of the catchment basin or sub-basin. Four principal objectives should be pursued, as follows: (a) To promote a dynamic, interactive, iterative and multisectoral approach to water resources management, including the identification and protection of potential sources of freshwater supply, that integrates technological, socio-economic, environmental and human health considerations; (b) To plan for the sustainable and rational utilization, protection, conservation and management of water resources based on community needs and priorities within the framework of national economic development policy; (c) To design, implement and evaluate projects and programs that are both economically efficient and socially appropriate within clearly defined strategies, based on an approach of full public participation, including that of women, youth, indigenous people and local communities in water management policy-making and decision-making; (d) To identify and strengthen or develop, as required, in particular in developing countries, the appropriate institutional, legal and financial mechanisms to ensure that water policy and its implementation are a catalyst for sustainable social progress and economic growth" (para. 18.9).
3. "Freshwater is a unitary resource. Long-term development of global freshwater requires holistic management of resources and a recognition of the interconnectedness of the elements related to freshwater and freshwater quality. There are few regions of the world that are still exempt from problems of loss of potential sources of freshwater supply, degraded water quality and pollution of surface and groundwater sources. Major problems affecting

the water quality of rivers and lakes arise, in variable order of importance according to different situations, from inadequately treated domestic sewage, inadequate controls on the discharges of industrial waste waters, loss and destruction of catchment areas, ill-considered siting of industrial plants, deforestation, uncontrolled shifting cultivation and poor agricultural practices. This gives rise to the leaching of nutrients and pesticides. Aquatic ecosystems are disturbed and living freshwater resources are threatened. Under certain circumstances, aquatic ecosystems are also affected by agricultural water resource development projects such as dams, river diversions, water installations and irrigation schemes. Erosion, sedimentation, deforestation and desertification have led to increased land degradation, and the creation of reservoirs has, in some cases, resulted in adverse effects on ecosystems. Many of these problems have arisen from a development model that is environmentally destructive and from a lack of public awareness and education about surface and groundwater resource protection. Ecological and human health effects are the measurable consequences, although the means to monitor them are inadequate or non-existent in many countries. There is a widespread lack of perception of the linkages between the development, management, use and treatment of water resources and aquatic ecosystems. A preventive approach, where appropriate, is crucial to the avoiding of costly subsequent measures to rehabilitate, treat and develop new water supplies" (para. 18.35).

To fashion a BMP for water resources, Agenda 21 further recommends that "three objectives will have to be pursued concurrently to integrate water-quality elements into water resource management: (a) Maintenance of ecosystem integrity, according to a management principle of preserving aquatic ecosystems, including living resources, and of effectively protecting them from any form of degradation on a drainage basin basis; (b) Public health protection, a task requiring not only the provision of safe drinking-water but also the control of disease vectors in the aquatic environment; (c) Human resources development, a key to capacity-building and a prerequisite for implementing water-quality management" (para. 18.38). Agenda 21 also denotes critical principles for water management:

Key strategic principles for holistic and integrated environmentally sound management of water resources in the rural context may be set forth as follows: (a) Water should be regarded as a finite resource having an economic value with significant social and economic implications reflecting the importance of meeting basic needs; (b) local communities must participate in all phases of water management, ensuring the full involvement of women in view of their crucial role in the practical day-to-day supply, management and use of water; (c) Water resource management must be developed within a comprehensive set of policies for (i) human health; (ii) food production, preservation and distribution; (iii) disaster mitigation plans; (iv) environmental protection and conservation of the natural resource base; (d) It is necessary to recognize and actively support the role of rural populations, with particular emphasis on women." (para 18.68)

These are only a few of Agenda 21's recommendations for developing national and local laws on integrated water management.[15] Other important recommendations, such as the International Law Association's Berlin Rules on Water Resources, have been set forth by international NGO experts.[16]

An interesting illustration of how these soft-law instruments are followed in a hard-law international environmental agreement involves the Zambezi River watershed. This watershed is the subject of an agreement made in 1987 in Harare, Zimbabwe—the Action Plan for the Environmentally Sound Management of the Common Zambezi River System (ZACPLAN). This international agreement details management of the entire Zambezi River basin across nations. It states that a "diagnostic study identified the following main problems relating to the environmentally sound management of the river basin which should be dealt with through selected activities as part of the Zambezi Action Plan." These problems include most of the issues that international environmental law broadly addresses.[17] The ZACPLAN agreement then structures how each national government in the river catchment area will adapt its national environmental laws, providing that "the key to sustainable, environmentally sound development is proper management of the resource base. Such management should take into account the assimilative capacity of the environment, the development goals as defined by national authorities and the economic feasibility of their implementation. The following activities may be undertaken to strengthen the ability of Governments to adopt appropriate environmental management practices for water and natural resources." The plan then discusses the BMPs at length.[18] Many of the Zambezi River legal provisions remain to be implemented because the governments of the region lack the institutional capacity to do so, and because they are coping with other social and economic crises (such as levels of HIV/AIDS at epidemic proportions across the region). Still, the relationship of the MDGs to the implementation of international environmental law is well demonstrated in this instance.

IMPLEMENTING BMPs AND MANAGEMENT SYSTEMS THROUGH ENVIRONMENTAL IMPACT ASSESSMENT

Given the reality of the degradation of grasslands, arid lands, and freshwater resources, and the complexity and urgency of the international environmental law provisions restated and quoted here, what is a nation to do? How can the evidently weak provisions of international environmental law be made robust and effective as nations implement them? How can nations that lack the resources to do so comply with these international norms across all of their territory or sectors?

One principal means exists by which nations can observe and implement provisions of international environmental law on a decision-by-decision basis. This is the technique of environmental impact assessment (EIA). The technique was first developed in the United States under the National Environmental Policy Act of 1969, and in the 1970s and 1980s it began to be widely emulated and adopted in many other nations. The European Union adopted its EIA directive in 1985.[19] In article 206 of UNCLOS, nations agreed to use EIAs for actions that they took affecting the marine environment; this was reiterated in several other hard-law agreements. By the time of the 1992 UNCED meeting, the EIA was believed to be a technique that could evaluate any national action in terms of how applicable international or national environmental rules could be applied to avoid and avert or mitigate any adverse environmental impacts associated with the proposed governmental action.

These assessments are intended to support the goals of environmental protection and sustainable development; to integrate environmental protection and economic decisions at the earliest stages of planning an activity; to predict environmental, social, economic, and cultural consequences of a proposed activity; to assess plans to mitigate any adverse impacts resulting from the proposed activity; and to provide for the involvement of the public, local government, and national government in the review of the proposed activities. Without EIAs the negative impacts of a project or activity on the landscape, water and soil resources, and biodiversity could not be quantified, avoided, or mitigated.

The benefits of incorporating the EIA as an essential element of sustainable development is that it implements a best management approach, follows the precautionary principle, allows evaluation of competing resource uses, identifies project alternatives, and preserves ecosystem services. There are two types of environmental assessment methodologies: strategic environmental assessment (SEA) and EIA. Because they take place at different stages of planning and development, they should be used in conjunction. The EIA evaluates the environmental impacts of specific projects, while the SEA incorporates environmental considerations into policies, plans, and programs.

International environmental law requires the use of EIAs, and the national environmental laws in many nations also require the use of EIAs. These assessments are an integral part of the obligations of state parties under many MEAs. They should be conducted at the earliest possible stage of the planning process, and they must involve the participation of all stakeholders, including local and indigenous communities. They should include analysis of alternatives as well as mitigation plans. Additionally, when the environmental impacts of a project or activity have the potential to have transboundary or multi-jurisdictional effect, an EIA must address these transboundary impacts. The host jurisdiction must notify all affected jurisdictions of the results of the EIA and should consult with those jurisdictions as to alternatives and mitigation. The EIA must also be integrated with other tools such as sustainable forest management, integrated coast zone management, and integrated water resources management.

The fact that nations still perform EIAs in a mechanistic and superficial manner indicates that the use of EIAs is still at an early phase. The duty under international environmental law is clear under many instruments, and national governments need to evaluate how to undertake the assessments more effectively. The CBD provides clear elaboration of EIA duties:

> Each Contracting Party, as far as possible and as appropriate, shall: (a) Introduce appropriate procedures requiring environmental impact assessment of its proposed projects that are likely to have significant adverse effects on biological diversity with a view to avoiding or minimizing such effects and, where appropriate, allow for public participation in such procedures; (b) Introduce appropriate arrangements to ensure that the environmental consequences of its programs and policies that are likely to have significant adverse impacts on biological diversity are duly taken into

account; (c) Promote, on the basis of reciprocity, notification, exchange of information and consultation on activities under their jurisdiction or control which are likely to significantly affect adversely the biological diversity of other States or areas beyond the limits of national jurisdiction, by encouraging the conclusion of bilateral, regional or multilateral arrangements, as appropriate; (d) In the case of imminent or grave danger or damage, originating under its jurisdiction or control, to biological diversity within the area under jurisdiction of other States or in areas beyond the limits of national jurisdiction, notify immediately the potentially affected States of such danger or damage, as well as initiate action to prevent or minimize such danger or damage; and (e) Promote national arrangements for emergency responses to activities or events, whether caused naturally or otherwise, which present a grave and imminent danger to biological diversity and encourage international cooperation to supplement such national efforts and, where appropriate and agreed by the States or regional economic integration organizations concerned, to establish joint contingency plans. (article 14)

The UNCLOS convention also requires EIA performance, stating that "when States have reasonable grounds for believing that planned activities under their jurisdiction or control may cause substantial pollution of or significant and harmful changes to the marine environment, they shall, as far as practicable, assess the potential effects of such activities on the marine environment and shall communicate reports of the results of such assessments in the manner provided in article 205" (article 206).

There are two treaties that specifically address to the duty to undertake EIAs. First, the Convention on Environmental Impact Assessment in a Transboundary Context is one codification of Principle 21 of Declaration of the UN Conference on the Human Environment, which holds that no state should act so as to harm the environment of another nation or the international commons. Second is the Espoo Convention, which provides that:

Environmental impact assessment means a national procedure for evaluating the likely impact of a proposed activity on the environment. Impact means any effect caused by a proposed activity on the environment including human health and safety, flora, fauna, soil, air, water, climate, landscape and historical monuments or other physical structures or the interaction among these factors; it also includes effects on cultural heritage or socio-economic conditions resulting from alterations to those factors; Transboundary impact means any impact, not exclusively of a global nature, within an area under the jurisdiction of a Party caused by a proposed activity the physical origin of which is situated wholly or in part within the area under the jurisdiction of another Party. (article 1(vi)–(viii))

Under the Espoo Convention, nations are to "either individually or jointly, take all appropriate and effective measures to prevent, reduce and control significant adverse transboundary environmental impact from proposed activities" (article 2(1)). The party of origin shall "provide, in accordance with the provisions of this Convention, an opportunity to the public in the areas likely to be affected to participate in relevant environmental impact assessment procedures regarding proposed activities and shall ensure that the opportunity provided to the public of the affected Party is equivalent to that provided to the public of the Party of origin" (article 2(6)). Moreover, the party of origin shall, after completing the EIA, ""without undue delay enter into consultations with the affected Party concerning, *inter alia*, the potential transboundary impact of the proposed activity and measures to reduce or eliminate its impact. Consultations may relate to: (a) Possible alternatives to the proposed activity, including the no-action alternative and possible measures to mitigate significant adverse transboundary impact and to monitor the effects of such measures at the expense of the Party of origin; (b) Other forms of possible mutual assistance in reducing any significant adverse transboundary impact of the proposed activity; and (c) Any other appropriate matters relating to the proposed activity" (article 5).

The Espoo Convention specifies that the EIA document is to contain nine elements:

> (a) A description of the proposed activity and its purpose; (b) A description, where appropriate, of reasonable alternatives (for example, locational or technological) to the proposed activity and also the no-action alternative; (c) A description of the environment likely to be significantly affected by the proposed activity and its alternatives; (d) A description of the potential environmental impact of the proposed activity and its alternatives and an estimation of its significance; (e) A description of mitigation measures to keep adverse environmental impact to a minimum; (f) An explicit indication of predictive methods and underlying assumptions as well as the relevant environmental data used; (g) An identification of gaps in knowledge and uncertainties encountered in compiling the required information; (h) Where appropriate, an outline for monitoring and management programmes and any plans for post-project analysis; and (i) A non-technical summary including a visual presentation as appropriate (maps, graphs, etc.). (appendix II)

The states that ratified the Espoo Convention later elaborated its terms in the Protocol on Strategic Environmental Assessment to the Convention on Environmental Impact Assessment in a Transboundary Context, also known as the Kiev Protocol. This protocol addresses "strategic environmental assessment [which] means the evaluation of the likely environmental, including health, effects, which comprises the determination of the scope of an environmental report and its preparation, the carrying out of public participation and consultations, and the taking into account of the environmental report and the results of the public participation and consultations in a plan or program" (article 2(6)). The protocol also specifies requirements for SEAs:

> 1. Each Party shall ensure that a strategic environmental assessment is carried out for plans and programs referred to in paragraphs 2, 3 and 4 which are likely to have significant environmental, including health, effects.
> 2. A strategic environmental assessment shall be carried out for plans and programs which are prepared for agriculture, forestry, fisheries, energy, industry including mining, transport, regional development, waste management, water management, telecommunications, tourism, town and country planning or land use, and which set the framework for future development consent for projects listed in annex I and any other project listed in annex II that requires an environmental impact assessment under national legislation.
> 3. For plans and programs other than those subject to paragraph 2 which set the framework for future development consent of projects, a strategic environmental assessment shall be carried out where a Party so determines according to article 5, paragraph 1.

4. For plans and programs referred to in paragraph 2 which determine the use of small areas at local level and for minor modifications to plans and programs referred to in paragraph 2, a strategic environmental assessment shall be carried out only where a Party so determines according to article 5, paragraph 1. (article 4)

The second hard-law treaty concerning EIAs is the UNECE Convention on Access to Information, Public Participation in Decision-Making and Access to Justice in Environmental Matters, signed in Aarhus, Denmark, in 1998, and better known as the Aarhus Convention.[20] The convention details how the public is to be informed and consulted in environmental decision making, namely: "The public concerned shall be informed, either by public notice or individually as appropriate, early in an environmental decision-making procedure, and in an adequate, timely and effective manner, *inter alia*, of ... the fact that the activity is subject to a national or transboundary environmental impact assessment procedure" (article 6(2)(e)). This convention has been ratified widely in the former U.S.S.R. (except for Russia) and in several western European nations.

International organizations also require EIAs. For instance, the World Bank's Operational Policy 4.01: Environmental Assessment provides that "environmental assessment (EA) is a process whose breadth, depth, and type of analysis depend on the nature, scale, and potential environmental impact of the proposed project. EA evaluates a project's potential environmental risks and impacts in its area of influence; examines project alternatives; identifies ways of improving project selection, siting, planning, design, and implementation by preventing, minimizing, mitigating, or compensating for adverse environmental impacts and enhancing positive impacts; and includes the process of mitigating and managing adverse environmental impacts throughout project implementation. The Bank favors preventive measures over mitigatory or compensatory measures, whenever feasible" (para. 2). In the World Bank's procedures:

> EA takes into account the natural environment (air, water, and land); human health and safety; social aspects (involuntary resettlement, indigenous peoples, and cultural property); and transboundary and global environmental aspects. EA considers natural and social aspects in an integrated way. It also takes into account the variations in project and country conditions; the findings of country environmental studies; national environmental action plans; the country's overall policy framework, national legislation, and institutional capabilities related to the environment and social aspects; and obligations of the country, pertaining to project activities, under relevant international environmental treaties and agreements. The Bank does not finance project activities that would contravene such country obligations, as identified during the EA. EA is initiated as early as possible in project processing and is integrated closely with the economic, financial, institutional, social, and technical analyses of a proposed project. (para. 3)

The World Bank also requires that during the assessment process for certain categories of projects, the borrower "consult project-affected groups and local nongovernmental organizations (NGOs) about the project's environmental aspects and take their views into account. The borrower initiates such consultations as early as possible. For Category A projects, the borrower consults these groups at least twice: (a) shortly after environmental screening and before the terms of reference for the

EA are finalized; and (b) once a draft EA report is prepared. In addition, the borrower consults with such groups throughout project implementation as necessary to address EA-related issues that affect them" (para. 14).

The EIA is also required among the regional development banks. For instance, the ADB Operations Manual: Environmental Considerations in ADB Operations provides that:

> Environmental assessment is a generic term to describe a process of environmental analysis, management, and planning to address environmental impacts of development policies, strategies, programs, and projects. ADB requires environmental assessment of all project loans, program loans, sector loans, sector development program loans, financial intermediation loans, and private sector investment operations. ADB's environmental assessment process starts as soon as potential projects for ADB financing are identified, and covers all project components whether financed by ADB, cofinanciers, or the borrower. While the scope of the environmental assessment covers the project in a broad sense, the specification of detailed mitigation measures to be implemented by the borrower will focus on components for which the borrower and therefore ADB has some influence. Environmental assessment is ideally carried out simultaneously with the prefeasibility and feasibility studies of the project. Environmental assessment, however, is a process rather than a one-time report, and includes necessary environmental analyses and environmental management planning that take place throughout the project cycle. Important considerations in undertaking environmental assessment include examining alternatives; identifying potential environmental impacts, including indirect and cumulative impacts, and assessing their significance; achieving environmental standards; designing least-cost mitigation measures; developing appropriate environmental management plans and monitoring requirements; formulating institutional arrangements; and ensuring information disclosure, meaningful public consultation, and appropriate reporting of results. (para. 4)

At the 1992 UNCED meeting, the EIA process was mandated in the Rio Declaration on Environment and Development: "Environmental impact assessment, as a national instrument, shall be undertaken for proposed activities that are likely to have a significant adverse impact on the environment and are subject to a decision of a competent national authority" (principle 17).

Regional EIAs across nations is required in many instances. The EU Directive on EIA is matched in North America by the North American Agreement on Environmental Cooperation (signed in 1993, and also known as the NAFTA Environmental Side Agreement), which holds as follows: "Recognizing the significant bilateral nature of many transboundary environmental issues, the Council shall, with a view to agreement between the Parties pursuant to this Article within three years on obligations, consider and develop recommendations with respect to: (a) assessing the environmental impact of proposed projects subject to decisions by a competent government authority and likely to cause significant adverse transboundary effects, including a full evaluation of comments provided by other Parties; (b) and persons of other Parties; (c) notification, provision of relevant information and consultation between Parties with respect to such projects; and (d) mitigation of the potential adverse effects of such projects" (article 10.7).

REGIONAL STREAMLINING OF INTERNATIONAL
ENVIRONMENTAL LAW

One way to cope with the complexity of international environmental legal norms is to recodify them on a regional basis. The guidance can then be focused upon, understood, and adapted to regional geographic conditions. This is being done across the European Union,[21] as well as across the nations in ASEAN.[22] One useful model for such integration is the African Convention on the Conservation of Nature and Natural Resources (signed in Algiers, Algeria, in 1968 and revised at Maputo, Mozambique, in 2003). IUCN's Environmental Law Program assisted the African Union (formerly the Organization of African Unity) with studies based on the norms of international environmental law and BMPs, and the African nations negotiated agreement on a recodified treaty in 2003. This contemporary convention offers an important example of how to integrate the many duties under international environmental law into a specific regional framework.

Instead of taking a purely utilitarian approach to natural resources conservation, the African Convention on the Conservation of Nature and Natural Resources acknowledged the principle of common responsibility for environmental management and called for the conservation and rational use of natural resources for the welfare of present and future generations. Considered the most forward-looking regional agreement of the time, it significantly influenced the development of environmental law in Africa. Twenty-five years later, developments in international environmental law made it necessary to revise the treaty, update its provisions and enlarge its scope, in particular to include the establishment of institutional structures to facilitate compliance and enforcement. This was undertaken under the auspices of the African Union, and the revision was adopted by its heads of state and government in July 2003.

The convention spells out three objectives, which also correspond to key elements of a sustainable development approach: the achievement of ecologically rational, economically sound, and socially acceptable development policies and programs. In keeping with these broad objectives, the parties to the convention are required to enhance the protection of the environment, foster the conservation and sustainable use of natural resources, and coordinate and harmonize polices in these fields (see article II). Articles VI, VII, and VIII consider fundamental rules for the conservation and sustainable management of land resources: land, soil, and water. Article VI addresses the issues of land degradation and directs the parties to take measures for the conservation and sustainable management of land resources, including soil, vegetation, and related hydrological processes. The measures are spelled out in some detail and include the implementation of improved agricultural practices and agrarian reforms, particularly on soil conservation, sustainable farming and forestry practices, and pollution control. The commitments in this provision reflect those contained in the UNCCD.

Article VII of the African Convention compels parties to address the management of water resources—whether underground, surface, or rainwater—in a way to maintain them at the highest possible quantitative and qualitative levels and to ensure the protection of human health. To this end, it requires that measures be taken to control pollution, water-borne diseases, and excessive abstraction. Parties must also ensure that people have access to sufficient and continuous supply of water. In article VIII, the convention directs the parties to take measures for the conservation, sustainable use, and rehabilitation of vegetation cover, and thus to adopt measures to soundly manage forests, woodlands, and wetlands. To this end, states must take concrete steps for the control of forest exploitation and fires and for land clearing for cultivation and grazing by domestic and wild animals, as well as invasive species. Forest reserves must also be established, and reforestation programs carried out.

Article IX deals with species and genetic diversity. It calls for their conservation in situ and for the sustainable use of harvestable plants and animals, whether terrestrial, freshwater, or marine. It requires the parties to conserve their habitat and take ex situ conservation measures. Moreover, it provides for the preservation of as many cultivars or domestic varieties of animals and plants as possible, as well as the control of both intentional and accidental introductions of exotic species and of genetically modified organisms. These provisions are in line with the requirements of the CBD.

In article XII(1) the Convention directs the parties to establish, maintain, and extend terrestrial and marine conservation to ensure the long-term conservation of biological diversity. Conservation areas are to be dedicated specifically for conservation and management purposes. They are classified according to the six main categories of protected areas that are acknowledged under IUCN's Guidelines for Protected Areas Management Categories (1994).

Article XIII, "Processes and Activities Affecting the Environment and Natural Resources," commits parties to take measures to prevent, eliminate, and reduce the adverse effects of radioactive, toxic, and other hazardous substances and waste. In order to achieve this objective, parties must harmonize their policies in a manner consistent with their international legal obligations. In particular, they should "establish, strengthen and implement national standards, including for ambient environmental quality, emission and discharge limits, as well as process and production methods and products quality." They are also required to collaborate with the competent international organizations over this matter.

In addition to the substantive issues outlined above, the African Convention contains procedural and institutional provisions to strengthen public participation (article XVI), promote research, education, and capacity-building (articles XVIII and XX), protect traditional knowledge (article XVII), and establish institutional mechanisms for dispute settlement, implementation, and compliance (articles XXI, XXIII, and XXIX). The convention embodies a comprehensive and integrated regional approach to environmental protection and sustainable development. It reflects a renewed perception of resource management that reconciles nature and culture. People-centered and advocating integrated resource management, it provides an excellent case study for international legal principles and best practices in the area of combating land degradation and natural resources management.

CONCLUDING REFLECTIONS

International environmental law awaits adoption and application of strong national measures that can ensure its implementation. Regional measures can promote such action. However, even when the African Convention is ratified, it will have little effect in guiding national environmental protection measures unless the nations of that region build the capacity and will to do so. On one hand, international environmental law today has achieved the status of a persuasive and important set of norms. On the other hand, its beneficial results are realized across nations in a very inconsistent and incomplete way. It awaits a more serious pattern of implementation.

International environmental law arises from and largely depends upon the environmental sciences. Ecology has refined what Darwin discovered, but international law is still enmeshed in sovereignty of nations and artificial legal fictions that each nation can manage its territory by itself. Even Agenda 21 is premised on promoting common actions by independent states, rather than requiring all authorities across a shared resource to follow the same rules. Regional steps, such as the federal rules for the United States or the supranational rules for the European Union, can compel adherence to the same rules across large regions. This regional approach is seen in Chinese law, in the action plans of the ASEAN, in Brazil's federal laws, and in the Union List rules in India. Its promise is seen in the African Convention. Less promising is the current inaction in the Russian Federation or in Eurasian nations of the former Soviet Union, or in the Hispanic nations of South America. From a global perspective, much regional integration remains to be addressed. Nonetheless, international environmental law is a recognized force, and the fate of nations lies in realizing its norms.

Review of the international legal norms on land stewardship reveals a sound body of international rules and recommendations. All the attention that has been given to preparing these norms must now be matched by efforts to explain them to national constituencies and secure their implementation. Greater focus is needed on the methodologies and tools for doing so, such as the techniques of ecosystem management or environmental impact assessment. Perhaps most important is the urgent need for national leaders—judicial, executive, and legislative—to embrace the norms of international environmental law. Without institutional and political support, the norms of international environmental law will remain in the realm of what is called *lex ferenda*, the law we envision and wish to be, rather than *lex lata*, the law as it is in fact.

The advance of international environmental law is an evolutionary process for human society, learning how to live in the natural systems of the earth. It is a work in progress, at an early stage in its development.

ACKNOWLEDGMENTS: The author acknowledges the assistance of Sarah Burt, of the Environmental Law Institute, in the research for parts of this chapter. The research on land degradation issues discussed in this chapter was also provided in an earlier version to Wuhan University's Research Institute for Environmental Law, for use in the development of its education programs on combating land degradation in China (done in cooperation with the Asian Development Bank and several Chinese provincial and central government ministries).

Notes

1. This argument suggests that environmental law is embedded in environmental science—that it is not a purely normative policy construct. It also suggests that environmental law is neither strictly international nor national in scope and application, but instead must be seen as a coherent set of rules to guide human society in its interaction

with natural systems. Since nature does not exist as a construct of national governments, it is the scientific understanding of how humans exist in nature that informs how environmental law must function. This is entirely alien to the rules of international law, in which only nations (and some international intergovernmental organizations of nations) determine what it international law, and national authorities are free to posit whatever national laws they wish. Environmental law is constrained to follow the knowledge of the environment. See, for example, O. Kolbasov (1985).

2. The Bonn Convention provides that member states need to take action to avoid any migratory species from becoming endangered. In particular, the parties (a) should promote, cooperate in, and support research relating to migratory species; (b) should endeavor to provide immediate protection for migratory species included in appendix I of the convention; and (c) should endeavor to conclude agreements covering the conservation and management of migratory species included in appendix II of the convention.

3. The Basel Convention requires that each member state shall take the following measures.

(a) Ensure that the generation of hazardous wastes and other wastes within it is reduced to a minimum, taking into account social, technological and economic aspects; (b) Ensure the availability of adequate disposal facilities, for the environmentally sound management of hazardous wastes and other wastes, that shall be located, to the extent possible, within it, whatever the place of their disposal; (c) Ensure that persons involved in the management of hazardous wastes or other wastes within it take such steps as are necessary to prevent pollution due to hazardous wastes and other wastes arising from such management and, if such pollution occurs, to minimize the consequences thereof for human health and the environment; (d) Ensure that the transboundary movement of hazardous wastes and other wastes is reduced to the minimum consistent with the environmentally sound and efficient management of such wastes, and is conducted in a manner which will protect human health and the environment against the adverse effects which may result from such movement. (article 4(2))

4. For more information on global trends in these areas, see UNEP's information database, which is called the Global Environmental Outlook (GEO), at www.unep.org. See also UNEP GEO (2006).

5. For instance, national implementation of international norms to develop and apply a cross-sectoral approach could include the following steps. (a) It is necessary to define the management scale beyond the boundaries of a single habitat type, conservation area, or political or administrative unit to encompass an entire ecosystem. (b) Because the needs of human beings play a major role in the disturbance of ecosystems, natural resource management programs should integrate economic and social factors into ecosystem management goals. (c) Because ecosystems are dynamic, management planning should be flexible and adaptive so that management strategies can be adjusted in response to new information and experience. (d) International principles for land degradation control should be intersectoral and participatory and should acknowledge the importance of the role of women in combating land degradation. (e) Such principles should make use of scientific and technical knowledge as well as local knowledge and experience. (d) Local and regional as well as national authorities need clear rules for combating land degradation.

6. Both soft law and a small body of hard law exist to guide governments in enacting and implementing land use laws across sectors. See, for example, references provided by John Nolon (2005).

7. On issues of property law reform, see the Bathurst Declaration on Land Administration for Sustainable Development (1999). Section 3 provides that the "lack of

secure property rights in the land will inhibit investments in housing, sustainable food production and access to credit, hinder good governance and the emergence of civic societies, reinforce social exclusion and poverty, undermine long term planning, and distort prices of land and services. Without effective access to land and property, market economies are unable to evolve and the goals of sustainable development cannot be realized." Section 4 further notes:

> Given that more than half the people in most developing countries currently do not have access to secure property rights in land and given the concerns about the sustainability of development around the globe and the growing urban crisis, the Bathurst Workshop *recommends* a global commitment to: (a) Providing effective legal security of tenure and access to property for all men and women, including indigenous peoples, those living in poverty and other disadvantaged groups; (b) Promoting the land administration reforms essential for sustainable development and facilitating full and equal access for men and women to land-related economic opportunities, such as credit and natural resources; (c) Investing in the necessary land administration infrastructure and in the dissemination of land information required to achieve these reforms; and (d) Halving the number of people around the world who do not have effective access to secure property rights in land by the year 2010.

8. Forests ecosystem management measures were expressly not endorsed at UNCED in 1992, although nonbinding principles were announced (see the Statement of Principles for a Global Consensus on the Management, Conservation and Sustainable Development of All Types of Forests [1992]), and subsequently the UN General Assembly has hosted a forum on forests as a means to debate and formulate a consensus on forests. A treaty on tropical timber production is premised on an assumption that sustained yields are possible (see the International Tropical Timber Organization's Web site at http://www.itto.or.jp/live/index.jsp), but this is more a trade agreement to promote tropical timber sales than a sustainable development agreement. For more information, see the following sources: Statement on Criteria and Indicators for the Conservation and Sustainable Management of Temperate and Boreal Forests ("Santiago Declaration"); Statement of Principles for a Global Consensus on the Management, Conservation and Sustainable Development of All Types of Forests; and chapter 11, Agenda 21. Other helpful sources include the UN Forum on Forests Web site (http://www.un.org/esa/forests/ipf_iff.html), the CBD Forest Biodiversity Programme Web site (http://www.biodiv.org/programmes/areas/forest/default.asp), the International Union of Forestry Researchers' sites on NGO expert bodies recommendations (http://www.iufro.org/) and principles and issues (http://www.iufro.boku.ac.at), and the Forest Stewardship Council Web site (http://www.fsc.org/en/).

Forests policies are provided as soft law in Agenda 21's chapter 11:

> Forests worldwide have been and are being threatened by uncontrolled degradation and conversion to other types of land uses, influenced by increasing human needs; agricultural expansion; and environmentally harmful mismanagement, including, for example, lack of adequate forest-fire control and anti-poaching measures, unsustainable commercial logging, overgrazing and unregulated browsing, harmful effects of airborne pollutants, economic incentives and other measures taken by other sectors of the economy. The impacts of loss and degradation of forests are in the form of soil erosion; loss of biological diversity, damage to wildlife habitats and degradation of watershed areas, deterioration of the quality of life and reduction of the options for development...The present situation calls for urgent and consistent action for conserving and sustaining forest resources. (paras. 11.10–11.11)

Further policy guidance is contained in the Statement on Criteria and Indicators for the Conservation and Sustainable Management of Temperate and Boreal Forests: "The development of criteria and indicators for the conservation and sustainable management of temperate and boreal forests is an important step in implementing the UNCED Forest Principles and Agenda 21, and is relevant to the UN conventions on biodiversity, climate change and desertification. Six criteria were identified for sustainable management of temperate and boreal forests." This soft-law statement in article 3 expressly states that "no priority or order is implied in the alpha numeric listing of the criteria and indicators." The criteria are as follows.

Criterion 1: Conservation of biological diversity. Biological diversity includes the elements of the diversity of ecosystems, the diversity between species, and genetic diversity in species.

Criterion 2: Maintenance of productive capacity of forest ecosystems

Criterion 3: Maintenance of forest ecosystem health and vitality

Criterion 4: Conservation and maintenance of soil and water

Criterion 5: Maintenance of forest contribution to global carbon cycles

Criterion 6: Maintenance and enhancement of long-term multiple socioeconomic benefits to meet the needs of societies

Criterion 7: Legal, institutional, and economic framework for forest conservation and sustainable management and associated indicators relate to the overall policy framework of a country that can facilitate the conservation and sustainable management of forests.

These soft-law policies have not been widely endorsed by nations, and they remain outside the realm of hard, treaty-based international environmental law.

9. Agricultural resource management has long been the responsibility of the FAO, and thus a greater body of international environmental law exists. New issues, such as the unwanted accumulation of waste nitrogen from fertilization, are also being addressed through new soft-law recommendations. For more information, see the references in Agenda 21, chapter 14. See also Web sites for the FAO Plant Production and Protection Division (http://www.fao.org/ag/AGP/Default.htm), the FAO Land and Water Development Division (http://www.fao.org/ag/AGL/default.stm), FAO Good Agricultural Practices (http://www.fao.org/ag/AGP/AGPC/doc/themes/5g.html), the International Initiative for the Conservation and Sustainable Use of Pollinators (http://www.biodiv.org/programmes/areas/agro/pollinators.aspx), the International Nitrogen Initiative (http://initrogen.org/), and the Alternative Farming Systems Information Center (http://www.nal.usda.gov/afsic/agnic/agnic.htm#definition).

10. For further soft law and policy analysis of grasslands, see the Web database for the FAO Crop and Grassland Service (AGPC) (http://www.fao.org/ag/AGP/AGPC/doc/Default.htm). See also the CBD Dry and Sub-humid Lands Biodiversity Programme and the CBD Official Chinese Text.

11. Sinks are resources that soak up more carbon than they emit. The concept of carbon sinks is based on the natural ability of trees, other plants, and the soil to soak up carbon dioxide and temporarily store the carbon in wood, roots, leaves, and the soil.

12. Further information on the legal authorities for these BMPs can be found in the following legal instruments: Convention on Wetlands of International Importance (see especially material on waterfowl habitat) and Agenda 21, chapter 12. See also the CBD Inland Waters Biodiversity Programme Web site (http://www.biodiv.org/programmes/areas/water/default.asp), the FAO Soil Biodiversity Portal (http://www.fao.org/ag/agl/agll/soilbiod/index_en.stm), the Web site of the UNEP strategy on land use management and

soil conservation (http://www.unep.org/pdf/UNEP-strategy-land-soil-03–2004.pdf), and the European Soil Bureau Web site (http://eusoils.jrc.it/esbn/Esbn_overview.html).

13. See, for example, the Great Lakes Water Quality Agreement between Canada and the United States, the Convention on the Protection and Use of Transboundary Watercourses and International Lakes, and the UN Convention on Non-Navigational Use of International Waters.

14. See, for example, the IUCN Water and Nature Initiative Web site (http://www.iucn.org/themes/wani/), the Global Water Partnership Web site (http://www.gwpforum.org/servlet/PSP), and the UNEP Water Policy and Strategy Web site (http://www.unep.org/dpdl/water/).

15. For instance, Agenda 21 also makes the following recommendations.

The development of new irrigation areas at the above-mentioned level may give rise to environmental concerns in so far as it implies the destruction of wetlands, water pollution, increased sedimentation and a reduction in biodiversity. Therefore, new irrigation schemes should be accompanied by an environmental impact assessment, depending upon the scale of the scheme, in case significant negative environmental impacts are expected. When considering proposals for new irrigation schemes, consideration should also be given to a more rational exploitation, and an increase in the efficiency or productivity, of any existing schemes capable of serving the same localities. Technologies for new irrigation schemes should be thoroughly evaluated, including their potential conflicts with other land uses. The active involvement of water-users groups is a supporting objective. (para. 18.72)

The objectives with regard to water management for inland fisheries and aquaculture include conservation of water-quality and water-quantity requirements for optimum production and prevention of water pollution by aquacultural activities. The Action Programme seeks to assist member countries in managing the fisheries of inland waters through the promotion of sustainable management of capture fisheries as well as the development of environmentally sound approaches to intensification of aquaculture. (para. 18.74)

The objectives with regard to water management for livestock supply are twofold: provision of adequate amounts of drinking-water and safeguarding of drinking-water quality in accordance with the specific needs of different animal species. This entails maximum salinity tolerance levels and the absence of pathogenic organisms. No global targets can be set owing to large regional and intra-country variations. (para. 18.75)

16. In 2004 the ILA adopted its Berlin Rules on Water Resources, which provide in part that (a) "States shall use their best efforts to manage surface waters, groundwater, and other pertinent waters in a unified and comprehensive manner" (article 5); (b) "States shall use their best efforts to integrate appropriately the management of waters with the management of other resources" (article 6); (c) "Basin States shall in their respective territories manage the waters of an international drainage basin in an equitable and reasonable manner having due regard for the obligation not to cause significant harm to other basin States. In particular, basin States shall develop and use the waters of the basin in order to attain the optimal and sustainable use thereof and benefits therefrom, taking into account the interests of other basin States, consistent with adequate protection of the waters" (article 12); (d) "Every individual has a right of access to sufficient, safe, acceptable, physically accessible, and affordable water to meet that individual's vital human needs" (article <number>); (e) "States shall ensure the implementation of the right of access to water on a non-discriminatory basis. States shall progressively realize the right of access to water by:

[1] Refraining from interfering directly or indirectly with the enjoyment of the right; [2] Preventing third parties from interfering with the enjoyment of the right; [3] Taking measures to facilitate individuals access to water, such as defining and en-forcing appropriate legal rights of access to and use of water; [4] Providing water or the means for obtaining water when individuals are unable, through reasons beyond their control, to access water through their own efforts. [5] States shall monitor and review periodically, through a participatory and transparent process, the realization of the right of access to water" (article 17); (f) "States shall compensate persons or communities displaced by a water program, plan, project, or activity and shall assure that adequate provisions are made for the preservation of the livelihoods and culture of displaced persons or communities" (article 21); (g) "States shall take all appropriate measures to protect the ecological integrity necessary to sustain ecosystems dependent on particular waters" (article 22); and (h) "States shall take all appropriate measures to ensure flows adequate to protect the ecological integrity of the waters of a drainage basin, including estuarine waters" (article 24).

17. The ZACPLAN-identified problems are as follows: (a) inadequate monitoring and exchanges of information with regard to climatic data and water quantity and quality, including pollution control; (b) soil erosion and inadequate soil and water conservation and floodplain management; (c) deforestation due to population growth and pressure on land; (d) lack of adequate drinking water supply and proper sanitation facilities; (e) insufficient community participation, especially on the part of women as "end users" of water, in planning, construction, and maintenance of water supply and sanitation systems; (f) inadequate health education for the public, especially for women; (g) inadequate land-use and river basin planning in general; (h) inadequate human resources development; (i) inadequate coordination and consultation both at national and river basin levels; (j) degradation of the natural resources base; (k) degradation of flora and fauna; (l) inadequate information on environmental impacts of water resources and related development projects, such as hydropower irrigation; (m) inadequate dissemination of information to the public; and (n) inadequate protection of wetlands.

The objective of ZACPLAN is to overcome the problems listed above and thus to promote the development and implementation of environmentally sound water resources management in the whole river system. It will contribute to the incorporation by the river basin states of environmental considerations in water resources management while increasing long-term sustainable development in the river basin. See annex I (14–15).

18. At annex I (29), ZACPLAN provides for:

(a) Strengthening or expansion of the relevant ongoing development activities that demonstrate sound environmental management practices; (b) Improvement of drinking water supply, sanitation and human health through strengthening of sector institutions, drinking water supply and sanitation programs; (c) Development of water quality control programs based on a uniform water monitoring system; (d) Encouraging "end users" of water, women in particular, who are actually in charge of making use of water in daily life, to participate in planning, construction and maintenance of water distribution, purification and sanitation systems; (e) Co-operation in preparedness for pollution emergencies and water-related natural hazards, and measures to prevent them and/or mitigate their consequences; (f) Environmentally sound development of water resources to meet the demand for water for industries, mines, irrigation, hydropower, navigation, drinking water supply, etc.; (g) Co-operation in the application of existing international measures to reduce and control the degradation and wasteful use of the natural resource base, to combat the vast problem of desertification, and to co-ordinate efforts concerning the problems of land-use practices in relation to flood and drought control and management and pollution control; (h) Formulation of regionally and locally applicable programs including guidelines and standards for the management and control of domestic, agricultural and industrial waste water, including the development of principles governing the treatment and discharge of such wastes; (i) Integration of environmental management components in decision-making on water and water-related projects; (j) Harmonization of policies on the management of wildlife, genetic resources, natural habitats and landscapes; (k) Co-operation in the establishment and management of protected rivers, lakes (natural and man-made), coastal areas of the river basin and its related marine habitats, such as wetlands, nurseries, breeding grounds and mangroves, including the training of technical personnel and managers in the conservation of wildlife and habitats; (l) Co-operation in devising land-use practices, watershed management, soil conservation and development patterns appropriate for conditions in the river basin and its related marine regions, including improvement of national capabilities to assess the environmental impact of development; (m) Co-operation in the preparation of measures to conserve wood resources, and to increase its supply on a sustainable basis which may reduce the rate of deforestation. In this context, improvement of biomass fuel processing and combustion techniques should be investigated; (n) Co-operation in the assessment and utilization of fisheries to achieve the highest rational utilization on a sustainable basis; (o) Development of a river basin planning process based, *inter alia*, on sound environmental management practices; (p) Studies of the environmental, social and cultural effects of tourism, and elaboration of environmentally sound strategies for tourism development; (q) Implementation of intensive human resources development programs to support the above measures and provision of environmental education and training in order to develop the knowledge of human resources in all basin countries.

19. A good general source of EU environmental laws and information on EIAs can be found on the EUROPA (Gateway to the EU) Web site's environment page (http://europa. eu.int/comm/environment/eia/home.htm).

20. A page on the UNECE Web site (http://www.unece.org/env/pp/documents/chinese. pdf) allows readers to access different language translations of the Aarhus Convention in different language translations.

21. See, for example, the European Union Framework Directive on Waste Disposal, as well as the European Union Integrated Pollution Prevention and Control directive.

22. See, for example, the ASEAN Agreement on the Conservation of Nature and Natural Resources.

References

AARHUS Convention [The UNECE Convention on Access to Information, Public Participation in Decision-making and Access to Justice in Environmental Matters]: http://www.unece.org/env/pp/treatytext.htm (accessed November 13, 2007).

Action Plan for the Environmentally Sound Management of the Common Zambezi River System: http://www.fao.org/docrep/W7414B/w7414b0j.htm

The ADB Operations Manual: Environmental Considerations in ADB Operations: http://www.adb.org/Documents/Manuals/Operations/OMF01_25sep06.pdf (accessed November 13, 2007).

African Convention on the Conservation of Nature and Natural Resources. Revised: http://www.intfish.net/treaties/africa2003.htm

Agenda 21, the Rio Declaration on Environment and Development: http://www.un.org/documents/ga/conf151/aconf15126-1annex1.htm (accessed November 13, 2007).

Asian Development Bank, Operations Manual: Environmental Considerations in ADB Operations, available at www.adb.org

Association of Southeast Asian Nations, Agreement on the Conservation of Nature and Natural Resources: http://sedac.ciesin.columbia.edu/entri/texts/asean.natural.resources.1985.html.

Basel Convention on the Transboundary Movements of Hazardous Wastes and Their Disposal: http://www.basel.int/text/documents.html (accessed November 13, 2007).

Bathurst Declaration on Land Administration for Sustainable Development: http://www.fig.net/pub/figpub/pub21/figpub21.htm (accessed November 13, 2007).

The Bonn Convention on the Conservation of Migratory Species of Wild Animals: http://www.undp.org/biodiversity/biodiversitycd/frameCMS.htm (accessed November 13, 2007).

Bruch, C., and E. Mrema. 2006. *Manual on compliance with and enforcement of multilateral environmental agreements.* Nairobi, Kenya: United Nations Environment Programme.

CBD Dry and Sub-humid Lands Biodiversity Programme: http://www.cbd.int/drylands/pow.shtml (accessed: 13 November, 2007).

CBD's Subsidiary Body on Scientific, Technical and Technological Advice (SBSTTA): http://www.cbd.int/convention/sbstta.shtml (accessed November 13, 2007).

Convention on Biological Diversity: http://www.cbd.int/convention/convention.shtml (accessed: November 13, 2007).

Convention on International Trade in Endangered Species of Wild Fauna and Flora: http://www.cites.org/eng/disc/text.shtml (accessed November 13, 2007).

The Convention on the Protection and Use of Transboundary Watercourses and International Lakes: http://www.unece.org/env/water/text/text.htm (accessed November 13, 2007).

Convention on Wetlands of International Importance (RAMSAR): http://www.ramsar.org/index_very_key_docs.htm (accessed November 13, 2007).

Craig, D., N. Robinson, and K. K. Lian. 2002. *Capacity building for environmental law in the Asian and Pacific region.* 2 vols. Manila, Philippines: Asian Development Bank.

The Environmental Law Institute (ELI): http://www2.eli.org/index.cfm (accessed November 13, 2007).

The Environmentally Sound Management of the Common Zambezi River System (ZACPLAN): http://www.fao.org/docrep/W7414B/w7414b0j.htm (accessed November 13, 2007).

The (Espoo) Convention on Environmental Impact Assessment in a Transboundary Context: http://www.unece.org/env/eia/documents/conventiontextenglish.pdf.

European Union Framework Directive on Waste Disposal: http://europa.eu.int/scadplus/leg/en/lvb/121197.htm.

European Union Integrated Pollution Prevention and Control Directive: http://europe.eu.int/scadplus/leg/en/lvb/128045.htm.

General Agreement on Tariffs and Trade: http://www.wto.org/english/docs_e/legal_e/ursum_e.htm#General (accessed November 13, 2007).

Global Environmental Outlook (GEO): http://www.unep.org/geo/ (accessed November 13, 2007).

Hardin, G. 1968. Tragedy of the Commons. *Science* 162: 1243–1248.

Holdgate, M. 1999. *The green web: A union for world conservation.* London: Earthscan.

The International Federation of Surveyors—FIG: http://www.fig.net/indexmain.htm (accessed November 13, 2007).

The International Framework for Environmental Compliance and Enforcement (INECE): http://www.inece.org/index.html (accessed November 13, 2007).

The International Law Association (ILA): http://www.ila-hq.org/ (accessed: November 13, 2007).

The International Law Association's Berlin Rules on Water Resources: http://www.ila-hq. org/pdf/Water%20Resources/Final%20Report%202004.pdf (accessed November 13, 2007).

International Union for the Conservation of Nature and Natural Resources (IUCN), "Guidelines for Protected Areas Management Categories," available at www.iucn.org/ themes/wcpa

International Year of Deserts and Desertification: http://www.unccd.int/publicinfo/ iyddlogo/menu.php (accessed November 13, 2007).

IUCN's Guidelines for Protected Areas Management Categories: http://www.iucn.org/ themes/wcpa/pubs/guidelines.htm (accessed November 13 2007).

IUCN Statutes: http://www.iucn.org/members/statutes.htm (accessed November 13, 2007).

Johannesburg Declaration on Sustainable Development: http://www.un.org/esa/sustdev/ documents/WSSD_POI_PD/English/POI_PD.htm (accessed November13 , 2007).

Kolbasov, O. S. 1985. *Ecological law.* Moscow: Progress Press.

Kyoto Protocol to the United Nations Framework Convention on Climate Change: http:// unfccc.int/essential_background/kyoto_protocol/items/1678.php (accessed: November 13, 2007).

Leopold, A. 1949. *A Sand County almanac, and sketches here and there.* New York: Oxford University Press.

Nolon, J. R. 2006. *Compendium of land use laws for sustainable development.* Cambridge: Cambridge University Press.

North American Agreement on Environmental Cooperation: http://www.cec.org/pubs_ info_resources/law_treat_agree/naaec/index.cfm?varlan=english (accessed November 13, 2007).

Protocol on Strategic Environmental Assessment to the Espoo Convention: http://www. unece.org/env/eia/documents/protocolenglish.pdf.

Rio Declaration on Environment and Development: http://www.un.org/documents/ ga/conf151/aconf15126–1annex1.htm and at http://www.unep.org/Documents. multilingual/Default.asp?DocumentID=78&ArticleID=1163.

Robinson, N. A., ed. 1992. *Agenda 21 and the UNCED proceedings.* 6 vols. New York: Oceana Publications.

Robinson, N. 2004. Forest Fires. *Pace Environmental Law Review.*

Robinson, N. A. 2004. *Strategies toward sustainable development: Implementing Agenda 21.* New York: Oceana Publications.

SBSTTA Recommendation III/1: http://www.cbd.int/recommendations/default.shtml?m= SBSTTA-03&id=7004&lg=0 (accessed November 13, 2007).

Speth, J. G. 2004. *Red sky at morning: America and the crisis of the global environment.* New Haven, CT: Yale University Press.

Statement of Principles for the Sustainable Management of Forests: http://www.un.org/ documents/ga/conf151/aconf15126-3annex3.htm (accessed November 13, 2007).

Statement of Principles for a Global Consensus on the Management, Conservation and Sustainable Development of all Types of Forests: http://habitat.igc.org/agenda21/ forest.html (accessed: November 13, 2007).

Statement on Criteria and Indicators for the Conservation and Sustainable Management of Temperate and Boreal Forests ("Santiago Declaration"): http://www.fs.fed.us/land/sustain_dev/sd/sfmsd.htm.

Tolba, M. K., and I. Rummel-Bulska. 1998. *Global environmental diplomacy: Negotiating environmental agreements for the world, 1973–1992.* Cambridge, MA: MIT Press.

United Nations Convention to Combat Desertification: http://www.unccd.int/ (accessed November 13, 2007).

UNCLOS: http://www.un.org/Depts/los/convention_agreements/texts/unclos/unclos_c.pdf.

UNFCCC. http://unfccc.int/2860.php.

UNCED. 1992. Statement of Principles for a Global Consensus on the Management, Conservation and Sustainable Development of All Types of Forests. Rio de Janeiro.

United Nations Convention on the Law of the Sea: http://www.un.org/Depts/los/convention_agreements/convention_overview_convention.htm (accessed November 13, 2007).

United Nations Convention on Non-Navigational Use of International Watercourses: http://untreaty.un.org/ilc/texts/instruments/english/conventions/8_3_1997.pdf (accessed November 13, 2007).

The United Nations Development Goals (MDGs): http://www.un.org/millenniumgoals/ (accessed November 13, 2007).

U.N. Environment Programme, Global Environmental Outlook, Report for 2006, available at www.unep.org/geo/yearbook/yb2006.

United Nations Long-Range Transboundary Agreement and Protocols: http://www.unece.org/env/lrtap/lrtap_h1.htm; http://www.unece.org/env/lrtap/status/lrtap_s.htm (accessed November 13, 2007).

Weiss, E. B. 1995. New directions in international environmental law. In *Capacity building for environmental law in the Asian and Pacific region,* eds. D. Craig, N. Robinson, and K. K. Lian, 13. Manila, Philippines: Asian Development Bank.

World Bank's Operational Policy 4.01: Environmental Assessment: http://wbln0018.worldbank.org/Institutional/Manuals/OpManual.nsf/8d1a4edd930ec36-6852567fa00106d34/9367a2a9d9daeed38525672c007d0972?OpenDocument (accessed November 13, 2007).

The World Heritage Convention: http://whc.unesco.org/en/conventiontext/ (accessed November 13, 2007).

World Commission on Environment and Development. 1987. *Our common future.* Oxford: Oxford University Press.

22

The Evolution of Sustainability in Forest Management Policy

V. Alaric Sample

Many of the chapters in this volume describe a transformation from "conservation," described as a controlled drawing down of the supply of a limited resource, to "sustainability," characterized by the careful utilization and continual replenishment of a resource in perpetuity. In the science and practice of forestry, sustainability has always been a central concern. Indeed, it was concern that forest resources would be inadvertently depleted, leading to unacceptable social and economic impacts, that first gave rise to the systematic study of forests and a scientific approach to the long-term management of forests as renewable resources. What has changed most in recent years, and formed the basis for "sustainable forest management" as we define it today, is the range and diversity of resources that forests are seen to represent, and the societal values associated with the protection and perpetuation of this array of resources.

Originating with the simple aim of avoiding local timber shortages, sustained-yield forest management evolved into a highly technical process of modeling growth, mortality, and risk in order to set timber removals at a level that theoretically could be maintained in perpetuity. Changing scientific understanding of the ecological functioning of forest ecosystems has challenged the notion that a sustained yield of timber is equivalent to sustaining all the components and natural processes necessary to maintain the long-term health and productivity of these ecosystems. Continuing uncertainty over what is socially and economically acceptable, as well as ecologically sustainable, will make optimality in forest management much more difficult to achieve than in the past.

There is an ongoing public debate, both in the United States and abroad, as to what actually constitutes sustainable forest management. This chapter will briefly summarize the historical evolution of sustainability in forest management, examine the ways in which natural resource policy development in the United States has promoted forest conservation, and explore additional possibilities for policy development to achieve sustainable forest management at the local level, and in the global context.

ORIGINS OF SUSTAINED-YIELD FOREST MANAGEMENT

Sustainability in forest management began as a socioeconomic as well as biological concept. Early forest managers developed an understanding of natural forestry productivity—and how it might be enhanced through silviculture—to maintain a continuous supply of wood, game, and other products for human use and consumption. The concept was fundamentally driven by the desire to avoid the social and economic disruption associated with shortages of timber, whether for local use or as the basis for a community export economy. Forest products clearly held the potential of being a perpetually renewable resource, and foresters undertook the responsibility of making this so.

The origins of sustained-yield forest management can be found in late-medieval Europe (Heske 1938). The lack of well-developed systems for transportation and communication at this time resulted in a system of small, independent political units with high customs barriers that prevented any significant degree of regional trade (Waggener 1977). Local consumption was almost entirely dependent on local production, and communities had to be largely self-sufficient. There was a distinct possibility of exhausting local timber resources unless collective use was strictly controlled, and the production and consumption of forest products became highly regulated. This applied not only to the cutting of timber and fuelwood, but to the gathering of leaf litter and grazing of livestock, both of which were understood to affect long-term soil productivity in forests. It has been argued that the concepts of secure land tenure for private property owners, mutual coercion by mutual consent under common law, and government intervention in free markets to protect the broader public interest—principles basic to the development of a constitutional democracy—had their origins in communities such as these (Adams 1993), seeking to avoid a "tragedy of the commons" (Hardin 1968). Perhaps because of the opportunities it afforded for stable employment and income in rural communities, this approach to sustained-yield forestry persisted long after improved transportation and communication systems had reduced the need for local self-sufficiency and turned wood into a widely traded economic commodity.

It was in this context that the concept of the "regulated forest" came into being. A regulated forest is one managed to yield a regular, periodic, and sustainable harvest of timber. The objective of sustained-yield management by itself does not indicate a single specific harvest level, since a forest can be sustained at a range of different management intensities. However, the objective of maximizing the sustainable volume of the timber harvest does generate a unique result. For even-aged stands, such an approach sets the length of rotation according to a biological rule, "culmination of mean annual increment," which determines that harvest occur when the mean of the annual increment of growth in the stand reaches a maximum (Smith 1962). This approach recognizes that as trees increase in size they add volume at an increasing rate, until at maturity the annual increase in volume falls below the average growth rate calculated over the life of the tree (Davis 1966). The culmination of mean annual increment rule gives the rotation age at which the sustainable harvest volume from the forest will be maximized. The harvest level is determined using Hanzlik's formula, which divides the net growth over the entire area of the economic enterprise by the rotation length and indicates the average annual volume of timber that can be removed on a sustainable basis.

This harvesting rule was complicated as far back as the mid-nineteenth century, when careful observers recognized that commercial timber production and harvesting would have a financial, not a strictly biological, objective. Although harvesting at the culmination of mean annual increment maximized the physical volume of the harvest, it almost never maximized the financial returns from the forest. Martin Faustmann ([1849] 1995) showed that, in a world with positive interest rates, the optimal financial rotation length was shorter than the biological rotation length. With a profit objective, the sustainability of timber production is a by-product of achieving financial maximization. On the assumption that the highest-value use of the land is timber production, forest management will involve a regime of harvests and regeneration that maximizes the financial returns from the forest as an economic asset. Similarly, the level of additional investments of labor and capital, such as thinning or fertilization, requires that the marginal return from each activity exceed its marginal cost.

SUSTAINED-YIELD FORESTRY IN THE UNITED STATES

These European concepts of sustained-yield forest management were transplanted to the United States in the late nineteenth century at a time of growing concern over the possibility of a timber famine—nationally as well as locally. Forests in the United States had been regarded as both an inexhaustible resource and an obstacle to the westward expansion of agriculture. At the time, wood was still the major building material and the predominant source of fuel. Vast areas of forest had been cleared but not reforested, and there was a very real concern that a timber shortage would begin to seriously limit the prospects for future economic growth.

Forest policy in the United States began with a reservation approach—withdrawing and protecting the remaining forests from private exploitation, to be used at some time in the future for public purposes (Adams 1993). Under a reservation-oriented policy, forests were treated as nonrenewable resources, their supply conserved and stretched over as much time as possible, with little regard for their dynamic nature. Federal forestry reserves were established in the United States by the Creative Act of 1891, securing nearly thirty-nine million acres out of public domain lands, mainly in the western states and territories. At about the same time, however, the basic notions of sustained-yield forest management—indefinitely replenishing forests through strictly controlled timber harvests, reforestation, and protection from insects, disease, and fire—were making their way from Europe to North America with the immigration of the first European professional foresters, and the education of the first American professional foresters at European universities.

As introduced in the United States by Bernhard Fernow, Gifford Pinchot, and others at the end of the nineteenth century, forestry was largely a technical undertaking. It was broadly assumed that by maintaining a continuous supply of timber and protecting the basic productivity of soils and watersheds, the broader set of forest uses and values would automatically be protected for the American people as a whole.

The forest reserves were symbolically transferred in 1905 from protection under the Department of the Interior to active management under the new Forest Service in the Department of Agriculture. The national forests, as they were now called, grew to comprise more than 194 million acres by 1910. With forester Gifford Pinchot as its first chief, the Forest Service was given the charge of managing the national forests to provide "the greatest good for the greatest number in the long run" (Pinchot 1947, 261). Sustainability in a broad sense—ecological, economic, and social—had been established as the fundamental concept underlying forest policy in the United States.

CUSTODIAL MANAGEMENT

Management of the national forests was largely custodial until the mid-1940s. Preventing theft and wildfire was the major managerial activity in the national forests of the western United States. In the East, large areas of land deforested and abandoned during the Dust Bowl years were acquired by the Forest Service and gradually restored. Conversion of forest to other land uses was generally prohibited. Little timber was cut in the national forests during this period, due in part to political pressure from timber companies seeking to minimize competition in the wood products industry and maintain favorable prices for private timber. Management of public forests emphasized maintaining the land in its native forest cover and relying upon natural regeneration following disturbances. The underlying biological and ecological systems were not well understood, however, as evidenced by the way wildfire was viewed at the time. Rather than recognizing that wildfire was part of a natural disturbance regime integral to the functioning of the forest ecosystem, the policy was to eliminate fire whenever and wherever it occurred in the forest. Thus, even custodial management requires a thorough understanding of natural disturbance regimes and other complexities of forest ecosystem functions.

By the 1940s, many private timber companies had also come to embrace the idea of sustained-yield forest management. Previously, the standard practice had been to acquire forest land, liquidate the timber assets, and abandon it—an approach often referred to as "cut out and get out." With the leadership of corporate pioneers such as Frederick Weyerhaeuser, private timber companies began to recognize the benefit of holding land, reforesting it, and harvesting the timber on a continuing basis. The management of many private forest lands in the United States reflected the sustained-yield forestry of nineteenth-century Europe, although utilizing modern technology that facilitated timber harvesting at a far greater scale.

MULTIPLE-USE FORESTRY

With wood supplies from private forests lands largely depleted during World War II, the federal forest reserves became a major supplier of timber for economic expansion and the suburban housing boom in the late 1940s and 1950s. Increased leisure time and improved transportation systems brought more Americans into contact with the national forests, increasing demand for recreation, wildlife, and other noncommodity resource values. With growing frequency, large-scale timber harvesting activities came into conflict with these other uses, challenging the operational utility of the traditional concept of sustained yield as the maximization of timber yield constrained only by the biophysical limits of the land itself.

The Multiple-Use Sustained-Yield Act of 1960 was an important turning point in foresters' interpretation of their responsibility for sustainable forest management. It defined sustained yield as "the achievement and maintenance in perpetuity of a high level annual or regular periodic output of *the various renewable resources of the national forests* without impairment of the productivity of the land" (§ 528; emphasis added). It has long been recognized that forests generate a host of goods and services. Medieval forests were commonly valued for their game and forest foods, as well as wood for both fuel and construction (Westoby 1989). Even when forests are managed for timber, other values are commonly produced as by-products. Wildlife, recreation, water and water quality, and other outputs are commonly generated incidentally to the production of timber.

The Multiple-Use Sustained-Yield Act provided the statutory basis for the application of this approach to U.S. public forests. Public controversies over the Forest Service's implementation of multiple-use forestry have led to additional statutory direction for sustainable management of the national forests. The Forest and Rangeland Renewable Resources Planning Act of 1974 required the development of periodic national assessments of the supply and demand for a large array of resource uses and values, and a strategic plan detailing how the Forest Service intended to address all demands simultaneously. The agency's response to Congress was to project significant increases in funding. Substantially higher investments in intensive resource management would allow the Forest Service to accommodate all the uses of and demands for the national forests—several of which competed and conflicted with one another—and sustain the forests indefinitely (Sample 1990a).

In the decade following the passage of the Multiple-Use Sustained-Yield Act, the public grew increasingly dissatisfied with the balance the Forest Service had struck in balancing these competing goals. The predominant focus on timber production that had developed in the agency during the 1950s persisted. Criticism from the scientific community as well as concerned citizens suggested that such high levels of timber removal not only imposed unacceptable impacts on the nontimber resources, but threatened the long-term sustainability of timber production as well (LeMaster 1984). The Forest Service's indomitable optimism and increasing estimates of the level of timber harvesting that could be sustained in the national forests were based on technical assumptions that included a vigorous program of reforestation and silvicultural treatments aimed at increasing tree growth rates. Year after year, however, Congress funded higher timber sale levels but did not adequately fund the kinds of reinvestments in forest management needed to support this level of harvesting (Hirt 1994). Local public challenges over issues such as large-scale clear-cutting eventually boiled over into a national controversy over the Forest Service's management of the national forests. Several successful legal challenges brought timber harvesting on the national forests to a virtual standstill, forcing Congress to take legislative action (LeMaster 1984). Policy changes proposed by the Forest Service would provide a new statutory basis for timber harvesting from the national forests, but public confidence in the Forest Service's ability to manage these public resources sustainably and in the broader public interest had been severely shaken. Congress determined that more sweeping changes were needed.

NON-DECLINING EVEN-FLOW

In 1976, the National Forest Management Act (NFMA) placed numerous additional statutory limits on timber harvesting in the national forests, and required the development of detailed management plans with ample opportunity for public involvement in national forest management decision making. Many of these limitations were aimed at reducing the impacts of timber harvesting on nontimber resources. But concern over the sustainability of timber production itself led Congress to add a new wrinkle to its definition of sustained yield, specifying that the sale of timber from each national forest be limited to "a quantity equal to or less than a quantity which can be removed from such forest annually in perpetuity" (§ 13(a)). This so-called non-declining even-flow constraint was criticized by some economists as inherently inefficient in managing the extensive areas of native forest "old-growth" that remained in many western national forests at the time (Clawson 1983).

Previously, policies aimed at promoting sustainable forestry were stated primarily in terms of staying within the limits of biological and physical resources. But in practice, considerations of socioeconomic sustainability have been implicit and intertwined (Dana 1918). The non-declining even-flow constraint was intended to meter out the remaining volume of old-growth until forest areas harvested and regenerated decades earlier had reached maturity, so that there would be no significant interruption in the supply of timber to communities where the local economy revolved around the processing of wood from national forests (Schallau 1974). For many forest products companies, however, even a small decline in timber supply in the near term was less acceptable than the prospect of a larger decline in the more distant future, and the NFMA included a provision allowing departures from the non-declining even-flow requirement (§ 1611).

A decade later, additional concerns over endangered species brought sudden and immediate court-imposed reductions in timber supply in many national forests. In the Pacific Northwest in particular, this resulted in the determination of near-term decreases in timber harvest levels that, even though less severe than could be expected in a continuation of the boom-and-bust approach, still resulted in severe economic disruptions at the local and regional scale. This has led to a political impasse and a fundamental reexamination of what forest managers are to sustain, for whom, and to what purpose (Shands, Sample, and LeMaster, 1990).

RECENT EVOLUTION IN THE CONCEPT OF SUSTAINABLE FOREST MANAGEMENT

This reexamination is leading to a further evolution in the definition of sustainability in forest management, one that explicitly rather than implicitly includes social and economic—as well as biological—objectives. A key tenet of an ecosystem-based approach to forest resource management—often abbreviated to "ecosystem management"—is that it must be not only ecologically sound but also economically viable and socially responsible (Aplet et al. 1993). If it is lacking in any one of these three areas the system will sooner or later collapse. Each of these three considerations represents a circle of possible options. Where all three circles overlap with one another delineates the subset of options that define sustainable forestry.

IS SUSTAINABILITY BIOCENTRIC OR ANTHROPOCENTRIC?

From one perspective, this approach to sustainability is overly biocentric. Forest condition has become the dominant objective rather than forest outputs. Since management is defined as the "judicious pursuit of means to accomplish an end," it is impossible to manage without identifying specific objectives of management required (Sedjo 2000). An ecosystem-based approach to forest management may involve "the abandonment of the dominant management objective of a stable flow of wood from the land" and its replacement by "management of whole systems for a variety of purposes" (Gordon 1994, 6). Ecosystem management means "thinking on a grander scale than we're used to . . . [I]t means sustaining forest resources over very long periods of time . . . and from that will flow many goods and services, not just timber" (Steen 2004, 267).

From another perspective, if one focuses on the ends rather than the means, the ecosystem management approach is ultimately anthropocentric. It can be argued that the greatest beneficiaries of this approach are human societies, whether of this or future generations. The greater attention to cumulative environmental effects over time and larger spatial scales is aimed primarily at sustaining the ability of natural ecosystems to meet human wants and needs, now and in the future. To some degree, concerns over threatened or endangered animals and plants reflect an expanded ethical consideration for the intrinsic rights of nonhuman species to survive, regardless of their utility to human societies (Sample 1990b). It can be argued, however, that the most successful attempts to date to rescue endangered species habitats have been motivated overwhelmingly by an anthropocentric focus on maintaining genetic potential for new pharmaceuticals and crop strains, or preserving wildlands for a variety of human uses.

Societal limitations on the current rate of forest resource use and consumption reflect a broader scientific understanding of the full effects of human activities on natural ecosystems, and the recognition that there is substantial continuing uncertainty in this area (Grumbine 1994). A more cautious, conservative approach serves as a sort of insurance policy to increase the likelihood that productive, functionally intact forest ecosystems will still be there to meet the needs of human societies for the foreseeable future (Callicott 1991). Thus, even though an ecosystem-based approach to sustaining forest ecosystems might focus operationally on guaranteeing hospitable environments for the diversity of nonhuman species, the ultimate objective is to guarantee the continued functioning of forest ecosystems as a basic life support system for *Homo sapiens*. Coal miners may be intensely interested in the health and well-being of their canary, but the canary is not their ultimate concern.

Ultimately, the question of whether sustainable forest management is biocentric or anthropocentric is rather moot. Despite the built environments in which most of us live, humans are an integral component in the natural environment. Our fate is inextricably intertwined with that of every other life-form on Earth. Sustaining forests, along with all the other terrestrial and aquatic biomes, is to sustain life itself, human and otherwise.

FOREST SUSTAINABILITY IN A GLOBAL CONTEXT

During the past decade, an important shift has taken place in the discussion of sustainability in forest management. The discussion itself has moved from the local arena to the world stage. Relatively localized debates over specific forest management practices such as clear-cutting now take place in the context of global issues such as biodiversity conservation and mitigation of global climate change. More than at any other time in history, forest policy discussions in the United States and around the world explicitly recognize that forests are essential to the long-term well-being of local populations, national economies, and the earth's biosphere as a whole.

What is emerging from this debate is a remarkable global convergence of views on a few basic precepts for sustainable forest management. These basic precepts are gradually forming the basis for (1) assessing current forest conditions and trends in terms of their sustainability, (2) determining the adequacy of existing forest management activities for moving closer to the goal of sustainability, (3) identifying particular actions needed to improve upon the existing management activities, and (4) monitoring progress on implementation of these actions.

Most important, these generally accepted principles for sustainable forest management are now being used to guide this kind of assessment-action-monitoring process not only at the national level, but at the local "forest management unit" level on individual public, private, and communal forests in the United States and in developed and developing nations around the world.

CRITERIA FOR ASSESSING FOREST SUSTAINABILITY AT THE NATIONAL SCALE

The controversies that arose in forestry in the United States from the 1970s onward were but a microcosm of emerging global issues in natural resource conservation and environmental protection. These conflicts are often characterized as "jobs versus the environment," particularly in the context of protecting habitats for threatened or endangered species in regions where local economies were largely centered on resource extraction.

To address this apparent conflict between the interests of economic development and the interests of environmental protection, the United Nations in 1983 appointed an international commission to propose strategies for improving human well-being in the short term without threatening the local and global environment in the long term. Norwegian prime minister Gro Harlem Brundtland chaired the commission, and its report, *Our Common Future,* became widely known as the Brundtland report (World Commission on Environment and Development 1987). The report helped popularize the term sustainable development, which it defined as "development that meets the needs of the present without compromising the ability of future generations to meet their own needs." This landmark report helped trigger a wide range of actions, including the UN Conference on Environment and Development (also known as the Earth Summit) in 1992 in Rio de Janeiro, Brazil.

The single most important outcome of the Earth Summit in terms of defining sustainable forestry was a set of Forest Principles, with linkages to several other new international conventions on biodiversity, climate change, and desertification (UN General Assembly 1992). The Forest Principles, in turn, served as the basis for several subsequent important developments in international forest policy. At a conference in Helsinki, Finland, in 1994, most of the tropical timber consumer nations made a joint commitment to maintain, or achieve, sustainable management of their forests by 2000 (International Institute for Sustainable Development 1994). At a 1993 meeting in Montreal, many of the world's industrialized nations developed a comprehensive set of "Criteria and Indicators for the Conservation and Sustainable Management of Temperate and Boreal Forests" which, following their ratification in Santiago, Chile, in 1995, became what are now commonly referred to as the Montreal Process Criteria and Indicators (Castañeda, Palmberg-Lerche, and Vuorinen, 2001), or simply the C&I. The seven criteria are as follows:

1. Conservation of biological diversity
2. Maintenance of productive capacity of forest ecosystems
3. Maintenance of forest ecosystem health and vitality
4. Conservation and maintenance of soil and water resources
5. Maintenance of forest contribution to global carbon cycles
6. Maintenance and enhancement of long-term multiple socioeconomic benefits to meet the needs of societies
7. Legal, institutional and economic framework for forest conservation and sustainable management (Castañeda, Palmberg-Lerche, and Vuorinen, 2001)

The seven criteria and sixty-seven indicators in the C&I provide an internationally agreed-upon framework for nations to measure their progress toward sustainable forest management—a major achievement. In 2001, the United States published its first forest assessment using the C&I framework (U.S. Department of Agriculture, Forest Service, 2001). In 2004, it published the more comprehensive *National Report on Sustainable Forests 2003,* reflecting the involvement of a broad and diverse array of public, private, and nongovernmental organizations (NGOs) in its evaluation of conditions and trends in U.S. forests (U.S. Department of Agriculture, Forest Service, 2004). Several state governments within the United States have also now issued assessments of conditions and trends in their own forests, utilizing the C&I as their framework.

Forest assessments based on the C&I are not action oriented. The C&I simply provide a basis for assessing current conditions and trends; they do not in themselves provide a basis for evaluation of the adequacy of current efforts to conserve forests, nor are they a basis for identified specific actions to correct any perceived shortcomings in current efforts. There is a parallel international process for accomplishing this, however. In the years following the Earth Summit, an array of developed and developing nations requested that the United Nations serve as convener for an Intergovernmental Panel on Forests (IPF), whose purpose was to develop "Proposals for Action"—specific activities aimed at moving nations forward toward achieving sustainable forest management. The IPF and its successor, the Intergovernmental Forum on Forests, which represented a broader array of participants, eventually produced a set of 270 Proposals for Action. Each of the Proposals for Action can be linked to specific indicators of sustainable forest management in the C&I (Washburn and Block 2001). In 2000, the UN Economic and Social Council (ECOSOC) voted to create a new office, the UN Forum on Forests (UNFF), with a five-year (2001–2005) program of work to facilitate nations' monitoring, assessment, and reporting of progress in implementing the Proposals for Action (ECOSOC 2000).

The U.S. national report to the UNFF details all of the current policies, programs, and activities in the United States that address each of the Proposals for Action (United States of America 2003). Unlike many of the reports from other countries, which describe only the actions by their central governments, the U.S. report details the actions by its federal agencies—but also those of state and local governments, industry, tribes, NGOs and private forest owners—all of which make their own contributions toward the collective U.S. response to the Proposals for Action, and progress toward sustainable forest management. This unique approach to the development of the U.S. report for the UNFF is intended to enhance the role that this report can play in facilitating a national dialogue within the United States to discuss the adequacy of our current efforts to achieve sustainable forest management, possible further actions that may be needed, and how we might prioritize among these potential actions so that the most urgent are addressed as soon as possible (Sample and Kavanaugh 2003).

So although the C&I themselves do not require nations to take specific actions to achieve sustainable forest management, the national-level and other assessments utilizing the C&I do help to promote a common understanding of the current status of a nation's forests. Indirectly this helps to stimulate timely and effective action, so that the next assessment that is conducted in the years ahead will serve to document substantial progress to the sustainable forestry goals that the United States and other countries have set for themselves.

STANDARDS FOR ASSESSING FOREST SUSTAINABILITY AT THE FOREST LEVEL

Progress toward sustainable forest management at the national level depends upon the collective progress made in the management of individual forests—on federal and state lands, on forest industry timberlands, and perhaps most important in the private forests that constitute nearly two-thirds of all the forest land in the United States. The gradual convergence on a set of generally accepted principles for sustainable forest management, facilitated by the increasing familiarity with and use of the C&I, is leading to a greater understanding by forest managers of all kinds of practices that will lead to better management of the forests under their stewardship.

Improved forest management at the forest management unit level provides the building blocks for progress toward sustainable forest management at the national level and ultimately on a global scale.

One of the more interesting mechanisms to develop out of this process is forest certification based on independent third-party assessments, using generally accepted principles of sustainable forest management as the yardstick by which current forest management practices are evaluated. What distinguishes certification from previous rule-based approaches to improving forest management practices is that it is private rather than governmental, voluntary, and aims to provide a positive incentive for sustainable forest management rather than punishment for violations (Cashore, Auld, and Newsome, 2004). Certification originally grew out of a consumer-led ban on the import of tropical timber into several European countries, aimed at reducing the rate of tropical deforestation (Sample 2000). At the request of tropical nations that were making substantial investments toward replacing forest exploitation with sustainable forestry practices, independent third-party certification was developed as a means to differentiate sustainably harvested tropical timber and to ensure its continued access to consumer markets in Europe.

The 1992 Earth Summit helped stimulate interest on the part of forest industry and private forest owners, as well as environmental NGOs, in articulating a set of principles and criteria describing forest management practices that are ecologically sound—but also financially viable and socially responsible. The Forest Stewardship Council (FSC) was established in 1993 by a diverse group of representatives from environmental NGOs, forest industry, indigenous peoples' organizations, and community forestry groups from twenty-five countries (Upton and Bass 1996), based on such a set of principles and criteria:

Principle #1: Compliance with Laws and FSC Principles
> Forest management shall respect all applicable laws of the country in which they occur, and international treaties and agreements to which the country is a signatory, and comply with all FSC Principles and Criteria.

Principle #2: Tenure and Use Rights and Responsibilities
> Long-term tenure and use rights to the land and forest resources shall be clearly defined, documented, and legally established.

Principle #3: Indigenous Peoples' Rights
> The legal and customary rights of indigenous peoples to own, use, and manage their lands, territories, and resources shall be recognized and respected.

Principle #4: Community Relations and Workers' Rights
> Forest management operations shall maintain or enhance the long-term social and economic well-being of forest workers and local communities.

Principle # 5: Benefits from the Forest
> Forest management operations shall encourage the efficient use of the forest's multiple products and services to ensure economic viability and a wide range of environmental and social benefits.

Principle #6: Environmental Impact
   Forest management shall conserve biological diversity and its associated values,
      water resources, soils, and unique and fragile ecosystems and landscapes, and,
      by so doing, maintain the ecological functions and the integrity of the forest.

Principle #7: Management Plan
   A management plan—appropriate to the scale and intensity of the operations—
      shall be written, implemented, and kept up to date. The long-term objectives
      of management, and the means of achieving them, shall be clearly stated.

Principle #8: Monitoring and Assessment
   Monitoring shall be conducted—appropriate to the scale and intensity of
      forest management—to assess the condition of the forest, yields of forest
      products, chain of custody, and management activities and their social and
      environmental impacts.

Principle # 9: Maintenance of High Conservation Value Forests
   Management activities in high conservation value forests shall maintain or
      enhance the attributes which define such forests. Decisions regarding high
      conservation value forests shall always be considered in the context of a
      precautionary approach.

Principle # 10: Plantations
   Plantations shall be planned and managed in accordance with Principles and
      Criteria 1–9, and Principle 10 and its Criteria. While plantations can provide
      an array of social and economic benefits, and can contribute to satisfying the
      world's needs for forest products, they should complement the management
      of, reduce pressures on, and promote the restoration and conservation of
      natural forests. (FSC 2000, 3–8)

   Other certification organizations have also developed since then, each with its
own principles and procedures, but most following the same general model of independent third-party review of forest management practices, and evaluation against a
set of criteria and specific standards. In the United States, these organizations include
the American Forest & Paper Association (which provides the Sustainable Forestry
Initiative program), the American Forest Council (which provides the Tree Farm
program), and the National Woodland Owners Association (which provides the
Green Tag program) (Rana, Price, and Block, 2003). The U.S. programs are generally not international, although the American Forest & Paper Association has agreed
to mutual recognition of programs similar to its own in several other countries.
   Another important aspect of forest certification programs in the management
of public forests is that they establish a clear set of principles, and directly or indirectly offer a number of positive incentives for adhering to these principles. Through
much of the controversy over forest management in the United States, especially
that relating to management of forests on federal and other public lands, the public
response has been to protest specific forest management activities that were felt to
be unacceptable. Environmental NGOs in particular have made effective use of the
judicial system to halt timber harvesting and related activities, from the forest level

to the national scale. Forest managers were often left with a rather spotty picture of what was regarded as unacceptable, but determining what was actually *acceptable* involved a significant amount of guesswork and trial and error. The articulation of a comprehensive set of principles, criteria, and specific standards for what constitutes sustainable forest management has made forest managers' jobs much easier, particularly given that the standards were consciously developed to support the financial viability of a forestry enterprise as well as its ecological soundness. Most of the public forestry agencies in the United States that have sought certification have discovered that the certification principles, criteria, and standards are generally congruent with the statutes and other public policies with which they are already required to comply (Sample, Price, and Mater, 2003). While certification may or may not result in higher values for the wood these public agencies sell, many have found that the resulting decrease in public challenges and controversy has reduced their operating costs and freed financial resources for other important needs (Mater et al. 1999).

The pathway to defining sustainability in forest management in the United States has been long and often difficult, but it has been productive. Fears on the part of environmental activists and forest industry alike that one or the other would dominate in a winner-take-all outcome have proven largely unfounded. Influenced strongly by approaches represented by efforts such as the Brundtland report, the Earth Summit, the World Business Council for Sustainable Development, and others, there has been a gradual convergence toward a basic set of generally accepted principles of sustainable forest management that are both practical and effective. It is important to note that none of these various efforts to describe sustainable forest management, whether in national-level assessments based on the C&I or forest-level assessments based on certification standards, assume that all forest lands will be managed in the same way for the same objectives. On the contrary, there is growing recognition that some forest areas with unique or significant conservation values will need to be largely protected, and that other areas can be well suited to management for a moderately high level of sustained wood production.

SUSTAINABLE FORESTRY AND THE PARTICULAR CHALLENGE OF BIODIVERSITY CONSERVATION

Sustainable forest management as we understand it today can be seen as a logical extension and further evolution of the multiple-use forest management approach that developed in the United States in the mid-twentieth century (Fedkiw 2004) and was codified in law by the Multiple-Use Sustained-Yield Act of 1960. Multiple-use forest management has been a versatile, flexible, and largely successful approach in the United States, and has become the prevailing approach on private forest lands as well as public. Focusing initially on securing a sustained supply of wood, multiple-use forest management has expanded its scope to protecting watersheds, wildlife, recreation, and grazing and even wilderness.

However, the need to conserve biological diversity—and especially to protect habitats for threatened and endangered species—represents a fundamentally different challenge to the multiple-use model of forest management. Scientific uncertainty as to just where the limits of sustainability lie, and the degree of resilience of sensitive species should these limits be exceeded, has resulted in a conservative approach to biodiversity conservation. In many instances, this precautionary approach means that even a modest level of human manipulation in the ecosystem may exceed the limits of what can be sustained. With the boundaries of sustainability thus so tightly drawn, it is difficult for forest managers to discern a future pathway by which biological diversity can be conserved within the context of actively managed forests.

ACCOMMODATING BIODIVERSITY

We are now in an era in which the downward trend in biodiversity, and the potential of forest protection to slow that decline, is seen by many as sufficient reason to cease any and all forest management activities that potentially interfere with that objective. This presents a particular challenge to defining and practicing sustainable forest management in a developed nation like the United States, one of the largest per capita consumers of wood products and a net importer of wood and wood fiber. The determination of what constitutes sustainable forest management in the United States must consider not only the nation as a whole, but how the United States interacts with other regions of the world in the global forest sector.

Many of the most recognized and respected biologists in the world believe that we are now in the midst of a biodiversity crisis, with extinctions of animal and plant species taking place at a rate not seen since the dinosaurs were wiped out sixty-five million years ago. Harvard biologist Edward O. Wilson (1992) has estimated the current rate of species extinctions at approximately 27,000 per year, or an average of seventy-four each day, out of a worldwide total of perhaps ten million species. The normal "background" extinction rate is about one species per one million species a year (Raup and Sepkoski 1984). More than 20,000 taxa are globally rare or threatened, and as many as 60,000 face extinction by the middle of this century (IUCN—The World Conservation Union 1988). According to Wilson, "Human activity has increased extinction between 1,000 and 10,000 times over this level . . . clearly we are in the midst of one of the great extinction spasms in geological history" (1992, 280).

The world's greatest concentration of biological diversity in forest ecosystems—and the greatest threats to conserving that diversity—is in the tropics (Raven 1987). Because of the means by which tropical rainforests cycle their nutrients, these seemingly lush and irrepressible forests are much more vulnerable to ecological damage than most temperate-zone forests, and much slower to recover from deforestation (Wilson 1992). The galloping losses of forest area in the tropics are the single greatest threat to global biodiversity, a trend that is exacerbated by population growth rates in many tropical nations that far exceed those in most temperate-zone nations. "An awful symmetry binds the rise of humanity to the fall of biodiversity: the richest nations preside over the smallest and least interesting biotas, while the poorest nations, burdened by exploding populations and little scientific knowledge, are stewards of the largest" (Wilson 1992, 272).

The importance of conserving biological diversity in forest ecosystems has generated policy proposals aimed at minimizing the conversion and fragmentation of the remaining large areas of native forests, and preventing the diminishment of remaining biological diversity by development for commodity production. Wilson (1992) estimates that the 4.3% of the world's land surface currently under legal protection should be expanded to 10%. Many eminent biologists and other scientists support a proposal to set aside 50% of the North American continent as "wild land" for the preservation of biological diversity. The largest grassroots environmental organization in the United States is actively working to ban all commercial timber harvesting on federal public lands, and signed up nearly a quarter of the members of the 106th Congress as sponsors of legislation to accomplish this (Sierra Club 1999).

Many conservation biologists today point to the need to think beyond "the reserve mentality" in designing strategies for conserving biological diversity (Brussard, Murphy, and Noss, 1992). But it is also clear that reserves will continue to be a major component of any successful biodiversity conservation strategy (Hunter and Calhoun 1996), particularly with regard to species endemic to late-successional forest ecosystems (Spies and Franklin 1996).

PROTECTION AND PRODUCTION: DUAL CONSERVATION RESPONSIBILITIES

The global nature of the biodiversity crisis points up the need for a strategy that integrates the management of temperate, tropical, and boreal forests with world demand for wood. Current global industrial wood demand is estimated at 1.6 billion cubic meters per year, and is expected to rise to 2.5 billion cubic meters per year by 2050 (Food and Agriculture Organization of the United Nations [FAO] 2000). Industrialized nations account for a disproportionate share of this global demand (figure 22.1). Even among the developed nations, the United States stands out as one of the world's largest consumers of wood. The U.S. per capita consumption of major wood products (lumber, plywood, and paper) is about double that of Germany, seven times that of Brazil, and fifteen times that of China (FAO 2000). The United States has one of the lowest average population densities among the developed nations (e.g., Oregon has a population of less than three million people; Germany, with a geographic area slightly larger than that of Oregon, has a population of more than eighty-two million), and some of the most productive forests. In spite of this, the United States continues to import more than a quarter of its wood—114 million cubic meters in1997—from harvesting in both tropical and boreal forests (Howard 1999).

While temperate forests are comparatively less biologically diverse, hot spots with extraordinary concentrations of species diversity exist, particularly where there are large, contiguous areas of largely undisturbed native forest (Ricketts et al. 2000). For wealthy, temperate-forest nations like the United States to support a credible and ethical program for biodiversity conservation in the poorer tropical nations, their own policies for sustainable forest management must encompass a two-pronged strategy of (1) protecting their own biodiversity hot spots where they exist, even when it means sacrificing economic values that could have been derived through resource development, and (2) sustainably utilizing productive forest areas of relatively low biodiversity value to help alleviate the pressure on tropical and boreal forests to meet global needs for wood fiber and other renewable resources.

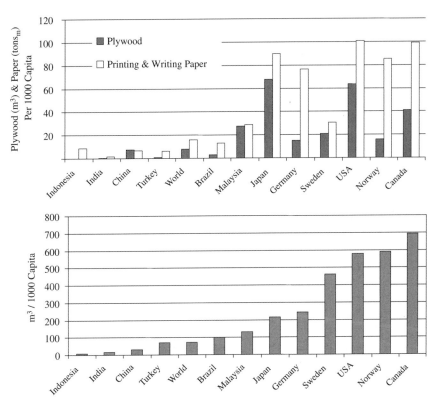

Figure 22.1 National per capita consumption of selected categories of forest products. (*Source:* Food and Agriculture Organization of the United Nations 2000).

U.S. forest policy appears to presume that all forests are to be managed to provide a wide range of uses and values, with only the particular mixture of these uses varying from place to place and ownership to ownership. We have quested after the holy grail of sustainable forest management as if there were a one-size-fits-all formula—a single set of standards that could be applied equally well to forests everywhere. Multiple-use forest management has proven enormously durable in many different circumstances. This flexible, adaptive approach has for the most part allowed forest managers to balance a wide variety of demands on forests while keeping within the bounds of sustainability. But most "all-purpose" tools, though convenient, are of limited value in any specific task, particularly when compared with other, more specialized tools developed for that particular application. Multiple-use forest management is an all-purpose tool in a world in which the demands on forests are also requiring the development of more specialized tools with greater precision and more direct application.

Today, the clear need to greatly improve our conservation of biological diversity in forests worldwide, while at the same time managing these renewable resources to help meet the material needs of an expanding human population, demands recognition in both policy and practice that (1) all forests are *not* equally suited to the same intensity of management and (2) there are important forest uses and values that are clearly *not* compatible with one another, and cannot be adequately protected under management aimed at accommodating a wide range of commodity and noncommodity uses.

A SYSTEMS APPROACH TO SUSTAINABLE FOREST MANAGEMENT

The necessity of simultaneously increasing both biodiversity conservation and wood production is accelerating the evolution toward three separate and distinct types of forest management (Hunter and Calhoun 1996):

- Commercial forest plantations intensively managed for the production of wood and wood fiber–based commodities—what Aldo Leopold (1949) alluded to as "Group A" forestry. This approach will likely be centered on highly productive private lands with relatively low value or potential value as habitats for rare or sensitive species due to their small tract size and/or history of past land use; these primarily private lands would be largely exempt from federal requirements for biodiversity conservation, particularly where the plantations derive from the afforestation of lands reclaimed from nonforest uses.
- Forests managed at a moderate or low intensity for a wide variety of goods, services, and natural values, not unlike the New England "working forest" concept, or Leopold's "Group B" forestry. These working forests would provide habitats primarily as a function of being maintained in forest land use; these lands, both public and private, would encompass the majority of the forest area of the United States, with the broad diversity of management approaches on individual tracts of varying size providing an accompanying diversity of habitats in terms of age, successional stage, vegetative composition, climate, and landform.
- Native forest reserves managed first and foremost for conservation and restoration of biological diversity—what Leopold might have termed "Group C" forestry. Management of these forests would be centered on identified biodiversity hot spots of global and national significance, and would likely encompass most of the remaining large tracts of undeveloped native forest on federal public lands, some state parks, and private lands where this style of management is consistent with landowner goals and objectives.

A systems approach to defining sustainable forest management must encompass all three categories—bioreserves, plantations, and working forests managed for multiple values and purposes (see figure 22.2).

Reinforcing the scientific foundation for sustainable forest management will require a continued high level of research activity in support of management in all three categories of forests. Substantial resources have been devoted to understanding all aspects of bioreserve delineation, consolidation, and management, from the relative advantages of large and small reserves (Diamond and May 1976; Robinson and Quinn 1992; Soulé and Simberloff 1986) to the rates and size of disturbance needed to maintain the ecological characteristics of old-growth forests (Spies and Franklin 1991). Decades of traditional forestry research in the United States have focused

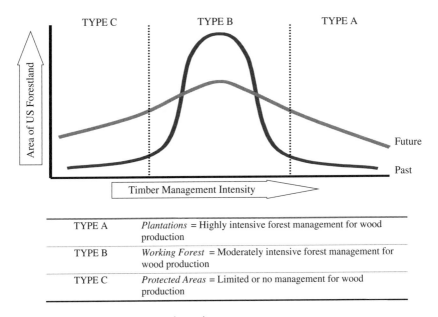

Figure 22.2 Forest management intensity spectrum.

on maximizing wood and fiber yields from intensively managed plantations, but it has been only recently that scientists have recognized the contributions that even industrial timberlands can make to biodiversity conservation (O'Connell and Noss 1992; Pimentel et al. 1992). Perhaps the greatest need—and greatest opportunity—for creating new knowledge in biodiversity conservation is on lands in "Group B," those public and private lands managed for multiple uses in an almost infinite variety of combinations.

It is important that public forest management agencies remain fully engaged in this kind of management. In spite of a century of experimentation, scientists today are keenly aware of the inadequacy of our current understanding of forest ecosystems and our limited ability to predict the outcomes of human interventions in these ecosystems (National Research Council 1992). The need is greater than ever for public forestry agencies to conceptualize, facilitate, and conduct research relating to the management of developed forest areas, experiment with different approaches in a variety of biophysical and socioeconomic settings, and provide a model for continuously improving forest stewardship on both public and private lands, in the United States and abroad.

A NATIONAL POLICY FRAMEWORK THAT FACILITATES SUSTAINABLE FORESTRY

More than any other forest use or value, biodiversity conservation has narrowed the bounds within which forest managers can accommodate all society demands within the limits of sustainability. In the case of the national forests in the United

States, measures to protect biodiversity have greatly constrained the long-standing multiple-use mandate for the management of these resources. This may not be the dilemma it has seemed to be. We are beginning to recognize that practicing forestry in more or less the same way everywhere should not necessarily be the most desirable goal—that such a one-size-fits-all approach sacrifices important forest values that can only be achieved with a more specialized approach to forest landscapes and forest ecosystems.

In forest areas characterized by extraordinary biodiversity values, particularly large contiguous areas of native forests primarily on public lands, we are likely to see management essentially as a bioreserve. In areas of relatively low biodiversity value, but with high productivity for meeting societal demands for wood and fiber, we are likely to see intensively managed forest plantations constrained by little more than basic principles of good land stewardship and protection of water quality. And in the majority of public and private forests, we are likely to see an infinitely varied array of approaches to multiple-use forest management that will produce moderate levels of wood and fiber while protecting a range of ecological values, including habitats for rare, sensitive, threatened, or endangered animal and plant species.

None of these three elements alone can be regarded as sustainable forestry. It is the overall system—with all its elements represented at the national, regional, and local levels—that will constitute sustainable forest management in the future. There is no single set of standards to define how forestry should be practiced in every location and in every circumstance. Any set of standards purporting to describe a system of sustainable forestry must take into account the need for bioreserves and intensively managed forest plantations as well as "working forests" managed to provide an array of forest values, renewable resources, and ecological services (see figure 22.3).

A recent report by the World Wildlife Fund (WWF) suggests that a significant expansion of the area of intensively managed forest plantations could allow the world's major forest products companies to meet a substantial share of the global demand for industrial wood from a relatively small proportion of the world's forest area, as well as open up new opportunities to provide outright protection to high conservation value forests, particularly those with globally significant biodiversity

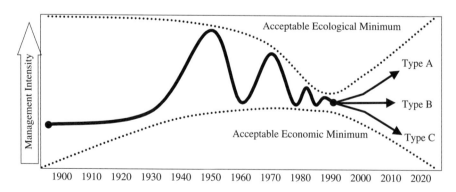

Figure 22.3 Forest management and the changing bounds of sustainability.

values (Howard and Stead 2001). The WWF (2001) is so convinced of the value of this approach that they have called upon the world's ten largest forest products companies to collectively increase the area of intensively managed forest plantations by five million hectares per year—for the next fifty years. With this level of investment, the WWF estimates that as much as 80% of the world demand for industrial wood in 2050 can be met from less than 20% of the world's forests. Furthermore, the WWF asserts this can all be done consistent with the FSC criteria for forest certification, meaning that much of that 20% will be new planted forests on retired marginal crop and pasture land, rather than plantations created by converting natural forests.

What would such an approach mean for the United States? In some ways, we are already moving in this direction, with wood production shifting increasingly to industrial timberlands and other private forests, and biodiversity conservation becoming a primary management goal in many public forests. Nevertheless, significant policy and political barriers remain to achieving either of these objectives efficiently or effectively. We are perhaps within reach of a new political consensus—one in which both the forestry community and the environmental community actively support the idea that intensively managed forest plantations *and* protected areas in high conservation value forests have an essential place in a comprehensive strategy for sustainable forest management.

Policy makers have an opportunity—and a responsibility—to further develop this potential for broad public consensus on forests and forestry, and to shape a policy framework that will support and facilitate this kind of practical approach to accomplishing sustainable forest management. It has been suggested (Binkley 2002) that a consensus agreement might include considerations such as:

- Devotion of 20–30% of the land base of plantation projects to ecological services
- Strict control of off-site impacts of plantation-based timber production, especially the movement of silt, fertilizer, or herbicides into waterways or groundwater
- Agreed-upon limits on the use of yield-enhancing chemicals such as fertilizers and herbicides, focused on minimizing use and maximizing impact
- Agreed-upon limits on the use of genetically modified organisms to instances in which it can be demonstrated that gene flow out of the plantation is not possible
- A commitment not to log old-growth forests

Numerous opportunities exist to create a policy framework that enables and encourages public and private forest land managers to make rational choices that will tend to be consistent with and supportive of this general approach. Developed temperate-forest nations like the United States, particularly those with high per capita consumption of wood, have a dual conservation responsibility to fulfill. The have an obligation to protect their remaining hot spots of biological diversity—and bear their share of the local, short-term economic effects of doing so—and at the same time meet their share of the demand for renewable wood and fiber that they themselves generate, without shifting an undue burden onto biologically rich forests in other regions of the world.

TOWARD A SHARED VISION OF SUSTAINABILITY
IN FOREST MANAGEMENT

Sustainable forest management involves a simultaneous pursuit of ecological, economic, and social objectives that, rather than mutually exclusive, are in fact mutually *dependent*. The modern definition of sustainable forest management requires meeting three conditions simultaneously; it must be ecologically sound, economically viable, and socially responsible. Reflecting a difficult lesson learned in developing countries around the world, conservation interests are recognizing that it is not possible to get long-term protection of forest ecosystems without incorporating the economic and social needs of the local people into conservation strategies. Economic development and commercial interests are recognizing that ensuring the ecological soundness of their activities not only helps to assure raw material supplies for the future, but also helps maintain essential social and political support (Schmidheiny 1992). Communities are no longer willing to accept the social disruptions and family dislocations that have always accompanied a boom-and-bust approach. They are recognizing that government policy makers alone cannot lead the way toward stable, resilient, and economically diverse communities—that there is an important role for partnerships between business interests and the communities themselves in finding a new basis for sustainable resource use and sustainable communities.

These stark realities are causing conservation interests and business interests alike to reconsider their previous adversarial approaches to one another. Without setting out to do so, these two segments of society are finding themselves on new paths that are no longer divergent, but in many ways are converging toward one another. This convergence—a new sense of common purpose and goals among environmental interests, communities, and the business sector—holds the potential for forming a strong working consensus for conservation such as has not been seen in the United States for at least a generation. This is beginning to illuminate a new array of rational, implementable policy options that offer us hope for finding a way out of the current political and legal impasse over forest conservation.

Sustainability in forest management is a dynamic, evolving concept, reflecting changing social values and changes in our scientific understanding of the effects of human activities on the functioning of forest ecosystems. As an increasingly broad cross-section of forestry interests comes to accept that truly sustainable forestry must reflect ecological, economic, and social objectives, the most challenging trade-off for policy makers may be between short-term needs and long-term assurances.

The central idea behind sustainable development—that is, meeting the needs of present human society without unduly compromising the capacity of future human societies to meet their needs (World Commission on Environment and Development 1987)—is not materially different from the basic motivating concept behind sustained-yield forestry in medieval Europe or sustainable forest management in twentieth-century America. From a policy-making and operational management perspective, the sustainability challenge will always be to protect the long-term productivity of forest ecosystems, to the best of our biological, social, and economic understanding, without unduly limiting the utilization of forests to meet current needs. From an analytical perspective, it is to operate as close to the margin as is socially and politically acceptable, neither exceeding ecological capacities nor leaving significant ecological capacity unutilized. How conservative a margin for error is incorporated is as much a political decision as a scientific one. But the question "forests for whom and for what?" (Clawson 1975) can perhaps never be answered once and for all, nor will the answer be the same for all forests everywhere. It must be revisited periodically as societal needs and conditions change, and as we come to a more complete knowledge of what is needed to sustain the regenerative capacity of forest ecosystems to meet current and anticipated needs.

References

Adams, D. A. 1993. *Renewable resource policy: The legal-institutional foundations.* Washington, DC: Island Press.

Aplet, G., N. Johnson, J. Olson, and V. A. Sample. 1993. *Defining sustainable forestry.* Washington, DC: Island Press.

Binkley, C. S. 2002. Forestry in the long sweep of history. In *Forest policy for private forestry: Global and regional challenges,* ed. L. Teeter, B. Cashore, and D. Zhang, 1–7. Wallingford, UK: CABI.

Brussard, P. F., D. D. Murphy, and R. F. Noss. 1992. Strategy and tactics for conserving biological diversity in the United States. *Conservation Biology* 6(2): 157–159.

Callicott, J. B. 1991. Conservation of biologic resources: Responsibility to nature and future generations. In *Challenges in the conservation of biological resources: A practitioner's guide,* ed. D. J. Decker, M. E. Krasny, G. R. Groff, C. R. Smith, and D. W. Gross, 33–42. Boulder, CO: Westview Press.

Cashore, B., Auld, G., and Newsome, D. 2004. *Governing through markets: Forest certification and the emergence of non-state authority.* New Haven, CT: Yale University Press.

Castañeda, F., Palmberg-Lerche, C., and Vuorinen, P. 2001. Criteria and indicators for sustainable forest management: A compendium. Forest Management Working Paper 5, Forestry Department, Food and Agriculture Organization of the United Nations, Rome. http://www.fao.org/docrep/004/AC135E/ac135e00.HTM.

Clawson, M. 1975. *Forests for whom and for what?* Baltimore, MD: Johns Hopkins University Press.

Clawson, M. 1983. *The federal lands revisited.* Baltimore, MD: Johns Hopkins University Press.

Creative Act of 1891. U.S. Statutes at Large 26:1103, codified at *U.S. Code* 16, § 471. [Also referred to as the *Forest Reserve Act* or the *General Land Law Revision Act.*]

Dana, S. T. 1918. *Forestry and community development.* U.S. Department of Agriculture Bulletin, no. 638. Washington, DC: Government Printing Office.

Davis, K. 1966. *Forest management: Regulation and valuation.* New York: McGraw-Hill.

Diamond, J. M., and R. M. May. 1976. Island biogeography and the design of nature reserves. In *Theoretical ecology: Principles and applications,* ed. R. M. May, 163–186. Philadelphia: W. B. Saunders.

Faustmann, M. [1849] 1995. Calculation of the value which forest land and immature stands possess for forestry. Repr. *Journal of Forest Economics* 1(1): 7–44.

Federal Regulations Pursuant to the National Forest Management Act. *Federal Register* 65:218, 67567–67581, codified at *Code of Federal Regulations* 36:219 and *U.S. Code* 16, § 1600 (note).

Fedkiw, J. 2004. Sustainability and the pathway hypothesis. In *Pathway to sustainability: Defining the bounds on forest management*, ed. J. Fedkiw, D. MacCleery, and V. A. Sample, 8. Durham, NC: Forest History Society.

Food and Agriculture Organization of the United Nations. 2000. *Global outlook for the future wood supply from forest plantations*. Rome: FAO.

Forest and Rangeland Renewable Resources Planning Act of 1974. Public Law 93–378. *U.S. Statutes at Large* 88:476, amended; codified at *U.S. Code* 16, § 1600–1614.

Forest Stewardship Council. 2000. *Principles and criteria for forest stewardship*. Washington, DC: FSC. http://www.fscus.org/images/documents/FSC_Principles_Criteria.pdf.

Gordon, J. 1994. *The new face of forestry: Exploring a discontinuity and the need for a new vision*. Pinchot Distinguished Lecture. Washington, DC: Pinchot Institute for Conservation.

Grumbine, M. 1994. What is ecosystem management? *Conservation Biology* 8(1): 27–38.

Hardin, G. 1968. The tragedy of the commons. *Science* 162(3859): 1243–1248.

Heske, F. 1938. *German forestry*. New Haven, CT: Yale University Press.

Hirt, P. 1994. *A conspiracy of optimism: Management of the national forests since World War II*. Lincoln: University of Nebraska Press.

Howard, J. L. 1999. *U.S. timber production, trade, consumption, and price statistics 1965–1997*. General Technical Report FPL-GRT-116. Madison, WI: U.S. Department of Agriculture, Forest Service, Forest Products Laboratory.

Howard, S., and J. Stead. 2001. *The forest industry in the 21st century*. London: World Wildlife Fund/World Wide Fund for Nature.

Hunter, M. L., and Calhoun, A. 1996. A triad approach to land-use allocation. In *Biodiversity in managed landscapes: Theory and practice*, ed. R. A. Szaro and D. W. Johnston, 477–491. Oxford: Oxford University Press.

International Institute for Sustainable Development. Web site presenting "European criteria and most suitable quantitative indicators for sustainable forest management." Sponsored by the Ministerial Conference on the Protection of Forests in Europe Liaison Unit in Helsinki; adopted by the first expert-level follow-up meeting of the Helsinki Conference, Geneva, June 24, 1994. http://www.iisd.ca/forestry/indicat.html.

IUCN—The World Conservation Union. 1988. *Plant conservation programme*. Gland, Switzerland: IUCN.

LeMaster, D. 1984. *Decade of change: The remaking of Forest Service statutory authority during the 1970s*. Westport, CT: Greenwood Press.

Leopold, A. 1949. *A Sand County almanac, and sketches here and there*. New York: Oxford University Press.

Mater, C., V. A. Sample, J. Grace, and G. Rose. 1999. Third-party performance-based certification: What every public forest land manager should know. *Journal of Forestry* 97(2): 6–12.

Multiple-Use Sustained-Yield Act of 1960. Public Law 86–517. *U.S. Statutes at Large* 74:215, codified at *U.S. Code* 16, § 528–531.

National Forest Management Act of 1976. Public Law 94–588. *U.S. Statutes at Large* 90:2949, amended; codified at *U.S. Code* 16, § 472a, 476, 476 (note), 500, 513–516, 518, 521b, 528 (note), 576b, 594–592 (note), 1600 (note), 1601 (note), 1600–1602, 1604, 1606, 1608–1614).

National Research Council. 1992. *Forestry research: A mandate for change*. Report of the National Research Council Panel on Forestry Research. Washington, DC: National Academy Press.

O'Connell, M. A., and R. F. Noss. 1992. Private land management for biodiversity conservation. *Environmental Management* 16(4): 135–151.

Pimentel, D., U. Stachow, D. A. Takacs, H. W. Brubaker, A. R. Dumas, J. J. Meaney, J. A. S. O'Neil, D. E. Onsi, and D. B. Cornelius. 1992. Conserving biological diversity in agricultural/forested ecosystems. *BioScience* 42(5): 354–362.

Pinchot, G. 1947. *Breaking new ground.* New York: Harcourt, Brace.

Rana, N., W. Price, and N. Block. 2003. *Forest management certification on private forest-lands in the U.S.* Washington, DC: Pinchot Institute for Conservation.

Raup, D. M., and J. J. Sepkoski. 1984. Periodicity of extinctions in the geologic past. *Proceedings of the National Academy of Sciences* 81(3): 801–805.

Raven, P. H. 1987. We're killing our world: Preservation of biological diversity. *Vital Speeches of the Day* (May 15): 472–478.

Ricketts, T. H., E. Dinerstein, D. M. Olson, and C. J. Loucks. 2000. *Terrestrial ecoregions of North America: A conservation assessment.* Washington, DC: Island Press.

Robinson, G. R., and R. F. Quinn. 1992. Habitat fragmentation, species diversity, extinction, and design of nature reserves. In *Applied population biology*, ed. S. K. Jain and L. W. Botsford, 147–159. Dordrecht, Netherlands: Kluwer.

Sample, V. A. 1990a. *The impact of the federal budget process on national forest planning.* Westport, CT: Greenwood Press.

Sample, V. A. 1990b. *Land stewardship in the next era of conservation.* Washington, DC: Pinchot Institute for Conservation.

Sample, V. A. 2000. Forest management certification: Where are we now, and how did we get here? *Forest History Today* (Spring): 27–30.

Sample, V. A., and S. Kavanaugh. 2003. *U.S. interim assessment report: UNFF4.* Washington, DC: Pinchot Institute for Conservation. .

Sample, V. A., W. Price, and C. Mater. 2003. Certification on public and university lands: Evaluations of FSC and SFI by the forest managers. *Journal of Forestry* 101(8): 21–25.

Schallau, C. 1974. Forest regulation II: Can regulation contribute to economic stability? *Journal of Forestry* 72(6): 214–216.

Schmidheiny, S. 1992. *Changing course: A global business perspective on development and the environment.* Cambridge, MA: MIT Press.

Sedjo, R. 2000. Does the Forest Service have a future? A thought-provoking view. In *A vision for the U.S. Forest Service: Goals for its next century*, ed. R. Sedjo, 176. Washington, DC: Resources for the Future.

Shands, W. E., V. A. Sample, and D. LeMaster. 1990. *National forest planning: Searching for a common vision.* Washington, DC: U.S. Department of Agriculture, Forest Service.

Sierra Club. 1999. Statement on National Forest Protection and Restoration Act.

Smith, D. 1962. *The practice of silviculture.* New York: Wiley.

Soulé, M. E., and D. Simberloff. 1986. What do genetics and ecology tell us about the design of nature reserves? *Biological Conservation* 35(1): 19–40.

Spies, T. A., and J. F. Franklin. 1991. The structure of natural young, mature, and old-growth Douglas-fir forests in Oregon and Washington. In *Wildlife and vegetation of unmanaged Douglas-fir forests*, ed. L. F. Ruggiero, K. N. Aubry, A. B. Carey, and M. H. Huff, 91–110. General Technical Report PNW-GTR-285. Portland, OR: U.S. Department of Agriculture, Forest Service.

Spies, T. A., and J. F. Franklin. 1996. The diversity and maintenance of old-growth forests. In *Biodiversity in managed landscapes: Theory and practice*, ed. R. A. Szaro and D. W. Johnston, 296–314. Oxford: Oxford University Press.

Steen, H. K. 2004. *Jack Ward Thomas: The journals of a Forest Service chief.* Durham, NC: Forest History Society.

UN Economic and Social Council. 2000. *Report of the fourth session of the Intergovernmental Forum on Forests*. E/2000/L.32. New York: United Nations.

UN General Assembly. 1992. *Non-legally binding authoritative statement of principles for a global consensus on the management, conservation and sustainable development of all types of forests*. Report of the United Nations Conference on Environment and Development (Rio de Janeiro, 3–14 June 1992), Annex III, **A/CONF.151/26 (vol. III)**. New York: United Nations. http://www.un.org/documents/ga/conf151/aconf15126-3annex3.htm.

United States of America. 2003. *Preliminary national report to the Third Session of the United Nations Forum on Forests: Progress and issues related to the implementation of the IFF/IPF proposals for action*. Washington, DC: U.S. Department of State. http://www.un.org/esa/forests/pdf/national_reports/unff3/usa.pdf.

Upton, C., and S. Bass. 1996. *The forest certification handbook*. New York: St. Lucie Press.

U.S. Department of Agriculture, Forest Service. 2001. *2000 RPA assessment of forest and range lands*. FS-687. Washington, DC: USDA.

U.S. Department of Agriculture, Forest Service. 2004. *National report on sustainable forests 2003*. FS-766. Washington, DC: USDA.

Waggener, T. R. 1977. Community stability as a forest management objective. *Journal of Forestry* 75(11): 710–714.

Washburn, M., and N. Block. 2001. *Comparing forest management certification systems and the Montreal process criteria and indicators*. Washington, DC: Pinchot Institute for Conservation.

Westoby, J. 1989. *Introduction to world forestry: People and their trees*. Oxford, UK: Blackwell.

Wilson, E. O. 1992. *The diversity of life*. New York: W. W. Norton.

World Commission on Environment and Development. 1987. *Our common future*. Oxford: Oxford University Press.

World Wildlife Fund. 2001. *Top ten companies can help save the world's forests*. Washington, DC: WWF.

PART VII

LOOKING TO THE FUTURE

Larry L. Rockwood & Ronald E. Stewart

In part VII, three prominent ecologists turn their attention to the future of ecology as a science, the future of the global environment, and the prospects for sustainability in the face of climate change.

Daniel B. Botkin begins chapter 23 by outlining important advances in the science of ecology. These include the fundamental recognition that ecological systems are not steady state, but instead vary over time and space. Some of the advances were made through "big science" projects—for example, the large-scale ecosystem studies at Oak Ridge National Laboratory, Brookhaven National Laboratory, Hubbard Brook Experimental Forest, and Coweeta, among others. The National Science Foundation has also recognized the value of long-term ecological research through its Long-Term Ecological Research program.

Ecological research, however, has failed in many ways and Botkin finds much to critique. The primary problem, he asserts, is the failure of ecologists to settle upon central questions. Although authors such as Robert E. Ricklefs (1993, 1997) and Peter Turchin (2001) have attempted to articulate general principles of ecology and population ecology, respectively, Botkin prefers that ecologists identify central questions and themes. Referring to a paper that he coauthored (Belovsky et al. 2004), Botkin describes ten problems that ecological science needs to address. He also notes that one of the major reasons why ecologists have not formulated a set of major questions is the commonly held opinion within the field that ecological generalizations are simply not yet possible: what is true here and now is not necessarily true in the next county, or next year. But Botkin points out that through a better knowledge of the literature, and through better methodology, we can begin to address fundamental questions (and articulate general principles).

Finally, Botkin discusses the relationship between basic and applied ecological research and concludes that we need to eliminate the idea that basic science always precedes applications. Hence, applied research has as much intrinsic value as does basic research.

In chapter 24, George M. Woodwell suggests that the environment must be elevated to a position of prominence, not secondary to war and terrorism, as environmental issues "are now the core issues of governmental purpose." According to Woodwell, for this to happen, scientists must be more effective in communicating environmental concerns so that the public consciousness can be galvanized into action. While the current focus on protection of biodiversity and the establishment of parks and reserves is important, it is inadequate by itself in preserving functional ecosystems.

This chapter is a passionate call to all citizens of the world, not just ecologists or environmentalists, to place conservation on the center stage of the public arena in order to ensure a future for civilization on the earth. Woodwell argues that our post-9/11 focus on terrorism has distracted us from the real international crisis, the ravaging of the global human environment. Many more lives (as well as monetary resources) have been lost due to environmental degradation in all its forms—deforestation, loss of agricultural land due to drought and climate change, and the spread of diseases such AIDS, malaria, and schistosomiasis—than have been lost to terrorism. While scientists have begun speculating that the future of our civilization is at stake, citizens and governmental leaders seem unconcerned.

Raising the power of science and conservation requires, in Woodwell's view, "new insights on key issues, a new cadre of powerful individuals as exponents in key places around the world, [and] the emergence of…new institutions whose purpose is focused intensively" on these new insights. "The concept that is basic to [this] entire undertaking," he continues, "is that the world, our only habitat, is a biotic system that is failing rapidly and requires major efforts at restoring and stabilizing its functional integrity." Woodwell identifies six key issues as particular challenges to conserving our planet: climatic disruption, energy and the environment, biotic impoverishment, toxification, agriculture, and the human population.

To Woodwell, a "new environmentalism" that acknowledges the seriousness of the emergent challenges of environment is essential. We must restore and protect the functional integrity of landscapes. A crucial first step lies in "quickly moving the world away from its reliance on fossil fuels." The second step is restoration and permanent protection of the global human habitat. For this to happen, "nothing short of a renaissance of science in the public realm" must take place.

The chapter concludes with a plan of action. Woodwell contends that the view that the world is large and capable of dealing with any human intrusion must be debunked (for there exists a viewpoint among many people that humankind is not actually capable of seriously damaging the earth). He also believes that we must redefine what public welfare and public interest mean in the modern world. Essentially, the preservation of our common environment is a prerequisite for the continuation of human civilization itself.

In chapter 25, Thomas E. Lovejoy examines the meaning of sustainability in the face of climate change. As he points out, the average temperature of the earth has been remarkably stable for the last 10,000 years, but it is now clear that we can no longer count on that stability. Indeed, climate change is already upon us. Despite a few skeptics who have scientific credentials (such as Patrick Michaels and Frederick Singer), the scientific community overwhelmingly affirms the fact that the earth's ecosystems are being affected by global warming. As Lovejoy notes, the average temperature of the planet is 0.7°C warmer than in preindustrial times, with an inevitability of further warming due to the lag time between the production of greenhouse gases, the trapping of heat, and the rise in atmospheric temperatures. The general public in the United States—as evidenced by the popularity of the 2006 book and movie *An Inconvenient Truth,* featuring Al Gore, and by the radical rises in gasoline prices—is receptive to decreasing our dependence on carbon-based energy sources. But is it already too late? Glaciers around the globe and the tundra in the Northern Hemisphere are melting. As the tundra melts it is releasing methane, a greenhouse gas with the potential to trap heat over thirty times that of carbon dioxide. Methane may already account for 20% of global warming. And one of the "sinks" for excess carbon dioxide, the oceans, may begin releasing it as pH and chemical equilibria change. As Lovejoy puts it, "Basically, human activity has reached such magnitude that it is changing the essential physics and chemistry of the planet."

Relating all of this to a goal of sustaining biological populations and the functioning of ecosystems is difficult in the extreme. Since we cannot predict how the climate will change, and how ecosystems or communities of plants and animals will react, what can we do to promote sustainability? Lovejoy suggests the following course of action. (1) Reduce the stress, particularly human-generated stress (e.g., pollutants), on ecosystems. (2) Restore connectivity in landscapes to promote successful dispersal among habitat patches. This recommendation implicitly acknowledges the fact that natural areas have been reduced in size around the world, that extinction is a common phenomenon in the small populations found in small habitat patches, and that some species depend for their existence on regular dispersal from one habitat patch to another. This is, of course, the metapopulation approach to conservation. Connectivity allows higher rates of successful dispersal and, in theory, promotes the survivorship of the metapopulation. (3) Pay special attention to the coastal regions of the world. As sea levels rise, present-day coastal areas will become inundated. Yet we need functioning coastal ecosystems, whether estuaries, sand dunes, marshes, or mangrove swamps, to promote biodiversity and biological productivity, and to prevent damage to ecosystems further inland. (4) Acknowledge that human populations living at or near sea level (whether in New Orleans or Bangladesh) are in serious jeopardy and that the "short-sightedness, vested economic interests, and perverse incentives" provided by government must be eliminated before serious planning can begin.

Lovejoy also discusses how we can begin stemming and then reducing the tide of carbon dioxide entering the atmosphere. One possible action is to pay landowners to grow trees and maintain the carbon stock of a forest, while allowing them to use the forest for anything else that would not affect its carbon content. What is interesting from the viewpoint of sustainability is that, in addition to addressing climate change, this would maintain biodiversity, and it would also provide local and regional ecosystem services such as watershed protection and erosion prevention. Such a program already exists in Costa Rica, though it faces many bureaucratic problems. (Namely, it tends to favor large versus small landowners, and the verification process is flawed.)

Lovejoy concludes with the sobering thought that when we discuss sustainability in the face of climate change, we are not just talking about sustainability of golden lion tamarins, for example, but about the sustainability of human civilization itself. "If we are serious about sustainability of the human species, we need to address climate change with greater urgency than any other problem at any other time in history. It is clear from the energy/climate change perspective alone that we are long overdue for a global reckoning on consumption per capita and consumption in total. Without that, sustainability will be unattainable except in a degraded sense."

Both Woodwell and Lovejoy agree: without serious action on environmental issues we face, as described by Jared Diamond (2005), yet another example of a society that collapses and disappears. But the collapse this time would not be limited to some isolated and largely unknown civilization. It would be collapse of civilization on a global scale.

References

Belovsky, G. E., D. B. Botkin, T. A. Crowl, K. W. Cummins, J. F. Franklin, M. L. Hunter Jr., A. Joern, D. B. Lindenmayer, J. A. MacMahon, C. R. Margules, et al. 2004. Ten suggestions to strengthen the science of ecology. *BioScience* 54(4): 345–351.

Diamond, J. 2005. *Collapse: How societies choose to fail or succeed.* New York: Viking.

Gore, A. 2006. *An inconvenient truth: The planetary emergency of global warming and what we can do about it.* Emmaus, PA: Rodale.

Ricklefs, R. E. 1993. *The economy of nature: A textbook in basic ecology.* 3rd ed. New York: W. H. Freeman.

Ricklefs, R. E. 1997. *The economy of nature: A textbook in basic ecology.* 4th ed. New York: W. H. Freeman.

Turchin, P. 2001. Does population ecology have general laws? *Oikos* 94(1): 17–26.

23

The Future of Ecology and the Ecology of the Future

Daniel B. Botkin

Beginning in the 1970s, while working with my colleagues on the Marine Mammal Protection Act, I began to pose questions about the worldview that many conservation scientists held. This led me to explore the relationship between major prescientific beliefs in Western civilization about man and nature, resulting in a book, *Discordant Harmonies: A New Ecology for the Twenty-First Century* (1990). I have continued to consider this worldview and the way that research is conducted in ecology. In recent years this has led me to some fundamental questions.

The questions I wish to discuss here are as follows: What path has ecology followed in the past fifty years? And where should it go in the future? We can point to a number of advances in the past half century of which we can be proud. But we must also acknowledge failures in the advancement of ecology as a science that are still with us, and for which there are both scientific and societal needs for improvement in the future.

ADVANCES

There have been numerous advances in the science of ecology during the last fifty years, including the recognition that ecological systems are not steady state, that time and space within them vary, and that such non-steady-state systems are open to analysis. Ecology also gained popularity as a science during this period, especially in the early 1960s (after Sputnik, of course), spurred by growing societal concern with the environment. Among the first and classic "big science" developments in ecology were:

1. Three projects on the effects of irradiation on ecosystems—one at Brookhaven National Laboratory in New York, one at Oak Ridge National Laboratory in Tennessee, and one in Puerto Rico—conducted by the (then) Atomic Energy Commission (now the Department of Energy)
2. Watershed studies at Hubbard Brook, New Hampshire; Coweeta, Georgia; and the Andrews Experimental Forest in Oregon

3. The beginnings of formal, long-term ecological research and monitoring, through the National Science Foundation's Long-Term Ecological Research (LTER) program
4. The extension of ecological research to global earth system science, both through applied problems such as global warming and through fundamental, agency-sponsored research—for example, the National Aeronautics and Space Administration's Mission to Earth and Earth System Science programs and the National Center for Atmospheric Research, as well as the Environmental Protection Agency's support for global climate models including biological systems
5. The acceptance in the science of ecology of computer models of complex systems

LACK OF ADVANCES

Even as technology and basic scientific methods have enabled advances in ecology over the past half century, other areas within the field have lagged. Some delays and failures have been:

1. The fact that much basic and applied work is still based on the idea and the mathematical statement of steady-state systems (this is true, e.g., for fisheries, forestry, endangered species, restoration ecology, and disturbance ecology)
2. Failure of the science of ecology to develop agreed-upon central questions
3. Failure to establish comparative long-term monitoring (i.e., comparison of different geographic regions over time variation), which would be a natural extension of the National Science Foundation's LTER program
4. Failure to integrate observation and theory
5. Failure to measure key variables
6. Continued dependency on plausibility rather than true science as a way to answer "scientific" questions—a continuation of the problem I discussed originally in *Discordant Harmonies*

WHAT ECOLOGY NEEDS

Similar concerns to those I have expressed above are shared by other ecologists, a group of whom published some of their concerns in an article titled "Ten Suggestions to Strengthen the Science of Ecology" that addresses the question, how can the science of ecology be improved in the future (Belovsky et al. 2004)? As defined by this group, the problems that continue to beleaguer ecology are:

- *Fashionability:* Scientific issues tend to go in and out of fashion in ecology, without scientific resolution.
- *Literature:* Connected to the first problem, there is lack of appreciation of past literature. On one hand, this is part of the reason why ecologists have been fickle about determining the central issues in the field; conversely, the lack of

familiarity with the field's literature ensures that ideas will be rediscovered, lost, and rediscovered again.

- *Theory:* Formal mathematical and computer modeling theory has been inadequately integrated with empirical research. There has been plenty of ecological theory, but much of it has been accepted without adequate testing, and sometimes accepted in spite of contrary empirical information.
- *Valid experiments:* Observations in the field, both qualitative and quantitative, have been inadequately integrated with formal, statistically valid experiments.
- *Complexity:* In spite of the recognition of nature's great complexity, ecological theory has tended to be based on an implicit belief that there are single causes of complex events.
- *Dynamic nature:* The idea that nature achieves a constant, fixed state unless affected by human actions has been an overriding belief, one that has surprising dominance in ecology. As a result, there is an inadequate analysis and understanding of equilibria and disequilibria in ecological systems.
- *Replication:* Too often, something demonstrated once—in one place for one species in one habitat—is accepted as universally true. Replications of studies, over both time and space, are needed. Comparative experiments are inadequate.
- *Rigor:* Even where there is replication, studies are often done differently, with different methods and nonstandard conditions, so that data are not compatible. There is a general lack of rigor in obtaining data.
- *Methods as methods:* Ironically, partially in recognition of methodological failures, ecologists have tended to let methods (especially statistical methods) become driving forces rather than tools. The result is a kind of blind analysis leading to blind acceptance of what is generated by a computer.
- *Applications:* In ecology, as in other sciences, fundamental knowledge and applied knowledge must be integrated. Applications force the fundamental science to develop; the fundamental science is valuable to environmental issues. But often these two are seen as distinct and unrelated rather than complementary.

WHAT CAN BE DONE

Central to correcting these failings is the identification of central questions and themes. Most sciences have a set of readily identified major leading questions. This is not the case with ecology. Over the years, I have kept asking my ecological colleagues to name the five or ten leading questions in ecology, or in their part of ecology; a common reaction is, "I never thought about that." Perhaps because of the great complexity of ecological systems and processes, ecologists have tended to operate under the assumption that generalizations are not possible, and so the field has tended to proceed in different directions based on specific questions of interest to an individual scientist.

The establishment of central questions in ecology will help us avoid the fashionability trap, will lead to better appreciation of the scientific literature, will improve the integration of theory and observation, and will aid in acceptance of the need for better and more extensive replications—especially performing the same kind of experiments in different biomes, ecosystems, and habitats, with different sets of species.

The inadequate integration of natural history and experimentation will also be improved by the determination of central questions and themes. But ecologists still need to come to terms with the role of natural history observations within a formal science. There is a curious kind of love-hate relationship between these two aspects of ecology. Many ecologists have come into the field because of their love of natural history, and good naturalists often provide insights that serve as valuable hypotheses. There is a tendency to dismiss good qualitative observations as totally useless, though, and ecologists miss valuable insights that could help pose the central issues (Botkin 2001).

Improving how we deal with complexity requires two additional steps. First, ecologists need to follow Occam's razor—use the simplest explanation consistent with observations—which has been too often ignored. Elaborateness has sometimes been selected for its own sake. Second, ecologists have tended to develop single-variable theory where it is inappropriate, unsupported, or contradicted by observations. Modern computer numerical methods, including simulation, along with a variety of modern mathematics, including stochastic processes, can help ecologists think in terms of more than one variable and make multivariable theory more tractable (Botkin 1993).

In order to improve the ecological concept of stability, further steps must be taken. Ecologists seem of two minds about stability in ecological systems: it has become fashionable since the 1990s to agree that ecological systems are generally not characterized by a steady state; however, when ecologists develop theory or attempt to solve real-world problems, the tendency is to revert to steady-state concepts. One possible solution is to reconsider the fundamental properties of stability in complex systems within an ecological context. Once again, modern computer and mathematical approaches can help. For although ecologists have tended to believe that non-steady-state systems dynamics are intractable, many nonequilibrium systems are tractable, and have been shown to be (Botkin 1993).

An idea of what the central questions of the field are will also help us deal with the dilemma of methodology, so that methods no longer substitute for goals, questions, themes, or solutions.

The integration of basic and applied ecological science is a slightly thornier problem. This is to some extent because the two are quite naturally integrated in natural resource management issues—they raise both fundamental and applied questions—but the integration is not always recognized as such. Perhaps ecologists have been caught up in the fashionable twentieth-century argument that true science is only basic science, and that basic science always precedes its applications. This is a modern myth about the scientific method. In fact, applied problems have frequently led to recognition of the need to improve some aspect of fundamental scientific understanding, subsequently leading to advances in basic science.

Consider, for example, the nineteenth-century development of thermodynamics, which arose in large part because of the desire to develop more efficient engines. The result of this desire was the development of the Carnot cycle, as well as the first and second laws of thermodynamics. These, in turn, led to the invention of the diesel engine. Another example is the Wright brothers' invention of the airplane. In the process of making this invention, they had to invent, in an integrated fashion, aeronautical science and aeronautical engineering (Combs and Caidin 1979).

We need to eliminate the outdated and inaccurate scientific myth that basic science always precedes applications.

My colleagues and I were not the first to point out these needs (Belovsky et al. 2004), but we believed it was time to restate them and to urge ecologists to consider them as a package. We hoped that these recommendations would aid in the improvement of both fundamental and applied ecology in the future. In reviewing these needs here, I maintain that same hope.

Author's Note: From the late twentieth century through today, Lee Talbot has been, in my opinion, the most successful scientist in the United States in actually solving environmental problems through government action in a constructive way that is good for people, nonhuman life-forms, and the environment. He has been an important stimulus for the ideas I have presented here. He has excelled in getting our federal government to pass laws and establish policies that have profoundly affected how our nation approaches the conservation of its living resources. Although Talbot might not take the credit as completely as I see it, he is responsible for the composition and the passing of the federal Endangered Species Act (1973) and the Marine Mammal Protection Act (1972). These achievements alone are sufficient to qualify him as one of the prime figures in our time for conservation of living resources. But he has also worked in more than 120 nations, helping with the conservation of their living resources, and he has affected government by working for the Council on Environmental Quality, understanding how to get things done in the arcane world of Washington politics. He has done all of this while maintaining a solid basis in science, an objective view of the issues, and a concern for people as well as the rest of nature. Moreover, his applied efforts have stimulated new research in ecology. I, for one, have been led to do new ecological research because of his inventive activities.

To be specific, Talbot inserted wording into the Marine Mammal Protection Act that its primary goal was to maintain the health and stability of marine ecosystems, and that its secondary goal was to maintain an optimal sustainable population of marine mammals. Soon after the act was passed, the members of the commission created by the act contacted me and said that they could not understand what "optimum sustainable populations" could mean, either scientifically or legally. Under contract with them, I worked with my colleague Matthew J. Sobel on a report that made use of ecological science and applied mathematics to explain what the law, as written, allowed (Botkin and Sobel 1975).

While working on this report, I spoke with Talbot and voiced my opinion that none of the terms—health, stability, marine ecosystem, or optimum sustainable populations—had an agreed-upon scientific meaning. I believe that that was his intent. He had placed these terms into the act to stimulate what he hoped would be important and constructive discussions, so that the terms would become meaningful.

References

Belovsky, G. E., D. B. Botkin, T. A. Crowl, K. W. Cummins, J. F. Franklin, M. L. Hunter Jr., A. Joern, D. B. Lindenmayer, J. A. MacMahon, C. R. Margules, et al. 2004. Ten suggestions to strengthen the science of ecology. *BioScience* 54(4): 345–351.

Botkin, D. B. 1990. *Discordant harmonies: A new ecology for the twenty-first century.* New York: Oxford University Press.

Botkin, D. B. 1993. *Forest dynamics: An ecological model.* New York: Oxford University Press.

Botkin, D. B. 2001. *No man's garden: Thoreau and a new vision for civilization and nature.* Washington, DC: Island Press.

Botkin, D. B., and M. J. Sobel. 1975. Stability in time-varying ecosystems. *American Naturalist* 109(970): 625–646.

Combs, H., and M. Caidin. 1979. *Kill Devil Hill: Discovering the secret of the Wright brothers.* Boston: Houghton Mifflin.

24

A New Environmentalism: Conservation and the Core of Governmental Purpose

George M. Woodwell

No longer can the public interests of environment be relegated with impunity to the backseat of public policy, secondary to war and terrorism. These issues are now the core issues of governmental purpose. They must be elaborated far more effectively than heretofore by the scientific community and forced into the public consciousness ten to a hundred times more powerfully than either science or conservation has been willing or able to do. Protection of biodiversity and the establishment of parks and reserves is a worthy goal, but totally inadequate in preserving a functional biosphere. Responsibility for these advances lies now with the scientific and conservation communities, which must rise to the challenge. Continuing on the present course guarantees the destruction of civilization through rapidly cumulative environmental impoverishment, which is already well advanced. It is time to seek a reorientation of environmental science and conservation and their nongovernmental supporters to ensure a future for this civilization.

TERRORISM AND ENVIRONMENTAL IMPOVERISHMENT

While public opinion has been galvanized in favor of extraordinary steps to combat terrorism, the global human environment has been ravaged. The erosion of human welfare, including the loss of life quite apart from the effects of war, has been unprecedented and continues. Environmental degradation—be it climatic disruption; loss of biological diversity and denudation of world forests; the spread of diseases such as AIDS, malaria, and schistosomiasis; or loss of potential agricultural lands in vast regions affected by drought—has involved far greater costs in money and lives than losses due to terrorism, as grim as some of those continuing tragedies have been. The bellicosity engendered in the response to terrorism is the cause of further environmental erosion and waste of resources, quite apart from the commitments of hundreds of billions of dollars in widening this wasteland through war.

The scientific community, acutely aware of major changes in the global human environment, is alarmed by its obvious failure in drawing the attention of policy

experts and governmental leaders to such discrepancies in governmental performance. The question arises as to whether there is an overt decision by governments
to use short-term political and economic crises to mask the costs of environmental
impoverishment, thereby protecting long-standing economic constituencies such as
those benefiting from subsidies supporting the fossil fuel and agricultural industries.
What might be done to defend the larger public interest in these issues? How can we
distill, from the abundance of information that we have in terms of losses of human
lives and damage to the economy, just where the greater public interests lie? And
what forum might we use for science to develop and publicize this information? The
topic has been examined in various forms in many books and treatises, yet the follies
continue.[1]

The war on terrorism, while agreed to in principle by all, is commonly held as a
special responsibility of the United States, which has successfully defined the "war"
for the world. But the set of arguments used in advancing "national security," narrowly defined, is never applied to the continuing, cumulative, environmental catastrophes of far greater cost and potential for cumulative damage to national security
than acts by vandals. Moreover, the intensification of demands on the environment
requires continued advances in the understanding of basic ecology, the biophysics of
the earth, to provide the insights needed in conservation and government. There is
no exaggeration in the observation that scientists and many others are raising serious questions as to the future of this civilization over the next decades (Diamond
2005). Sir David King, chief scientist in the United Kingdom's Blair administration, has recently pointed to climatic disruption as the most serious issue facing us
in this century and beyond. Despite such sentiments, widely shared in the scientific
community, the United States has taken an official position opposed to any action;
it has also succeeded in imposing its will on nearly 200 other nations represented at
the Tenth Conference of the Parties to the UN Framework Convention on Climate
Change meeting in Buenos Aires, Argentina, in early December 2004, and again at
the Eleventh Conference meeting (held jointly with parties to the Kyoto Protocol) in
Montreal in December 2005.

THE WORLD: SMALL AND FULL

Over the last sixty years, within the lives of many now living, the rapid intensification of human influences has changed the whole earth. Three phases have been clear.
During the 1940s through the 1960s, the bomb, its fallout, and the potential effects
of both shook the civilized world and made it clear that the earth had become small
in proportion to real and potential human influences. From the 1960s through the
early 1990s, the earth shrank further as industry expanded and industrialized agriculture loosed its toxins into the same atmospheric circulation that distributed radioactive fallout over our planet. The new suite of agricultural poisons and other toxins
appeared in the far corners of the earth, including the Antarctic, the depths of the
oceanic abyss, and in the tissues of virtually all life-forms, including people. Concern
about effects, both environmental and personal, was well placed. Now, from the
1980s through the present, a rising wave of global biotic impoverishment—caused

by climatic disruption, toxification, and the expansion of all chronic disturbances throughout the biosphere—has been joined by the political and economic chaos that is engendered and perpetuated by environmental chaos to shake an increasingly unstable and uncertain world. The issues of conservation have expanded well beyond the challenges of preserving interesting plants and animals, views, or parks to the conservation of the human environment and the continuity and further evolution of this civilization (Woodwell 1990). Failure to correct these trends is producing a surge of global biotic and human impoverishment that anticipates a new Dark Age of political, economic, and social chaos, including reductions in the human population, as the effects of environmental impoverishment explode over the earth. The 2004 national election in the United States has virtually assured the world that major new increments of destruction are in store as that industrialized country attempts to prolong as long as possible the expansion of the fossil fuel–based industrialized world.

Such considerations are now, for the first time in history, global. They have, moreover, a last-chance element as resources considered essential disappear, irretrievably. They present, whether acknowledged or not, a major challenge to governments, to the academic community, to conservation, and to science. They also offer an opportunity to establish some new and constructive trends in a world already in flux.

A "new environmentalism" is urgently needed, and it must be advanced vigorously into the public realm to become an intrinsic part of the governmental purpose. It is especially important that environmental issues be moved back to the central political stage in the United States. The scientific message of uncertainty about the future of the planet must be replaced by lucid descriptions of likely future scenarios of a human population that has surpassed the capacity of the planet's life-support systems. The antiscientific, anti-intellectual, antienvironmental, xenophobic lobbies must be countered by an aggressive, persistent campaign that presents the state of the planet in such clear and compelling ways that common citizens will be moved to action in defense of common interests—not only to ensure a habitable earth, for people can and will survive on a much impoverished planet, but to preserve the opportunity for the further evolution and development of this civilization.

We see a need to redefine conservation as the definition, restoration, and indefinite protection of the global human habitat. Emerging now as the central governmental purpose globally is this overwhelmingly urgent mission to project that must gain the power to displace, or at least to compete, with bellicosity and active war in the public realm. There is in fact no place or time for war in this shrunken, last-chance world. Unfortunately, environmental pressures have never been a major *causa belli*, and this fact makes even more urgent a well-developed plan for countering a global trend toward impoverishment.

How can we establish environmental issues as sufficiently powerful politically to compete with the political and economic crises that have historically attracted statesmen and scoundrels alike to dominate all governmental functions? How can we make the protection and stewardship of the biosphere, including its full complement of biodiversity, the core of governmental purpose for the next decades?

There is no simple or final solution with a satisfactory outcome. There is only a series of hard-fought battles, such as those that have brought the great social and political advances of the past. It took a lifetime, seventy years of postconstitutional experience and the Civil War, to abolish slavery in the United States, despite the fact that those engaged in writing the Constitution knew well what must happen. It took the destruction of two Japanese cities and global radioactive contamination to restrict the use of nuclear weapons to underground tests, and it took further contamination and extraordinary arguments, experience, and data to ban testing altogether. It took half a century of struggle to eliminate some long-lived agricultural poisons; we still see total failure in spurring the revolution in agriculture required to eliminate the routine contamination of human food and water and air with commercially produced agricultural poisons as blatantly undesirable as methyl bromide, directly toxic to all and indirectly toxic through destruction of high-altitude ozone.

We do not have decades or even years to reverse the frightfully progressive impoverishment of the earth. For examples of where we are headed, we need not look far. Nearby Haiti has been driven so far into environmental decay that there is no internal economy capable of supporting a government. The country's only hope lies in an external infusion of money and insight that can restore a functional landscape, a water supply, a stable indigenous agriculture, forests in the uplands to hold soils and to stabilize water flows, and water quality, as well as provide a resource base to support both a government and a viable economy. A stable landscape that is functioning normally to support life is the sine qua non of both government and business. And it does not emerge from an unregulated market. It requires governmental leadership, in this case outside leadership, for there is no government and none possible internally. The decay of Haiti has passed the point of no return. Both a vision as to what will work and outside financial help are now necessary before a plan can be implemented. It could be a U.S. effort, a model of success for the world, if we could have agreement as to the vision and find the people and resources to implement the transition.

The progress that has been made in recognizing the global problem has emerged largely from science, as the scientific community has produced both data on damaging effects and paths to solutions that have become common knowledge. It is this process, the development of new information, insights, and their diffusion into the public realm, that is the key. The new information includes a new model and new data, all focused on what will work in restoring a functional and durable human environment. What is clean water? Clean air? What is a viable city? More important than that is a fundamental understanding of what governments are for, what they do, what taxes are for, why we have public schools, and why we respect and build governmental purpose in the common interest. These issues emerge ever more sharply in a world pressed by expanding demands of an expanding population (Diamond 2005). Nothing short of a renaissance of science in the public realm will work. How to proceed?

THE OBJECTIVE

Raising the power of science and conservation requires new insights on key issues, a new cadre of powerful individuals as exponents in key places around the world, the emergence of a matrix of new institutions whose purpose is focused intensively on the insights, and the data and people required to make the transition. And it depends on a new intensity of public interest and discussion of the public welfare in a world too small for war.

The concept that is basic to the entire undertaking is that the world, our only habitat, is a biotic system that is failing rapidly and requires major efforts at restoring and stabilizing its functional integrity. The topic has been elaborated extensively over decades but has not emerged as an overwhelming public purpose, the core of governmental function. The time for its emergence is upon us now.

KEY ISSUES

The issues continuously evolve. At this moment there is little question that the following will dominate and require continuous elaboration in detail and in concept.

Climatic disruption: The causes, effects, and potential course for the future—
and, most of all, cures and how to implement them—are the key issues.
Although the core of these issues is scientific, the definition of the problem
and recognition of patterns of corrective action are well advanced, even to
the point of a global treaty, the UN Framework Convention on Climate
Change, which has been universally ratified. Progress has fallen victim to
politics. Progress must resume. Continuing on the present course will lead
to biotic and economic impoverishment of the earth. The topic is urgent and
immediate and requires profound changes, which carry with them trivial costs
and obvious benefits that range from corrections of trade imbalances through
major contributions to the stabilization of climates globally.

Energy and environment: Deflection of the climatic disruption sets clear limits
on the extent to which we can exploit fossil fuels for the further development
of technology. The transition away from fossil fuels will bring not a simple
replacement, but a new world in which we shall use different sources of energy
far more efficiently and in different ways. The transitions will be as profound
and continuing as the computer revolution.

Biotic impoverishment: Chronic disturbances of all types (chemical, physical,
biotic) contribute to the loss of biodiversity. The transitions are from the
complex to the simple; from forest to shrubland to barren ground; from
agriculture to salinized wasteland; from a working landscape to eroding
gravel; from a rich fishery to a dead zone; from a pristine valley to a dump
for mining wastes. We are moving from a world that works to a planet that is
warming rapidly, on which glaciers are melting, sea level is rising, forests are
burning, and storms are increasingly violent and costly. Scientists are saying
that even larger changes may be entrained already as they discover that no
corner of the planet has escaped the influences of human-induced chronic
disturbance (Diamond 2005).

Toxification: This is the introduction of long-lasting xenochemicals with which
life has had no evolutionary experience. All are major contributions to the
continuing processes of impoverishment. The problem is huge, with large
economic consequences for industrial interests and even larger consequences

for public health, welfare, and security. One important issue is toxic air that affects the morbidity and mortality of plants, including trees.

Agriculture: The current model of industrial agriculture is not sustainable and is inconsistent with a finite, biotically mediated human habitat. It requires rebuilding within the new vision of a closed-system, finite world. In the short term, we must elevate to the scale of a major global crisis the great agro-industrial transition, through which virtually all new areas of grain production are now carved out of the world's tropical rainforests and woodlands—a transition that will be hastened by the imminent dismantling of the agricultural subsidy programs of the industrial nations.

The human population: The earth is saddled with 6.3 billion humans whose numbers will increase over the next decades by at least 25–50%, possibly more. These increases ensure that the problems already recognized in environment, health, and government will be intensified. The problem is not simply a problem of total numbers. It is complicated by migration pressures, by shifts in age groupings, by mores and religions, and by the emergent aspirations of nations. Our effort is designed to offer insights into what can be expected and what can be done to deflect the obvious problems of current trends.

What is called for here is recognition of the seriousness of the emergent challenges of environment, the fact of our residence within the fold of a biotic complex in which each of us is but one of the actors, guests who have taken over the house and now must maintain it, all of it in self-preservation. The key is recognizing, defining, restoring, and protecting the functional integrity of landscapes. The first step lies in quickly moving the world away from its reliance on fossil fuels in the interest of stabilizing the composition of the atmosphere. In the absence of governmental leadership, major efforts through scientific, conservation, and nonprofit communities are appropriate. The process is the "jazz" of James Gustave Speth (2004), and the sooner it emerges from every element of the nonprofit and the commercial world, the better off we shall all be.

In a world in which the United States can deflect the core purpose of two Conferences of the Parties to the UN Framework Convention on Climate Change, there is little hope that we can find support in the currently dysfunctional U.S. administration for such a sea change in perspective. The alternative lies in local action, innovations bubbling up from the bottom and engaging the remarkably versatile and powerful U.S. nongovernmental community, including private foundations (Speth 2004). Success will require a major reorientation of public purpose, a maturity of approach currently quite foreign to most foundations, and at least a ten-year commitment to the transition.

A PLAN OF ACTION

The time for this reorientation of the scientific and conservation communities and their supporters is now. The objective is to advance and implement two concepts. First, we must transition away from the common worldview that the earth is large and resilient, capable of supporting virtually any human intrusion, to the new realization that the world is in fact small and, by comparison with the world of our fathers, fragile and already in need of restoration from cumulative increments of impoverishment. The implications of this change are profound, for they require a rapid (a few years at most) shift away from fossil fuels to a quite new world of closed industrial systems of high energy efficiency based on renewable sources of energy. Defining the details of that new world is the immediate challenge, and the money and interest must come initially from the foundation community and the nonprofit scholarly community. In the end, it must be the dominant governmental purpose in a world too busy with rebuilding the human environment to allow the distractions of war.

Second, the transition will require definition and redefinition of the public welfare and the public interest in the largest sense. What is a viable ocean? A viable landscape? There will be many answers, different for different places and intensity of human activities. And the perspectives will change with time and require continuous local review. The business of defining the circumstance and measuring the details of the quality of the habitat is the business of all. It should become a part of the grammar school curriculum, as well as a major issue for science programs throughout universities and colleges. For it is everyone's business. In the end, we shall be living in a park under new rules that require the preservation of the environment first, for there is no other way to guarantee the continuity of a human habitat and an industrial civilization. The blueprint is emerging from several sources cited above in the works of Paul Hawken, Amory Lovins, and L. Hunter Lovins (1999), Lester Brown (2006), Paul R. Ehrlich and Anne Ehrlich (2004), James Speth (2004) Peter H. Raven and Tania Williams (2000), and myself (2000), along with a score of others.

To speed the process we need models and examples, financed by the private foundation community. These can be comprehensive programs throughout the country and beyond. Success might even include enlisting larger sums from the World Bank, gradually persuading governments of their central role of environmental action.

Impossible? Not at all. The cost of failure is civilization, and even life itself.

Note

1. Experience in defining how the world works, both as social, economic, and political system and as a biophysical system, is comprehensive. The central political issue for the capitalistic world was defined compellingly by Garrett Hardin in his 1968 article "The Tragedy of the Commons" and in the barrage of articles and books that his provocative statement triggered. The biophysical experience accumulated as the scientific community recognized global contaminations with radioactivity (Tykva and Berg 2004) and, later, the pesticides and other industrial poisons so brilliantly set forth as human and environmental hazards by Rachel Carson in 1962, an issue still unresolved despite long-term recognition and cumulative experience (U.S. Environmental Protection Agency 2004). See recent books by Speth (2004), Ehrlich and Ehrlich (2004), Hawken, Lovins, and Lovins (1999) and Brown (2006).

References

Brown, L. 2006. *Plan B 2.0: Rescuing a planet under stress and a civilization in trouble.* New York: W. W. Norton.

Carson, R. 1962. *Silent spring.* Boston: Houghton Mifflin.

Diamond, J. 2005. *Collapse: How societies choose to fail or succeed.* New York: Viking.

Ehrlich, P. R., and A. Ehrlich. 2004. *One with Nineveh: Politics, consumption, and the human future.* Washington, DC: Island Press.

Hardin, G. 1968. The tragedy of the commons. *Science* 162(3859): 1243–1248.

Hawken, P., A. Lovins, and L. H. Lovins. 1999. *Natural capitalism: Creating the next industrial revolution.* Boston: Little, Brown.

Raven, P. H., and T. Williams, eds. 2000. *Nature and human society: The quest for a sustainable world: Proceedings of the 1997 Forum on Biodiversity.* Washington, DC: National Academy Press.

Speth, J. G. 2004. *Red sky at morning: America and the crisis of the global environment.* New Haven, CT: Yale University Press.

Tykva, R., and D. Berg. 2004. *Man-made and natural radioactivity in environmental pollution and radiochronology.* Dordrecht, Netherlands: Kluwer.

U.S. Environmental Protection Agency. 2004. *DDT: Persistent Bioaccumulative and Toxic (PBT) Chemical Program.* Washington, DC: PBT Chemical Program.

Woodwell, G. M. 1990. The earth under stress: A transition to climatic instability raises questions about biotic impoverishment. In *The earth in transition: Patterns and processes of biotic impoverishment,* ed. G. M. Woodwell, 3–7. New York: Cambridge University Press.

Woodwell, G. M. 2000. Science and the public trust in a full world: Function and dysfunction in science and the biosphere. In *Nature and human society: The quest for a sustainable world: Proceedings of the 1997 Forum on Biodiversity,* eds. P. H. Raven and T. Williams, 337–346. Washington, DC: National Academy Press.

25

Climate Change and Prospects for Sustainability

Thomas E. Lovejoy

The challenges of sustainability are tough and complicated, and to them we must now add the mega-challenge of climate change. The average temperature of the earth has been remarkably stable for the last 10,000 years—that is, for all of human recorded history and some of our prehistory. Prior to this period it varied fairly significantly, with swings between glacial and interglacial periods. The entire human enterprise as we know it, however, has been based on the assumption of a stable climate.

Today it is clear that we have departed from that stability. The average temperature of the planet is 0.7°C warmer than in preindustrial times; even if we stopped adding to the problem, another 0.7°C increase is inevitable because of lag time between the addition of greenhouse gases and the resultant increase in atmospheric temperatures.

Signs of physical response in nature to the 0.7°C increase already abound. Arctic ice has retreated. All tropical glaciers will be gone in twenty years, including the famous snows of Kilimanjaro. Records of ice on lakes in Europe and North America show later onset and earlier breakup year after year. There is a greater frequency of higher-intensity storms, personified by the tragedy of Hurricane Katrina, and including the first hurricane-intensity storm in recorded South American history. Additionally, the increased wildfire activity in the American Southwest has now been demonstrated to be driven by climate change.

As predicted, the warming is greater at higher latitudes, leading to serious concerns about melting of the Greenland glaciers. Sea level is rising due to the thermal expansion of water, with about one meter predicted by 2100; the sea level rise will be even larger should the land-based ice of Greenland and Antarctica be added to it. (Arctic ice, in contrast, essentially floats and thus does not add to sea level rise as it melts).

The climate change has been caused by an increase in greenhouse gases from the burning of fossil fuels and biomass burning (principally in the tropics). Elevated CO_2 levels have also caused a change in ocean acidity of 0.1 pH, with grave implications for marine ecosystems.

Before considering what climate change portends for sustainability, we need to recognize that climate change itself is an indicator of unsustainable development.

Basically, human activity has reached such magnitude that it is changing the essential physics and chemistry of the planet. Similar changes in human blood chemistry, noted during an annual physical, would be considered serious symptoms of disease.

If the question had been raised at the onset of the Industrial Revolution as to whether it would be a good idea to change the global climate, it is doubtful whether anyone would have replied in the affirmative. Today, though, while the biggest and most ignored policy question is how much climate change to allow, we face the additional question of how to live with it—that is, how to achieve sustainability with the shifting base of climate change.

One element of dealing with the issue, of course, is to try and understand impending climate change. Most climate models are still on a coarse scale, so the development of more detailed models should be a priority. In some cases the existing models agree, as they do for the northeastern United States, where they show a climate unfavorable for sugar maple and the associated legendary autumn foliage. In others they do not; for example, the United Kingdom's Met Office Hadley Centre model projects massive change and drying in the Amazon, while five other models do not.

Another difficulty with the models—and, indeed, much of the discussion about climate change—is that they generally assume gradual and linear change. That is clearly an incorrect assumption. Paleoclimatic data are replete with examples of abrupt and rapid change. One of the most worrisome involves the major ocean current often referred to as the "conveyor belt." It transports warm water from the tropical Pacific west across the Indian Ocean and around the Cape of Good Hope, northward into the Atlantic until finally, between Europe and Greenland, it becomes cool and dense enough to sink, fold under itself, and flow back to rise again in the Pacific. This current is a major driver of global climate and has been known to shut down in the past. Already there are signs of possible weakening, as freshwater from melting ice makes the waters less dense. Whether the weakening is imminent or a prospect in the relatively near future is very difficult for science to determine. Nonetheless, this implies quite clearly that the best way to avoid the prospect of a current shutdown or any other abrupt change is to stabilize atmospheric concentrations of greenhouse gases and minimize consequent climate change as soon as possible. In a sense it is quite simple: as climate change increases, the probability of abrupt change increases even faster.

There are rising numbers of responses by living nature to the climate change that has already taken place. Many temperate and arctic birds are migrating earlier and nesting earlier. Similarly, many plant species are blooming earlier; the grapevines in Bordeaux, for instance, are now blooming three weeks earlier than in the past. Some Northern Hemisphere butterfly species are changing their geographical distributions, some moving up in altitude as well as northward. Many arctic species adapted to ice environments, including polar bears, are encountering serious problems as the ice retreats. In Alaska and parts of western Canada, elevated nighttime temperatures have fostered serious infestations of bark beetles and massive coniferous tree mortality.

In the oceans, elevated water temperatures have led to frequent coral bleaching events in which coral animals evict the alga that serves as symbiotic partner. Two coral species are now on the U.S. endangered species list. Independent of climate change, the increased acidity of the oceans from elevated CO_2 concentrations has serious complications for the tens of thousands of species that build skeletons of calcium carbonate. For as atmospheric CO_2 concentrations continue to grow, so will acidity. Calcifying organisms' ability to build shells depends on an equilibrium that is pH dependent; if the trend to acidification continues, by century's end some organisms' shells will go back into solution while they are alive. Science has only begun to investigate the implications of ocean acidification, but the change will surely ripple through marine ecosystems.

Climate change is not new to the history of life on earth, so past change can inform us to some degree what we might expect. What is quite clear is that biological communities do not move as a unit. Rather, each species tracks its required conditions and moves in its own direction and at its own rate. Consequently, we can anticipate that ecosystems will disassemble and the constituent species will assemble into novel ecosystems. That in itself is a challenge for conservation. But the challenge is made bigger yet by the extensive human alteration of landscapes, which has resulted, in most cases, in a virtual obstacle course for many species. Combined with reduced population sizes for numerous species, this is the principal factor contributing to the projections of major extinctions from climate change.

In the end, sustainability is to a great degree dependent on the biological underpinnings of human society, whether it is for direct use (e.g., fisheries) or ecosystem services (e.g., dependable water supplies; maintenance of the rainfall system east of the Andes in southern South America; disaster prevention such as the necessity of intact mangrove forests to protect against storm surges on ocean coastlines, or forested hillsides to diminish the likelihood of floods and landslides). So the logical question is, what can be done to minimize the impact of climate change? An obvious goal is to reduce the stresses on ecosystems. For example, in the case of coral reefs, this means eliminating siltation from terrestrial sources that cover corals and impede photosynthesis.

Restoring connectivity in landscapes would clearly go a long way toward minimizing the impact of climate change and would enhance the probability of successful dispersal. However, this is harder to do in some parts of the world than others, just because of sheer human numbers and pressure on the landscape. Approaches such as restoring natural vegetation along watercourses, which is often important for other reasons (here, reducing soil erosion and water pollution as well as fostering normal stream temperatures), can nonetheless contribute to enhancing connectivity. One thing is certain: isolated protected areas (bits of nature surrounded by human-dominated landscapes) will lose much of their diversity, especially in an era of climate change, and will do little to prevent escalating extinction rates. Even if climate change was not a real threat, restoring the biodiversity and ecological function of landscapes is important on its own merit. At first consideration, existing protected areas may seem unable to fulfill the purposes behind their creation; with increased landscape connectivity they become even more important, though, because they will be the safe havens from which species will disperse and create the new biogeographic patterns of a climate-changed world.

The coastal regions of the world are worthy of special attention. Sea level rise, even on its own, will affect wetlands and other coastal ecosystems. The Nature

Conservancy is undertaking an interesting project in North Carolina's Albemarle Sound aimed at anticipating the effect of sea level rise and enabling coastal ecosystems to move inland (presumably in some sort of analogy to how that must have happened with the rising sea levels after the last glacial period).

Our greatest challenge is the massive concentration of human population and development in the coastal regions throughout the world. In Louisiana, for example, half the population lives below or within a meter above sea level. A seemingly impervious Gordian knot of shortsightedness, vested economic interests, and perverse incentives—flood insurance, often provided by government, that leads to rebuilding after severe storms in almost willful ignorance of the increasing likelihood of a more severe storm—seems to keep present coastal use locked in place and frozen in time.

The fundamental issue, of course, is addressing climate change itself—drastically slowing it, halting it, and maybe even reversing it. This goes to the very heart of modern society, namely its energy base. Simply speaking, it means phasing out fossil fuel use and replacing it with other forms of energy as rapidly as possible. This feat is not as impossible as it might seem at first, but it clearly cannot be achieved sufficiently rapidly by depending on the private sector alone, with too many fixed costs. Such an achievement will require a public-private partnership with incentives and subsidies for (1) conversion to climate-friendly energy and (2) research and development.

There is no single energy alternative; our problems can be resolved only through a combination of alternate energy sources: renewable (solar, wind, tidal, geothermal), biofuel (renewable if done correctly), hydrogen (depending on how the hydrogen is generated), and, in my view, perhaps nuclear, in some situations. Improvements in energy efficiency and conservation have much to contribute as well.

Another part of the possible solution is carbon sequestration. One method of sequestration involves using biological systems (e.g., reforestation, plantations) to fix CO_2 as carbon accumulates in growing trees. Another method (which could make the coal industry part of the solution rather than an impediment) is to burn coal in a way that pumps CO_2 belowground for permanent storage. This is already being done successfully in a limited number of examples.

Biofuels are currently generating a lot of interest and certainly will be part of the overall solution. Like any technology, biofuels can be employed for better or worse, and the details matter enormously. Ethanol from corn is hardly a good fuel option because so much energy goes into fertilizer, pesticides, harvesting, and ethanol production itself. Much more favorable is cellulose-based ethanol, which can be derived from almost any source of cellulose, including grasslands. Biodiesel from soy is likely to be as costly in energy terms as ethanol from corn, whereas other sources, such as palm oil, are likely to be less costly. Production of biofuels also raises environmental issues—that is, concerns not about what or how the fuels are planted, but rather where.

Agriculture can contribute to the solution if, for example, it is conducted in ways that increase carbon stored in soils. This is presently being explored, and it is likely that payments to farmers for carbon sequestration can soon replace the commodity price supports that continue to rankle free trade negotiations.

Little attention has been paid to date to the other large source of greenhouse gases accumulating in the atmosphere, namely deforestation and burning of biomass, principally in the tropics. About 20% of annual CO_2 emissions come from this source, so finding ways to reduce that to a minimal number is an important part of the solution. Fortunately the potential of carbon trading—currently restricted under the Kyoto Protocol only to energy, reforestation, and plantation forestry—to contribute to this is significant. Details of how this would work remain to be resolved, but they are only details. At the December 2005 meeting of the Conference of the Parties to the UN Framework Convention on Climate Change, a group of tropical forest nations including Papua New Guinea and Costa Rica initiated a proposal on "avoided deforestation."

The appeal of such a mechanism is that a forest owner could receive payment for maintaining the carbon stock of a forest, but he or she could use the forest for anything else that would not affect its carbon content—essentially receiving multiple income streams. From the sustainability perspective, in addition to addressing climate change, this would maintain important biodiversity stocks as well as local and regional ecosystem services. It would certainly be valid to view this approach's role vis-à-vis the global carbon cycle as yet another ecosystem service. The approach is already functioning on a national scale in Costa Rica. All that remains to make it work on, say, an Amazonian scale, is a workable system and sufficient funds— namely sufficient demand in carbon markets.

All of these solutions depend on humanity taking the sustainability/global climate change issue sufficiently seriously. This, in turn, relates to the biggest environmental question we face, alluded to above: What is the acceptable level of greenhouse gas concentration in the atmosphere? That is, how much climate change should be allowed? As long as the debate is about emissions, this critical question is largely being avoided. It is a difficult question because there are multiple tipping points, and because we are in the middle of the process of the climate change (with lags in the system that postpone daily recognition of the consequences). But it must be faced.

The UN Framework Convention on Climate Change is based on concerns about economics, agriculture, and ecosystems. It is clear from the stirrings in nature that ecosystems constitute the most sensitive of the three. There is some talk about 2°C increases in global average temperature being a safe or acceptable limit. That corresponds to 450 ppm of CO_2 (compared to 280 ppm at the onset of the industrial age, and 379 ppm currently). It seems virtually impossible that we will be able to stop at 450 ppm, yet the changes we are already seeing in the Arctic would suggest that that limit is too generous.

The last thing that makes sense is to discover the safe level of greenhouse gas concentrations for climate change after we have already exceeded it. If we are serious about sustainability of the human species, we need to address climate change with greater urgency than any other problem at any other time in history. It is clear from the energy/climate change perspective alone that we are long overdue for a global reckoning on consumption per capita and consumption in total. Without that, sustainability will be unattainable except in a degraded sense.

Index

Page numbers followed by *t* and *f* indicate tables and figures; those followed by *n or nn* refer to endnotes.